Biotic Communities

Biotic Communities
Southwestern United States
and Northwestern Mexico

David E. Brown

Editor

University of Utah Press
Salt Lake City

Originally published as *Biotic Communities of the American Southwest—United States and Mexico*, in *Desert Plants* vol. 4, nos. 1–4, 1982, by the Boyce Thompson Southwestern Arboretum.

A full-color, 60" × 48" map, "Biotic Communities of the Southwest," is also available for $20.00, plus shipping, from The University of Utah Press, 1795 E. South Campus Drive, Salt Lake City, Utah, 84112.

(800) 773-6672   www.upress.utah.edu

5 4 3

08 07 06 05 04 03 02 01

LIBRARY OF CONGRESS CATALOGING-IN-PUBLICATION DATA

Biotic communities : southwestern United States and northwestern Mexico / David E. Brown, editor.
    p. cm.
    "Originally published as Biotic communities of the American Southwest—United States and Mexico, in Desert plants vol. 4, nos. 1–4, 1982, by the Boyce Thompson Southwestern Arboretum"—T.p. verso.
    Includes bibliographical references (p.).
    ISBN 0-87480-459-0
    1. Biotic communities—Southwestern States. 2. Biotic communities—Mexico. I. Brown, David E. (David Earl), 1938–
QH104.5.S6B56   1994
574.5'247'0979—dc20   94-20777

Contents

Preface, 1994

Biotic Communities of the American Southwest — United States and Mexico was first published in 1982 as a special issue of *Desert Plants*, the journal of the Boyce Thompson Southwestern Arboretum. This multi-authored volume was an immediate success and was widely used by resource managers, biologists, teachers, and those interested in natural history assessment. Accordingly, even though ten thousand copies had been printed, the book was soon out of print along with the accompanying map, *Biotic Communities of the Southwest*, published as General Technical Report RM–78 by the Rocky Mountain Forest and Range Experimental Station of the U.S. Forest Service. Even after twelve years, hardly a week goes by without an inquiry as to the availability of one or both of these publications. Now, thanks to the University of Utah Press, such requests can be fulfilled.

This printing of *Biotic Communities* differs little from its predecessor. Thanks to an exemplary editing job by Frank Crosswhite, we were able to reproduce the original text and photos with only minor corrections and revisions. The scale on the drainage map on page 227 has been corrected, and some plant and animal names have been changed or added, mostly in Appendix II. In this regard, Gale Monson graciously provided a list of characteristic and indicator birds for the Sinaloan deciduous forest, and Randy Babb has done the same for this biotic community's herptiles. Such lists were lacking in the earlier Appendix II due to a lack of biological information on this still poorly known biome.

Because of the major revision required, no attempt was made to incorporate the numerous taxonomic changes in plants and animals that have accrued since 1982. Some of the most dramatic of these changes in nomenclature have been in fishes, the most up-to-date references in taxonomy of which may be found in Mayden (1992) cited below. Also, because of cost considerations, we have had to substitute black and white photos for the five color frontispieces that introduced the formation-classes. Otherwise, except for the new color cover and publication credits, the book and map are essentially the same as the earlier versions.

One of the principal values of the original work was its extensive bibliography on Southwest environments. Such a compendium could of course now be greatly expanded. A list of some of the more significant recent books dealing with the region's natural history and naturalists follows.

Alcock, J. 1985. *Sonoran Desert Spring*. University of Arizona Press, Tucson.

___. 1990. *Sonoran Desert Summer*. University of Arizona Press, Tucson.

Barbour, M. G., and W. D. Billings. 1988. *North American Terrestrial Vegetation*. Cambridge University Press.

Betancourt, J. L., T. R. Van Devender, and P. S. Martin, eds. 1990. *Packrat Middens*. University of Arizona Press, Tucson.

Bowers, J. E. 1988. *A Sense of Place: The Life and Work of Forrest Shreve*. University of Arizona Press, Tucson.

___. 1992. *The Mountain Next Door*. University of Arizona Press, Tucson.

Brown, D. E., ed. 1983. *The Wolf in the Southwest*. University of Arizona Press, Tucson.

Brown, D. E. 1984. *The Grizzly in the Southwest*. University of Oklahoma Press, Norman.

___. 1984. *Arizona Tree Squirrels*. Arizona Game and Fish Department, Phoenix.

___. 1985. *Arizona's Wetlands and Waterfowl*. University of Arizona Press, Tucson.

___. 1989. *Arizona Game Birds*. University of Arizona Press, Tucson.

___. 1994. *Vampiro: The Vampire Bat in Fact and Fantasy*. High-Lonesome Books, Silver City, New Mexico.

Brown, D. E., and N. B. Carmony. 1991. *Gila Monster*. High-Lonesome Books, Silver City, New Mexico.

Brown, D. E., and N. B. Carmony, eds. 1990. *Aldo Leopold's Wilderness*. Stackpole Books, Harrisburg, Pennsylvania.

Carmony, N. B., and D. E. Brown, eds. 1993. *The Wilderness of the Southwest: Charles Sheldon's Quest for Desert Bighorn Sheep and Adventures with the Havasupai and Seri Indians*. University of Utah Press, Salt Lake City.

Day, G. I. 1985. *Javelina Research and Management in Arizona*. Arizona Game and Fish Department, Phoenix.

Dick-Peddie, W. A. 1993. *New Mexico Vegetation: Past, Present, and Future*. University of New Mexico Press, Albuquerque.

Hartmann, W. K. 1989. *Desert Heart: Chronicles of the Sonoran Desert*. Fisher Books, Tucson, Arizona.

Hocutt, C. H., E. O. Wiley, eds. 1986. *The Zoogeography of North American Freshwater Fishes*. John Wiley and Sons, New York.

Hoffmeister, D. F. 1986. *Mammals of Arizona*. University of Arizona Press, Tucson.

Lee, D. S., C. R. Gilbert, C. H. Hocutt, R. E. Jenkins, D. E. McAllister, and J. R. Stauffer, Jr. 1981. *Atlas of North American Freshwater Fishes*. North Carolina State Museum of Natural History, Raleigh.

Lowe, C. H., C. R. Schwalbe, and T. B. Johnson. 1986. *The Venomous Retiles of Arizona*. Arizona Game and Fish Department, Phoenix.

Mayden, R. L., ed. 1992. *Systematics, Historical Ecology, and North American Freshwater Fishes*. Stanford University Press, Palo Alto, California.

Minckley, W. L., and J. E. Deacon, eds. 1992. *Battle Against Extinction: Native Fish Management in the American West*. University of Arizona Press, Tucson.

Ramamoorthy, T. P., R. Bye, A. Lot, and J. Fa, eds. 1993. *Biological Diversity of Mexico: Origins and Distribution*. Oxford University Press, New York.

Rinne, J. N., and W. L. Minckley. 1991. *Native Fishes of Arid Lands: A Dwindling Resource of the Desert Southwest*. Gen. Tech. Rep. RM–206, Rocky Mountain Forest and Range Experimental Station, USDA Forest Service, Fort Collins, Colorado.

Rosenberg, K. V., R. D. Ohmart, W. C. Hunter, and B. W. Anderson. 1991. *Birds of the Lower Colorado River Valley*. University of Arizona Press, Tucson.

Russell, S. M., and G. Monson. 1995. *The Birds of Sonora*. University of Arizona Press, Tucson.

Schoenherr, A. A. 1992. *A Natural History of California*. University of California Press, Berkeley.

Sigler, W. F., and J. W. Sigler. 1987. *Fishes of the Great Basin: A Natural History*. University of Nevada Press, Reno.

Stebbins, R. C. 1985. *A Field Guide to Western Reptiles and Amphibians*. Houghton Mifflin Co., Boston.

Sublette, J. E. , M. D. Hatch, and M. Sublette. 1990. *The Fishes of New Mexico*. University of New Mexico Press, Albuquerque.

Swift, C. C., T. R. Haglund, M. Ruiz, and R. N. Fisher. 1993. *The Status and Distribution of the Freshwater Fishes of Southern California*. Bulletin of the Southern California Academy of Science 92:101–167.

W. L. Minckley assisted with the editing and provided the revised stream map. Special thanks also to Barry Spicer of the Arizona Game and Fish Department for sacrificing a clean original for photo reproduction, and to Raymond M. Turner for safely guarding the plates to the map for all these years. It is because of their foresight, and University of Utah Press editor Jeff Grathwohl's enthusiastic support that these publications are now again available.

David E. Brown

Acknowledgements

In 1974 the Rocky Mountain Forest and Range Experiment Station (USDA Forest Service) undertook a job at the request of Forest Service Region-3 headquarters to compile and store all available published information concerning wildlife species and their habitat for Arizona and New Mexico. David R. Patton accepted responsibilities as Project Leader for the Forest Service, working at the USFS-RM offices on the Arizona State University campus at Tempe. David E. Brown of the Arizona Game and Fish Department became not only a major collaborator but agreed to author and edit material for publication. As part of this effort the Arizona Game and Fish Department and Rocky Mountain Station published a detailed map (1978) on the Biotic Communities of the Southwest by David E. Brown and Charles H. Lowe.

The purpose of the map was to tie wildlife to recognized biomes to meet local assessment needs and for use by management at the regional level. As a result of the map there were many requests to document, in a publication, the physical and ecological description of the identified biotic communities. The effort to do this was officially started in 1977.

David E. Brown organized a group of experts to write chapters dealing with the various biomes and became chief author and editor for the unified work. He has authored or co-authored several major chapters and filled in the gaps between sections written by others. Throughout *Biotic Communities of the American Southwest — United States and Mexico* he is the author of all text not specifically attributed to other authors. Aside from authors in the Forest Service (David R. Patton and Charles P. Pase), collaboration came from Arizona State University (W.L. Minckley and James P. Collins), Desert Botanical Garden (Howard Scott Gentry), University of Arizona (Charles H. Lowe), and the U.S. Geological Survey (Raymond M. Turner).

A work of this nature is the product of numerous individuals aside from the authors. Special thanks go to Robert P. Winokur and his staff at the Rocky Mountain Forest and Range Experiment Station for final review and editorial expertise. Matthew E. Alderson, Arizona Game and Fish Department, Phoenix, prepared the line drawings for the text and David Daughtry, Dale R. Dundas and Todd Pringle processed most of the photographs used. Other photographs not taken by the authors were provided by Roland Wauer, National Park Service, Washington, D.C. and J.D. MacWilliams, U.S. Forest Service, Riverside, California. Vern Booth, Phoenix, prepared the map base and original color map plates — a truly monumental job. JoAnn Alwin, Omega Communications, Tucson, later provided additional base enhancement and edited the color separates. The efforts of several typists, particularly Roseann Wishner who worked on a voluntary basis, are much appreciated.

On the map part of the manuscript, we are particularly indebted to Reid V. Moran, Curator of the Herbarium, San Diego Natural History Museum, Balboa Park, San Diego (information on Baja California); Tom Wendt and David H. Riskind, University of Texas, Austin (information on Texas, Chihuahua, Coahuila); Barton Warnock, Sul Ross University, Alpine, Texas (information on Texas); Tom Zapatka, New Mexico Game and Fish Department, Santa Fe and Dale R. Jones, U.S. Forest Service, Washington, D.C. (information on New Mexico) and Killian Roever, Chemargo, Phoenix, Arizona for sharing their special knowledge. Craig B. Jones, Albuquerque, New Mexico, was especially helpful in the editing and correction of the New Mexico portion of the map.

Other important changes to earlier works were prompted by Frank Reichenbacher, Arizona Natural Heritage Program, Tucson, Arizona; Neil B. Carmony, Tucson; and W.L. Minckley, Arizona State University, Tempe, Arizona. Paul S. Martin, Laboratory of Paleoenvironmental Studies, University of Arizona, Tucson, Patricia M. Bergthold, Office of Economic Planning and Development, Phoenix, Rodney Engard, Desert Botanical Garden, Phoenix, and especially Raymond M. Turner, U.S. Geological Survey, Tucson, helped generally and gave needed support and suggestions.

Seymour Levy, Tucson, contributed greatly in assigning the proper birds to the biomes, as did John S. Phelps, Arizona Game and Fish Department, Tucson, for the reptiles.

The manuscript was subjected to review by the following experts who greatly assisted with improvement or comments: Earl F. Aldon, USFS; Robert G. Bailey, USFS; Janet F. Barstad, W.D. Billings, Duke University; Ardell J. Bjugstad, USFS; W. Glen Bradley, University of Nevada; Tony Burgess, U.S. Geological Survey; James Collins, Arizona State University; Frank S. Crosswhite, Boyce Thompson Southwestern Arboretum; Richard S. Driscoll, USFS; Arthur Gibson, James R. Griffin, Hastings Natural History Reservation; Alden R. Hibbert, USFS; John Hubbard, New Mexico Game and Fish Department; Elbert Little, Smithsonian Institution; John Marr, University of Colorado; S. Clark Martin, University of Arizona; Harold Mooney, Stanford University; Phil Ogden, University of Arizona; Duncan T. Patten, Arizona State University; Dixie R. Smith, USFS; Tom Van Devender, Arizona Heritage Program; Frank C. Vasek, University of California; Neil West, Utah State University; Frank Ronco, USFS.

All-important inspiration was provided by David R. Patton of the Rocky Mountain Station, Paul M. Webb, Arizona Game and Fish Department, and Charles H. Lowe, University of Arizona. Financial and administrative support and assistance by the U.S. Forest Service through Richard S. Driscoll, Richard G. Krebill and David R. Patton and by the Arizona Game and Fish Department through Paul M. Webb were essential for completion of this publication and for which we are truly grateful.

This is a publication of the wildlife habitat research project, David R. Patton, Project Leader, of the U.S. Forest Service Rocky Mountain Forest and Range Experiment Station research work unit at Tempe, Arizona in cooperation with Arizona State University.

Substantial Funding was provided through Arizona Federal Aid to Wildlife Project W-53-R.

Preface, 1982

When I first "discovered" the Southwest some 20 years ago, I began what has become a continuous Southwest adventure. My first 6 months with the Arizona Game and Fish Department were spent traveling throughout the State on a "getting acquainted basis" for which I will be forever grateful. I was continuously impressed by the variety and juxtaposition of Arizona's natural landscapes. I found much that resembled. my native California — chaparral, stands of relict cypresses, streamside woodlands of sycamore, cottonwood, willows, walnut, and alder. Not only were the familiar Great Basin, Mohave, and California deserts also represented in Arizona, I "found" the Arizona and Chihuahuan deserts. Equally impressive were the montane and subalpine forests and meadows— southward extensions of the Rocky Mountains. Like many newcomers to the Southwest, I discovered Mexico in southeast Arizona, with its evergreen pine-oak woodlands, trogons, and Mearns' Quail. It was also here, in the San Rafael Valley of southeast Arizona, that I first saw the Great Plains. Truly, Arizona and the Southwest is the meeting place of North American biomes.

Discovery, however, is only the beginning of the real knowledge that comes with understanding. Three events in the mid-1960's changed my perspective of the Southwest and made each new discovery so much more rewarding. The first was an introduction to Dr. Charles H. Lowe. I was mapping the vegetation of my game management unit and on the suggestion of one of his students, I arranged a meeting with Lowe at the University of Arizona. Although the purpose of this meeting was to discuss the distribution and occurrence of ironwood trees, the most memorable outcome of that session for me was a complimentary copy of his *The Vertebrates of Arizona*. By 4:00 a.m. the next day I had read the "Landscape and Habitats" section several times. It was, in writing, a complete treatment of Arizona's natural environment. The book's bibliography also provided an excellent introduction to the literature of the natural history of the Southwest.

Dr. Lowe was a great influence on my life, both on my appreciation of Southwestern natural history and on a larger personal basis. I consider myself fortunate to have worked with him on a number of projects—the most recent being this publication. This presentation and its companion map (Brown and Lowe, 1980) were his idea—an expanded treatment of Lowe's *Arizona's Natural Environment* for the entire Southwest.

Also in the mid-1960's, Dr. James R. Hastings and Dr. Raymond M. Turner published *The Changing Mile*. It wasn't just the changed and unchanged landscapes that were so impressive, — it was something basic, yet cryptic. Only slowly did the significance of their contribution become clear, — they had captured evolution in progress. Changes and evolutionary responses were not only happening, they were literally taking place before our eyes. Daniel Axelrod was right, the Southwest was a dynamic place—not just 10,000 years ago, but today—during our lifetimes.

A third memorable event was a presentation by Dr. W.L. Minckley at a departmental school on Arizona's native aquatic habitats and endemic fishes. Before this, I had considered Southwestern wetlands to be modified by man so as to be almost cultural features. But this was not the case. Our aquatic environments had their own evolutionary history and distinctive flora and fauna. Our wetlands took on a new significance.

We are fortunate that Charles H. Lowe, Raymond M. Turner, and Wendell Minckley have authored chapters of this publication. Their participation is moreover evident throughout this work, and their contributions go far beyond their authored passages. Their continued encouragement for this and all aspects of natural history inquiry has been of prime importance in the completion of this work. The other authors—Charles P. Pase, David R. Patton, James P. Collins and Howard Scott Gentry—are established authorities on their subjects. Their assistance, and the inspiration of Paul M. Webb, and David R. Patton, who steadfastly supported and administered our effort, made this publication a reality.

Although the Southwest is no longer a wilderness, many of its native landscapes remain. It is still a dynamic showplace of natural habitats undergoing evolution, and is at the same time both old and young.

Although the other authors and I hope to enjoy our adventure for some time to come, our initial discovery period is essentially over. We hope that for others our text and map will be of some use. If this is so, and if we are permitted to allow our modest understanding of the Southwest's biotic communities to become a part of their discovery experience, our purpose in writing this will have been satisfied. Perhaps, too, they will see and love the Southwest as we do.

David E. Brown
January 1982

Management Applications of Biotic Community Data

David R. Patton

USDA Forest Service

The need to summarize and map plant and animal information received new emphasis in the 1970's with a number of environmental laws enacted by Congress—the National Environmental Policy Act, National Forest Management Act, Endangered Species Act, and Resources Planning Act. These laws are forcing a systematic approach to land use planning through resource inventory and assessment. Such activities are greatly facilitated when resource information is in an hierarchical structure similar to the Linnaean taxonomy of plant and animal systematics developed in the 18th Century. A systematic hierarchy for vegetation has the great advantage of having data ordered from the general to the specific so that it can be used for different levels of abstraction—national, regional, and/or local.

Mapping The Southwest Landscape

One obvious advantage of mapping natural vegetation is to illustrate and thereby measure biological and cultural potential. Land use managers, biologists, agronomists and others concerned with renewable resources can obtain valuable insights into the biotic capabilities of a given area and its actual and potential uses. Biological and cultural limitations are then also often apparent. Especially important, mapping of vegetation (=cover) enables wildlife, forest, and range managers to stratify and sample populations in any given land area efficiently (Leopold, 1933). Statistically valid surveys then can be used to measure and predict an area's wildlife density, timber potential, and range capability (see e.g. Brown and Smith, 1976). Areas to be assessed can range in size from a few acres to subcontinental units.

The uses of a biotic community (or vegetation) map to the student of natural history are also readily apparent. Plant and animal distributions can be plotted and determinations made as to their actual and expected occurrence. Knowledge of an organism's distribution in relation to natural communities can provide insights into factors responsible for or influencing a species' occurrence, density and limitations. So important is this knowledge that the study and mapping of plant distribution (phytogeography), animal distribution (zoogeography), and biome distribution (biogeography) have become recognized sciences in their own right.

Biogeographic maps are essential to those interested in the identification, study, acquisition, and preservation of natural areas (see e.g. Dasmann, 1976; Franklin, 1978; Udvardy, 1975). Their use greatly facilitates the comparison and selection of representative habitats through the delineation of representative candidate areas. Similar areas can then be sampled and quantified before a choice is made. Perhaps, equally or more important, types not heretofore protected can be identified.

The chapters of this publication describe the biotic communities outlined on a map of the Southwest (Brown and Lowe, 1980). The first edition of this map (Brown, Lowe, and Pase, 1978) was a major contribution in bringing together information at the biome level. The revised map (Brown and Lowe, 1980) and the hierarchical classification system in Appendix I provides an ecological base for the location of plant and animal communities for the American Southwest and parts of Mexico.

As a result of the increased emphasis to develop systems for vegetation classification and related habitat factors (Chambers, 1974; Driscoll et al., 1978; Donart el al., 1978a; Brown et al., 1980), land managers in Arizona and New Mexico are beginning to manage in terms of ecological units (Montane Conifer Forest, Sonoran Desert, etc.) instead of purely geographic areas (Mogollon Rim, Kaibab Plateau, etc.) and descriptive terms (salt marsh, cold desert, etc.). The ecological unit has meaning to plant geographers, ecologists, and biologists because of its common evolutionary history; a geographical or descriptive area may or may not have a shared flora and fauna with other areas of the same name.

Biotic community data consisting of plant form (tree, shrub, etc.) and structure (seedling, sapling, etc.) and animal profile (use of form and structure for feeding and breeding) is being collected for storing in local and regional wildlife data bases. These data bases are being used to provide information to write environmental analysis reports and environmental impact statements, to evaluate land management practices, and to develop habitat improvement projects. The application of biotic community data for decision making will increase as biologists in state and federal agencies accumulate better information on the management of all species in an ecosystem.

Introduction
Charles H. Lowe

Department of Ecology and Evolutionary Biology
University of Arizona

and David E. Brown

Arizona Game and Fish Department

Table 1. Map legend for Biotic Communities of the Southwest (scale 1:1,000,000), Brown and Lowe, 1978, 1980.

Tundra Formation
    Alpine Tundra

Forest Formation
    Petran Subalpine Conifer Forest
    Sierran Subalpine Conifer Forest
    Petran Montane Conifer Forest
    Sierran Montane Conifer Forest
    Sinaloan Deciduous Forest

Woodland Formation
    Great Basin Conifer Woodland
    Madrean Evergreen Woodland
    Californian Evergreen Woodland

Scrub Formation
    Great Basin Montane Scrub
    Californian Chaparral
    Californian Coastalscrub
    Interior Chaparral
    Sinaloan Thornscrub

Grassland Formation
    Subalpine Grassland
    Plains and Great Basin Grassland
    Californian Valley Grassland
    Semidesert Grassland

Desertscrub Formation
    Great Basin Desertscrub
    Mohave Desertscrub
    Chihuahuan Desertscrub
    Sonoran Desertscrub
        Lower Colorado subdivision
        Arizona Upland subdivision
        Plains of Sonora subdivision
        Central Gulf Coast subdivision
        Vizcaino subdivision
        Magdalena subdivision

Wetlands

This publication is one part of a two-part work. This part accompanies the revised color map (scale 1:1,000,000)—Biotic Communities of the Southwest, (Brown and Lowe, 1980) and is coordinated with the map so that the table of contents corresponds to the map legend (Table 1). Also, wetlands are included.

This publication is explanatory and descriptive, not exhaustive. Selected references are provided without presenting an extensive literature review. Major references are included for the states of Arizona and New Mexico as well as for the half of the Southwest which occurs in Mexico in the states of Sonora, Chihuahua, Coahuila, and Baja California del Norte.

The chapters on the biotic communities tend to be more uniform than diverse in matters of content and style, with greater emphasis on community ecology than physiological ecology. Both floristics and faunistics are used in characterizing the biotic communities. Thus, in addition to information on the vegetation and flora, authors have made an effort to provide information on the animal taxa relevant to the biomes for the area shown in Figure 1—the North American Southwest. In addition, summaries of scientific and common names for characteristic plant and animal taxa discussed in each biome chapter are given in a separate appendix.

Insofar as possible the scientific names for plants and animals conform with those accepted by recent authorities. These authorities include the American Ornithologist's Union (1973, 1975, 1976), Birkenstein and Tomlinson (1981), Collins et al. (1978), Correll and Correll, (1972, 1975), Correll and Johnston (1970), Critchfield and Little (1966), Gentry (1972), Harrington (1964), Hitchcock and Chase (1971), Jones et al. (1975), Kearney and Peebles (1960), Lehr (1978), Lehr and Pinkava (1980), Martin and Hutchins (1980), Munz and Keck (1968), Nickerson et al (1976), Patton (1978), Peterson and Chalif (1973), Shreve and Wiggins (1964), Smith and Smith (1976), Standley (1920-1926), Taylor and Patterson (1980), White (1949) and Wiggins (1980).

Following zoological custom common names for animals when derived from personal names generally have a possessional ending ('s), e.g. "Lucy's Warbler" or "Gambel's Quail." On the other hand, following botanical custom for plants with similar derivation from personal names, possessional endings are not used, e.g. "Jeffrey Pine" or "Gambel Oak." Common names in English or Spanish are capitalized when they refer to a specific plant or animal at the species, subspecies or variety level but are not capitalized when used in a general sense, e.g. "Goodding Willow" and "Fremont Cottonwood" are capitalized whereas "willows" and "cottonwoods" are not.

Unless otherwise indicated, photographs, captions, and introductory passages are by David E. Brown.

The System

A comparative evolutionary approach to plant communities is consistent with an open-ended hierarchical system of classification (in contrast to a classless system). Moreover, it is capable of producing more stable and meaningful classification over time by permitting orderly accumulation of more information (Brown and Lowe, 1974a; 1974b; Brown, Lowe, and Pase, 1979, 1980; Brown, 1980). A summary of this classification system as it pertains to North America and the Southwest is presented in Appendix I.

This hierarchical system for North American biomes is

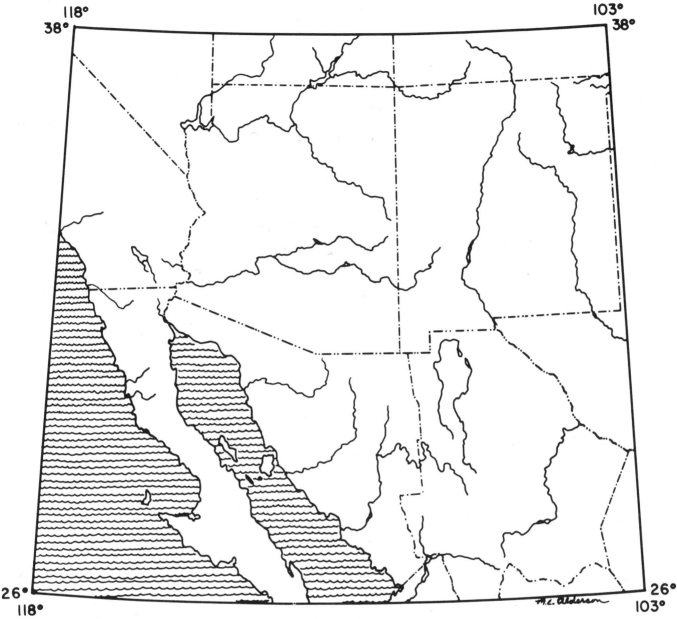

Figure 1. *The North American Southwest. In delineating a natural Southwest region, approximately one half of the area falls in the Republic of Mexico and one half in the United States; the U.S. States of Arizona and New Mexico constitute less than half of the "American Southwest." Parts or all of the following states are included: Arizona, Baja California Norte, Baja California Sur, Chihuahua, Coahuila, Colorado, Nevada, New Mexico, Sonora, Texas, and Utah.*

ecosystem-based and is concordant with the life-zone concept of Merriam (1890, 1898; Figure 2) and with the more encompassing geography-based system of North American biotic provinces (Fig. 2 in Appendix I); this is reflected in the legend of the map units in Table 1. However, the structure of biomes is based on vegetation, and they are not provinces *per se,* which are biotic, faunistic, or floristic in structure, function, or other aspects (Shreve, 1915; Dice, 1943; Pitelka, 1941, 1943; Miller, 1951; Munz and Keck, 1959; Whittaker and Niering, 1964; Daubenmire and Daubenmire, 1968; Udvardy, 1975; Barbour and Major, 1977; Franklin, 1977; Bailey, 1978[1]). Part or all of 11 biotic provinces occur in the

Southwest, as delimited here (Fig. 3).

The Biome Approach

Because biotic provinces have distinctive evolutionary histories, their occurrence is genetically based. Biomes (biotic communities) are natural formations, characterized by a distinctive vegetation physiognomy, within a biotic province. They are plant and animal community responses to integrated climatic factors, more or less regional in scope.

The limits of a particular biome or biotic community are determined by climate (i.e., minimum seasonal temperatures, minimum seasonal precipitation). The actual boundaries, therefore, are often tenuous and commonly determined by local phenomena—elevation, longitude, slope exposure, geomass, temperature inversions, cold air drainages, soil

---

[1]*Bailey (1976, 1978) has recently reworked Dice's original concepts and contributions (biotic pronvices) and proposes them as "ecoregions".*

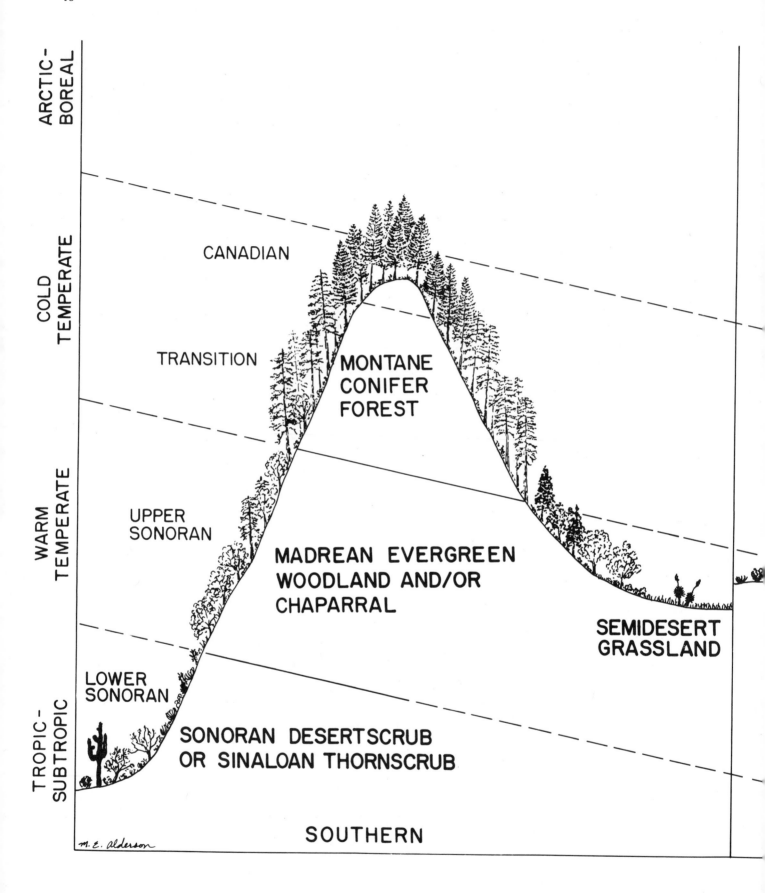

ARCTIC-BOREAL

COLD TEMPERATE

CANADIAN

TRANSITION

MONTANE CONIFER FOREST

WARM TEMPERATE

UPPER SONORAN

MADREAN EVERGREEN WOODLAND AND/OR CHAPARRAL

SEMIDESERT GRASSLAND

TROPIC-SUBTROPIC

LOWER SONORAN

SONORAN DESERTSCRUB OR SINALOAN THORNSCRUB

SOUTHERN

m. E. alderson

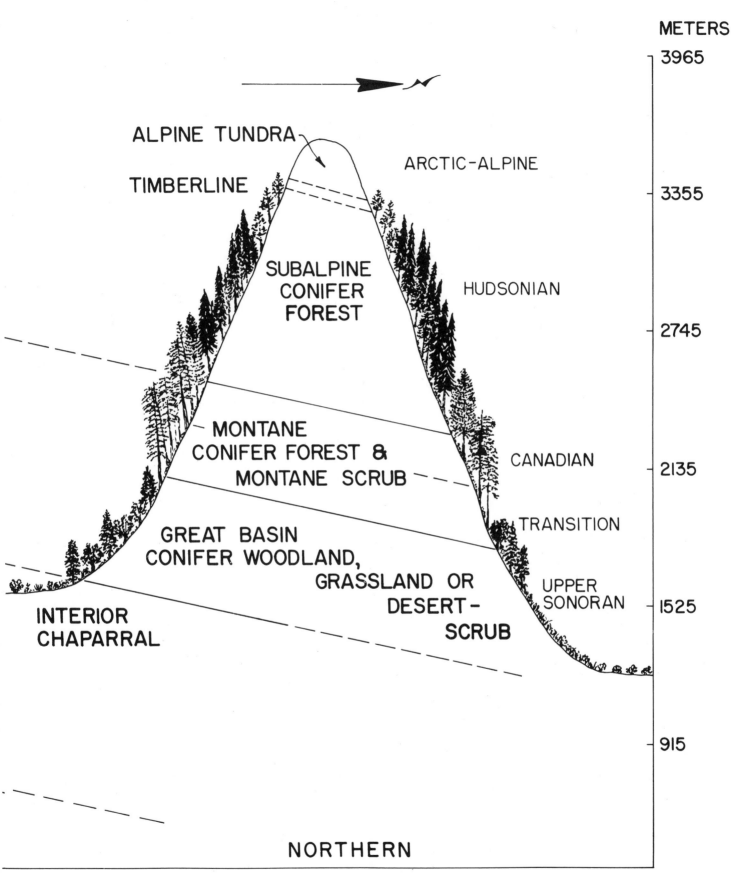

Figure 2. *Diagrammatic profiles of hypothetical mountains in the southern and northern portions of the Southwest indicating the vertical climatic zonation and concordance of biotic communities with Merriam's (1890, 1898) Life-zones.*

12

porosity, etc. (Lowe, 1964:84–91). Accordingly, local micro-climates may result in the unusual occurrence of one or more biomes in an area, contributing to the overall diversity.

The "Southwest"

The latitudinal limits for the Southwest in Figure 1 are 26°N at the south, and 38°N on the north. The area for the vegetation color map (Brown and Lowe, 1980) is slightly less—27°N and 37°30'N, respectively. The expanded area for the Southwest addressed here, both north and south of the USA–Mexico border, gives a still too restricted geographic and ecologic representation for this subcontinental region, especially in time as well as in space. The landscape evolution of the American Southwest during the past 100 million years derived out of the major Tertiary Geofloras over a much greater subcontinental area. Even the most pristine conditions within the derivative biomes are, for the most part, remnants of once-greater biotic communities of a greatly expanded geographic "Southwest."

Southwest Landscape Evolution

Important evolutionary sequences in the Cenozoic history of southwestern landscapes are now reasonably accurately inferred from the paleobotanical record. California and Nevada especially have yielded abundant fossil-bearing rocks of Tertiary age, permitting greater detail in evolutionary analysis there than possible further east (Axelrod, 1950, 1956, 1957, 1958a, 1958b, 1966, 1967, 1970, 1972, 1973, 1975, 1976, 1979a, 1979b).

During the early Tertiary, the vegetation of North America was composed of three great Geofloras—roughly equivalent to grand subcontinental provinces—(a) a mesophytic broad-leafed evergreen (Neotropical-Tertiary) Geoflora in the south half of the continent; (b) a temperate conifer and mixed-deciduous Arcto-Tertiary Geoflora at the north, and between them; (c) an emerging sclerophyllous and microphyllous Madro-Tertiary Geoflora, appearing on drier sites within and bordering the Neotropical Tertiary Geoflora. Southwest forests of relict conifers, montane conifers, subalpine conifers, and riparian deciduous trees are today relatively simplistic and depauperate modern derivatives out of the more generalized and diverse temperate Arcto-Tertiary Geoflora.

At the other end of the present regional moisture-temperature gradient, Southwest deserts are relatively simplistic modern derivatives out of the newer and enormously more generalized and ecologically diverse Madro-Tertiary Geoflora. Madro-Tertiary evolution has yielded Southwest evergreen woodlands, evergreen sclerophyllous scrublands (=chaparral), microphyllous desertscrub, as well as the drought-deciduous thornscrub and subtropical deciduous forest out of which thornscrub was derived. The dominant species in Southwest semidesert grasslands appear ecologically polyphyletic; Madro-Tertiary dry-tropic shrubs and grasses share dominance with Arcto-Tertiary grasses.

Subsequent to the Eocene epoch, dry climates expanded throughout the second half of the Tertiary, culminating in their greatest geographic area and severity during Mio-Pliocene time. By the close of the Tertiary, more than 75 million years after the period began, the dry-adapted Madro-Tertiary Geoflora had spread over southwestern North America. This was accomplished with concomitant retreats, under expanding aridity, of the northern temperate Arcto-

Tertiary Geoflora and the southern mesophytic Neotropical-Tertiary Geoflora from which Madro-Tertiary species and communities evolved. The major families, genera, and species in Southwest communities of essentially modern aspect were in place at the end of the Tertiary—at the beginning of the Quaternary and Pleistocene events (2 million years before present). For reconstruction of Quaternary Southwest environments see Wells (1976, 1978, 1979) and Van Devender et al. (1977) and Van Devender and Spaulding (1979).

Current mapping and reporting involves natural vegetation in situ during the current Holocene interglacial following the close of the Pleistocene at 11,000 ybp. The mixing of species compositions into present biotic communities under strong secular climatic changes characterized the Pleistocene and Holocene in the Southwest (2 mybp to present). During this time, there were significant biogeographic shifts of taxa in both elevation and latitude. Quaternary glacial periods were on the order of 100,000 years duration, and interglacials on the order of 10,000 to 20,000 years. There is much speculation and some knowledge on their Pleistocene and early to middle Holocene whereabouts of the elements present and available for floristic and physiognomic characterization of modern communities in the Southwest. Although questions remain on what species stayed or went, and to where, there is wide consensus on many points.

Alpine tundra and present subalpine forests reached their greatest extent in the recent Pleistocene. Although it remained too warm in most of the Southwest for glaciation, many smaller glaciers captured the high mountains. Lower temperatures reduced evapotranspiration and accompanied a southward extension of the polar air mass. Storms and precipitation were greater than at present. Modern boreal and cold temperate conifer forests, which are primarily disjunctive island-like features in the Southwest, reached their present elevated positions during post-Wisconsinan time, sometime during the past 10,000 years.

Southwest riparian deciduous forests and woodlands are modern water-controlled relicts once part of the western late Tertiary mixed deciduous forests. The populations of large winter-deciduous trees in the dominant gallery stratum, and the subdominant trees and shrubs in lower strata in these riparian communities, have been conspecific taxa throughout the Southwest for several million years—since well before the close of the Tertiary. The species in these unique communities are among the most misunderstood taxa, partially because of insufficient ecological investigation. For example, the nomenclature in such plant manuals as Kearney and Peebles (1960) is characteristic in not adequately reflecting the paleo-botanical evidence (see Little, 1950 for further detail on the botanical synonyms of species and subspecies in *Juglans, Platanus,* et al.).

Southwestern evergreen woodlands were derived out of a more generalized Madro-Tertiary woodland vegetation before the Pleistocene. In the late Pleistocene, during middle to late Wisconsinan time (40,000 to 11,000 ybp), mesophytic evergreen woodland communities predominated across landscapes presently in the Chihuahuan, Sonoran, and Mohave Deserts, before, during, and after the glacial maximum (22,000 to 17,000 ybp). This environmental domination by winter climate did not provide critical summer requirements for germination, establishment, and growth for many subtropical taxa that now characterize modern derivative com-

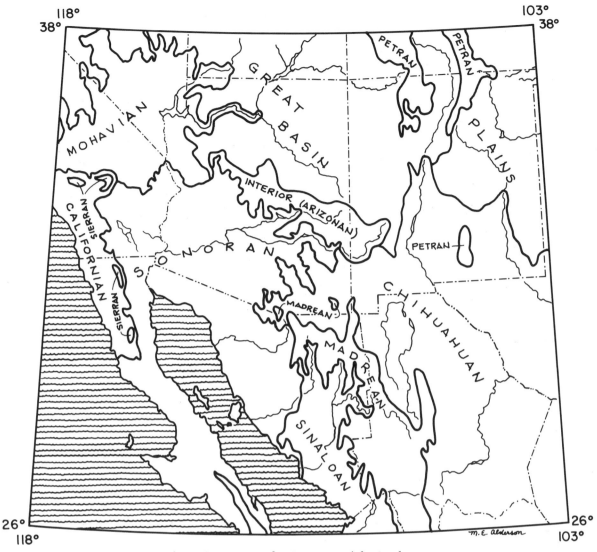

Figure 3. *Biogeographic Provinces of the Southwest.*

munities in Southwest subtropical ecosystems. Such subtropical components, depending on moderate temperatures and significant summer rainfall, were isolated in oases south of the retreating woodland perimeter, or pushed into geographic areas occupied today by Sinaloan deciduous forest on the Pacific side and homologous Tamaulipan communities on the Atlantic side of the continent.

By the early Holocene (11,000 to 8,000 ybp), decreasing temperatures and precipitation minima had resulted in the widespread persistence of xeric pinyon-juniper woodlands and included many Southwest landscapes formerly and presently occupied by vegetation established under dominant warm-season monsoon rainfall. By the middle to late Holocene (8,000 ybp to present), further significant reductions in summer as well as total precipitation resulted in the present assemblages of diverse subtropical and warm-temperature scrub communities inclusive of present Great Basin, Mohave, Sonoran, and Chihuahuan desertscrub. Rapid northward and upward deployment of floral and faunal elements into modern subtropical and warm temperate desertscrub assemblages was accelerated by melting of the ice sheets and stronger development of the Azores (Bermuda) high—with increased and expanded summer precipitation—favored by warmer global temperatures. The result is the Southwest landscape visible today—relict conifer woodlands, subalpine and montane forests, warm temperate grasslands, evergreen woodlands, sclerophyll chaparral, and most recently, North American deserts.

Maps depicting biotic communities are based primarily on natural vegetation. Although animal constituents are an important factor in the determination and classification of biomes, it is the vegetative structure and components of biomes that provide the readily observable and, therefore, measureable manifestation of these natural ecosystems. Delineation of biomes requires knowledge of their identity; mapping biotic communities, therefore, requires the recognition and delineation of classified vegetation. It is a taxonomic effort, separating one evolutionary-derived entity from others of the same rank.

Even when one recognizes prescribed units of natural vegetation, it may be difficult to draw a line separating them. Moreover, it soon becomes apparent that the various classifications of vegetation often form broad ecotones, intergrading over a considerable area. Disturbances, past and present, may make it difficult to recognize an area's potential natural vegetation (=pnv). Each classification effort requires the interpretation of criteria. The delineation of biomes is, therefore, somewhat subjective; although not usually deter-

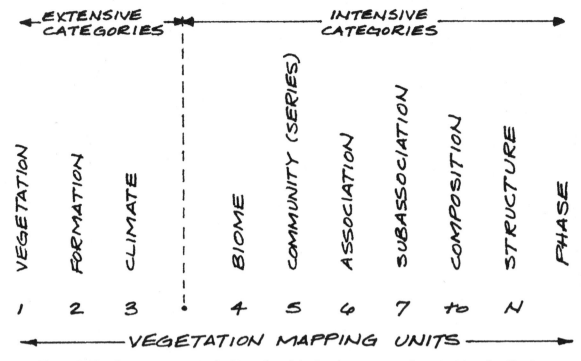

Figure 4. *The digit component in the hierarchy of the Southwest system for vegetation classification. Vertical dash line indicates the position of the decimal point that anchors the system.*

mined through measured criteria, the communities depicted are based on natural criteria and are subject to quantitative assessment.

The 27 biotic communities identified here are those in the digitized hierarchical classification system developed by Brown and Lowe (1974a,b), primarily for Southwest ecosystems. The multiple levels and open-ended arrangement of hierarchical components in the system provides for sensitivity to scale (Fig. 4). The system, therefore, permits classification and mapping at any scale, the classification was recently expanded for continental North America (Brown, Lowe, and Pase, 1977, 1979, 1980). The classification system is a workable blueprint for world ecosystems; the fourth level of the classification is the basis for the Southwest color map at scale 1:1,000,000.

The delineation of biotic communities in the Arizona portion of the Southwest was largely based on an earlier effort (Brown, 1973a) that incorporated several revisions. The 1973 Arizona map was itself greatly facilitated by the earlier vegetation map of A.A. Nichol (1937, 1952), and the Arizona range maps of R.A. Darrow (1944) and R.R. Humphrey (1950, 1953, 1960). Maps and publications of E. Little (1950) and E.F. Castetter (1956) were particularly helpful for New Mexico. Although the useful map of potential natural vegetation in New Mexico by Donart et al. (1978a, 1978b) has only recently been available, the Brown and Lowe (1980) map is in agreement with its classifications and major delineations. The excellent 1935 map by Morris in Gregg (1963) provides much of the basis for mapping that portion in Colorado. Smith's *Range Types of Utah* (in Vallentine, 1961) was the major source of consultation for that state and was used along with Utah Department of Natural Resource Game Range Resource Inventories.

West Texas, southern Nevada, and Baja California were without suitable statewide reference; the national treatments

of Küchler (1964) and of Flores Mata et al. (1971), and the regional maps of Buechner (1950), Bradley (1964), Allred et al. (1963), and others, were consulted and modified where deemed appropriate and when based on fieldwork. The valuable map produced by the Comisión Técnico Consultiva para la determinación Regional de los Coeficientes de Agostadero (1974) in conjunction with Shreve's (1951) classic treatment of the Sonoran Desert greatly assisted the mapping of Sonora. Of the published maps of Chihuahua, those by Brand (1936), Lesuer (1945), and Shreve (1939) proved most useful. Information for Coahuila was obtained from Muller (1947), Flores Mata et al. (1971), and from field work there, as in Chihuahua, Sonora, and Baja California.

Jensen's (1947) map of California, while of value, was at too large a scale to be of great benefit. Particularly useful for southern California and worthy of special mention are the truly remarkable vegetation maps produced between 1934 and 1940 under the direction of E.A. Wieslander (Wieslander, 1935). These maps, while showing vegetation unequal in rank (desertlands and grasslands are not differentiated further), accurately depict forest, woodland, and chaparral vegetation at the equivalent of association level. Küchler's (1977) work on California natural vegetation, while published too late to be incorporated here, shows a marked similarity in the difficult delineation of the Sonoran-Mohave Desert boundary, as well as between some other biomes. The major difference with Küchler for California is with chaparral and evergreen woodland communities, and is a result of interpretive differences of climax vegetation.

Other regional mapping efforts were also consulted. Those found particularly helpful were Rasmussen's (1941) map of the biotic communities of the Kaibab Plateau, Grinnell's (1908) life-zone map of the San Bernardino Mountains, Grinnell and Swarth's (1913) life-zone map of the San Jacinto Mountains, Hall's (1946) life-zone map of Nevada, and Storer

Figure 5. *Routes traveled by the editor.*

and Usinger's (1974) vegetation map of the Sierra Nevada Mountain region.

The lack of maps illustrating vegetation in sufficient detail to enable the interpretation necessary for the delineation of the biomes, required extensive travel and field mapping (Fig. 5). This was especially important for the delineation of communities not heretofore mapped in parts of Mexico, Texas, and New Mexico. The ordinary high altitude imagery available is not greatly useful without on the ground corroboration. Its limitations have indeed inhibited vegetation and biome mapping below formation rank. This has resulted in the "method" dictating the "means" for aerial based classifications, systems, and maps. Aerial photographic techniques, therefore, were rarely used here for interpretive purposes except for delineating formations of known biomes and when used in conjunction with verified ground data.

Alpine tundra (and alpine grassland) was mapped as those treeless areas above approximately 3,500 ± 150 m as indicated on other vegetation maps and/or as described in the literature (Part 1). No attempt was made at 1:1,000,000 to delineate the

relatively small areas of alpine scrub and subalpine krummholz. Similarly, no attempt was made to differentiate those relatively minor treeless areas adjacent to alpine tundra that are in early successional stages and/or under edaphic control.

Subalpine and montane forests were mapped in large part from other maps, except in Mexico. Because commercial and resource interest in these biomes is high, their boundaries could often be determined correctly from U.S. Forest Service habitat-type maps, as well as from Wieslander's detailed maps for California, and from several of the more general map sources cited above. In Mexico and outside of the U.S. National Forests, field investigations and literature review aided determination of upper and lower elevational limits, which vary significantly with longitude as well as latitude. Generally, subalpine forests in the Southwest are restricted to north of Mexico and above 2,400-2,600 m, depending on slope exposure. The lower Rocky Mountain, Madrean, and Sierran montane forests come in at elevations from 1,650 m to as high as 2,300 to 2,600 m (Part 2). The lower elevational limits of forest communities were mapped at the elevations of ecotone at which dominant forest tree species (yellow pines, deciduous oaks) were quantitatively more important than characteristic woodland species (pinyons, junipers, evergreen oaks) or otherwise gained major prominence in the community. For mapping purposes, open, woodland-like communities of relatively stunted ponderosa pines were treated as montane forest.

Woodland and chaparral biomes were mapped on the presence of characteristic physiognomic dominants. Many of these characteristic taxa (e.g., species of scrub oak, mountain-mahoghany, pinyon, juniper, evergreen oak) determine the physiognomic structure as well as floristics of the vegetation (Parts 2 and 3). An exception is some more or less open woodlands of single-seed junipers; these communities, where fully identified, were most often considered disclimax grass-land (Part 4).

Montane and piedmont communities are commonly subject to strong edaphic control, in addition to being dependent on the precipitation gradient; therefore, they are readily differentiated over relatively short distances from adjacent valley vegetation (e.g., chaparral versus grassland). Such is not necessarily the case between desertscrub and grassland communities in the inland Southwest, both of which may intergrade over broad areas and with many reversions. Desertland and grassland are the two "base" formations, encompassing more than 70% of the Southwest. Slight changes in elevation, slope exposure, available soil moisture, or differences in grazing history often determine the local presence of one or another of these biotic communities.

Particularly difficult to resolve, and to delineate for mapping purposes, are the ecotones (hence "lines") between (i) semidesert grassland and plains grassland, (ii) semidesert grassland and Chihuahuan desertscrub, (iii) Sonoran and Mohave desertscrub, and (iv) Great Basin desertscrub and grassland. These difficulties are resolved in many cases by drawing an arbitrary line through the approximate center of the discontinuous phase between these biomes. Some decisions are necessarily based on judgment as to the reversibility of scrub invasions that are today greatly advanced across wide expanses; the presence of Snakeweed and Burroweed indicate disclimax grassland; Creosotebush, Tarbush, and scrub Mesquite hummocks indicate desertscrub. Indicator plant species were also used to delineate other desertscrub biomes and their natural subdivisions (Part 5). At the fourth digit (biome) level in mapping, floristic criteria are especially important (see e.g., Hastings, Turner, and Warren 1972). Thus, elevation contours determined from field investigation and the literature were used in conjunction with knowledge on the ecology and distribution of indicator plant species and communities, as the basis for mapping the various desertscrub biomes at scale (1:1,000,000).

The most complex, least known, and certainly most misunderstood vegetation and flora in the North American Southwest are tropic-subtropic deciduous forest and thorn-scrub in northern Mexico. Sinaloan thornscrub and Sinaloan deciduous forest on the Pacific side (Sinaloa-Sonora) are in the mapped area. Their Tamaulipan homologs on the Atlantic side (Tamaulipas-Texas) are not. More equable climate with greater warm-season precipitation increases southward on the complex continental gradient, determining the associated vegetation continuum of increasing stature and of structural and species diversities. On the Pacific gradient southward,

the more recently evolved subtropical desertscrub (Sonoran Desert) is derived out of Sinaloan thornscrub, which, in turn, is derived out of Sinaloan deciduous forest (thornforest). Ordination does not resolve the mapping problems encountered on the continuum, where multiple floristic and physiognomic criteria were used for their determination.

Because of the scale involved on the Southwest map, relict conifer sites and many small discontinuous outlier stands in all of the biomes had to be omitted. With some important exceptions, this type of exclusion also pertains to wetland formations, most of which are small in extent even when shown; generally, therefore, wetlands are not further differentiated on the map. It is the primary purpose of the map to provide general location, geographic and areal extent, relative sequential relationship, and other aspects of the biomes of the Southwest. To provide an illustrated record of each biome's actual occurrence would require numerous maps and efforts at different scales over a wider geographic area.

The choice of colors was carefully considered. A color scheme was adapted from Gaussen (1953, 1955) who instituted the use of color for phytocartographic representation of ecological relationships. Arid habitats are presented by light colors. For wetter communities the colors become darker; very wet habitats are represented by dark solid colors. Cold habitats are dull, cold colors—black, grays, cold blues, and purples. As communities become warmer, the colors representing them become brighter until the brilliant warm colors of the tropics—yellows, oranges, reds, magentas—are reached. The color selected for each biome was a result of the combination of the colors representing these two basic environmental gradients—temperature and moisture. Although some selections had to be compromised in some degree to give needed contrast between adjoining biomes, and to keep the number of color plates from being excessive, Gaussen's color principles were not violated.

The only suitable maps covering the Southwest at the same scale both north and south of the border were provided by the 1:1,000,000 aeronautical charts. A base map without aeronautical and navigational enhancement was constructed from these charts, provided by the U.S. Department of Commerce. Both the scale and the base proved ideal. A scale of 1:1,000,000 permits a general overview of the biotic communities of the Southwest region, while allowing enough detail to identify specific areas. It is a convenient scale where 1 mm = 1 km; this simplification for distance with latitude and longitude intersections greatly facilitated plotting the locations of towns, highways, climate stations, mountain ranges, and other landscape features. The scale is increasingly used for natural resource maps (e.g., Comisión Técnico Consultiva Para La Determinación Regional De Los Coeficientes De Agostadero, 1974; Küchler, 1977; Donart et al., 1978b; Brown, Carmony, and Turner, 1979). The aeronautical charts used for base map construction provide much topographic detail including mountains, dune fields, wetlands, and, importantly, 1,000-foot (305 m) elevation contours. Such features greatly increased the serviceability of the base map in facilitating the delineation of biotic communities.

The field maps used were U.S. Geological Survey maps at scale 1:500,000 for Arizona, New Mexico, and southern California. The states of Texas, Colorado, Utah, Nevada, and the Mexican states were mapped on 1:1,000,000 aeronautical charts without navigational enhancement.

Historical Background to Southwestern Ecological Studies

David E. Brown
Arizona Game and Fish Department

W. L. Minckley
Department of Zoology
Arizona State University

and James P. Collins
Department of Zoology
Arizona State University

The Southwest is justifiably famous as a focal area for North American natural history inquiry and discovery. Many of America's great conservationists and naturalists, President Theodore Roosevelt, Charles Sheldon, John Wesley Powell, and Aldo Leopold, were strongly influenced by its spectacular and diverse environments. In addition, many eminent biologists and ecologists were lured to work in the Southwest, C. Hart Merriam, E. W. Nelson, E. A. Goldman, Joseph Grinnell, Homer Shantz, Forrest Shreve, F. E. Clements, and more recently Walter P. Taylor, Carl L. Hubbs, Robert MacArthur, and Donald Tinkle. Four presidents of the Ecological Society of America worked and made their homes in Arizona (Shantz, Shreve, Charles Vorhies and Taylor; Burgess, 1977). As a result, much understanding of the continent's biogeography originated here with the work of Merriam, Shreve, Lee R. Dice, and their associates, students and successors.

The main reason the Southwest has contributed so greatly to knowledge of natural history is the juxtaposition of North American biomes so obvious in the region. Communities of plants and animals are often clearly demarcated by elevational gradients. Consequently, habitats in vastly different biomes may be readily sampled during a day's fieldwork. Reaching a similar diversity of habitats could literally require days of travel in other parts of the continent.

A second reason favoring Southwestern studies is the excellent historical record of this region, especially the extensive natural information obtained from relatively continous fossil records. There are Tertiary floras complete enough for interpretation of past elevations, climates, and other ecological features, as well as the lineages of existing species and even communities (Axelrod, 1958b, 1979). Fossil to recent pollen in sediments of closed drainage basins have added another paleobotanical technique for interpretations of change, or lack of it (Martin, 1963; Martin and Mehringer, 1965; Meyer 1973, 1975; many others). Complementary data are available from such diverse sources as ancient woodrat middens (Wells, 1978), vertebrate remains and other materials in cave deposits (Van Devender and Worthington, 1978; Van Devender et al. 1978), tree-ring chronologies (Fritts, 1974; Fritts et al., 1979), and archaeological sites (Mehringer, 1967). Aridity of the region has preserved a remarkable amount of these data, and a vast amount of information is undoubtedly yet to be revealed. Another advantage to understanding Southwest environments is that we have descriptions of how things were prior to invasion of technological culture. This was made possible by the Southwest's relatively late and low density of occupation by Hispanic cultures, and an even later acquisition of its northern half by Anglo-Americans.

The chapters which follow illustrate that Southwestern ecosystems are distinctive because of the particular evolutionary history of their geographic locations. These communities and their component plant and animal species, however, are similar to their ecological equivalents in other parts of the world (Mabry et al., 1977; Simpson, 1977; Orians and Solbrig, 1977). It is not surprising, therefore, to find that ecological studies have contributed not only to a record of the natural history of the region, but also to an understanding of general ecological principles of worldwide application. Thus, some investigations describe the natural history of organisms such as Saguaro Cactus (Niering et al., 1963) and/or processes such as arroyo cutting (Cooke and Reeves, 1976), that are either peculiar to, or particularly well illustrated in, the Southwest. Other, more general investigations in the region have illustrated or suggested basic ecological concepts or patterns for the first time,—concepts such as the life zones of Merriam (1890, 1898). Southwestern ecological studies have also incorporated theoretical and methodological biases prevalent in contemporary ecological studies conducted in areas outside the region. Late 19th and 20th Century ideas on predator control and wildlife management, for example, were applied to the Kaibab Plateau of Arizona with disastrous results from 1920 to 1939 (Rasmussen, 1941).

Historical development of our knowledge of the ecology of Southwestern ecosystems recapitulates a pattern frequently seen in development of ecological knowledge in other parts of the world. In general, Southwestern ecological studies fall into three overlapping periods of development: 1) *Exploration*, 2) *Inventory*, and 3) *Synthesis*. Those who studied and described the natural environment during each of these periods often were distinctive individuals, and formed the basis for much of the legend and lore of the region (see for example Corle, 1951). Their occupations, like those of their counterparts at other times and in other parts of the world, typically reflected their motives for being in the Southwest. Thus, natural history reports during the period of exploration (from 1807 to 1900) were prepared by trappers, personnel of military surveys, and physician-naturalists assigned to such surveys.

The inventory period (from 1850 to 1940) saw hunter-naturalists and collectors beginning the task of cataloging and preserving representatives of Southwestern plant and animal life. During this period, a movement with its origins in sportsmen's social clubs of the Northeastern United States spawned a group of wildlife advocates known as "conservationists" who also reported on the natural history of the West. Even as the cataloging process was just seriously beginning, these men fought to obtain laws and lands to protect wildlife species and wilderness areas that were already vanishing in the 1800s.

The synthetic period (from 1890 to the present) has been characterized by studies and publications of professional ecologists from academic, governmental and private institutions. Many of these professionals have made their homes in the Southwest, sharply distinguishing them from the largely transient researchers of previous periods. These scientists contributed greatly to maturation of ecology as a science. More recently, the work they started has been reflected in species inventories and habitat descriptions that typically accompany modern-day "environmental impact statements." Contemporary students of Southwestern ecology are trying to understand the causes of patterns initially described by these workers. Ecologists now use long- and short-term population, community, and ecosystem studies, as well as comparative and experimental approaches, to demonstrate how distribution and abundance of plants and animals are affected not only by variation in abiotic factors, but also by processes such as competition, predation, and nutrient availability.

The objective of this chapter is first to summarize contributions of investigators characteristic of each of the major periods of ecological development. Incidentally, it will be noted how their work reflected theoretical and methodological biases of the science of ecology at the time they conducted their research. Secondly, we summarize ways in which these studies contributed to our knowledge of natural history of the Southwest, and perhaps how they influenced in a more general way ideas concerning basic ecological principles beyond the region.

Exploration: Trappers, Military Surveys, and Physician-Naturalists.

Beaver trappers were important among the first people of Northern European extraction to enter the Southwest (Davis, 1982). Between 1823 and 1846, these so-called "mountain men" found an unsettled land sparsely occupied by people at a Neolithic stage of development. Except for a few outposts, pueblos or presidios at Santa Fe, Janos, Taos, Tubac, San Diego, and at the slightly more numerous missions, Hispanic presence was local and, with the important exception of introduced livestock, largely passive. So it also was with the beaver trappers who, disdainful of Spanish and later Mexican laws against their vocation, and even more of farming, mining, ranching, and other sedentary pursuits, left the wilderness and its inhabitants largely intact. Some, like James Ohio Pattie (1962), kept journals or narrated memoirs, thereby leaving valuable accounts of a soon-to-vanish wilderness.

As the United States expanded its boundaries westward, newly acquired lands had to be surveyed, mapped and described. This job fell to the military. Most survey parties had in attendance a physician who doubled as a botanist, zoologist, or naturalist. Although most early expeditions were limited to collecting and cataloging specimens, this in itself was a great contribution. Explorer-naturalists provided initial descriptions of species, upon which later, more detailed studies were based.

As early as 1807, military sorties had begun to penetrate the Southwest. In that year Maj. Zebulon Montgomery Pike (1810) "wandered" into Spanish territory by traveling through Plains grasslands up the Arkansas River to the Rocky Mountains, then south to Santa Fe on what was termed exploration. Pike and his party were detained, then escorted by Spanish militia from the region *via* the Rio Grande to El Paso, thence to Ciudad Chihuahua and eastward to be released. This trip revealed a vast, somewhat unexpected area, plus potential military routes west to California.

Stephen Harriman Long's Army Corps of Engineers returned in 1820 from an expedition to the Plains and Rocky Mountains through what is now New Mexico. Santa Fe was then the Southwest's trading center and capitol. Accompanied by such naturalists as Thomas Say (now honored by Say's Phoebe, *Sayornis sayi*) and Edwin James (*Hilaria jamesii*, *Eriogonum jamesii*, *Jamesia americana*) this expedition recorded a number of important new plants and animals including Limber Pine (*Pinus flexilis*) and the Rocky Mountain Mule Deer (*Odocoileus h. hemionus*). Biological discovery of the Southwest was underway.

The great soldier-explorer John C. Fremont led the next major expedition along the Old Spanish Trail. He moved from the west, and entered the Mohave Desert at Tehachapi Pass in 1844. Leading a detachment of topographical engineers and accompanied by Christopher "Kit" Carson, Fremont crossed southern Nevada, camping at what is now Las Vegas before continuing up the Virgin River into the Great Basin Deserts of southwestern Utah. Fremont was a true naturalist and collected numerous plants enroute including *Ambrosia dumosa, Coleogyne ramosissima, Pinus monophylla,* and *Populus fremontii.* Although much of his plant collection was later lost in winter snow and summer flood, extensive material remained. The plants were sent to John Torrey at Columbia University, who described them and named the cottonwood after Fremont. It was Fremont who appropriately coined the term "Great Basin" for much of the Intermountain West.

In the mid-1840's the United States acquired what is now the northern half of the American Southwest, Texas, southern California, and New Mexico (the last of which included what was to become Arizona), and military surveys began in earnest.[1] In 1846, while war with Mexico was in progress, Lt. William Hemsley Emory, a red-headed engineer assigned to Col. Stephen W. Kearney's "Army of the West," was commissioned to map from the "Rio Grande to the Pacific", with 14 topographical engineers. It was on this expedition that the Saguaro Cactus (*Cereus giganteus*) was first collected and made known to science. The Roundtail Chub (*Gila robusta*) from the Gila River was accurately illustrated in Emory's (1848) journal, although it was technically described from material collected elsewhere. Emory's journey was followed by three consecutive boundary surveys after the end of war in 1848, the last in 1855 again under Emory, now a major. A. L. Heerman, C. B. R. Kennerly, and Arthur Schott were scientists accompanying these surveys. They collected mammals, fishes, and plants for the U.S. National Museum to be housed at the newly built (1852) Smithsonian Institution. Journals were kept, botanical and zoological reports were submitted, and Congress authorized the printing of 3,000 copies of the reports (1859).

After collecting in Texas, Dr. S. W. Woodhouse, physician and naturalist, accompanied Capt. Lorenzo Sitgreaves' survey of northern New Mexico and Arizona in 1851. It was on that trip that Woodhouse described and named Abert's Squirrel (*Sciurus aberti*) after Lt. James William Abert who was John C. Fremont's brother-in-law and head of the Corps of Topographic Engineers. Abert himself had explored northeastern New Mexico, in 1845 prior to joining Emory's expedition to New Mexico, Arizona, and California in the following year. Woodhouse's travels took him up the Rio Grande from El Paso to the Rio Puerco and on to Laguna. The Sitgreaves party thence proceeded down the Zuñi River to Arizona. Along this last route two western big-river fishes, the Bonytail Chub and Roundtail Chub (*Gila elegans* and *G. robusta*) were discovered and named (Baird and Girard, 1853).

In 1853, C. B. R. Kennerly was again an attending physician and naturalist with Lt. A. W. Whipple's expedition attempting to find a railroad route along the 35th parallel. Accompanied by the German artist-naturalist Heinrich B. Möllhausen, they collected a vast quantity of material for the U.S. National Museum along the trail from Albuquerque to the Colorado River. Both Kennerly and Möllhausen later explored and

collected along the Little Colorado and Colorado rivers under Lts. Beale and Joseph C. Ives (1857-1858), the former of Camel Corps fame.

Later military surveys attempted to be even more specific in nature. The G. M. Wheeler Survey of 1871-74 reported on geology of the region, as did the civilian John Wesley Powell Expeditions (Powell, 1875) to the Colorado Plateau and Colorado River and the F. V. Hayden Expedition (Hayden, 1873) to the north. Henry W. Henshaw, an ornithologist with the Wheeler party, collected extensively, especially in New Mexico, as did Joseph T. Rothrock, the U.S. Army botanist assigned to the survey. The botanist George Vasey made an extensive plant collection during the Powell survey, and he later became curator at the U.S. National Herbarium.

The last, and in many ways the most productive survey with a primary geographic mission, was the U.S. and Mexican Boundary Survey of 1892-93. It had as its physician and naturalist Edgar Alexander Mearns. He was primarily interested in mammals and published a treatise (Mearns, 1907) on the natural history of the Southwestern boundary with Mexico, and a descriptive catalog of its mammals. His collections of other organisms added substantially to our knowledge of the overall Western biota (*e.g.* fishes, Snyder, 1915; plants, Britton, 1889). Mearns continued as a collector and investigator in the United States and Mexico until his death from diabetes at age 61.

Although the primary purpose of these surveys was the pragmatic task of determining boundaries, finding railroad routes, and ascertaining geological wealth, the government also valued documentation of the region's living resources. Naturalists accompanying these surveys kept journals as did many of their commanders and engineering colleagues. Given the pristine conditions they encountered, their recordings and descriptions are now invaluable. These surveys became the forerunners of the Geological Survey of the U.S. Department of the Interior and the Biological Survey of the U.S. Department of Agriculture. The latter was first designated in 1889 as the Division of Economic Ornithology and Mammalogy of the USDA.

The surveys had increasingly changed from those of an exploratory nature to scientific expeditions, requiring their participants to have an ever greater scientific expertise. The army was not the best organization to direct detailed scientific endeavors, and after the Wheeler Survey it became apparent that science would best be served by multidisciplined teams of civilian professionals, like the Powell Surveys. Although a few major operations were organized to explore remote areas, the period of general exploration *per se* thus came to a close by about 1880. Surveys henceforth were limited to restricted geographic areas, selected not for their military or political importance, but because there was little biological knowledge of the area. Thus began the collecting period.

Inventory: Hunter-Naturalists, Collectors, and Conservationists.

Even before 1800, expeditions from Europe had been seeking out the world's remote places with the objective of locating new plants and animals. Many of these expeditions were led by big game hunters, some by retiring botanists, others by royalty, but all their findings were eagerly awaited by scientists of all kinds. Little was contributed to North American zoology by North American scientists prior to 1800

---

[1] *A most important exception to the military efforts was physician, ornithologist and botanist, William Gambel, who in 1841 discovered and collected many new species on a wagon trip along the Old Spanish Trail from Aliquin to California. Gambel was a friend of the great naturalist Thomas Nuttall, who described and named several important species for him, including Gambel Oak (*Quercus gambelii*) and Gambel's Quail (*Lophortyx gambelii*).*

Figure 6. *Clinton Hart Merriam about 1910 after he published* Life-zones of the San Francisco Mountains, Arizona. *Merriam was a collector and naturalist when he became the first Chief of the U.S. Office of Economic Mammalogy and Ornithology (later to become the U.S. Biological Survey) from 1879 to 1885. After 1885 he conducted numerous biological and anthropological investigations including a study of the status of the Fur Seal on the Pribiloff Islands and his work on the San Francisco Peaks. This latter work resulted in his famous classification of North American life-zones,—a system still in use in the Southwest today.*

(Myers, 1964). Formation of the Academy of Natural Sciences of Philadelphia in 1812 started to change this condition, and the vast biota of Eastern North America began to be studied in detail. With exploration of the Southwest, collecting increased, and institutions such as the Museum of Comparative Zoology at Harvard University espoused a policy of preserving large collections of organisms. Reports of discovery and description of new forms constituted much of the scientific literature in this period.

A large part in development of this phase of descriptive science in North America was played by the prominent Swiss zoologist Louis Agassiz. Although Agassiz scarcely dealt with Southwestern animals, his influence after emigrating still is reflected in the works of his academic descendants. As

pointed out by Hubbs (1964), "few of the more active ichthyologists of the present day are unable to trace their lineage back to Jordan, and through him to Agassiz," and analogies hold for other biological disciplines as well.

Studies in the American Southwest were enhanced by establishment of the Smithsonian Institution in 1846, with its National Museum, and eventually (in 1890) the National Herbarium. That institution and the prestigious Academy of Natural Sciences of Philadelphia sponsored a large percentage of the published literature in the period from 1850 to 1875. Collections of animals and plants made during exploratory expeditions and surveys were housed in those repositories. Important collections of fishes for example, including much of the Southwest's known fauna, were described by Charles Fredrick Girard, alone or in co-authorship with Spencer Fulton Baird. The first of these men emigrated from France to study American fishes with Louis Agassiz. The second was selected as Assistant Secretary of the Smithsonian Institution in 1850, and brought in 1871 an early and vigorous beginning to the U.S. Fish Commission. That agency gave rise in turn to the Bureau of Fisheries, which was ultimately combined with the Bureau of Biological Survey to become the Fish and Wildlife Service. Edward Drinker Cope played much the same role as Baird and Girard in describing numerous reptiles and amphibians from the surveys (Cope, 1866), and in studying not a few fishes (Cope and Yarrow, 1875).

The policy of assignation of collectors to field stations by the Smithsonian Institution resulted in Dr. Elliot Coues' mission to Fort Whipple, Arizona, in 1864. His assignment was to collect and prepare specimens of wildlife from the Rio Grande to the Colorado River. This he did, and his *List of Birds of Fort Whipple, Arizona,* published by the Academy of Natural Sciences of Philadelphia in 1866, and his *Quadrupeds of Arizona,* published by the American Museum of Natural History in 1867, are the first scientific papers on Southwestern wildlife other than species descriptions. These works are even more impressive when one remembers that much of the material was obtained in unsettled territory inhabited only by military personnel and Indians. Coues was the first to report Bobwhite Quail (*Colinus virginiana*) in Arizona. The diminutive Coues' Whitetail Deer (*Odocoileus virginiana couesi*) inhabiting Arizona, New Mexico, Sonora, and Chihuahua, was appropriately named for him.

Although many collectors were primarily concerned with discovering and naming new species, all believed that they contributed greatly to the knowledge of natural history, and so they did. Although taxonomists are dependent on collections for their studies, ecologists, zoogeographers, phytogeographers, paleobiologists, and evolutionary biologists also depend on these resources for their research. David Starr Jordan (whose long and successful career included the latter part of the 19th and the early part of the 20th centuries) lauded these contributions in dedicating the 13th edition of the *Manual of Vertebrate Animals* in 1929 "To five of my early students, 'brought up on the Manual of Vertebrates' ... and to five others, equally gifted, who lost their lives while engaged in field work." We owe a great deal to these early naturalists who frequently financed, organized, led, and died on their expeditions.

After the Civil War, settlement of the Southwest proceeded more rapidly. Although some like George Thurber began collecting plants around what is now Silver City, New

Mexico, in the 1850's, the 1870's saw several collecting trips to trans-Pecos Texas, and into New Mexico. A few like John G. Lemmon (and later his wife, Sara Plummer Lemmon) collected plants in southern California, the Mohave, Great Basin, and Sonoran deserts, and throughout Arizona when these areas were still relatively unsettled (Crosswhite, 1979). It was not until the 1880's, however, with suppression of the Apaches, that museum expeditions could be sent out in earnest for Arizona, Sonora, and Chihuahua. This was the peak of the survey and collecting period. Identification and description of subspecies as well as species proceeded actively. For the next 60 years racial characters as well as animal distributions were considered necessary parts of an understanding of the evolution and taxonomy of our biota. Investigators traveled by train and hired horse and wagon or else hiked to reach collecting sites; the most eminent scientists of the day participated.

The Death Valley Expedition under the Division of Economic Ornithology and Mammalogy and composed of no less than C. Hart Merriam, V. Bailey, Edward W. Nelson and others, covered the vast deserts of southern Nevada and southeastern California in 1891. Hence the names of White River Springfish (*Crenichthys baileyi*), Ash Meadows Poolfish (*Empetrichthys merriami*) and Desert Bighorn (*Ovis canadensis nelsoni*). Nelson, accompanied by Edward A. Goldman, initiated studies of mammals and birds in Mexico in 1892, also under the Division, and ended it in 1905-06 with a trip by pack animals through the entire length of Baja California (Goldman, 1951). Goldman went on to collect extensively throughout the Southwest and elsewhere in Mexico, and for the next 40 years his name was associated with more than 100 published works dealing with Southwestern animals and landscape descriptions. Nelson also collected throughout the Southwest. He first collected the native elk in Arizona (the so-called Merriam Elk), and later proved to be a most able administrator of the Biological Survey. In subsequent years, Barton W. Evermann and Cloudsley M. Rutter reported on fishes of the Colorado River basin in 1895, and Seth Eugene Meek began his studies of the fishes of Mexico (Meek, 1902, 1903, 1904), under the auspices of the Field Columbian Museum, now the Chicago Natural History Museum.

Collecting and compilation of descriptive information was not limited to the Field Museum, the U.S. National Museum and the Philadelphia Academy. The Museum of Vertebrate Zoology at Berkeley, the San Diego Natural History Museum, the University of Michigan Museum of Zoology, and as years passed, the Museum of Natural History at the University of Kansas, and others, sponsored collectors and housed specimens. Private collectors and collections flourished. A. W. Anthony collected in New Mexico in 1886 and again in 1889 for the Merriam collection. H. H. Rusby (1889) collected plants under pharmaceutical industry sponsorship in southwestern New Mexico and Arizona. Anthony's specimens went to the National Museum as a result of Merriam's employment as first head of the U.S. Biological Survey.

Most private collections eventually found their way to a university. This is fortunate in lieu of the fact that the Southwest has never had a regional or even a state-based depository for natural history material. Some of the most significant contributions were the D. R. Dickey collection of vertebrates. It was Dickey who financed the important trips of A. J. van Rossem in Sonora and southern Arizona, and some

early travels of L. M. Huey in Baja California. The Herbert Brown (editor for the Tucson Citizen newspaper) Ornithological Collection (now at the University of Arizona) was another important contribution of the later 19th and early 20th centuries (Phillips *et al.*, 1964). W. H. Burt (1932, 1938) reported on his and other collections of mammals in Baja California and Sonora, respectively.

Vernon and Florence Merriam Bailey (C. H. Merriam's daughter) collected throughout New Mexico for the Bureau of Biological Survey from 1889 to 1924. Results of this long and fruitful effort are V. Bailey's *Mammals of New Mexico* (1931; recently reprinted as *Mammals of Southwestern United States*) and Florence Bailey's (1928) *Birds of New Mexico*. Scattered descriptions and new information on the Southwest's fishes were summarized in the monumental *Synopsis of Fishes of North America* by Jordan and Gilbert (1883), and in the even greater *Fishes of North and Middle America* by Jordan and Evermann (1896-1901). In the case of fishes, these works suppressed research for a number of years, with their completeness overshadowing needs for further studies (see *e.g.*, Hubbs, 1964). Results of the remarkably extensive biological exploration in Mexico by Nelson and Goldman did not appear until far later (*Lower California and Its Natural Resources* by Nelson, 1922; *Plant Records of an Expedition to Lower California* by Goldman, 1916, *Mexican Tailless Amphibians...*by Remington Kellogg, 1932; *Birds of North and Middle America* by Robert Ridgeway and Herbert Friedmann, 1901-1950; and *Biological Investigations in Mexico* by Goldman, 1951).

The extensive collections of E. O. Wooton, now in the U.S. National Herbarium, were gathered mostly from New Mexico in the late 1890's and early 1900's. They with other collections at the National Herbarium and New Mexico State University, were the basis fore the pioneer work by Wooton and Standley, entitled *The Flora of New Mexico* published in 1915. Meanwhile, P. C. Standley was collecting in Mexico preparatory to his five-part opus, *Trees and Shrubs of Mexico*, published between 1920 and 1926 by the Smithsonian Institution, while he was Associate Curator at the U.S. National Herbarium.

Other early Biological Survey collectors of note were the entomologists T. D. A. Cockerell and C. M. Barber (who amid myriad accomplishments sampled in the Sacramento Mountains and Mesilla Park areas of New Mexico), J. A. C. Rehn, and Walter P. Taylor. The latter is probably best known for his editorship of *The Deer of North America* (Taylor, 1956). Cockerell, like E. A. Goldman, Vernon Bailey, and other collectors, was also greatly interested in biogeography.

Synthetic Period: The Biogeographers.
In 1890, Clinton Hart Merriam (Fig. 6) published his *Results of a Biological Survey of the San Francisco Mountain Region and Desert of the Little Colorado, Arizona*, thereby ushering in a unique life-zone concept of delineating habitats. This system, which equated elevational zones with biogeographic realms within the North American continent (Fig. 3) proved extremely useful in the Southwest where it was developed. Initially used by Merriam's Bureau of Biological Survey colleagues, it has been generally applied in descriptions of habitats for more than 80 years (see *e.g.* Lowe, 1964). Development of a common language to describe habitats facilitated efforts at additional ecological surveys by the

Figure 7. *Forrest Shreve about 1940. Shreve truly provided the key to our understanding of the North American deserts. A prolific as well as intuitive botanist, Shreve's work on the effects of climate on plant distribution made him one of the great ecologists of our time. Photograph courtesy of Mrs. Margaret Shreve Conn.*

Bureau, resulting in Cockerell publishing *Life-Zones in New Mexico* (1897, 1898), and V. Bailey publishing a *Biological Survey of Texas* (1902), *Life-Zones and Crop-Zones of New Mexico* (1913) and other works, in and outside of the Southwest. Other worthy surveys tying terrestrial vertebrates to prescribed life-zones were accomplished by Joseph Grinnell, another Merriam advocate working with the University of California. In our region these surveys resulted in a publication entitled *The Biota of the San Bernardino Mountains* (1908), an account of the San Jacinto Mountain area with H. S. Swarth (1913) and an account of the lower Colorado River (1914). E. Raymond Hall also published information on the life-zones of Nevada in his monumental *Mammals of Nevada* (1946).

These and numerous other early efforts first resulted in a correlation of plants and animals occurring in Southwestern environments. By the 1920's, correlations with climatic variables were recognized on a continental basis (Merriam, 1898; Shreve, 1917; Shantz and Zon, 1924; Shelford and Shreve, 1926). It now remained only to continue to implement the use of vegetative descriptions within the life-zone system on a more intensive basis.

This endeavor was largely underway by the 1930's. A. A. Nichol (1937) described and mapped Arizona's natural environments, Tharp (1939) summarized those in Texas, and Melvin Morris (1935) mapped Colorado's natural vegetation.

Gloyd (1937) dealt with reptiles in southern Arizona. Swarth (1929) and Phillips (1939) designated "faunal areas" on the basis of birds. Studies reviewed by Lowe (1964) dealt with geographic regions based on congruent distributional ranges of several animal groups. Furthermore, the Mexican half of the Southwest was being inventoried, mapped, and described in the context of vegetational zones or biomes. The Carnegie Institution of Washington funded Forrest Shreve's (Fig. 6) great work on arid lands (1934, *et seq.*) in the United States and Mexico. Similarly, H. S. Gentry (1942) was cataloging and describing the tropical and sub-tropical vegetation of southern Sonora and Sinaloa. D. D. Brand's (1936, 1937) descriptions and maps of northwest Mexico, Harde LeSueur's (1945) work on the vegetation of Chihuahua, E. G. Marsh's largely unpublished investigations of Cuatro Ciénegas and elsewhere in Coahuila (Hubbs and Miller, 1965; Minckley, 1969; Pinkava, 1979), C. H. Muller's (1939, 1947) studies in Nuevo Leon and Coahuila, and Elzada U. Clover's (1937) work on the lower Rio Grande in Texas, all were accomplished in the 1930's. This was also a period of great productivity in the fields of natural history and ecology in general. The successional and climax principles of Weaver and Clements (1938), the biome concepts of Shelford (1932, 1945), Pitelka (1941), and Lee R. Dice (1939, 1943) used in this book, were also products of that decade. Wieslander's (1932-1940) remarkable vegetation maps of California at 1:125,000, Forest Shreve's (1951) brilliant treatment of the Sonoran Desert, E. Lucy Braun's (1950) treatise on the Eastern Deciduous Forest, and Weaver and Albertson's (1956) analysis of grasslands of the Great Plains were all products of the 1930's.

Study and interpretation of wetlands and their biota in the Southwest lagged behind similar terrestrial pursuits. However, an event in 1894 in Williams, Arizona, the birth of Carl Leavitt Hubbs (Fig. 8) was to insure additional search for knowledge of Southwestern fishes despite the doldrums following publication of *Fishes of North and Middle America*. After a childhood in southern California, Hubbs was influenced by Loye Holmes Miller and George Bliss Culver to enter Stanford University, where under the tutelage of David Starr Jordan and C. H. Gilbert he became one of the leading naturalists in North America, a position he retained even after his death in 1979 (Norris, 1974; Pister, 1979). Hubbs' efforts along with those of his student, son-in-law, and colleague, Robert Rush Miller resulted in one of the largest collections of freshwater fishes in the World at the University of Michigan Museum of Zoology, a large percentage of which are from the American Southwest. Their work on fishes contributed greatly to our knowledge of biogeographic relations in Western deserts and elsewhere. Hubbs' early collecting in streams and lakes of the arid West with John

Figure 8. *Carl L. Hubbs about 1954. Although Hubbs' contributions to fish management, systematics, and research were continentwide, his interest in palaeoclimate resulted in a great contribution to understanding Southwest fish evolution. Photograph by Lewis W. Walker.*

Otterbein Snyder led him back to the region. Hubbs, Miller, and their families traced ancient shorelines, delineated former watercourses, and studied the remarkable relict fish populations of isolated pools and streams, deriving hydrographic interpretations of climate, geology, and biology, which stand as a challenge to even the most astute geologists of today (Hubbs and Miller, 1948; Miller, 1948, 1959; Hubbs et al., 1974; and many others). Their dedication to systematics, conservation, and natural history of Western fishes was exceptional, yet the Southwest's wetland environments went largely unnoticed until very recently—mostly until the last two decades. Indeed, the wetland section presented in this volume is the first holistic treatment of these valuable and neglected resources.

The late 1930's and early 1940's was a high-water-mark period of general natural history studies. The job of resource inventory must, however, continue to better interpret and understand our natural history heritage, and to communicate information and ideas of what is and what has been. This is the purpose of this publication and its companion map—to communicate the accumulated knowledge of Southwestern ecology and biogeography.

Part 1. Tundras

*Alpine tundra on San Francisco Mountain,
Coconino National Forest, Arizona. Here, in this
barren rock-fell community at an elevation of
about 3860 m, the only plants are scattered sedges,
lichens, and grasses in what is essentially a desert
landscape within an arctic-boreal environment.*

Arctic and Alpine Tundras

The low, treeless vegetation of the polar regions and high mountains is commonly designated "tundra." Actually, tundra is a general term properly used to describe any treeless landscape in the arctic, the antarctic, or on mountains above timberline (alpine landscapes). Tundra may consist of grassland, scrubland, marshland (wet tundra), or desertland in an Arctic-Boreal climatic zone. Accordingly "polar desert" and other open herbaceous, alpine ecosystems are considered tundra, and tundralands are regarded as the Arctic-Boreal equivalent of desertlands.

Whether arctic or alpine, all tundras have a short growing season with a low mean air temperature (often at or near 0°C). Such short-season moisture availability precludes the establishment of all but short-stemmed perennial herbaceous plants, prostrate schrubs, lichens, and mosses (Billings, 1973a; Thilenius, 1975). Aside from these characteristics, many of the floras of the world's tundras are more distinctive than similar, despite the natural tendency to equate them because of certain shared growth forms, plant genera, and physical parameters of their environments.

Tundras in North America extend from sea level, at latitude 83°N, on Ellesmere Island, southward to above 4,875 m on the volcanos of central Mexico (frontispiece photo for chapter). Disjunct alpine tundras in the United States occur in the Sierra-Cascades, Great Basin, Rocky Mountains, and northern Appalachians. The southernmost *Rocky Mountain alpine tundra* reaches the Southwest in the San Juan and Sangre de Cristo Mountains in southern Colorado and northern New Mexico; outliers occur on San Francisco Mountain, Arizona and on Sierra Blanca in the Sacramento Mountains in south-central New Mexico. Depauperate examples of *Great Basin* and *Sierra Nevadan alpine tundra* are found on the White Mountains of eastern California, isolated Mount Charleston in southern Nevada, and on Mount San Gorgonio and Mount San Jacinto in southern California. Further north, many of the basin ranges in Nevada (e.g., Toiyabe, Toquima, and the Grant Range) have well developed alpine ecosystems (Billings, 1978).

111.5 Alpine Tundra

Charles P. Pase

USDA Forest Service

Alpine tundra in the Southwest is often a small summit area above timberline. It is sometimes construed to include the krummholz, the twisted and gnarled subalpine scrub (elfinwood) of prostrate conifers—lying between the forestline and treeline—distorted by almost constant winds on these high mountains peaks, ridges, and slopes. Alpine tundra vegetation consists primarily of low-growing woody shrubs, diverse herbaceous plants, lichens and mosses, all adapted to a brief and often interrupted growing season. Every year all are subjected to severe subfreezing temperatures, severe physiological drought, and intense insolation.

Major and Taylor (1977) described a xerophytic or "Mediterranean" alpine tundra in southern California mountains. The more extensive alpine tundra of northern New Mexico, with outliers in southern New Mexico and central Arizona, more nearly fits the classical alpine type (Tomalchev, 1948). These communities represent extensions of the much larger alpine tundra of the main Rocky Mountain cordillera, which has provided a pathway for north-south migration during the colder periods of the Pleistocene (Billings, 1978). During ice-free periods, including the present one, many species which once were more generally distributed survive only on summit areas, giving unique if depauperate identities to Southwest alpine plant communities.

Floras of isolated mountain masses, such as those encountered in the Southwest, tend to show more relationship to some floras of surrounding lowlands, and the number of alpine species derived from adjacent lowland progenitors probably depends on the time available for evolutionary development, and the diversity of niches available (Chabot and Billings, 1972). In the case of southern California alpine summits, the proportion of locally derived species to far-ranging ones tends to be high, indicating long isolation from the great Sierra massif to the northeast. These isolated alpine tundras of southern California have even less floristic similarity to the Southern Rockies; the great desert barriers have effectively isolated them even during the glacial epochs of the Pleistocene, when the alpine zone probably extended some 1,200 m lower than at present (Moore, 1965; Billings, 1978). Consequently, the alpine floras of San Gorgonio and San Jacinto are impoverished; they include a higher percentage of endemics and species closely related to adjacent subalpine and high desert forms than is found either in the High Sierra or the Southern Rockies. The alpine tundra of San Francisco Mountains, Arizona, midway between the Southern Rockies and the Southern California summits, has provided an important refugium for tundra plants during warm interglacial periods (Billings, 1978).

Colorado and New Mexico

Here alpine tundra consists of relatively small, isolated areas above timberline, stretching through southern Colorado to 32 km northeast of Santa Fe, along the main crest of the Sangre de Cristo Mountains, and on high peaks above 3,500 m in the San Juan Mountains in southwestern Colorado. The major mountain peaks—Pecos Baldy, Santa Fe Baldy, Truchas Peak, Wheeler Peak, Conejos Peak, Montezuma Peak, etc.—range from 3,810 m to 4,011 m, with a substantial area of intervening high country above 3,500 m (Fig. 9). A small, marginal, alpine grassland area is also found at the 3,659-m summit of Sierra Blanca, on the Mescalero Apache Indian

Figure 9. *Walter Peak (right background) rises to 4,002 m elevation in the Wheeler Peak Wilderness of the Carson National Forest in northern New Mexico. The sedge-grass-forb community here is dominated by the low, mat-like growth of* Geum turbinatum, Festuca brachyphylla, *and species of* Carex.

Reservation, and represents the southernmost alpine community in the United States (Moir, 1967).

Few ecological studies have been made of New Mexico's alpine tundra, and recognition of series and associations is tentative. Two series are recognized: a *Lichen-Moss Series*, confined to scree slopes and rocky, soil-less sites, and an *Avens-Sedge Series* on more favored sites where soil has accumulated and the low, hardy alpine vascular flora has gained a toehold. The Avens-Sedge Series has been termed "Meadow Series" by the New Mexico Interagency Range Committee (1978).

The alpine tundra along the crest of the Sangre de Cristo and San Juan Mountains is floristically diverse for alpine floras. Some 143 vascular plant taxa have been reported there, although many of these are also found at lower elevations, especially in the subalpine forests and grasslands (Wooton and Standley, 1915; Tidestrom and Kittell, 1941; Harrington, 1954; and MacKay, 1970) and, therefore, are not exclusively alpine tundra species.

In a recent study of the relationship of the mountain flora of Wheeler Peak, New Mexico, with the flora of Blanca Peak, 120 km to the north in southern Colorado, MacKay (1970) observed that the alpine zone on Wheeler Peak (elevation 4,011 m) has 91 species of vascular plants, Blanca Peak (elevation 4,378 m) has 89 species, with 63 species common to both areas. Using Sorensen's index of similarity as suggested by Southwood (1966), the quotient of similarity (QS) between the two areas at the species level was 0.70. This high value suggests relatively easy migration during cooler, moister periods, recent isolation from a common flora, or both. Some 142 species of vascular plants have been reported from the alpine zone in New Mexico (Wooton and Standley, 1915; Harrington, 1954; Tidestrom and Kittell, 1941, MacKay 1970); this list is probably not complete.

Few shrubs are found in Rocky Mountain alpine tundra, although a number of root-perennial forbs may have woody or semi-woody bases. Two wetland species of *Salix* (*S. petrophila* and *S. planifolia*) mingle with species of *Carex* to form low, almost mat-like communities where adequate soil moisture can collect in microsites and where there is at least limited relief from the constant wind. *Salix petrophila* produces long, trailing stems that hug the ground; the short ascending vertical branches seldom exceed 15 cm. The latter willow may rise to 60 cm or 90 cm, although at lower elevations it forms a robust shrub 1.5 m or 1.8 m tall. *Ribes montigenum* may also grow in favored sites, seldom more than 60 cm tall, but it may provide limited but valuable wildlife cover.

Figure 10. *Avens-Sedge series above timberline in the Wheeler Peak Wilderness, Carson National Forest. Photograph by Clait Braun.*

Often a grass-like cover is present—sometimes enough to qualify as an alpine grassland. These communities are usually dominated by the ubiquitous *Carex*, which is an important component of almost every vascular plant community above timberline. Of the six species in New Mexico alpine areas (*C. aquatilis, C. ebenea, C. foena, C nova, C. petasata,* and *C. scopulorum*), three also occur in Arizona's alpine zone, but none in California's southern mountains. Intermingled with the sedges are varying amounts of Tufted Hairgrass (*Deschampsia caespitosa*), Alpine Fescue (*Festuca ovina* var. *brachyphylla*), Arctic, Greenland, Nodding, and Timberline Bluegrass (*Poa arctica, P. glauca, P. reflexa,* and *P. rupicola*), and Spike Trisetum (*Trisetum spicatum*). Other grasslike plants include the rushes *Juncus drummondii, J. albescens,* and *J. castaneus,* Millet Woodrush (*Luzula parviflora*), and Kobresia (*Kobresia myosuroides*).

Low mat and cushion forbs are common, and lend a characteristic appearance to the vegetation. Golden Avens (*Geum turbinatum*) is an important forb, dominating many mixed forb-sedge-grass communities from 3,350 m to 4,000 m (Fig. 10). Other characteristic forbs include Yarrow (*Achillea lanulosa*), *Erigeron simplex, E. formosissimus, Hymenoxys brandegei, Potentilla nivea, Pseudocymopterus montanus,* starworts (*Stellaria calycantha* and *S. umbellata*), Bigroot

Springbeauty (*Claytonia megarrhiza*), saxifrages (*Saxifraga* spp.), clovers (*Trifolium* spp.) including *T. nanum* common up to 4,000 m, and Rockjasmine (*Androsace carinata* and *A. septentrionalis*). Among the more showy forbs are Colorado Primose (*Primula angustifolia*), Snowy Erysimum (*Erysimum nivale*), Hayden Paintedcup (*Castilleja haydenii*), and the ubiquitous Bluebell (*Campanula rotundifolia*), which in this area may be found from robust forms at less than 2,450 m to the more depauperate but showy forms near the upper limit of vascular plant growth.

Few mammals live and breed in the alpine tundra—the forbidding climate is too severe for all but the most hardy or specialized species. One of the most characteristic is the Pika or Cony (*Ochotona princeps*), which may be seen scurrying among the rocks storing hay for winter use, or more likely identified only by its unmistakable shrill whistle or bleat. Other mammals breeding within the alpine zone, or sharing the border zone with subalpine forest and meadow, include the Colorado and Least Chipmunks (*Eutamias quadrivittatus* and *E. minimus*), Yellow-bellied Marmot (*Marmota flaviventris*), and the Vagrant and Masked Shrews (*Sorex vagrans* and *S. cinereus*). Fresh mounds of dirt within the forb-grass meadows indicate the presence of the Northern Pocket Gopher (*Thomomys talpoides*), and the mounds often become "gopher

Figure 11. *Above timberline in the Wheeler Peak Wilderness Area, a horseback party approaches the crest of the Sangre de Cristo Mountains.*

gardens," favorite sites for those species that flourish on raw, disturbed soils (Zwinger and Willard 1972). Although this area once was the grazing and lambing habitat of the Rocky Mountain Bighorn (*Ovis canadensis*), few of these animals can now be found in the Southwest's alpine tundra. The last native bighorn in the Pecos Baldy area was taken in 1902 (Johnson, 1979). The present small population was introduced in the early 1970's, and a small herd of 100 to 200 animals now again use the alpine areas of the Carson and Santa Fe National Forests.[1] Domestic Sheep (*Ovis ovis*) and their introduced diseases have essentially eliminated bighorns from most other high mountains of the Southwest.

Perhaps the best known of the alpine birds is the White-tailed Ptarmigan (*Lagopus leucurus*), a shy, grouse-like bird that breeds and winters above timberline. While just a few years ago the effects of habitat destruction (grazing) had substantially reduced populations, increased protection of the willows necessary for food and cover has permitted an increase of this recently rare bird in the Southwest (Hubbard, 1979).

Other nesting birds in the alpine zone in New Mexico

include the Water Pipit (*Anthus spinoletta*), White-crowned Sparrow (*Zonotrichia leucophrys*), and occasionally the Broad-tailed Hummingbird (*Selasphorus platycercus*), Brown-capped Rosy Finch (*Leucosticte australis*), and Horned Lark (*Eremophila alpestris*). While other birds are present, most are migrants or late summer visitors such as the Mountain Bluebird (*Sialia currocoides*) and Killdeer (*Charadrius vociferus*).

The alpine zone receives substantial recreational use from hikers and horseback riders (Fig. 11). Earlier heavy livestock use, particularly by domestic sheep, and present heavy recreational use exerts considerable pressure of the flora and fauna on the mountain heights. Once damaged, healing of the soil and recovery of the native biota is apt to be a slow and uncertain process.

Arizona

San Francisco Mountain, an extinct volcanic cone rising to 3,862 m elevation, is the highest point in Arizona, and contains the only well developed alpine tundra in the State (Fig. 12). The "Peaks" are located within the Coconino National Forest, about 16 km north of Flagstaff. Some 5 square kilometers lie above timberline, which occurs here at approximately 3,500 m. Illustrated accounts are given in Little (1941), Lowe (1964), Moore (1965), and Lowe and Brown

---

[1]*Personal communication with Art Renfro, Wildlife Biologist, USDA Forest Service, Region 3, Albuquerque, New Mexico.*

Figure 12. *The alpine tundra of approximately 5 km² atop the peaks of San Francisco Mountain (to 3,862 m elevation), 16 km north of Flagstaff, in the Coconino National Forest, in north-central Arizona.*

(1973). A small area on Mt. Baldy (3,533 m), in the White Mountains of east central Arizona, consists of small, marginal, alpine grassland at the summit. These areas are separated from the nearest alpine areas in the southern Rockies by 400 km of Great Basin desertscrub and woodland.

As might be expected from their small size and relative isolation, the Arizona alpine areas are more species-poor than their New Mexico counterparts. Approximately 88 species and subspecies of vascular plants are reported from Arizona alpine areas, including a few timberline species that penetrate the alpine fringe. The degree of similarity to the southern Rockies, however, is high; 68 of these taxa (76%) also occur in New Mexican alpine areas. On this basis, Sorensen's QS = 0.59 is a rather high value considering the relatively small size and wide isolation of the alpine communities on the San Francisco Peaks and Mt. Baldy.

There are three major habitats on the San Francisco Peaks: (1) alpine meadow, (2) boulder field or felsenmeer, and (3) fellfield or talus. The alpine meadow occurs only where adequate soil has developed, and contains the richest assortment of vascular plants found above timberline. The boulder field consists of extensive areas of large, often layered and overlapping rocks in a matrix of finer rock debris that provide partial shade and protection from the drying winds. The

fellfield is the most severe environment, with little soil and limited protection against insolation, sub-freezing cold, and drying winds. Lichens and mosses may abound, but the site is too severe for most vascular plants (Shreve, 1942c; Lowe, 1964; Schaack, 1970).

Approximately 82 species of vascular plants have been collected from the alpine zone on San Francisco Mountain (Little, 1941; Kearney and Peebles, 1960; Schaack, 1970). At the lower elevation limits, a gnarled and twisted krummholz of Bristlecone Pine (*Pinus aristata*), Corkbark Fir (*Abies lasiocarpa* var. *arizonica*), and Engelmann Spruce (*Picea engelmanni*) forms islands in the tundra, or borders the neighboring subalpine forest (Fig. 13). The *Geum turbinatum-Carex* Association is the most prevalent within the Avens-Sedge Series of the alpine meadows, and contains most of the alpine vascular plants found here. Golden Avens is the characteristic forb, but present also throughout the season are the following forbs: *Sibbaldia procumbens*, Parry Lousewort (*Pedicularis parryi*), Wild Candytuft (*Thlaspi montanum* var. *fendleri*), Alpine Speedwell (*Veronica wormskjoldii*), Orange Sneezeweed (*Helenium hoopesii*), Strawberry (*Fragaria ovalis*), gentians (*Gentiana barbellata, G. tenella*, and *G. heterosepala*), *Erigeron simplex*, Fernleaf Fleabane (*Erigeron compositus*), Moonwort (*Botrychium lunaria*), Pacific Windflower

Figure 13. *Timberline within the Inner Basin on the San Francisco Peaks. An open, subalpine forest of Engelmann Spruce, Corkbark Fir, and Bristlecone Pine on the recent volcanic soils gives way to krummholz and eventually to alpine forbs, grasses, and sedges at 3,500 m on south-facing slopes.*

(*Anemone globosa*), *Ranunculus inamoenus*, Hairy Whitlow-grass (*Draba crassifolia*), Diamondleaf Saxifrage (*Saxifraga rhomboidea*), dandelions (*Taraxacum laevigatum* and *T. lyratum*), Chickweed (*Cerastium beeringianum*), Stonecrop (*Sedum rhodanthum*), *Oxyria digyna*, *Silene acaulis*, and Painted Alumroot (*Heuchera versicolor* forma *pumila*).

Shrubs are rare, and are mostly confined to the boulder field or felsenmeer. Gooseberry Currant (*Ribes montigenum*) is the most common shrub; Bearberry Honeysuckle (*Lonicera involucrata*) occurs only rarely.

Only two species of vertebrates are known to breed in the Arizona alpine tundra—the Water Pipit and the Deer Mouse (*Peromyscus maniculatus*) (Lowe and Brown, 1973).

Nevada

Above 3,500 m elevation on Charleston Peak (3,630 m), in the Spring Valley Mountains, in southern Nevada, there are approximately 400 ha of above-timberline vegetation (Bradley, 1964). The community was termed "pseudo-alpine" by Bradley because it apparently lacked a distinctive alpine biota. It may be properly considered a depauperate extension of Great Basin alpine tundra possessing several endemic taxa (Bradley and Deacon, 1965). Its only resident vertebrate is the deer mouse (Hall, 1946).

The following list of small sprawling woody plants and herbs is provided by Bradley (1964); see also Clokey (1951):

Aquilegia scopulorum	*Draba jaegeri*
Arabis pendulina	*Ivesia cryptocaulis*
Arenaria filiorum	*Phacelia hastata*
Astragalus platytropis	*Ribes montigenum*
Cerastium beeringianum	*Silene clokeyi*
Crepis nana	*Sitanion hystrix*

Eastern California

The White Mountains are on the California-Nevada border and are shown in part on the map; however, neither White Mountain Peak (4,342 m) in California nor Boundary Peak (4,005 m) in Nevada are within the mapped area. Geology, vegetation, and comparative floristics of the White Mountains are summarized with the relevant literature in Lloyd and Mitchell's (1973) recent flora of the White Mountains.

The modern ecological literature on the White Mountains is a fairly large one, and mostly postdates Schulman's (1954, 1958) announcement on the Bristlecone Pine there as the "oldest known living thing." Timberline elevations vary considerably throughout the mountains and according to different authors, but it is at approximately 3,500 m for most of the area which, with some local exceptions, is the

Figure 14. *Krummholz of Limber Pine near summit of Mt. San Gorgonio, near 3,200 m elevation, in the San Bernardino National Forest, southern California. Subalpine forest stands on the south-facing slope with tundra-like depression between peaks. Photograph courtesy of San Bernardino National Forest.*

uppermost distributional limit of Bristlecone Pine and Limber Pine (*Pinus flexilis*).

The surface area above 3,500 m in the isolated White Mountains is relatively large. Mitchell (1973) observes that, in floristic comparison to the high Sierra Nevada and Colorado Rockies, the presence of some 200 alpine species in the White Mountains indicates a relatively high floristic diversity despite smaller mountain mass and more recent orogeny.

Southern California

South of the Sierra Nevada, a depauperate alpine zone is limited to the three highest peaks in the San Bernardino and San Gabriel Mountains, and on Mt. San Jacinto (Fig. 14). Long isolation from the Sierra Nevada has resulted in a low degree of similarity to the great massif, and the barriers of the Mohave and Colorado deserts have effectively restricted migration between the isolated southern California summits and the Southern Rockies.

Major and Taylor (1977) note that not only is the flora depauperate, but it is fragmentary as well; such common alpine genera as *Saxifraga* and prostrate *Salix* are missing, and even *Carex* is poorly represented. *Castilleja*, a common Rocky Mountain alpine genus, is represented by only one species (*C. miniata*) as is *Gentiana* (*G. amarella*).

Alpine plants of xeric habitats listed by Major and Taylor (1977) include *Selaginella watsonii, Eriogonum umbellatum, E. saxatile, E. kennedyi* ssp. *alpigenum, Arenaria saxatile, Calyptridium monospermum,* Wax Current (*Ribes cereum*), Littleleaf Rockspirea (*Holodiscus microphyllus*), *Oreonana vestita, Monardella odoratissima, Carex mariposana,* Bottlebrush Squirreltail (*Sitanion hystrix*), Western Needlegrass (*Stipa occidentalis*), Alpine Fescue, (*Festuca brachyphylla*), Rockjasmine, *Arenaria rubella, Leptodactylon pungens, Phlox covillei, P. diffusa, Hulsea vestita,* and *Raillardella argentea.*

Mesic habitats are uncommon, but almost as many species occur here as on the much more extensive xeric habitats of exposed ridges and rocky slopes. Included are *Oxyria digyna,* Sitka Columbine (*Aguilegia formosa*), *Ranunculus eschscholtzii* var. *oxynotus,* Alpine Shooting Star (*Dodecatheon alpinum*), *Mimulus primuloides* and *M. tilingii, Carex mariposana, Cryptogramma acrostichoides, Cystopteris fragilis, Gentiana amarella, Juncus parryi, Phleum alpinum, Polygonum bistortoides, Sibbaldia procumbens,* starworts (*Stellaria calycantha, S. longipes* var. *laeta,* and *S. umbellata*), Horned Dandelion (*Taraxacum ceratophorum*), and Spike Trisetum.

Of 63 Southern California alpine taxa, most are western in distribution, occurring also in the Rocky Mountains and Pacific Coastal Mountain ranges, including the Sierra Nevada; 13% are circumpolar, and a relatively large 19% are endemic.

Part 2. Forests and Woodlands

Open woodland of ponderosa pines (Pinus ponderosa) about 2400 m elevation near Mount Tiptonatop, the Cerbat Mountains, Mohave County, Arizona. This "sky-island" of Rocky Mountain conifer forest on BLM land is now included within a congressionally designated wilderness area.

Boreal Forests and Woodlands

Much of the northern hemisphere's vast subarctic region is covered by post-Pliocene conifer forests of Arcto-Tertiary genera of *Picea, Abies, Larix, Tsuga,* and *Pinus.* This appropriately-called boreal (=northern) forest, or as it is known in Eurasia, "taiga," maintains a recognizable circumpolar integrity throughout a number of regional fasciations. Similar evolutionary derived and adapted—but ecologically distinct—forests extend southward throughout the Holarctic's subalpine regions.

Cold-adapted spruces, firs, larches, and hemlocks are usually actually or potentially dominant within wetter, cooler sites, although millions of acres of disturbed forest lands are covered by subclimax stands of aspen (*Populus*) and birch (*Betula*). Sunnier, drier, and windier sites and less fertile soils tend to be occupied by cold-tolerant species of pine. At treeline, many forest species form open woodlands and eventually scrubland (=krummholz).

In North America, a Canadian subarctic conifer forest[1] analogous to taiga extends southward to the Great Lakes region. Subalpine forests, also of spruce, fir, or pine, occupy higher elevations in the Appalachians and Adirondacks, the Rocky Mountains, the Sierra-Nevada and Cascades, mountains in the Great Basin, and a number of high volcanoes in central Mexico. Two of these biomes, the *Rocky Mountain* and *Sierra-Cascadean subalpine forests*, reach their southern terminus in the Southwest.

---

[1]*Not to be confused with the cold-temperate Canadian zone (=fir forest of Merriam (1898).*

121.3 Rocky Mountain (Petran) Subalpine Conifer Forest

Southwestern fasciations of this biotic community are contained within the Hudsonian life-zone of Merriam (1890, 1898), and, in the southern Rocky Mountains, are largely equivalent to spruce-fir forest. Included are the higher mountainous areas of Colorado, Utah, New Mexico and Arizona from above 2,450-2,600 m to timberline at ca. 3,500-3,800 m. Engelmann Spruce(*Picea engelmanni*) is the prevalent spruce throughout the Southwest (Fig. 15), where it commonly grows with either Subalpine Fir (*Abies lasiocarpa*) or its southern form, Corkbark Fir (*Abies lasiocarpa* var. *arizonica*). The southernmost stands of Engelmann Spruce are found on the summits and high slopes of the Pinaleño (=Graham) and Chiricahua mountains of Arizona and the Sacramento Mountains of New Mexico. Well-developed stands of Corkbark Fir occur in Arizona as far south as the Pinaleño and Catalina mountains, and on Sierra Blanca in New Mexico.

These forests are wet as well as cold, probably averaging from 635 mm to more than 1,000 mm of precipitation annually. Although much of this precipitation falls as snow, summer precipitation contributes substantially to the total, and afternoon thundershowers can be expected almost daily in late summer. The growing season is commonly less than 75 days, and even then subject to occasional nighttime frosts.

Virgin stands often exceed 25 m or more in height and are commonly layered with two or more age-classes of trees. Below 2,900 m, one or more of these classes may be composed solely of Aspen (*Populus tremuloides*); this tree is the principal successional pioneer after fire or other forest disturbance. Fire suppression has resulted in difficulty in the maintenance of aspen clones, causing a major problem in some areas for wildlife managers, because young stands of this short-lived species are a major browse species for deer and other herbivores.

Blue Spruce (*Picea pungens*) is sometimes present with Engelmann Spruce or forms small stands—alone or with Aspen. This is especially so in canyons, where it descends to lower elevations and may extend well into the lower montane forest. It is also commonly marginal to the small grassy mountain parks.

Wetter sites within the forest may have Rocky Mountain Maple (*Acer glabrum*), Bebb Willow (*Salix bebbiana*), Scouler Willow (*Salix scouleriana*), Blueberry Elder (*Sambucus glauca*), Thin-leafed Alder (*Alnus tenuifolia*), or Bitter Cherry (*Prunus emarginata*) present. Dry windy sites, such as ridges, may be occupied by the Limber and Bristlecone Pine (*Pinus flexilis* and *P. aristata*) (Fig. 16). At lower elevations, below 2,600 m, there is much intermingling with Douglas-fir (*Pseudotsuga menziesii*), White Fir (*Abies concolor*), and even Ponderosa Pine (*Pinus ponderosa*). Consequently, the lower contact with *Rocky Mountain montane conifer forest* is often indistinct and poorly differentiated, but usually takes place between ca. 2,450 and 2,900 m, depending on location, slope exposure, and mountain geomass.

Dwarf Juniper (*Juniperus communis*) is an understory shrub closely associated with spruce-fir forest. Otherwise shrubs are poorly represented, except in certain seral stages, natural openings, and at the edge of the forest. Species here may include Red Elderberry (*Sambucus microbotrys*), Creeping Mahonia (*Berberis repens*), currants (*Ribes* spp.), raspberries (*Rubus spp.*), snowberries (*Symphoricarpos* spp.), blueberry (*Vaccinium oreophilum*), Kinnikinnick (*Arctostaphylos uva-ursi*), Bearberry Honeysuckle (*Lonicera involucrata*), and Shrubby Cinquefoil (*Potentilla fruticosa*).

Where the forest is closed, considerable duff and debris

Figure 15. *Subalpine conifer forest of Engelmann Spruce (*Picea engelmanii*), and Aspen (*Populus tremuloides*) with some White Fir (*Abies concolor*), Subalpine Fir (*A. lasiocarpa*) and Douglas-fir (*Pseudotsuga menziesii*) on VT Ridge in the Kaibab National Forest, Arizona, ca. 2,600 m elevation. The shrub in the foreground is Dwarf Juniper (*Juniperus communis*)—another boreal species.*

may accumulate. In places few if any herbaceous plants occur, their places being taken by lichens and other fungi, sedges, mosses, and liverworts. Herbaceous species tend to be more abundant in aspen stands, or where the forest is opened after fire or logging. Any number of flowering herbs may be found there, such as vetches (*Vicia* spp.), fireweed (*Epilobium* spp.), Wild Pea (*Lathyrus arizonicus*), Daisy (*Erigeron formosissimus*), Strawberry (*Fragaria ovalis*), Mountain Dandelion (*Agoseris*), Skunk Cabbage or Falsehellebore (*Veratrum californicum*), Mountain Parsley (*Pseudocymopterus montanus*), Balsamroot (*Balsamorhiza*), groundsels (*Senecio*), primroses (*Primula*), clovers (*Trifolium*), violets (*Viola*), sneezeweed (*Helenium*), gentians (*Gentiana*), dandelions (*Taraxacum*), woodsorrels (*Oxalis*), and many others. Grasses include Spike Trisetum (*Trisetum spicatum*), fescue (*Festuca*), Nodding and Fringed Brome (*Bromus anomalus* and *B. ciliatus*), oat grasses (*Danthonia*), Alpine Timothy (*Phleum alpinum*), bluegrasses (*Poa*) and are mostly lesser quantities of those making up adjacent subalpine grassland.

In the drier mountains, such as the western portions of the Great Basin in the White, Panamint, Spring Valley (=Charleston), and Sheep Mountains, spruces and firs are absent, and boreal forests above 2,600-2,900 m are represented by open woodlands of one or both of two high elevation pines—Limber Pine and Bristlecone Pine (Fig. 16). Water Birch (*Betula occidentalis*) and Rocky Mountain Maple may

occur here in the moister sites. Aspens may or may not be present, while Curlleaf Mountain-mahogany (*Cercocarpus ledifolius*), Mountain Whitethorn or Snowbush (*Ceanothus cordulatus*), Fernbrush (*Chamaebatiaria millefolium*), Tansy (*Tanacetum canum*), and Big Sagebrush (*Artemesia tridentata*) may extend upward into the forest's lower, drier elevations. Understory associates in these more open woodlands may include a number of the same genera of herbs also present in the spruce-fir forest, e.g. alumroot (*Heuchera*), Scarlet Gilia (*Gilia aggregata*), raspberry, and hawksbeard (*Crepis*). Dwarf Juniper and Ninebark (*Physocarpus alternans*) are often common shrubs at higher elevations.

Of the distinctly boreal forest mammals, only the Snowshoe Hare (*Lepus americanus*), Least Chipmunk (*Eutamias minimus*), Gapper's Redbacked Mouse (*Clethrionomys gapperi*), and Marten (*Martes americana*) reach into the Southwest, and then usually only in those larger and more northern ranges with extensive spruce-fir forests. Other subalpine species, such as the Vagrant Shrew (*Sorex vagrans*), Dwarf Shrew (*S. nanus*), Nuttall Cottontail (*Sylvilagus nuttallii*), Golden-mantled Ground Squirrel (*Citellus lateralis*), and Red Squirrel (*Tamiasciurus hudsonicus*), appear to be able successfully to inhabit both subalpine and/or the cooler, wetter montane communities, at least in the Southwest. Still others, such as the Deer Mouse (*Peromyscus maniculatus*) and Bushy-tailed Wood Rat (*Neotoma cinerea*), may be well represented here as well as in a variety of montane and other temperate habitats. Subalpine forests with their adjacent meadows and aspen stands are also important fawning and summering habitats for Mule Deer (*Odocoileus hemionus*), and to a lesser degree, Elk (*Cervus elaphus*).

Nesting birds closely associated with Rocky Mountain subalpine conifer forests in the Southwest are the Blue Grouse (*Dendragapus obscurus*), Williamson Sapsucker (*Sphyrapicus thryoideus*) (pines), Northern Three-toed Woodpecker (*Picoides tridactylus*) (spruce-fir), Hammond's Flycatcher (*Empidonax hammondii*), Gray Jay (*Perisoreus canadensis*) (spruce-fir), Clark's Nutcracker (*Nucifraga columbiana*) (pines), Red-breasted Nuthatch (*Sitta canadensis*) (spruce-fir), Pygmy Nuthatch (*Sitta pygmaea*) (pines), Golden-crowned Kinglet (*Regulus satrapa*), Ruby-crowned Kinglet (*Regulus calendula*), and Cassin Finch (*Carpodacus cassinii*). Other nesting birds also occur such as the Broad-tailed Hummingbird (*Selasphorus platycercus*), Flicker (*Colaptes auratus*), Western Flycatcher (*Empidonax difficilis*), Olive-sided Flycatcher (*Nuttallornis borealis*), Violet-green Swallow (*Tachycineta thalassina*), Mountain Chickadee (*Parus gambeli*), Brown Creeper (*Certhia familiaris*), American Robin (*Turdus migratorius*), Hermit Thrush (*Catharus guttatus*), Mountain Bluebird (*Sialia currucoides*), Townsend Solitaire (*Myadestes townsendi*), Warbling Vireo (*Vireo gilvus*), Yellow-rumped Warbler (*Dendroica coronata*), Western Tanager (*Piranga ludoviciana*), Pine Grosbeak (*Pinicola enucleator*), Pine Siskin (*Carduelis pinus*), Red Crossbill (*Loxia curvirostra*), and juncoes (*Junco* spp.). Many of these, while well represented here, may also be equally or more at home in montane conifer forests and other environments.

Because of the severity of the cold temperatures in this biome, amphibians and reptiles are poorly represented, if not absent entirely. One notable exception is the Jemez Mountain Salamander (*Plethodon neomexicanus*) found in the Jemez Mountains above 2,600 m under moss, rocks, and logs, on north-facing slopes, in and near mixed forests of fir and spruce. Also, the open sunnier Limber Pine-Bristlecone Pine communities allow the Sagebrush Lizard (*Sceloporus graciosus*) to occur up to and even above 2,900 m elevation.

Figure 16. *Subalpine conifer woodland of Bristlecone Pine
(Pinus aristata) on San Francisco Mountain, Coconino
National Forest, Arizona ca. 3,475 m elevation. These pines
and their downslope associate, Limber Pine (P. flexilis),
occur here as a distinctive high elevation association on dry
west slopes within an overall more mesic spruce-fir forest
community.*

121.4 Sierran Subalpine Conifer Forest

Subalpine forests of Limber Pine (*Pinus flexilis*) and/or Lodgepole Pine (*P. contorta* var. *murrayana*) occur above ca. 2,450-6,660 m, on the highest peaks·of the San Gabriel, San Bernardino, and San Jacinto ranges (Fig. 17). These forests are generally single layered, open (=woodland), and of shorter stature than the montane forests downslope; the pines range in height from a few feet above ground near krummholz above 3,050 m on Mts. San Gorgonio and San Jacinto, to trees 18 to 23 m tall on mesic north slopes. Climatological data are lacking, but mean annual precipitation probably averages more than 635 mm—almost all of which falls as snow. The overall xeric appearance is a result of a short growing season of often only 50 to 75 days (and even then occasionally interrupted by freezing temperatures), wind desiccation, and poor soil development.

Many of the Limber Pines appear to be very old, and neither they nor the Lodgepole Pines are here relegated to the subclimax role these species play farther north. Missing altogether in the Southwest are characteristic and representative Sierran subalpine conifers such as Mountain Hemlock (*Tsuga mertensiana*), Western White Pine (*Pinus monticola*), Whitebark Pine (*P. albicaulis*), and Foxtail Pine (*P. balfouriana*). Aspen, so common in successional stands in the subalpine zone farther north, is also virtually absent in Southwestern Sierran subalpine forests.

At lower elevations, below 2,600 m, there is much intermingling of Lodgepole Pine with White Fir (*Abies concolor*), Western Juniper (*Juniperus occidentalis*), and Jeffrey Pine (*Pinus jeffreyi*) and, at least in southern California, these and other montane species may persist in decreasing numbers upward through most of the subalpine forest. Conversely, Lodgepole Pine extends downward into cool, upward levels of the mixed conifer series of cold-temperate montane forest (Canadian life-zone). It would probably be an exaggeration to consider these lower elevation communities, and the small stands of this pine in the Sierra San Pedro Mártir, as subalpine.

The understory tends to be open except for occasional clumps and patches of Mountain Whitethorn (*Ceanothus cordulatus*), Bush Chinquapin (*Castanopsis sempervirens*), Green Manzanita (*Arctostaphylos patula*), Curlleaf Mountain-mahogany (*Cercocarpus ledifolius*), currants (*Ribes* spp.), and Rockspirea (*Holodiscus microphyllus*). Grasses, except for Parish Needlegrass (*Stipa parishii*) and Squirreltail (*Sitanion hystrix*), are also poorly represented and the major understory species are forbs. According to Thorne (1977) these include wintergreen (*Pyrola* spp.), Coral Root (*Corallorhiza maculata*), Pine Drops (*Pterospora andromedea*), Rockcress (*Arabis platysperma*), sedges (*Carex* spp.), Draba (*Draba corrugata*), Bedstraw (*Galium parishii*), Sandwort (*Arenaria nuttallii*), Silene (*Silene verecunda*), Penstemon (*Penstemon* spp.), Mariposa (*Calochortus invenustus*), Alumroot (*Oreonana vestita*), Fleabane (*Erigeron breweri*), *Monardella cinerea*, Wild Onion (*Allium monticola*), *Sarcodes sanguinea*, Goosefoot Violet (*Viola purpurea*), and the buckwheats (*Eriogonum saxatile, E. umbellatum, E. kennedyi*). None of these is confined to, or particularly indicative of, subalpine forest, and they either extend upward into alpine tundra fellfields or downward into montane communities; an apparent exception is Small Mistletoe (*Arceuthobium cyanocarpum*), which Thorne (1977) records as occurring in subapine forests in the San Bernardino

Figure 17. *Subalpine conifer forest of Lodgepole Pines (*Pinus contorta var. murrayana*) in the San Gorgonio Wilderness, San Bernardino National Forest, California, ca. 2,650 m elevation. The forest is approaching timberline, the trees losing stature and becoming more open as they approach Mount San Gorgonio, elevation 3,506 m. U.S. Forest Service Photograph.*

and San Jacinto mountains.

The Sierran subalpine fauna is depauperate. Of the boreal mammals, only the Lodgepole Pine Chipmunk (*Eutamias speciosus*), Golden-Mantled Ground Squirrel (*Citellus lateralis*), Northern Flying Squirrel (*Glaucomys sabrinus*), and Long-tailed Meadow Mouse (*Microtus longicaudus*) reach one or more of these southern "islands." A race of Chickaree (*Tamiasciurus hudsonicus mearnsi*), isolated in the high Sierra San Pedro Mártir, is the most southwestern representative of this essentially boreal and cold-temperate forest squirrel.

Characteristic nesting subalpine birds are thinly represented in these forests by the Williamson Sapsucker (*Sphyrapicus thyroideus*), Clark's Nutcracker (*Nucifraga columbiana*), Red-breasted Nuthatch (*Sitta canadensis*), Ruby-crowned Kinglet (*Regulus satrapa*), and Cassin's Finch (*Carpodacus cassinnii*). Other species such as the Dusky Flycatcher (*Empidonax oberholseri*), Mountain Chickadee (*Parus gambeli*), Pine Siskin (*Carduelis pinus*), and Red Crossbill (*Loxia curvirostra*), while present, are equally or more indicative of Sierran montane conifer forests.

Cold-Temperate Forests and Woodlands

Communities of medium to large conifers and/or winter deciduous trees are found over large and widespread areas of the northern hemisphere, below the more cold adapted boreal and subalpine forests (see e.g. Waring and Franklin, 1979). Included are such diverse types as the mixed deciduous forests of eastern North America and Europe, the montane relict *Cedrus* forests of the Middle East and North Africa, *Cryptomeria* forests of Japan and China, and the majestic Coast Redwoods (*Sequoia sempervirens*) of California. Other independently evolved communities occupy more limited areas in the southern hemisphere, e.g., *Nothofagus* woodlands at the southern end of South America.

In North America, these biotic communities are represented by *Northeastern deciduous forest, Pacific Coastal* (=Oregonian) *conifer forest, Rocky Mountain* and *Sierra-Cascadean montane conifer forests,* and a *Great Basin conifer* (juniper-pinyon) *woodland.* Based on a general grouping of dominant overstory trees and considering understory and animal components, the Southwest's cold-temperate forests can be divided into a Sierran biome comprising the montane forests of the Transverse and Peninsular Ranges in California and Baja California Norte, and those of the Rocky Mountains and Sierra Madres,[1] which include the montane forests of southern Colorado, Utah and Nevada, Arizona, New Mexico, Chihuahua, and Sonora. Considering the proximity of these two floras, there are surprising differences, suggesting a rather long separation and evolutionary development. Faunas of the two areas are also distinct, with a number of vertebrates characteristic of each that are wholly missing from the other. The differences between Rocky Mountain or Petran and Madrean fasciations are less distinct, with a large number of shared dominants, and overlapping or closely related botanical and faunal elements. These suggest a more recent evolutionary history, and the delineation of Madrean and Petran biomes is more difficult.

Although far outstripped in commercial value by the great Douglas-fir forests of the Pacific coastal forest and the pine forests of the Southeastern mixed evergreen deciduous forest, both Sierran and Rocky Mountain-Madrean montane conifer forests are of great importance to the economy and well-being of their respective regions. Those in the mountain islands of California, Nevada, Arizona, and New Mexico are increasingly recognized for their value as recreation centers, wildlife habitats, and watersheds. The larger forests on the Sangre de Cristo, San Juan, Sacramento, Mogollon, and White Mountains, Defiance and Kaibab Plateaus, and the Sierra Madre Occidental are major lumbering and livestock grazing centers.

---

[1]*The Rocky Mountain division can be further divided to include a Great Basin element or fasciation. In the Southwest the element would include those arid Ponderosa Pine and White Fir montane forests in the Virgin, Sheep, Spring Valley, Clark, and other mountains.*

122.3 Rocky Mountain (Petran) and Madrean Montane Conifer Forests

Included here are both the pine (Transition zone) and fir (Canadian zone) forests of Merriam (1894a), Gentry (1942), Lowe (1964) and others. Appropriately named montane, these forests on high plateaus and mountains extend southward from the Rocky Mountains to the Southwest in Colorado and Utah through New Mexico and Arizona to the Sierra Madre Occidental and Sierra Madre Oriental and outlying mountains in Mexico. Some of these stands are extensive; others are confined to isolated summits and canyons. A number of disjunct ranges in adjacent states—the Spring Valley (Charleston) Mountains, Sheep Mountains and Virgin Mountains in Nevada; the Clark and Kingston mountains within the Mojave Desert in California; the Davis Mountains in extreme southwest Texas; and the Sierra del Pino and other high mountains in Coahuila also possess montane conifer forests at higher elevation. Elevations range from ca. 2,000 m at the colder, wetter sites (uncommonly lower to 1,700 m) to as high as 3,050 m on south slopes, and in the Sierra Madres. More often the forest comes in at 2,200-2,300 m, and in the Southwest reaches its best and most characteristic development between 2,300 and 2,650 m.

Mean annual precipitation ranges from a low of just above 460 mm to as high as 760 mm or more (Table 2). In contrast to the wetter Sierran montane forest, half or more of this precipitation falls during the growing season. This permits forests to exist on less than 640 mm per year, and some of these stands are among the driest forests in North America. All fasciations of this montane forest are uniformly cold, nighttime freezing temperatures usually beginning by mid-September and not ending until sometime in May.

Based on overstory dominants, both Rocky Mountain and Madrean fasciations can be conveniently divided into two major communities or series—a Ponderosa Pine forest (Fig. 18), generally at lower elevations; and at higher elevations, in canyons and on north slopes—a cooler mixed conifer forest of Douglas-fir (*Pseudotsuga menziesii*), White Fir (*Abies concolor*), Limber Pine (*Pinus flexilis*), and Aspen (*Populus tremuloides*) (Fig. 19). At its upper limits (ca. 2,600-2,900 m) in the northern Southwest, the mixed conifer series merges, and then gives way, to the Spruce-Alpine Fir and Bristlecone-Limber Pine series of the boreal Rocky Mountain subalpine forest. Although at least three summits in Chihuahua are reported to exceed 3,050 m, true subalpine forests are lacking in the Mexican half of the Southwest. These summits are capped by montane communities of Mexican White Pine (*Pinus ayacahuite*), with some Douglas-fir, Aspen, and Ponderosa Pine. At least one locale above 3,000 m, on the north slopes of Cerro Mohinora in southwestern Chihuahua, contains small stands of the relict and endemic spruce (*Picea chihuahuana*) found usually only further south in Durango.[1]

Generally from north to south, the lower limits of pine forest are typically in contact with pinyon-juniper woodland, interior chaparral, and Madrean evergreen woodland. Grassland (often now composed of sagebrush or invading composites) provides a major lower elevational contact almost throughout.

Ponderosa Pine Forest

This widespread yellow pine is the Southwest's most common montane tree and often grows in pure stands. On the

---

[1] *Personal communication with Dirk Lanning, Nature Conservancy, San Bernardino Ranch, P.O. Box 695, Douglas, Ariz.*

Table 2. Precipitation within or adjacent to Rocky Mountain and Madrean Montane Conifer Forest.

Station	Elevation (in m)	J	F	M	A	M	J	J	A	S	O	N	D	Total
Chama, NM 36°55′ 106°35′	2,393	44	29	39	36	29	26	47	65	45	40	28	43	471
Jacob Lake, AZ 36°43′ 112°13′	2,414	36	34	52	37	29	13	66	61	30	27	33	53	471
Chacon, NM 36°10′ 105°23′	2,590	25	19	28	30	37	39	79	97	33	32	21	25	465
Wolf Canyon, NM 35°58′ 106°46′	2,506	38	38	47	36	29	29	84	88	40	42	29	38	538
Fort Valley, AZ 35°16′ 111°44′	2,239	55	42	54	41	18	16	75	88	43	36	34	57	559
Williams, AZ 35°15′ 112°11′	2,057	47	40	50	35	18	13	71	93	45	32	36	59	539
Groom Creek, AZ 34°29′ 112°27′	1,859	54	48	49	32	12	15	97	113	47	31	32	63	593
Crown King, AZ 34°12′ 112°20′	1,829	76	57	67	35	11	12	94	129	49	36	45	91	701
Lakeside, AZ 34°12′ 109°59′	2,042	49	33	48	24	11	14	79	86	46	48	39	50	527
Greer, AZ 34°01′ 109°28′	2,588	55	42	42	21	12	22	115	108	51	46	28	67	609
Mt. Lemmon, AZ 32°27′ 110°45′	2,371	59	39	62	25	7	23	147	139	51	40	38	51	681
Ruidoso, NM 32°22′ 105°40′	2,084	27	27	33	17	22	50	109	76	64	27	17	40	509
Yecora, Son. 28°22′ 108°57′	1,662	69	34	27	8	10	87	301	272	125	38	22	48	1041
Creel, Chih. 27°45′ 107°38′	2,716	59	27	17	9	9	68	179	156	87	39	15	27	692
Guadalupe y Calvo, Chih., 26°06′ 106°59′	2,400	60	57	35	6	24	84	257	221	157	64	29	83	1076

mountains, plateaus, and mesas of central and northern New Mexico and Arizona, forests of Ponderosa Pine total about 3.4 million ha and provide the most important commercial timber (Choate, 1966; Spencer, 1966). There and in southern Colorado and Utah, one finds the typically three-needled and large cone Rocky Mountain form, *P. ponderosa* var. *scopulorum.* In southern Arizona *P. ponderosa* var. *scopulorum* is joined by the five-needled Arizona subspecies, *P. ponderosa* var. *arizonica*—with the latter usually dominating in the lower portions of the forest. *P. ponderosa* var. *arizonica* extends southward into the Sierra Madre where it is both the dominant montane conifer and commercial tree (Fig. 20).

Old growth Ponderosa Pine forests are often park-like, with the scattered yellow-barked older trees interspersed with occasional groups of their descendants. Otherwise, understories are grassy or at least herbaceous. Frequent light fires, to which older pines are relatively immune, probably kept the forests more open in pre-settlement times. Crown cover of the more open, mature stands may range from 50% to 75%, but that of younger well-stocked stands, sometimes with thickets of "dog-hair" (dense stands of stunted, young trees) may be much higher. Ponderosa Pine forests grow on a wide variety of soils and geologic parent materials, including andesite, basalt, granite, diabase, limestone, and sandstone (Pearson, 1931).

While Ponderosa Pine is the unquestioned dominant over most of the forest, such associated forest trees as Southwestern White Pine (*Pinus strobiformis*), Douglas-fir, White Fir, and Quaking Aspen are frequent participants in the forest at middle and higher elevations. In the Rocky Mountain

fasciation, Gambel Oak (*Quercus gambelii*) and the New Mexican Locust (*Robinia neomexicana*) are locally common and may dominate some of the lower and rockier locations. The deciduous Gambel Oak is of great importance and affects the distribution of several species of wildlife. White Fir is a frequent understory component at higher elevations within older undisturbed pine forests. Understory shrubs are few, rarely dense, and not especially common, but may include scattered populations and plants of Fendler Ceanothus (*Ceanothus fendleri*), Creeping Mahonia (*Berberis repens*), Smooth Sumac (*Rhus glabra*), Golden and Sticky Currant and Orange Gooseberry (*Ribes aureum, R. viscosissimum,* and *R. pinetorum*), Arizona Rose (*Rosa arizonica*), Blue Elderberry and Velvet Elder (*Sambucus cerulea* and *S. velutina*), Longflower, Mountain, Utah, and Roundleaf Snowberries (*Symphoricarpos longiflorus, S. oreophilus, S. utahensis,* and *S. rotundifolius*), Bush Rockspirea (*Holodiscus dumosus*), and Ninebark (*Physocarpus monogynus*) (Castetter, 1956).

Under more open stands, grasses and grass-like plants may prevail and include montane-centered grasses and sedges such as Mountain and Screwleaf Muhly (*Muhlenbergia montana* and *M. virescens*), Pine Dropseed (*Blepharoneuron tricholepis*), Nodding and Fringed Brome (*Bromus anomalus* and *B. ciliatus*), Arizona Fescue (*Festuca arizonica*), Prairie Junegrass (*Koeleria cristata*), Littleseed Muhly (*Muhlenbergia minutissima*), Bulb Panicum (*Panicum bulbosum*), Mutton and Kentucky Bluegrass (*Poa fendleriana* and *P. pratensis*), Squirreltail (*Sitanion hystrix*), Pringle Needlegrass (*Stipa pringlei*), Dryland Sedge (*Carex geophila*), and Fendler Flatsedge (*Cyperus fendlerianus*). Depending on location, some of the

Figure 18. *Ponderosa Pine (Pinus ponderosa var. scopulorum) forest near Chimney Rock on the San Juan National Forest east of Durango, Colorado, ca. 1,980 m elevation. The single-species dominance, interspersion of age classes and grassy understory is characteristic of thousands of acres of today's "multiple use" managed forests in Colorado, Utah, New Mexico, and Arizona. Increased timber sales are resulting in the gradual elimination of older age classes over increasingly large tracts, however.*

Figure 19. *Mixed conifer series (Canadian zone) of Douglas-fir (*Pseudotsuga menziesii*), White Fir (*Abies concolor*), Rocky Mountain Maple (*Acer glabrum*) and Quaking Aspen (*Populus tremuloides*) on trail to Flys Peak, ca. 2,770 m elevation, Chiricahua Mountains, Coronado National Forest, Arizona. This north slope community, while superficially resembling subalpine conifer forest (see Fig. 15), differs from that biome in the absence of numerous subalpine species (e.g., Engelmann Spruce and Alpine Fir) as well as in the presence of montane dominants.*

more characteristic and common forbs include Yarrow (*Achillea lanulosa*), Mountain Parsley (*Pseudocymopterus montanus*), New Mexico Groundsel (*Senecio neomexicanus*), Bracken Fern (*Pteridum aquilinum*), Pinedrops (*Pterospora andromeda*), Rusby Clover (*Trifolium rusbyi*), Mountain Bluebell (*Mertensia franciscana*), American Vetch (*Vicia americana*), Arizona and Grassleaf Peavine (*Lathyrus arizonicus* and *L. graminifolius*), Lupines (*Lupinus*), Trailing and Spreading Fleabane (*Erigeron flagellaris* and *E. divergens*), *Erigeron macranthus*, *E. concinnus*, and *E. formosissimus*, Dandelion (*Taraxacum officinalis*), Meadowrue (*Thalictrum fendleri*), Elegant Cinquefoil (*Potentilla concinna*), Sawatch Knotweed (*Polygonum sawatchense*), Purple Geranium (*Geranium caespitosum*), and Wild Strawberry (*Fragaria ovalis*).[1] In dog-hair thickets or under the high canopy of some mature stands, a heavy litter may develop, with little or no understory herbaceous cover.

In the more southern Ponderosa Pine forests in southeastern Arizona, southwestern New Mexico, and in the highlands of

Chihuahua and Sonora, aboreal associates commonly include Mexican White Pine and Douglas-fir at higher elevations. Apache Pine (*Pinus engelmannii* (=*P. latifolia*), Chihuahua pine (*P. leiophylla* var. *chihuahuana*), Alligatorbark Juniper (*Juniperus deppeana*), and evergreen oaks such as *Quercus fulva*, *Q. pennivenia*, *Q. arizonica*, *Q. grisea*, and *Q. viminea* enter the forest at the lower elevations adjacent to Mexican oak-pine woodland. Species such as Silverleaf and Netleaf Oak (*Quercus hypoleucoides*, *Q. rugosa*), as well as Buckbrush (*Ceanothus huichugore*) and Madrone (*Arbutus arizonica*), may occur locally on the drier sites.

Mixed Conifer (=Douglas-fir—White Fir) Forest

Douglas-fir dominated forests cover more than 600,000 ha in the Southwest, mostly in southern Colorado, in New Mexico, on the White Mountains and Kaibab Plateau in Arizona, and the higher slopes and canyons in the Sierra Madre Occidental (Choate, 1966; Spencer, 1966). General elevation range is usually from 2,450 to 2,900 m, often forming a discontinuous belt between the warmer, drier, more extensive pine forests below and the colder, wetter boreal spruce-fir forests above. Douglas-fir may occur in pure stands but more often it is mixed with firs and/or Engelmann Spruce (*Picea engelmannii*) near its upper limit and with Ponderosa and Mexican White Pine and Blue Spruce (*Picea pungens*) elsewhere.

Mature mixed conifer forests (the most common condition) are often dense, with high canopy cover and heavy litter accumulation that restricts undergrowth. Where openings in the canopy are caused by blowdowns, road construction, fires or other disturbances, a rather depauperate understory flora may develop, including such common species as Mountain Snowberry, Raspberry, Strawberry, Nodding and Mountain Brome (*Bromus marginatus*), Tufted Hairgrass (*Deschampsia caespitosa*), Rough Bentgrass (*Agrostis scabra*), and Figwort (*Scrophularia parviflora*). Where litter buildup is not too great, the undisturbed forest floor may contain such shade tolerant species as Feather Solomonseal (*Smilacina racemosa*), Pipsissewa (*Chimaphila umbellata*), Pyrola (*Pyrola virens*), Western Rattlesnake Plantain (*Goodyera oblongifolia*), alumroots (*Heuchera rubescens* and *H. versicolor*), Mountain-lover (*Pachystima myrsinites*), Hook and Canada Violet (*Viola adunca* and *V. canadensis*), and *Valeriana arizonica*.

Aspen Subclimax Communities

Quaking aspen is an important—though numerically minor—associate throughout the more mesic montane conifer forests of the Southwest. Close examination of the forest floor in even the most dense pine or Douglas-fir—White Fir stands usually reveals scattered greatly suppressed aspen sprouts, probably holdovers from a previous period of forest disturbance. The shade-intolerant aspen, which reproduces chiefly from root sprouts, produces a flourishing colony in such stands once the overstory conifers have been removed by fire, blowdown, or logging.

---

[1]*McLaughlin, Steve, and R. F. Wagle. Changes in herbaceous understory, productivity accompanying prescribed burning of Ponderosa Pine forests in Arizona. School of Renewable Natural Resources. University of Arizona, Tucson. Work done under a cooperative agreement with Research Work Unit RM-2108, Rocky Mountain Forest and Range Experiment Station. USDA Forest Service, Tempe, Ariz.*

Figure 20. *Ponderosa Pine (*Pinus ponderosa *var.* arizonica*) forest in the high Sierra Madre Occidental, Madero Campo Cinco Taco (5 Taco Lumber Camp), between Yepachic and Tutuaca, Chihuahua, ca. 2,280 m elevation. These "second growth" forests are more typical of today's pine forest—both north and south of the border. Such was not the case just one generation ago when the Southwest boasted the largest forests of virgin pine in the world. Photograph by R.L. Todd.*

Of the 200,000 ha of aspen in the United States portion of the Southwest, 75% is in New Mexico. Most of the remainder lies in the Mogollon Rim-White Mountain area of Arizona, with small colonies of mostly sterile individuals in the highest mountains of Chihuahua and Sonora (Standley, 1920-26), an alder (*Alnus firmifolia*) being the more prevalent successional tree (Lesuer, 1945).

Aspen stands are rich wildlife communities, providing abundant food and cover (when not heavily grazed) for a wide variety of mammals and birds (Patton and Jones, 1977). Common understory shrubs include gooseberries and currants, Arizona Rose, Mountain and Roundleaf Snowberry, and Arizona and Bearberry Honeysuckle (*Lonicera arizonica* and *L. involucrata*). The normally rich mixture of grasses and forbs include Nodding, Mountain, and Fringed Brome, wheatgrasses (*Agropyron* spp.), Kentucky Bluegrass, asters (*Aster* spp.), Bracken Fern, fleabanes (*Erigeron* spp.), Missouri and Few-flowered Golden-rod (*Solidago missouriensis* and *S. sparsiflora*),

Grassleaf Peavine (*Lathyrus graminifolius*), American Vetch, Rocky Mountain Iris (*Iris missouriensis*), lupines, Sneezeweed (*Helenium hoopesii*), Cutleaf Coneflower (*Rudbeckia laciniata*), Yarrow, Mintleaf Beebalm (*Monarda menthaefolia*), False Mountain-bluebell (*Mertensia* spp.) and geraniums (*Geranium* spp.). Poisonous plants include Carrotleaf Larkspur (*Delphinium tenuisectum*), and Columbia Monkshood (*Aconitum columbianum*).

A number of mammals make their home within Rocky Mountain montane forests including bats such as the Southwestern Myotis (*Myotis auriculus*), Long-eared Myotis (*M. evotus*), Long-legged Myotis (*M. volans*), and Big Brown Bat (*Eptesicus fuscus*). Species such as the Vagrant, Dwarf, and Merriam Shrews (*Sorex vagrans, S. nanus, S. merriami*); Chickaree (*Tamiasciurus hudsonicus*); and Nuttall's Cottontail (*Sylvilagus nuttalli*) reach their greatest abundance within mixed conifer communities, while others, such as the Tassel-eared Squirrel (*Sciurus aberti*) and Porcupine (*Erethizon dorsatum*), are most abundant in yellow pine forests. The former is particularly tied to these habitats, and the life cycles of this species is closely interwoven with that of Ponderosa Pine. Also indicative of these montane environments are several species of chipmunks—Colorado Chipmunk (*Eutamias quadrivittatus*), Gray-collared Chipmunk (*E. cinereicollis*), Gray-footed Chipmunk (*E. canipes*), Uinta Chipmunk (*E. umbrinus*), depending on regional fasciation. Also depending on locale is the presence of such montane denizens as the Montane Vole (*Microtus montanus*), Long-tailed Vole (*M. longicaudus*), and Mexican Vole (*M. mexicanus*). Other species centered elsewhere, but likely to be encountered in at least some of these Petran or Rocky Mountain forests are the Eastern Cottontail (*Sylvilagus floridanus*), Golden-mantled Ground Squirrel (*Citellus lateralis*), Deer Mouse (*Peromyscus maniculatus*), Mexican Woodrat (*Neotoma mexicana*), Long-tailed Weasel (*Mustela frenata*), and Mule Deer (*Odocoileus hemionus*). White-tailed Deer (*O. virginianus*) are generally found in the Madrean fasciation. Once restricted to a much smaller distribution in the Southwest, the Rocky Mountain Elk (*Cervus elaphus*) now occupies southwestern montane conifer forests in Colorado, New Mexico, and Arizona. The Merriam Elk (*C. e. merriami*), once found in the the the White Mountains of Arizona and the Mogollon Mountains of New Mexico, is now extinct. The Gray Wolf (*Canis lupus*), once widespread in these forests, is now extirpated; the last wolves on the North Kaibab were eradicated during the mid-1920's.

The list of characteristic nesting avifauna is large and contains such typically montane conifer forest species as the Goshawk (*Accipiter gentilis*), Flammulated Owl (*Otus flammeolus*), Pygmy Owl (*Glaucidium gnoma*), Spotted Owl (*Strix occidentalis*), Saw-whet Owl (*Aegolius acadicus*), Broad-tailed Hummingbird (*Selasphorus platycercus*), Western Flycatcher (*Empidonax difficilis*), Steller Jay (*Cyanocitta stelleri*), Pygmy Nuthatch (*Sitta pygmaea*), Brown Creeper (*Certhis familiaris*), Western Bluebird (*Sialia mexicana*), Townsend's Solitaire (*Myadestes townsendi*), Solitary Vireo (*Vireo solitarius*), Warbling Vireo (*V. gilvus*), Yellow-rumped Warbler (*Dendroica coronata*), Western Tanager (*Piranga ludoviciana*), Evening Grosbeak (*Hesperiphona vespertina*), Pine Siskin (*Carduelis pinus*), Red Crossbill (*Loxia curvirostra*), juncoes, and Chipping Sparrow (*Spizella passerina*); all of which extend southward to or through the Southwest. Southward these are complemented or replaced by Madrean species reaching northward; examples, several of which are analogous species, are the Wild Turkeys (*Meleagris gallopavo mexicana*, *M. gallopavo merriami*), Band-tailed Pigeon (*Columba fasciata*), Rivoli's Hummingbird (*Eugenes fulgens*), Coues' Flycatcher (*Contopus pertinax*), Pine Flycatcher (*Empidonax affinis*), Mexican Chickadee (*Parus sclateri*), Aztec Thrush (*Ridgwayia pinicola*), Brown-backed Solitaire (*Myadestes obscurus*), Olive Warbler (*Peucedramus taeniatus*), Grace's Warbler (*Dendroica graciae*), Red-faced Warbler (*Cardellina rufrifrons*), Hepatic Tanager (*Piranga flava*), and Yellow-eyed Junco (*Junco phaeonotus*). Several rare, and in some cases vanishing, species peculiar to Madrean montane forests occur in the Southwest—the Thick-billed Parrot (*Rhynchopsitta pachyrhyncha*), **Imperial Woodpecker** (*Campephilus imperialis*), **Eared Trogon** (*Euptilotus neoxenus*), **Russet Nightingale Thrush** (*Catharus occidentalis*), and **Red Warbler** (*Ergaticus ruber*).

Amphibians are limited to the Tiger Salamander (*Ambystoma tigrinum*), and locally in mixed-conifer forests within their mountain namesakes, the Jemez and Sacramento Mountain Salamanders (*Plethodon neomexicanus* and *Aneides hardyi*), and various frogs and toads. Skinks—in the southern Rockies, the Southern Many-lined Skink (*Eumeces multivirgatus epipleurotus*); in Great Basin fasciations, the Western Skink (*E. skiltonianus*) and in Madrean fasciations, the Mountain Skink (*E. callicephalus*)—are representative lizards. As with the birds, some of the reptiles occur throughout both Rocky Mountain and Madrean fasciations with only change in subspecies rank—Short-horned Lizard (*Phrynosoma douglassi*), Arizona Alligator Lizard (*Gerrhonotus kingi*), Ringneck Snake (*Diadophis punctatus*), Gopher Snake (*Pituophis melanoleucus*), and a number of subspecies of the Western Rattlesnake (*Crotalus viridis*). Madrean and Mogollon fasciations have a proportionally larger number of herptiles than more northern forests because of the warmer biomes downslope. The Bunchgrass Lizard (*Sceloporus scalaris*), Striped Plateau Lizard (*S. virgatus*), Yarrow's Spiny Lizard (*S. jarrovi*), Mountain Patch-nosed Snake (*Salvadora grahamiae*), Sonoran Mountain Kingsnake (*Lampropeltis pyromelana*), Western Terrestrial Garter Snake (*Thamnophis elegans*), and Twin-spotted Rattlesnake (*Crotalus pricei*) are all species that reach well into montane conifer forest within their regional distributions.

122.5 Sierran Montane Conifer Forest

In the higher Transverse and Peninsular Ranges, this forest is characterized by a well defined zone of pines above the chaparral, evergreen woodlands, mixed evergreen forest and grassland of the Californian Biotic Province. Especially well developed stands occur above 1,500 m elevation in the San Gabriel, San Bernardino, San Jacinto, Santa Rosa, Palomar, Cuyamaca, Laguna Juarez and San Pedro Mártir Mountains. While now isolated from each other and from their ecological metropolis in the Sierra Nevada by broad areas of low elevation chaparral, grassland, and desertscrub, these stands share many floristic and faunal constituents. They are, nonetheless, depauperate in a number of Sierran plants and animals.

Annual precipatation averages more than 635 m per year (Table 3), but because of the high elevations and winter rainfall pattern much of this precipitation falls as snow. These forests, therefore, must depend on snowmelt, and at least in some years the growing season is shortened by the lack of available moisture in late summer.

Two closely related yellow pines, Jeffrey Pine and Ponderosa Pine (*Pinus jeffreyi* and *P. ponderosa*) are the usual dominants over much of the lower elevation Sierran montane conifer forest.[1] Ponderosa Pine is most common on the wetter cismontane (Pacific side) mountain slopes, where it usually occurs between 1,300 and 2,140 m. It reaches its southern limit in the Cuyamaca Mountains of San Diego County. Jeffrey Pine is a common dominant above 1,200-1,500 m on both cismontane and transmontane slopes. The two forest-types share many common understory species, and indeed the open, often park-like stands of the two species look remarkably similar (Fig. 21). One can differentiate the two trees (and therefore the forests) by becoming familiar with their cones, needles, and bark. The prickle on the larger Jeffrey pine cone scales turns inward, those on Ponderosa Pines turn outward; and the bark of Jeffrey Pines is darker, more narrowly furrowed and smells like vanilla.

Ponderosa Pine Series

Near its lower altitudinal range, Ponderosa Pine associates include Coulter Pine (*Pinus coulteri*) and the deciduous California Black Oak (*Quercus kelloggii*) which is analogous to Gambel Oak (*Q. gambelii*) in the Rocky Mountains and *Q. gravesii* in the Sierra Madre Oriental. On moister sites and at higher elevations, common associates are White Fir (*Abies concolor*), Sugar Pine (*Pinus lambertiana*), and Big-cone Douglas-fir (*Pseudotsuga macrocarpa*), Incense-cedar (*Libocedrus decurrens*), and Canyon Live Oak (*Quercus chrysolepis*) of the mixed evergreen forest. Scattered shrubs in open Ponderosa Pine forest may include Eastwood and Pringle Manzanita (*Arctostaphylos glandulosa* and *A. pringlei* var *drupacea*), Deerbrush (*Ceanothus integerrimus*), Yellowleaf Silktassel (*Garrya flavescens*), Hairy Yerbasanta (*Eriodictyon trichocalyx*), California Buckthorn (*Rhamnus californica* var. *cuspidata*), and Purple Nightshade (*Solanum xanti*). A rich herbaceous flora may include California Glorybind (*Convolvulus occidentalis*), *Clarkia rhomboidea*, *Collinsia childii*, *Gilia splendens* spp. *grantii*, Mustang-clover (*Linanthus ciliatus*),

Figure 21. *Virgin pine forest of Jeffrey Pine (*Pinus jeffreyi*) near Vallecitos in the Sierra San Pedro Mártir, Baja California Norte, ca. 2,450 m elevation. The open understory, lack of young trees and herbaceous ground cover is characteristic of yellow pine forests throughout much of the Southwest.*

[1]*Not included are lower elevation stands dominated by Coulter Pine (*Pinus coulteri*) and in which evergreen oaks are typically present. These open woodlands are to be considered as an oak-pine community within Californian evergreen woodland.*

Table 3. Precipitation from 3 representative Stations within Sierran Montane Conifer Forest.

Station	Elevation (in m)	\multicolumn{13}{c}{Mean Monthly Precipitation in mm}
		J	F	M	A	M	J	J	A	S	O	N	D	Total
Mount Wilson, CA 34°14′ 118°04′	1,740	161	154	119	80	11	2	1	3	7	23	118	124	803
Lake Arrowhead, CA 34°15′ 117°11′	1,586	207	194	157	104	25	3	4	11	14	31	136	169	1055
Palomar Mountain Obser., CA 32°21′ 116°52′	1,692	123	118	119	64	10	2	9	12	10	20	78	116	681

Streptanthus bernardinus, and Goosefoot Violet (*Viola purpurea*). Common perennial grasses and grass-like plants include Slimleaf and Orcutt Brome (*Bromus breviaristatus* and *B. orcuttianus* var. *halli*), Coastrange Melic (*Melica imperfecta*), and Pine Bluegrass (*Poa scabrella*). Certain species enjoy a more general distribution and are common to both Ponderosa and Jeffrey Pine forests. These include Woolypod Milkweed (*Asclepias eriocarpa*), Penstemon, Castilleja martinii, Chaenactis santolinoides, Menzies Pipsissewa (*Chimaphila menziesii*), Lotus davidsonii, Mimulus johnstonii, Grinnell and Longlips Penstemon (*Penstemon grinnellii* and *P. labrosus*), Snowplant (*Sarcodes sanguinea*), Junegrass (*Koeleria cristata*), and Bottlebrush Squirreltail (*Sitanion hystrix*).

Jeffrey Pine Series

Jeffrey Pine tolerates a wide range of temperatures and soil moisture, and is rather more cosmopolitan than Ponderosa Pine in the southern California mountains and replaces it completely in Baja California. On cismontane slopes, it usually occurs at higher elevations than Ponderosa Pine, and often occurs to the exclusion of Ponderosa Pine on transmontane slopes bordering the Mojave Desert (Thorne, 1977). On the lower Mojave Desert border, it is commonly associated with One-needle Pinyon (*Pinus monophylla*), Coulter Pine, Sierra Juniper (*Juniperus occidentalis*), and Birchleaf and Curlleaf Mountain-mahogany (*Cercocarpus betuloides* and *C. ledifolius*). On higher, moister exposures, arboreal associates may include Sugar Pine, Incense Cedar, Bigcone Douglas-fir and White Fir. Canopies range from very open on steep south-facing slopes at lower elevations to nearly closed on mesic sites in more gentle slopes with well-developed soils. Crowns of mature dominant trees may reach a height of 46 m or more on better sites, but are commonly 23 to 30 m (Horton, 1960). Conspicuous understory shrubs may include Greenleaf Manzanita (*Arctostaphylos patula*), Wright Buckwheat (*Eriogonum wrightii subscaposum*) and Parish Snowberry (*Symphoricarpos parishii*). On transmontane slopes, Big Sagebrush (*Artemisia tridentata*), Mountain Whitethorn (*Ceanothus cordulatus*), Rubber Rabbitbrush (*Chrysothamnus nauseosus*), and Gray Horsebrush (*Tetradymia canescens*) are important shrubs.

On dry, rocky cismontane slopes, the Jeffrey Pine forest shares many subordinate herbaceous species with the Ponderosa Pine community. However, many common Jeffrey Pine associates are rare or absent from Ponderosa forests, including the following: *Arabis repanda, Cordylanthus nevinii, Eriogonum parishii, Fritellaria pinetorum,* Rock Melic (*Melica stricta*), Bridges and San Bernardino Penstemon (*Penstemon bridgesii* and *P. caesius*), and *Stipa parishii.*

Mixed Conifer Series

Sugar Pine and White Fir are important participants in a mixed conifer forest between elevations of 1,700 and ca. 2,600 m (Fig. 22). This series is, therefore, typically positioned above yellow pine forest and below the subalpine forest of Lodgepole and Limber Pine. Where this type occurs on south-facing slopes, especially at its lower elevations, the canopy is often very open, sometimes as low as 10%, and rarely exceeds 50%. Common associates on these droughty sites include Jeffrey Pine, Canyon Live Oak, and a wide variety of timberland chaparral species. On north and east exposures, and at high elevations, the moister conditions result in an increase in White Fir and a decrease in Jeffrey and Sugar Pine. Here arboreal associates may include Incense Cedar and Lodgepole Pine. While little herbaceous understory occurs, scattered understory shrubs may include Bush Chinquapin (*Castanopsis sempervirens*), Mountain Whitethorn Ceanothus, Greenleaf Manzanita, Sierra Currant (*Ribes nevadense*), Sierra Gooseberry (*Ribes roezli*), Western Thimbleberry (*Rubus parviflorus*), Coulter Willow (*Salix coulteri*), and Blueberry Elder (*Sambucus caerulea*) (Munz and Keck, 1959; Jaeger and Smith, 1966; Thorne, 1977).

Few endemic mammals are present; Merriam's Chipmunk (*Eutamias merriami*) is perhaps the most indicative. A limited number of others—Broad-footed Mole (*Scapanus latimanus*), Western Gray Squirrel (*Sciurus griseus*) Southern Pocket Gopher (*Thomomys umbrinus*), Deer Mouse (*Peromyscus maniculatus*), Long-tailed Weasel (*Mustela frenata*), and Mule Deer (*Odocoileus hemionus*), while found in this forest, are also well distributed throughout other series and biomes.

Birds are numerous, particularly during the summer months when migrant nesters are present. Nesting species generally restricted to Sierran montane forest—at least in the Southwest—are the Dark-eyed Junco (*Junco hyemalis*), White-headed Woodpecker (*Picoides albolarvatus*), Calliope Hummingbird (*Stellula calliope*), and Purple Finch (*Carpodacus purpureus*) while others such as the Band-tailed Pigeon (*Columba fasciata*), Flammulated Owl (*Otus flammeolus*)—pines, Hairy Woodpecker (*Picoides villosus*), and sapsuckers (*Sphyrapicus thyroideus, S. varius*), Yellow-rumped Warblers (*Dendroica coronata*), Western Wood Pewee (*Contopus sordidulus*), Steller Jay, Mountain Chickadee (*Parus gambeli*), Pygmy Nuthatch (*Sitta pygmaea*), American Robin (*Turdus migratorius*), Western Bluebird (*Sialia mexicana*), Hutton's Vireo (*Vireo huttoni*)—oaks, Western Tanager (*Piranga ludoviciana*), Pine Siskin (*Carduelis pinus*), and Brown Creeper (*Certhia familiaris*) are more or less indicative of montane conifer forests throughout the Southwest. Some, such as the Sharpshinned Hawk (*Accipiter striatus*), Spotted Owl (*Strix*

Figure 22. *Mixed conifer forest west of Big Bear Lake in San Bernardino National Forest, California, ca. 2,200 m elevation. Trees represented are White Fir (*Abies concolor*), Jeffrey Pine (*Pinus jeffreyi*), Sugar Pine (*P. lambertiana*), Incense Cedar (*Libocedrus decurrens*), Black Oak (*Quercus kellogii*) and Canyon Live Oak (*Q. chrysolepis*). Photograph by A.P. Gomez, courtesy of U.S. Forest Service.*

occidentalis), Pygmy Owl (*Glaucidium gnoma*), Townsend's Solitaire (*Myadestes townsendi*), and Red Crossbill (*Loxia curvirostra*), are rare and/or restricted to particular situations within the forest.

In the wetter, damper habitats, particularly at lower and warmer elevations, the Ensatina (*Ensatina eschscholtzi*) and Arboreal Salamander (*Aneides lugubris*) may be present. The representative skink is the Gilbert's Skink (*Eumeces gilberti*), and except for the widespread Western Fence and Sagebrush Lizards (*Sceloporus occidentalis* and *S. graciosus*)—which may be abundant in the more open habitats—lizards are un-

common. Snakes also are not prevalent, and, with some exceptions, their presence often indicates conditions reflecting lower biotic communities. This is not necessarily the case with the Southern Rubber Boa (*Charina bottae umbricata*), Ringneck Snake (*Diadophis punctatus*), Gopher Snake (*Pituophis melanoleucus*), California Mountain and Common Kingsnakes (*Lampropeltis zonata, L. getulus*), Western Terrestrial Garter Snake (*Thamnophis elegans*), and the Southern Pacific Western Rattlesnake (*Crotalus viridis helleri*), which if not centered in, are at least "at home" in these montane environments.

122.4 Great Basin Conifer Woodland

This cold-adapted evergreen woodland is characterized by the unequal dominance of two conifers—juniper (*Juniperus*) and pinyon (*Pinus*). These trees rarely, if ever, exceed 12 m in height and are typically openly spaced (woodland), except at higher elevations and other less xeric sites where interlocking crowns may present a closed (forest) aspect. The shorter, bushier junipers ("cedars") are generally more prevalent than pinyons, but either may occur as an essentially pure stand. Structurally, these juniper-pinyon woodlands are among the simplest communities in the Southwest.

This woodland has its evolutionary center in the Great Basin and is one of the most extensive vegetative types in the Southwest. It extends southward through Colorado, Utah, Nevada, southeastern California, northern Arizona, and New Mexico to mountainous area in Trans-pecos Texas, southern New Mexico, central Arizona, and northern Baja California Norte. Juniper-pinyon woodland covers extensive areas here between 1,500 and 2,300 m (extremes are 1,050 and 2,700 m) and reaches its greatest development on mesas, plateaus, piedmonts, slopes, and ridges.

Several species of juniper may assume or share dominance in the Southwest. Rocky Mountain Juniper (*Juniperus scopulorum*) is an important constituent in the higher and colder woodlands in Colorado, northern New Mexico and Arizona, and more locally in southern Nevada and Utah (Fig. 23). In northwestern New Mexico, western Colorado, Utah, northern Arizona, Nevada, and eastern California, the Great Basin, Utah Juniper (*J. osteosperma*) may be the more common. One-seed Juniper (*J. monosperma*) is the prevalent species in juniper-pinyon woodlands in west Texas, central and southern New Mexico, and much of sub-Mogollon Arizona (Fig. 24), as is the closely related *J. californica* in southern California and Baja California Norte. Rocky Mountain Pinyon (*Pinus edulis*) is the common pinyon pine almost throughout (Fig. 25) except that west of ca. longitude 113.5° it is largely replaced by the single needled form (*P. monophylla*) or farther south in Baja California Norte by the Four-leaved Pinyon (*P. quadrifolia*) (Fig. 26).

Not included as Great Basin conifer woodland species are Alligator-bark Juniper and Mexican Pinyon. Although the former species may be present in juniper-pinyon woodland with One-seed Juniper and Rocky Mountain Pinyon in central and southeastern Arizona and west-central New Mexico, both trees are species of the Madrean evergreen woodland farther south; they normally occur in the communities of oaks (*encinal*) and oaks and pine (oak-pine) that have their center of distribution in Mexico. This is also the case with Pinchott and Drooping Juniper east of the Sierra Madre Occidental and in the Big Bend region of Texas.

Precipitation ranges from 250 to 500 mm per year with extremes of 180 and 560 mm (Table 4). This sparse rainfall is more or less evenly spread throughout the year, and much of the winter precipitation falls as snow. Summer precipitation is of more importance in eastern juniper-pinyon woodlands than in the western portions where more than 80% of the precipitation falls during the late fall and winter. The unifying climatic feature of all these arid woodlands is cold winter minimum temperatures; freezing temperatures can be expected to occur about 150 or more days a year, precluding the participation of evergreen oaks and other warm-temperate forms.

Habitats tend to be rocky, with thin soils predominating. In

Figure 23. *Extensive juniper-pinyon woodland of Rocky Mountain Junipers (Juniperus scopulorum) with some Pinyon (Pinus edulis) and Alligatorbark Juniper (Juniperus deppeana) south of the Mogollon Rim, Coconino National Forest, Arizona, ca. 2,000 m elevation. This southern fasciation of Great Basin conifer woodland gives way to interior chaparral and semidesert grassland immediately downslope to the south.*

the central and eastern areas of the Southwest, the principal contact with Great Basin conifer woodland is grassland, and extensive landscapes there are characterized by parkland and savanna-like mosaics. The openness of these "cedar glades" depends on soil type, range history and condition. Here the understory is typically composed of grasses (e.g., *Bouteloua gracilis*) and shrubs, e.g. Threadleaf Groundsel (*Senecio longilobus*) and Snakeweed (*Gutierrezia sarothrae*) of the Plains grassland. Also well represented in many of these grass understories are Galleta Grass (*Hilaria jamesii*), Indian Ricegrass (*Oryzopsis hymenoides*), Western Wheatgrass (*Agropyron smithii*) and other grasses of the Plains grassland-Great Basin grassland transition. Other grasses locally common to abundant include several muhleys (*Muhlenbergia*

spp.), dropseeds (*Sporobolus* spp.), and Junegrass (*Koeleria cristata*).

Junipers have invaded large areas of former grassland (Humphrey, 1962). That this is also true for pinyons is less certain, and woodlands well stocked with pinyons are not to be considered as disclimax grassland—as numerous futile attempts to "reconvert" these areas to grass will attest. Junipers tend to be at lower elevations than pinyons and normally occupy the deeper soil sites below 2,000 m.

In the Great Basin, conifer woodland occurs on the mountain gradient above and within Great Basin desertscrub. Here Big Sagebrush is the principal and often the almost exclusive understory plant. Indeed, Big Sagebrush continues to be an important subdominant in juniper-pinyon woodlands

Figure 24. *Series of One-seed Juniper (*Juniperus monosperma*) and Rocky Mountain Pinyon (*Pinus edulis*) near Aurora, San Miguel County, New Mexico ca. 1,890 m elevation. An eastern fasciation with understory of Blue Grama (*Bouteloua gracilis*).*

south-westward to the Sierra Juarez in Baja California Norte. Other more or less Great Basin Desert associates of general or regional importance are rabbitbrush (*Chrysothamnus* spp.), Winterfat (*Ceratoides lanata*), Shadscale (*Atriplex confertifolia*), and Black Sagebrush (*Artemisia arbuscula* spp. *nova*). On those mountain ranges over 1,500 m elevation in and adjacent to the Mohave Desert, Blackbrush (*Coleogyne ramosissima*) is a common major understory component of the pinyon-juniper woodlands present there. In northwestern and central Arizona understory species of adjacent interior chaparral and even Sonoran desertscrub (Arizona Upland subdivision) may be important in the makeup (e.g., *Quercus turbinella*, *Rhamnus crocea*, *Garrya wrightii*, *Canotia holacantha*). Chaparral also intergrades with Great Basin conifer woodland in southern

California and Baja California Norte.

In the Sacramento, Guadalupe, Organ, Burro, Peloncillo, and other southern New Mexico mountains, in sub-Mogollon Arizona, and the Trans-Pecos region of Texas, Great Basin conifer woodland phases into the more southerly derived Madrean evergreen woodland. This transition is marked by the disappearance of *Juniperus osteosperma*, *J. scopulorum*, and *Pinus edulis*, and the appearance of *J. deppeana*, *Quercus emoryi*, *Q. grisea*, *Q. arizonica*, and *Pinus cembroides* with their respective floral and faunal associates. This replacement may be gradual or abrupt and is much influenced by slope exposure, elevation, and edaphic situation. Generally the warm-temperate and more moisture-requiring Madrean species first make their appearance on south slopes, pro-

Figure 25. *Pinyon (Pinus edulis) dominated Great Basin conifer woodland on Fish Tail Mesa, Kaibab National Forest, Arizona ca. 1,585 m elevation. The major understory species on this ungrazed site is Big Sagebrush (Artemisia tridentata). The shrub in right center of photo is Cliffrose (Cowania mexicana), a common constituent in Great Basin conifer woodlands and an important winter browse species for Mule Deer (Odocoileus hemionus).*

tected hillsides, and in drainages; the southernmost conifer woodlands are to be found on high north slopes and mesas. Also, particularly at lower elevation, there may be integration with interior chaparral, e.g., in the Organ and Burro mountains in New Mexico and in the Apache and other mountain ranges in Arizona.

The upslope contact with Great Basin conifer woodland is montane conifer forest, except north of Parallel 37° (and locally elsewhere as in the Sandia and Manzano mountains) where Great Basin montane scrubland makes its appearance. Here, and not uncommonly elsewhere at higher elevations within the conifer woodland, important plant associates are Gambel Oak (*Quercus gambelii*-in shrub form), mountain-mahoganies (*Cercocarpus montanus, C. ledifolius, C.*

intricatus), Skunkbush Sumac (*Rhus trilobata*), Saskatoon Serviceberry (*Amelanchier alnifolia*), snowberries (*Symphoricarpos* spp.), and currants (*Ribes* spp).

Other shrubs generally important as subdominant associates in juniper-pinyon woodland include Cliffrose (*Cowania mexicana*), Apache Plume (*Fallugia paradoxa*), Mormon-tea (*Ephedra viridis* and others), Barberry or Algerita (*Berberis fremonti* and *B. haematocarpa*), Fourwing Saltbush (*Atriplex canescens*), Small Soapweed (*Yucca glauca*), and Dátil (*Yucca baccata*): other associated species as Buffalo-berry (*Shepherdia* spp.), Antelope Bitterbrush (*Purshia tridentata*), and Fernbush (*Chamaebatiaria millefolium*) are of more local occurrence.

Herbs and grasses commonly encountered throughout much of the conifer woodland include gilias (*Gilia* spp.)

Figure 26. *Great Basin conifer woodland of Four-leaved Pinyon (*Pinus quadrifolia*), Singleleaf Pinyon (P. monophylla), and California Juniper (*Juniperus californica*) in the Sierra Juarez, Baja California Norte, ca. 1,400 m elevation. The sunlit shrub in lower right center is Big Sagebrush (*Artemisia tridentata*) and these woodlands possess a surprising physiognomic and biotic similarity through a number of widespread fasciations.*

buckwheats (*Eriogonum* spp.), Sego-lily (*Calochortus nuttalli*), penstemons (*Penstemon* spp.), several globemallows (*Sphaeralcea digitata, S. marginata, S. coccinea,* and others), Louisiana Sagebrush (*Artemisia ludoviciana*), lupines (*Lupinus* spp.), and bromes (*Bromus* spp.).

Several cacti have Great Basin conifer woodland as their center of distribution or are otherwise well represented here. Among these are the Red Hedgehog Cactus (*Echinocereus triglochidiatus* var. *melanacanthus*) and the hedgehogs *E. engelmanii* var. *variegatus* and *E. fendleri* var. *fendleri*; the prickly-pears *Opuntia erinacea, O. basilaris* var. *aurea, O. phaeacantha, O. macrorhiza, O. polyacantha, O. fragilis*; the chollas *O. whipplei* and *O. imbricata; Sclerocactus whipplei* var. *intermedius; Mammillaria wrightii; Pediocactus papyracanthus, P. simpsonii; Coryphantha vivipara* var. *arizonica* and *C. missouriensis.*

Only a few vertebrates are closely tied to Great Basin conifer woodland (e.g., Pinyon Mouse (*Peromyscus truei*), Pinyon Jay (*Gymnorhinus cyanocephalus*), Gray Flycatcher (*Empidonax wrightii*)) or are centered here (e.g., Bushy-tailed Woodrat (*Neotoma cinerea arizonae*), Gray Vireo (*Vireo vicinior*), Black-throated Gray Warbler (*Dendroica nigrescens*), Scott's Oriole (*Icterus parisorum*), and the Plateau Whiptail (*Cnemidophorus velox*)). A somewhat larger number of the more adaptable, and, therefore, more widely distributed species also may be found in these relatively recent environments. Furthermore, juniper-pinyon woodlands are more or less seasonal habitats for a number of montane and subalpine animals; as such they are often of great importance as winter range for Rocky Mountain Elk and Mule Deer.

Table 4. Precipitation within and adjacent to Great Basin Conifer Woodland.

Station	Elevation (in m)	Mean Monthly Precipitation in mm												
		J	F	M	A	M	J	J	A	S	O	N	D	Total
Orderville, UT 37°16' 112°38'	1,658	45	39	37	29	16	15	23	35	25	30	31	43	368
Mesa Verde N.P., CO 37°12' 108°29'	2,155	44	37	40	34	25	18	45	54	32	46	28	47	450
Trinidad F.A.A. CO 37°15' 104°20'	1,751	10	10	39	33	47	37	47	48	25	23	11	13	343
Conchas Dam, NM 35°24' 104°11'	1,294	7	9	14	22	34	38	65	63	36	27	6	11	332
Santa Rosa, NM 34°57' 104°41'	1,408	8	10	16	17	33	35	64	73	38	29	7	14	344
Elk, NM 3E 32°56' 105°17'	1,737	14	12	13	17	22	41	69	81	58	32	12	18	389
Mesacalero, NM 33°10' 105°48'	2,068	25	23	26	15	16	40	94	97	55	30	16	33	470
Santa Fe, NM 35°41' 105°56'	2,147	16	20	19	23	31	28	64	60	40	26	18	17	362
Mountainaire, NM 34°32' 106°15'	1,987	17	18	17	15	16	21	64	63	34	27	12	24	328
Cuba, NM 36°02' 106°58'	2,147	21	19	29	21	21	18	54	63	32	32	16	24	350
Fort Bayard, NM 32°48' 108°09'	1,872	21	16	15	7	6	21	70	83	50	24	12	24	349
Zuni R.S., NM 35°06' 108°47'	1,966	22	17	23	15	10	10	45	53	29	29	15	21	289
Blue, AZ 33°37' 109°06'	1,756	27	25	19	12	6	17	96	122	67	55	25	65	536
Lukachukai, AZ 36°25' 109°14'	1,987	23	16	32	21	11	8	34	38	24	30	20	21	278
Klagetoh, AZ 36°30' 109°32'	1,949	25	12	19	19	9	15	53	53	17	17	19	14	272
Pinedale, AZ 34°18' 110°15'	1,981	35	24	27	18	7	14	63	81	39	47	29	43	427
Cibecue, AZ 34°03' 110°29'	1,509	48	30	39	18	9	12	60	72	42	43	29	45	447
Betatakin, AZ 36°41' 110°32'	2,221	24	21	22	21	11	12	37	43	24	27	20	29	291
Pleasant Valley R.S., AZ 34°06' 110°56'	1,539	49	32	44	23	12	12	73	88	43	33	31	51	491
Walnut Can. N.M., AZ 35°10' 111°31'	2,038	42	29	47	27	16	13	62	64	40	28	32	48	448
Sedona R.S., AZ 34°52' 111°46'	1,317	43	39	42	30	14	12	48	61	38	29	33	44	434
Grand Canyon, AZ 36°03' 112°08'	2,125	34	32	37	25	14	12	38	54	32	27	21	40	366
Drake, AZ 34°58' 112°23'	1,417	32	28	26	23	9	9	40	62	29	23	14	35	330
Walnut Creek, AZ 34°56' 112°49'	1,551	34	28	31	19	9	12	68	78	34	23	22	36	394
Mt. Trumbull, AZ 36°25' 113°21'	1,707	21	18	26	18	10	11	43	52	24	20	20	25	288

Warm-Temperate Forests and Woodlands

Once covering extensive portions of southern Europe, North Africa, the Middle East, and southern North America, as well as parts of Asia, South America, eastern Australia, and the Cape region of Africa, these woodlands are now much altered and reduced because of their mild temperatures and otherwise favorable climate for human occupation. Those in the Mediterranean region, in particular, have been ravaged by more than 3,000 years of civilization and are now almost completely deforested. In the northern hemisphere these woodlands were, and are, typically composed of oaks (*Quercus* spp.), often accompanied by any of several genera of evergreen conifers (*Pinus, Juniperus,*

Cupressus, etc.) and/or other evergreen hardwoods (e.g., Arbutus). These communities are then largely evergreen or a mixture of evergreen and deciduous species (=mixed evergreen-deciduous).

Some of the best remaining, as well as least disturbed, examples of warm-temperate woodlands are to be found in the Southwest in the San Bernardino, Cleveland, and Coronado National Forests, Cuyamaca California State Park, Big Bend National Park, and in northern Mexico. Upland communities are here represented by *Madrean evergreen woodland, Californian mixed evergreen forest, Californian evergreen woodland,* and less commonly, *relict conifer forests.*

123.3 Madrean Evergreen Woodland

This mild winter-wet summer woodland is centered in the Sierra Madre of Mexico where it reaches northward to the mountains of southeastern Arizona (north-westward to Yavapai County), southwestern New Mexico, and Trans-Pecos Texas. At its lower elevations the woodland is typically open—sometimes very open. The trees are evergreen oaks from 6 to 15 m or more in height, or oaks, Alligator Bark and One-seed Junipers and Mexican Pinyon in unequal proportion (=encinal)[1] (Fig. 27).

Higher on the mountain gradient, a Mexican oak-pine woodland characteristically occurs above the encinal and below montane conifer forest (=pine forest) (Figs. 28, 29). Although this woodland has often been considered transitional to pine forest, it is so only in the geographic and physiognomic sense, because it is floristically distinct (see e.g., the definitive work by Marshall, 1957). The encinal oaks are usually accompanied or replaced by oak species characteristic of higher elevations as well as one or more Madrean pines, i.e., Apache Pine (*Pinus engelmannii*), Chihuahua Pine (*P. leiophylla*), Arizona Pine (*P. ponderosa* var. *arizonica*), Pino Triste (*P. lumholtzii*), Durango Pine (*P. durangensis*), and *P. cooperi.*

In the mountainous regions of sub-Mogollon Arizona as in the Chiricahua, Santa Rita, Baboquivari, Tumacacori, Huachuca, Catalina, Pinaleno and Pinal Mountains, the oaks most prevalent in the encinal are Emory Oak or Bellota (*Quercus emoryi*), Arizona White Oak (*Q. arizonica*), and south of the Gila River, Mexican Blue Oak (*Q. oblongifolia*) (Fig. 30). Emory Oak and Gray Oak (*Q. grisea*)—an oak closely related to Arizona White Oak—are the two common Madrean oaks in encinals farther east in the Peloncillo, Animas, Burro, Organ, Davis, Chinati, Chisos, and other mountains in New Mexico and southwest Texas. Silverleaf Oak (*Q. hypoleucoides*) and Netleaf Oak (*Q. rugosa*) are the characteristic oaks of the higher encinals and the restricted oak-pine zone in southeastern Arizona and extreme southwestern New Mexico.

In the foothills, bajadas, barrancas, and sierras of the Sierra Madre Occidental and its outlying ranges in Mexico, a large variety of oaks participate in both the encinal and oak-pine woodlands that cover hundreds of square miles of western Chihuahua and eastern Sonora. In both states, Chihuahua Oak (*Q. chihuahuensis*) is commonly the first oak encountered at the woodland's lower edge; further north in northeastern Sonora and southern Arizona this role is taken by Mexican blue oak. Other commonly found oaks within encinals in Sonora and western Chihuahua are *Quercus albocincta, Q. emoryi, Q. arizonica,* and the deciduous *Q. chuchiuchupensis* and *Q. santaclarensis.* Eastward from the Pacific drainage divide in central Chihuahua, Santa Clara Oak is an important woodland constituent as are *Q. emoryi, Q. chihuahuensis, Q. grisea,* and *Q.*

---

[1]*The term encinal is from Shreve (1915, and elsewhere) and is a Spanish designation to describe evergreen woodlands composed wholly or partially of oaks (encino = live oaks + al = place of). Darrow (1944) and Nichol (1952) termed these woodlands in Arizona oak woodland while Wauer (1973) used the designation of pinyon-juniper-oak woodland to describe these woodlands in the Chisos Mountains and elsewhere in Texas. Leopold (1950), Marshall (1957) and others have recognized and described pine-oak woodland, cypress-pine-oak woodland (Wauer, 1973) and Mexican oak-pine woodland (Lowe, 1964). Flores Mata et al. (1971) also refer to these communities in Mexico as bosques de encinos y bosques de pino-encino.*

Table 5. Precipitation within Madrean Evergreen Forest and Woodland.

Station	Elevation (in m)	J	F	M	A	M	J	J	A	S	O	N	D	Total
Cuauhtemoc, Chih. 28°22′ 106°50′	2,210	4	6	9	8	10	36	138	153	62	33	10	18	487
Mulatos, Son. 28°38′ 108°53′	1,524	28	15	11	6	5	66	198	162	72	23	8	39	633
Temosachic, Chih. 28°58′ 107°50′	1,858	9	4	7	7	12	27	109	128	64	30	15	40	452
Pilares de Nacozari, Son. 30°20′ 109°38′	1,409	37	26	24	13	4	31	148	128	53	49	16	49	578
Madera, Chih. 29°17′ 107°52′	2,079	17	17	5	7	5	15	80	96	43	22	16	40	363
Chisos Basin, TX 29°16′ 103°18′	1,615	15	14	10	12	37	46	92	77	82	47	15	15	462
Mount Locke, TX 30°40′ 104°00′	2,070	21	12	12	12	38	63	96	87	70	38	14	14	477
Bisbee, AZ 31°27′ 109°55′	1,631	29	31	25	12	6	17	105	113	50	26	20	33	467
Ruby, AZ 31°27′ 109°55′	1,212	44	32	24	12	4	16	112	118	45	18	25	32	482
Santa Rita Exp.Sta. (Florida Canyon) AZ 31°46′ 110°51′	1,311	43	40	31	15	7	16	106	107	48	19	28	35	495
Chiricahua N.M. AZ 32°00′ 109°21′	1,615	40	30	30	15	9	28	119	104	30	21	20	28	474
Oracle, AZ 32°36′ 110°44′	1,384	53	47	41	20	8	11	71	82	41	24	39	55	492
Pinal Ranch, AZ* 33°21′ 110°59′	1,378	80	72	64	27	10	11	71	87	50	32	47	76	627
Whiteriver, AZ** 1,609	1,609	41	35	41	25	12	12	71	77	45	30	30	36	455

* Woodland within interior chaparral
** At extreme northern edge of Madrean evergreen woodland

durifolia. Some of the important oaks at higher elevations (1,650-2,200 m) within oak-pine woodland in these parts of the Southwest are *Quercus viminea, Q. hypoleucoides, Q. pennivenia, Q. epileuca, Q. fulva,* and *Q. rugosa.* The principal pines are Apache Pine, Chihuahua Pine, Arizona Pine, Pino Triste, and Durango Pine. The latter two species are endemic to western Mexico. Another conifer, the relict Arizona Cypress (*Cupressus arizonica*) is largely confined to north-facing canyon slopes and drainages. Madroños (*Arbutus arizonica, A. texana*) are important and characteristic arboreal constituents within oak-pine woodland. Where air moisture is sufficient, the epiphytic bromeliad *Tillandsia recurvata* can be found clinging to the branches of oaks and other trees. Mexican Pinyon and Alligator-bark Juniper may occur within any of the fasciations in either oak-pine or encinal woodland.

On the eastern slopes of the Sierra Madre Occidental and in west Texas, New Mexico, and Cochise County, Arizona, the lower contact with Madrean evergreen woodland is with grassland (=plains and semidesert grassland) or rarely Chihuahuan desertscrub. This contact is apparently determined to a large extent by soil depth and type, since the woodland occupies much the same elevational range as grassland; the lower encinal communities here are at ca. 1,500-1,800 m along drainages, on rocky slopes, and on other thin-soiled habitats. To the northwest, in northern Sonora and south central Arizona, encinal woodland descends somewhat lower—to ca. 1,200-1,350 m to within semidesert grassland (in central Arizona also to interior chaparral).

Southwestward in central and southern Sonora, oak woodland drops downward to as low as 880-950 m where its contact with either subtropical deciduous forest or thornscrub may be remarkably abrupt (Fig. 31). Here the transition from oak-pine woodland to montane conifer forest is also lower, taking place at ca. 1,850 to 2,000 m; well-developed montane pine forest does not usually occur above oak-pine woodland elsewhere in the Southwest until 2,200-2,300 m.

At its northern limits in central Arizona, and in the Burro, Organ, and Guadalupe Mountains in New Mexico, the Big Bend region of Texas, and in the mountains in extreme northeastern Chihuahua and western Coahuila, Madrean evergreen woodland occurs above or within the drier interior chaparral, and below and along drainages within the drier and more cold tolerant Great Basin conifer woodland. The major contact throughout is with grassland, however, and grasses provide the major woodland understory.

The more prevalent grass species in this "savanna" zone may be any of a number of bunchgrasses centered here (e.g., the muhlys (*Muhlenbergia emersleyi, M. torreyi, M. porteri,* etc.), Woolspike (*Elyonurus barbiculmis*), and Cane Bluestem (*Bothriochloa barbinodis*) or, particularly at lower elevations, grassland species of wider distribution (e.g., Wolftail (*Lycurus phleoides*), Little Bluestem (*Schizachyrium scoparium*), Plains

Figure 27. *Madrean evergreen woodland. Encinal woodland in Juniper Canyon in the Chisos Mountains, Big Bend National Park, Texas, ca. 1,700 m elevation. The larger overstory trees are Pinyon* (Pinus cembroides), *some Alligatorbark Junipers* (Juniperus deppeana) *and Drooping Juniper* (J. flaccida) *(individual in lower right foreground). Less conspicuous are numerous young oaks* (Quercus grisea, Q. emoryi) *which promise a return to a more representative encinal community as existed prior to the drought of the 1950's (Whitson, 1974). The grass in immediate foreground is Bullgrass* (Muhlenbergia emersleyi).

Lovegrass (*Eragrostis intermedia*), Blue Grama (*Bouteloua gracilis*), Sideoats Grama (*B. curtipendula*), Hairy Grama (*B. hirsuta*), Tanglehead (*Heteropogon contortus*), and Green Sprangletop (*Leptochloa dubia*)). Herbaceous weeds, shrubs and forbs such as penstemons (*Penstemon*), lupines (*Lupinus*), bricklebushes (*Brickellia*), sages (*Salvia*), indigobushes (*Dalea*), buckwheats (*Eriogonum*), Louisiana Sagebrush (*Artemesia ludoviciana*), flatsedges (*Cyperus*), rose-mallows (*Hibiscus*), woodsorrels (*Oxalis*), beans (*Phaseolus*), and many others, while almost always present to some degree, may on occasion be so abundant on some of the steeper slopes as to present a "soft chaparral" aspect. Although many herbaceous and shrub species, such as Larchleaf Goldenweed (*Ericameria laricifolia*), increase with grazing, the usual aspect of over-grazed encinals is of a bareness of ground cover.

Many of the cacti and leaf succulents of the semidesert grassland extend well up into the Madrean evergreen woodland. These include the Rainbow Cactus (*Echinocereus pectinatus* var. *rigidissimus*), Barrel Cactus (*Ferocactus wislizeni*), Cane Cholla (*Opuntia spinosior*), Engelmann Prickly Pear (*O. phaeacantha*), Purple Prickly Pear (*O. violacea* var. *santarita*), Schott Yucca (*Yucca schottii*), Thornber Yucca (*Yucca baccata* var. *thornberi*), Palmer Agave (*Agave palmeri*), Parry Agave (*A. parryi*), and Sacahuista (*Nolina microcarpa*). Several cacti such as the Cream Cactus (*Mammillaria gummifera*), the Pincushion (*Mammillaria orestera*), the hedgehogs (*Echinocereus triglochidiatus* and *E. ledingii*) and the Hen and Chicks Cactus (*Coryphantha recurvata*), are largely centered in this biotic community.

The presence of scrubland species varies from that of an occasional plant within the woodland to the attainment of landscape dominance. Any of a number of either Madrean or

Figure 28. *Madrean evergreen woodland. Mexican oak-pine in the Sierra Madre Occidental off Durango-Mazatlan Highway ca. 1,980 m elevation. March, 1971. Such level habitats, common in the Sierra Madres east of the continental divide, are somewhat atypical of the steep slopes so characteristic of most oak-pine woodlands in the Southwest. While heavily grazed, the ground cover burns at intervals to provide an open understory. Note the number of Apache Pine (Pinus engelmannii) seedlings in opening in right foreground.*

interior chaparral species may be conspicuously present including:

Arctostaphylos pungens	Pointleaf Manzanita
Ceanothus huichagorare	—
Cercocarpus montanus	Alderleaf Mountain-mahogany
Cowania mexicana	Cliffrose
Garrya wrightii	Wright's Silktassel
Quercus toumeyi	Toumey Oak
Rhamnus betulaefolia	Birchleaf Buckthorn
Rhus choriophylla	Mearns' Sumac
R. trilobata	Skunkbush Sumac
Vauquelinia californica	Arizona Rosewood

These and other chaparral representative plants are especially prevalent on thin eroded soils, on limestone, and near the northern end and eastern range of Madrean woodland. At lower elevations and on rocky south slopes, certain species affiliated with thornscrub may be important associates and subdominants within the encinal. Some of the more common and widely distributed of these include Southwestern Coral

Bean (*Erythrina flabelliformis*), Kidneywood (*Eysenhardtia orthocarpa*), Senna (*Cassia leptocarpa*), Hopbush (*Dodonaea viscosa*), Wait-a-bit (*Mimosa biuncifera*), and Velvet-pod Mimosa (*Mimosa dysocarpa*).

Mean annual precipitation within Madrean evergreen woodlands usually exceeds 400 mm, with 200 mm or more falling during the summer growing season of May through August (Table 5). Annual extremes are 330-380 mm to as high as 890-1,020 mm. Freezing temperatures range from occasional in the south to an average of almost 150 days per year at the woodland's northern limits (e.g. near Whiteriver on the Fort Apache Indian Reservation).

Madrean evergreen woodland is the principal biotic community for the White-tailed Deer (*Odocoileus virginianus*) in the Southwest, and its oak-pine zone is a major habitat-type for the Coati (*Nasua nasua*). This woodland was also the home of the Mexican Grizzly (*Ursus horribilis*). Once widespread, this bear may now be extinct, no recent documentation of its existence having been received from its last strongholds in

Figure 29. *Madrean evergreen woodland of pines and oaks (Mexican oak-pine woodland) on west slopes of Sierra Madre Occidental, west of Durango, Mexico. Note the differences in slope physiognomy in this* **barranca** *habitat as compared to Fig. 28.*

Figure 30. *Madrean evergreen woodland. Encinal woodland of mostly Mexican Blue Oak (*Quercus oblongifolia*) and Emory Oak (*Quercus emoryi*); Potrero Canyon in the Pajarito Mountains on the Coronado National Forest near the Arizona-Sonora line, ca. 1,460 m elevation. November, 1970. Most of the understory grasses in left foreground are Little Bluestem (*Schizachyrium scoparium*) and Sideoats Grama (*Bouteloua curtipendula*).*

the Sierra del Nido of Chihuahua (Leopold, 1959). Other mammals well represented in or indicative of Madrean evergreen woodland include the Yellow-nosed Cotton Rat (*Sigmodon ochrognathus*), Southern Pocket Gopher (*Thomomys umbrinus*), Apache Squirrel (*Sciurus nayaritensis*), Bailey's Pocket Mouse (*Perognathus baileyi*), and the Eastern Cottontail (*Sylvilagus floridanus*).

It is the rich assortment of birds that make these woodlands so attractive to naturalists. A number of colorful inhabitants are found here such as the Coppery-tailed Trogon (*Trogon elegans*), Rivoli's Hummingbird (*Eugenes fulgens*), Violet Crowned Hummingbird (*Amazilia violiceps*), White-eared Hummingbird (*Hylocharis leucotis*), and Colima Warbler (*Vermivora crissalis*), all of which barely reach the United

States. A list of the most characteristic nesting species would include the Montezuma Quail (*Cyrtonyx montezumae*), Whiskered Owl (*Otus trichopsis*), Arizona Woodpecker (*Picoides arizonae*), Buff-breasted Flycatcher (*Empidonax fulvifrons*), Mexican Jay (*Aphelocoma ultramarina*), and Bridled Titmouse (*Parus wollweberi*). Other common and characteristic species such as the Acorn Woodpecker (*Melanerpes formicivorus*), Hutton's Vireo (*Vireo huttoni*), Bush Tit (*Psaltriparus minimus*), and Black-throated Gray Warbler (*Dendroica nigrescens*). The Western Bluebird (*Sialia mexicana*) is equally at home here and in Californian evergreen woodland. At higher elevations in oak-pine woodland one also may expect to find (or hear) some of the more montane Madrean species such as the Gould's Turkey (*Meleagris*

Figure 31. *Madrean evergreen woodland contact with Sinaloan deciduous forest near Santa Ana, Sonora, ca. 1,000 m elevation. It is springtime (April) and the brown, retained leaves of the Chihuahua Oak (*Quercus chihuahuensis*) present a sharp contrast to the leafless Sinaloan deciduous forest. One can literally step from one biome to the other.*

gallopavo mexicana), Band-tailed Pigeon (*Columba fasciata*), Mexican Chickadee (*Parus sclateri*), and Hepatic Tanager (*Piranga flava*).[1]

These woodlands are also the Southwestern metropolis for many terrestrial Madrean reptiles, including:

Crotalus lepidus	Rock Rattlesnake
C. pricei	Twin-spotted Rattlesnake
C. willardi	Ridgenose Rattlesnake
Elaphe triapsis	Green Rat Snake
Eumeces callicephalus	Mountain Skink
Lampropeltis pyromelana	Sonora Mountain Kingsnake
Phrynosoma ditmarsi	Ditmar's Horned Lizard
Salvadora grahamiae	Mountain Patchnose Snake
Sceloporus clarki	Clark's Spiny Lizard
S. jarrovi	Yarrow's Spiny Lizard
S. scalaris	Bunchgrass Lizard
S. virgatus	Striped Plateau Lizard
Tantilla wilcoxi wilcoxi	Huachuca Blackhead Snake
Thamnophis eques	Mexican Garter Snake

The Barking Frog (*Hylactophryne augusti*), while rarely seen or collected, is a characteristic terrestrial amphibian, extending northward to the Santa Rita Mountains. The Tarahumara Frog (*Rana tarahumarae*) is found with Madrean evergreen woodlands along permanent springs, streams, and ponds.

[1]*For an excellent discussion of the particular requirements of these and other evergreen woodland species, the reader is urged to consult Marshall (1957).*

123.4 Californian Evergreen Forest and Woodland

The center of these most Californian of biomes is the interior side of the Coast Ranges. Only the southernmost fasciations occur in southern California and northern Baja California Norte. The oak-pine associations of Blue Oak and Digger Pine, so common in the foothills above the Central Valley, do not reach the Southwest. The southernmost Blue Oaks and Digger Pines occur in northern Los Angeles Country. Other species important in adjacent woodlands but not in the defined boundaries for the Southwest include Valley Oak and California Buckeye.[1]

The lower elevation (60-1,050 m) woodlands (=encinals)— collectively termed southern oak woodland by Munz and Keck (1949)—are generally restricted to moister and cooler sites within sheltered valleys and foothills. At higher elevations (ca. 1,050-1,550 m), and in canyons, a mixed hardwood community, usually composed of evergreen and deciduous hardwoods with some conifers, may occupy mesic mountain slopes, "flat" and shoulder habitats. These associations are considered by Sawyer et al. (1977) to be the southernmost representatives of mixed evergreen forest (see also e.g. Shreve, 1927).

In addition to mixed hardwood forest, two major woodland communities—an Engelmann Oak (*Quercus engelmannii*) series and a Coast Live Oak (*Quercus agrifolia*) series—are recognized as occurring in the Californian Southwest (Griffin 1977). *Quercus agrifolia* is an important participant in all three communities. As elsewhere, these communities while cismontane, rarely face the sea.

Precipitation means range from ca. 300-380 mm per annum in the lower open woodlands to 635 mm or more in mixed hardwood forest (Table 6). This rainfall occurs in the typical Californian (=Mediterranean) pattern of October through April. In contrast with Madrean evergreen woodland, the summer period in Californian evergreen woodland is virtually rainless. Although from 200 to 350 days a year are frost free (Munz and Keck, 1949) winter temperatures of -6° C and even lower are also not uncommon.

Mixed Hardwood Series

This diverse group of associations is restricted in the Southwest to moist, favored sites in the Cuyamaca, Palomar, San Gabriel, San Bernardino, and other high southern California mountains. It is absent from the high mountains of Baja California. Here at its southernmost extremity there is much integration with Sierran montane conifer forest with which it shares a number of constituents, e.g., the deciduous Black Oak.

Although these associations are depauperate in species when compared to more northern associations, they nonetheless possess—at least locally—several characteristics and distinctive mixed evergreen forest species. These include Incense Cedar (*Libocedrus decurrens*), Pacific Madrone (*Arbutus menziesii*), California Bay (*Umbellularia californica*), and Coulter Pine (*Pinus coulteri*). Big-cone Douglas-fir (*Pseudotsuga macrocarpa*), a southern Californian endemic, is a locally important dominant (Fig. 32). Canyon Oak (*Q. chrysolepis*) and Coast Live Oak are the common participants and occur more or less throughout. There is considerable contact with the higher chaparral associations, and many of

Figure 32. *Californian mixed evergreen forest at Falling Springs, San Gabriel Mountains, Los Angeles National Forest, California ca. 1,220 m elevation. A moist canyon habitat dominated by Big-cone Douglas-fir (*Pseudotsuga macrocarpa*) with Canyon Live Oak (*Quercus* chrysolepis), Incense Cedar (*Libocedrus* decurrens) and Coast Live Oak (*Quercus* agrifolia) within Californian chaparral and adjacent to riparian deciduous forest.*

---

[1]*Another buckeye (*Aesculus parryi*) occurs—sometimes abundantly, in northwestern Baja California del Norte within coastalscrub.*

Figure 33. *Californian evergreen woodland of Coast Live Oak* (Quercus agrifolia) *near Dulzura, San Diego County, California ca. 370 m elevation.*

Figure 34. *Californian evergreen woodland of California Walnut* (Juglans californica) *in the San Jose Hills, west of Pomona, Los Angeles County, California ca. 340 m elevation. The dead branches are thought to be due to the prolonged drought experienced during the winters of 1975-76 and 1976-77. These interesting communities have suffered heavily from urbanization and without protection will soon be lost entirely.*

Figure 35. *Californian evergreen woodland of Engelmann Oak (*Quercus engelmannii*) at Oak Grove, Cleveland National Forest, San Diego County, California, ca. 840 m elevation. Heavy grazing precludes the establishment of young trees at this level alluvial site.*

Table 6. Precipitation within or adjacent to Californian Evergreen Forest and Woodland.

Station	Elevation (in m)	Mean Monthly Precipitation in mm												Total
		J	F	M	A	M	J	J	A	S	O	N	D	
San Gabriel Canyon, CA 34°09' 117°54'	227	116	100	85	53	10	4	.5	2	5	12	68	85	541
Pomona-Claremont, CA 34°04' 117°49'	226	86	74	70	41	6	1	.5	1	5	11	53	69	418
Henshaw Dam, CA 33°14' 116°46'	823	107	85	99	59	13	2	4	12	6	18	65	68	538
Cuyamaca, CA* 32°59' 116°35'	1,414	117	137	156	93	28	4	13	13	15	26	90	133	825
Barrett Dam, CA 32°41' 116°40'	495	73	64	69	45	12	1	2	5	5	12	43	66	397

* Californian mixed evergreen forest.

these species possess facultative and/or sprouting properties and are otherwise adapted at least to occasional fires.

No species of vertebrate appears to be particularly restricted to Californian evergreen woodland. The mammal inhabitants include both forest and tree-requiring species such as the Western Gray Squirrel (*Sciurus griseus*) and Raccoon (*Procyon lotor*) and a number of widespread influents, e.g., California Mule Deer (*Odocoileus hemionus californicus*), California Ground Squirrel (*Citellus beecheyi*) and Western Pocket Gopher (*Thomomys bottae*). Characteristic and representative birds are the Acorn Woodpecker (*Melanerpes formicivorous*), Nuttall's Woodpecker (*Picoides nuttallii*), Plain Titmouse (*Parus inornatus*), and Western Bluebird (*Sialia mexicana*). Several species more representative of the higher mixed evergreen forest include the Mountain Quail (*Oreortyx picta*), Hairy Woodpecker (*Picoides villosus*), White-breasted Nuthatch (*Sitta carolinensis*), and Band-tailed Pigeon (*Columba fasciata*). The Arboreal Salamander (*Aneides lugubris*) is largely found in Californian evergreen woodlands and the Ensatina (*Ensatina eschscholtzi klauberi*) and California Mountain Kingsnake (*Lampropeltis zonata*) are probably indicative of mixed evergreen forest. Some of the more commonly encountered herptiles include the Common Kingsnake (*Lampropeltis getulus*), Gopher Snake (*Pituophis melanoleucus*), Coast Horned Lizard (*Phrynosoma coronatum*), Western Fence Lizard (*Sceloporus occidentalis*), and Western Toad (*Bufo boreas*).

Coast Live Oak Series

While nowhere extensive, this is the woodland type most common in southern California. Here *Quercus agrifolia* dominates north slopes with deep soils, alluvial terraces, and most frequently, the recent alluvium of canyon bottoms (Fig. 33). Southward this species becomes increasingly confined to moister habitats within chaparral until it forms narrow riparian consociations such as those near San Antonio in Baja California Norte. Because the seedlings are intolerant of grazing, there is concern that these trees are gradually being eliminated both north and south of the border because nearly all evergreen woodlands are used to pasture livestock (Coyle and Roberts, 1975).

Coast Live Oak woodlands vary from open savanna-like landscapes with few or no woody associates to relatively dense woodlands in which Canyon Live Oak (*Quercus chrysolepis*), Engelmann Oak, and other trees participate. Toyon (*Heteromeles arbutifolia*) may be a conspicuous subdominant (Griffin, 1977). Manzanita (*Arctostaphylos* spp.), Squawbush Sumac (*Rhus integrifolia*), Sugar Sumac (*R. ovata*), and other chaparral shrubs may also be common associates, particularly on slopes and rocky soils. Because of overgrazing, some woodlands on slopes now have coastalscrub shrub understories. Understories are more typically herbaceous however, and annual grasses predominate.

An interesting coast live oak association with California Black Walnut (*Juglans californica*) present and locally dominant enters the Southwest from the northwest in Los Angeles County. This regional north slope association has suffered much from urbanization and the best remaining examples are disturbed populations in the Puente and San Jose Hills southeast of Covina (Wieslander, 1934; Griffin, 1977) (Fig. 34).

Engelmann Oak Series

Woodlands dominated by this semi-evergreen "white" oak appear to require relatively deep clay soils. Consequently, many—perhaps most—of these associations, which were always restricted in distribution, have been eliminated by agricultural and urban development. Wieslander (1934d, 1934f) shows remnants of Engelmann Oak woodland as occurring south of the San Gabriel Mountains in the vicinities of Pasadena and Pomona. No intact woodland remains there now (Griffin, 1977). Other associations still occur on the Santa Rosa Plateau, Mesa de Burro, Mesa de Colorado, etc., in the Santa Ana Mountains in Riverside County (Wieslander, 1938), in San Diego County, and in interior valleys south of Tecate in Baja California Norte. The most extensive stands of Engelmann Oak woodland remaining are probably those in the Ramona-Santa Ysabel area in north-central San Diego County (Wieslander, 1934c; Griffin, 1977) (Fig. 35).

This species characteristically forms an open to dense woodland where it usually grows with Coast Live Oak and less commonly other oaks, e.g., *Quercus kelloggii*. Because of the soil type and level habitats, chaparral species are poorly represented, and understories are relatively open and composed mostly of annual grasses and forbs.

123.5 Relict Conifer Forests and Woodlands

Small populations of cypress (*Cupressus*) and closed-cone pines (Bishop Pine [*Pinus muricata*] and Knobcone Pine [*P. attenuata*]) are locally distributed in the Southwest within warm-temperate scrublands (chaparral) and woodlands. These postclimax (=relict) communities are restricted to certain maritime situations and favorable inland sites—in canyons, along drainages and on suitable slopes—where both winter and summer temperatures are moderate. Both the cypresses and pines appear to require sterile soils where these species presumably have a competitive advantage over the dominant plants of adjacent communities.

Closed-cone pines get their name from the fact that their ovulate cones are persistent and may remain closed for many years until opened by fire or age. This is also more or less true for the cypress. Fire is an integral part of the life history and maintenance of all relict conifer communities (Bakker, 1971).

The fossil record indicates that the distribution of these woodlands was much more extensive before the Pleistocene (Mason, 1932; Axelrod, 1967). Decreasing minimum temperatures and increased aridity since the advent of the Ice Ages have now reduced these Arcto-Tertiary relics to only the most favorable temperate habitats. Their distribution is limited by narrow tolerances to climatic parameters, competition with more contemporary species, and in some cases disease. These communities appear at an evolutionary end.

Inland, several forms of *Cupressus arizonica* may be found in groves or "pockets" at mid elevations (450-2,200 m) within evergreen woodland or chaparral (Fig. 36). In southern Arizona, extreme southwestern New Mexico, the Chisos Mountains in Texas, and in the Sierra Madre and other mountain ranges in northern Mexico eastward to Nuevo Leon, Arizona Cypress discontinuously occupies favored canyon sites and drainage bottoms. Stands of Smooth-back Arizona Cypress (*C. glabra*) are found on north slopes and in canyons in sub-Mogollon Arizona within interior chaparral, sometimes adjacent to pinyon-juniper communities as near Sedona in Yavapai County. The Cuyamaca Cypress (*C. stephensonii*) and San Pedro Mártir Cypress (*C. montana*) occupy similar habitats in Californian chaparral in their respective mountain namesakes. The closely related Tecate Cypress (*Cupressus forbesii*) is found in or at the head of chaparral-covered canyons in the Santa Ana Mountains on Guatay, Otay, and Tecate Peaks in San Diego County, and in Baja California Norte.

Another inland-occurring species, Knobcone Pine, is naturally limited to widely separated locations in the San Bernardino Mountains, Santa Ana Mountains, and near Ensenada in Baja California (Critchfield and Little, 1966).

Small groves of Bishop Pine occur below a local chaparral community in two widely separated areas, in the central and northern parts of Isla Cedros, and on Punta San Quentin and in El Cañon de Pinitos in Baja California (Gentry, 1949). This latter population is unique in that it occurs in association with Tecate Cypress (Vogl et al., 1977). These maritime communities are the southern outliers of more extensive postclimax forests and woodlands interruptedly distributed just inland from the immediate coast in northern and central California (Munz and Keck, 1949; Ornduff, 1974). Another relict pine, Torrey Pine (*Pinus torreyana*), is restricted to a few canyons in coastalscrub near Del Mar in San Diego County, California.

None of the vertebrates are restricted to, or particularly representative of, relict conifer forests, and the associated fauna are largely the tree-requiring species from nearby communities.

Figure 36. *Postclimax community of Arizona Cypress (Cupressus arizonica) on north-facing slope of Bear Canyon in the Santa Catalina Mountains, Coronado National Forest, Arizona, ca. 1,660 m elevation. Relict stands (postclimax) of Arizona Cypress such as this one within Madrean evergreen woodland, are now restricted to cool mid-elevation sites within warm-temperate woodlands and chaparral from extreme southeastern Nevada through sub-Mogollon Arizona southeastward into Nuevo Leon. Other relict cypress forests occur as small populations in California and Baja California del Norte.*

Tropical-Subtropical Forests

Such forests possess an enormous variety of species. They include some of the world's richest and most diverse ecosystems (see e.g. Hubbell, 1979). They are of great importance in the landscape and biotic makeup of the Neotropical, Oriental, Ethiopian, and Australian realms and many oceanic islands. In addition to being differentiated on the basis of their biogeography, these forests may be structurally divided into tropical rain forests, seasonal evergreen forests, cloud forests, semi-deciduous forests, and deciduous (=drought deciduous) forests (see Walter, 1973; Mueller-Dombois and Ellenberg, 1974; DeLaubenfels, 1975). Examples of all five of these structural types are found in North America (see e.g. Leopold, 1959).

Two analagous tropic-subtropic deciduous forests extend northward on the cismontane sides of the Sierra Madre Oriental and Sierra Madre Occidental, — Tamaulipan semi-deciduous forest and Sinaloan deciduous forest. Although the former reaches the Lower Rio Grande Valley in Texas (Fig. 37), only its Sinaloan counterpart is included within our boundaries for the Southwest.

The Sinaloan Deciduous Forest is equivalent to the Short-Tree Forest of Gentry (1942) and is included within the Selva Baja Caducifolia and Selva Mediana Caducifolia of Flores Mata et al. (1971). It enters the Southwest in southeastern Sonora and extreme southwestern Chihuahua between ca. 300 m and 1,050 m and extends northward to the vicinities of Nuri and Movas, where its lower limit is somewhat higher at ca. 400 m. Elements occur in protected pockets on canyon slopes northward to the vicinity of Opodepe. Going upslope, the forest begins with the dominance of *Lysiloma watsoni*, the stature of which is above the Hecho (*Pachycereus pecten-aboriginum*) and other cacti. Mean annual precipitation exceeds 635 mm, 75% or more of which falls during the June through October growing season (Table 7).

Much of the subtropical deciduous forest's extent in the Southwest is within the watershed of the Río Mayo. This biotic community was originally described by H. Scott Gentry who spent considerable time in its exploration and study. The results of these studies, published by Gentry (1942), remain the most comprehensive treatment of an extremely interesting biotic area.

124.6 Sinaloan Deciduous Forest

The Sinaloan deciduous forest is a heterogenous drought-deciduous forest with a strong infusion of tropical elements. It also alternates periods of growth with periods of dormancy, relative to the two wet seasons and the two dry seasons. Though it intergrades with Sinaloan thornscrub, it is in the aggregate distinguished from it by greater height, by larger leafage, by a greater proportion of mesomorphic and hydromorphic elements, and by the relative infrequency of thorny and succulent plants. Larger trees are present, dominating lesser trees (Fig. 38). The variable but unbroken forest canopy is 10 to 15 m above ground, with a mixed perennial population in typical stands numbering 3,000 to 5,000 individuals per acre (7,500 to 12,300 per ha).

The Sinaloan deciduous forest lies principally within the canyons, is nearly confined to the barranca region, and is characterized by steep slopes. Its range in elevation is from about 300 to 1,050 m above sea level. It is bordered on the east by Madrean evergreen woodland. To the north it is like a tail pinching out in the northern barrancas. Its southern extensions are unknown, but it merges with semi-deciduous forest in the region of Jalisco. Its position is phytogeographically peculiar, being a long, tough transitional salient, by way of which the tropics have all but put a finger in northern climes.

The tropical element is represented in the Sinaloan deciduous forest by the hydromorphic *Ficus*, the lianas *Arrabidaea littoralis*, *Mardensia edulis*, and *Gouania mexicana*, and the scandent form of *Pisonia capitata*. Epiphytes are exemplified by the orchid *Oncidium cebolleta*, and the bromeliads by *Tillandsia inflata* and *Hechtia* sp. Other prominent plants of tropical distribution are: *Guazuma ulmifolia, Solanum verbascifolium, S. umbellatum, S. madrense, Cestrum lanatum, Drypetes lateriflora, Bursera grandifolia, B. stenophylla, Coutarea latiflora, Stemmadenia palmeri, Cassia emarginata, C. occidentalis, Trichilia hirta, Sassafridium macrophyllum, Vitex mollis, Urera caracasana*, and many others. Most of these plants have leaves or leaflets of comparatively large size, of a mesomorphic character, and without hirsute covering, epidermal thickening, or other features commonly found in arid environments. The largest leaf is that of *Solanum tequilense*, which is about 320 by 450 mm. Many of the leaves are deciduous, however, so that their existence is confined to the warm, moist summer.

The leaf of *Conzattia sericea* is interesting as displaying the probable maximum development of size in the leguminous pinnatifid type of leaf. Though the leaflets are small—8 by 24 mm—they number 500 to 600 and are spread over an area of nearly half a square meter. The entire leaf is about 45 by 75 dm in size, and forms a thin, lacy canopy 12 to 18 m above ground and overtopping the lesser forest trees.

Jarilla chocola, which may be endemic to Sinaloan deciduous forest of the Rio Mayo, admirably reflects the nature of its environment. It is a dioecious, tolerant forest underling, perennial from a crown of erectly placed tubers. It springs forth quickly with the summer rains, putting out leafy, turgid stems, which in several brief weeks attain heights of 60 to 90 cm. It fades quickly with the last dwindling rains, leaving its fruit to lie on the ground for an indefinite period. It is typical of the general behavior of plants of the barrancas in its rapid response to summer moisture and tropical temperatures, alternating with a long period of dormancy through the

Figure 37. *Tamaulipan semideciduous forest along the Rio Grande at Bentsen State Park, Texas. Present are Winged Elm* (Ulmus crassifolia), *Granjeno* (Celtis pallida), *Texas Ebony* (Pithecellobium flexicaule), *Hackberry* (Celtis reticulata), *Huisache* (Acacia smallii) *and Mesquite* (Prosopis glandulosa). *Spanish Moss* (Tillandsia usneoides) *grows on branches.*

months of drought and lower temperatures. In its location among the forest shrubbery it receives the double advantage of constant soil moisture and the reduced transpiration afforded by shade. Its ability to persist in the highly competitive "jungle" growth is also a characteristic of tropical vegetation. The two other known species of *Jarilla* occur in tropical or subtropical climate: *J. heterophylla* in southern Mexico, *J. caudata* from Baja California to Sinaloa and southern Mexico. Yet coupled with the hydromorphic activity of the three summer months are xeric adjustments, adapting *Jarilla chocola* to existence through the nine months of the year which are characterized by drought and higher temperatures. Besides the regular annual production of seeds, *Jarilla* forms each year of its mature life a new plant, rising from the old underground root crown.

Many other plants have root storage systems, providing them with the reserve which enables them to make a quick response to summer rains, and giving them the advantage of a longer growing season. Among the plants with subterranean reserves adjusting their activities to the wet and dry seasons of the Sinaloan deciduous forest are the following: *Ceiba acuminata* (tree; young plant with storage root), *Ipomoea arborescens* (tree; young plant with storage root), *Exogonium bracteatum* (vine with tuberous root), *Dioscorea convolvulacea* var. *grandifolia* (vine with tuberous root), *Hymenocallis sonorensis* (lily with bulb), *Phaseolus caracala* (vine with thickened root), *Amoreuxia palmatifida* (tuberous root), *Vincetoxicum caudatum* (tuber), *Manihot angustiloba* (tuber), *M. isoloba* (tuber), *Salpianthus macrodontus* (tuberous root), *Tigridia pringlei* (bulb).

Figure 38. *Sinaloan deciduous forest 1.8 km southwest of El Taymuco and 56 km northeast of Alamos, Sonora, ca. 460 m elevation. Summer (wet season) aspect on August 9, 1980. The cacti (*Pachycereus pecten-aboriginum, Stenocereus thurberi*) are now almost completely hidden in the leafy foliage of* Cassia emarginata, Brongniartia alamosana, Jatropha platanifolia, Alvaradoa amorphoides, Randia echinocarpa, Croton ciliato-glandulosa, Bursera *spp. and other deciduous components of the forest. Photograph by R.M. Turner.*

The forest stature is highly variable. The lesser trees, including *Bursera confusa, Coutarea pterosperma* and *Haematoxylum brasiletto*, rarely exceed 8 m in height, whereas among the tallest are *Conzattia sericea, Cochlosperum vitifolium, Ceiba acuminata, Bursera inopinnata,* and *Lysiloma watsoni*, rising 12 to 18 m above ground. Along the arroyos are leafy mounds of *Celtis, Guazuma ulmifolia, Montanoa rosei,* and many other shrubs, as well as the large spreading trees *Taxodium mucronatum, Platanus racemosa,* and special species of *Ficus*, rising to heights of 18 to 25 m. On the whole one may safely indicate the average height of the forest as about 12 m.

The Sinaloan deciduous forest lacks the broken canopy characteristics of much of the thornscrub. Except for the interruptions of clearing and arroyos, it presents an unbroken canopy of luxuriant leafage in the summer rainy season and a myriad of living sticks in the spring dry season. Though the dominant color of moist summer is green, it is highly varied between the pale, ashy leaf of *Manihot isoloba* and the deep green of *Ficus* species. In fall and winter it is broken and colored by trees in various stages of deciduation. Variegation of autumn coloring includes the continued green of *Tabebuia* (Amapa), the red and vermilion of *Caesalpinia platyloba*, the changing yellow leaves of *Ipomoea arborescens* (Palo Santo), and countless other hues determined by the pigmental variation of the plants. In winter its beauty is continued by the bright red flowers of *Tabebuia palmeri*, the intense yellow flowers of *Tabebuia chrysantha*, the white starlike canopies of *Ipomoea arborescens*, and other winter bloomers.

Table 7. Precipitation data from stations in or near Sinaloan Deciduous Forest.

Station	Elevation (in m)	J	F	M	A	M	J	J	A	S	O	N	D	Total
Alamos, Sonora*	389	39	17	8	2	1	35	173	173	92	48	10	42	640
Minas Nuevas, Sonora	518	31	14	6	1	1	46	188	182	95	45	10	45	664
Nuri, Sonora*	440	41	17	9	4	4	58	180	165	83	36	17	45	659
San Bernardo, Sonora*	308	46	14	8	5	2	50	179	144	76	34	8	52	618
Quiriego, Sonora*	252	28	13	6	3	2	28	187	181	107	38	12	35	640
Palo Dulce, Sinaloa	800	47	31	16	1	2	64	207	271	110	83	14	80	926

* Outside or at lower edge of subtropical deciduous forest.

In the long spring dry season the forest is a dreary scene: a naked infinite host of trunks and branches, spreading interminably over the volcanic hills and mesas, bared to the fiery sun, under which the last leaf seems to have withered and died (Fig. 39). He who walks this land in the month of May walks with a parched throat. The plants are waiting for the rains and their union with the soil, when like a piece of magic they turn the look of the dead into a fiesta of voluptuous growth, all in a few brief days. The sudden starting of new leaves with the first summer rain is aided greatly in many plants by the advanced development they have attained in the bud. Ten days is sufficient to bring the bare forest into green leaf, though it requires several weeks to bring the leaves to maturity.

Like the thornscrub area, the Sinaloan deciduous forest is composed of two major vegetative types: the deciduous uniform forest of the slopes, and the partially evergreen cover of the canyon bottoms and arroyo margins. Thus the Sinaloan deciduous forest proper is indented and irregularly striate with ribbons of green in the dry season. These are riparian plants of variable associations, which will be discussed below under "Wetlands." In general it is a mesic group less markedly affected by drought. Many of the plants have established root systems in the subterranean waters and are evergreen, as some of the species of *Ficus, Sassafridium macrophyllum,* and *Celtis iguanea,* or partially spring deciduous, as *Taxodium mucronatum, Guazuma ulmifolia, Montanoa rosei,* and others. In the hot spring months the arroyos are oases between the arid and dull hillsides. Small grasses and various flowers enliven the scene, cool air eddies down the deep channels, and shade and water may be found to refresh the traveler. Trails often take advantage of the clear strip offered by the stream beds and follow them for miles.

The Sinaloan deciduous forest tends to form three layers of foliage. The top-layer dominants are, however, usually scattered, with the most common members not adapted to colonial existence, hence the forest level is uneven in appearance. Dominant species forming the leafy mounds above the more uniform middle layer are the following: *Ceiba acuminata, Lysiloma watsoni, L. divaricata, Bursera inopinnata,* and *Cochlospermum vitifolium,* the tops of which are 12 to 15 m above ground. *Conzattia sericea* is the only high dominant forming colonies in itself, and its presence leads to the most definite three-layered forest.

Because of the wide and thin spacing of the leaves, *Conzattia sericea* forms a light shade which is a benefit rather than a deterrent in the light factor for many plants. The association is found only on steep slopes with a gradient of 40° to 60°. Annuals and herbaceous species are, however, greatly restricted because of the shade aggregate of the several arborescent forms. The association occurs high in the Sinaloan deciduous forest, above 600 m altitude, in the heart of the area.

Lists follow of prominent plants found on the Sinaloan deciduous forest slopes and in the arroyo and canyon bottoms.

Slopes (Deciduous)	Bottoms (Evergreen-deciduous)
Acacia coulteri	*Ambrosia ambrosioides*
Arundinaria longifolia	*Baccharis salicifolia*
Bursera fragilis	*Caesalpinia pulcherrima*
B. grandifolia	*Cassia emarginata*
B. epinnata	*C. occidentalis*
B. stenophylla	*Celtis iguanea*
Caesalpinia platyloba	*Cochlospermum vitifolium*
C. standleyi	*Ficus cotinifolia*
Calliandra rupestris	*F. padifolia*
Cassia biflora	*Guazuma ulmifolia*
C. emarginata	*Haematoxylon brasiletto*
Ceiba acuminata	*Hymenoclea monogyra*
Cephalocereus alensis	*Leucaena lanceolata*
Stenocereus thurberi	*Lysiloma divaricata*
Conzattia sericea	*Montanoa rosei*
Coutarea latiflora	*Opuntia* spp.
C. pterosperma	*Pachycereus pecten-aboriginum*
Croton fragilis	*Piscidia mollis*
Haematoxylon brasiletto	*Pisonia capitata*
Hybanthus mexicanus	*Pithecellobium dulce*
Ipomoea arborescens	*P. mexicanum*
Jatropha cordata	*P. undulatum*
J. platanifolia	*Platanus racemosa*
Lemaireocereus montanus	*Randia echinocarpa*
Lysiloma divaricata	*Sassafridium macrophyllum*
L. watsoni	*Solanum madrense*
Pachycereus pecten-aboriginum	*S. verbascifolium*
Pisonia capitata	*Stemmedenia palmeri*
Tabebuia chrysantha	*Taxodium mucronatum*
T. palmeri	*Tithonia fruticosa*
Willardia mexicana	*Urera caracasana*
Wimmeria mexicana	*Vitex mollis*

The fauna is distinctly Neotropical, and these communities are the northern terminus for a large number of tree and forest-requiring vertebrates of more southern distribution. The avifauna in particular is exceptionally rich and diverse—even by Southwestern standards, and this biotic community is deservedly famous as a "Mecca" for bird-watching. Although several of the mammalian inhabitants of the subtropical

Figure 39. *Sinaloan deciduous forest near Alamos, Sonora, ca. 550 m elevation. Spring season (March) aspect of the drought-deciduous "thorn forest" of southern Sonora and Sinaloa. Note that the height of the tree overstory* (Lysiloma watsoni, Acacia cymbispina) *exceeds the layers of scrub and cacti* (Pachycereus pecten-aboriginum) *contained within the short tree forest canopy.*

deciduous forest such as the White-tailed Deer (*Odocoileus virginianus*) and Coati (*Nasua nasua*) extend northward into wooded habitats in the Southwestern United States, other species such as the Ocelot (*Felis pardalis*), Sonoran Squirrel (*Sciurus truei*), and Mexican Cottontail (*Sylvilagus cunicularius*) are "exotic" to Norteamericanos. Although the Beaded Lizard (*Heloderma horridum*), Boa (*Constrictor constrictor*), and Iguana (*Iguana iguana*) are among its best known reptiles, they are but a small sample. Hardly and McDiarmid (1969) list around 70 species of reptiles and amphibians as being abundant to moderately abundant in their analogous *Tropical Semiarid Forest* in Sinaloa.

Part 3. Scrublands

Madrean fasciation of Interior Chapparal at an elevation of about 2460 m in the Sierra del Nido, Chihuahua. The dominant shrubs are species of an oak (Ouercus sp.) and a manzanita (Arctostaphylos sp.).

Arctic-Boreal Scrublands

Scrublands are limited in this climatic regime. The extreme latitudes and high elevations within this climatic zone usually allow for enough precipitation to permit the establishment of forest or woodland. It is only in the coldest, more rigorous locales that decreased temperatures and wind shear decrease plant water availability enough to preclude the establishment of trees. Conversely, at slightly higher elevations, or closer to the poles, these same factors preclude even the presence of scrub, and scrublands are limited to an ecotone along a minimum temperature gradient between taiga and tundra.

Communities of arctic-boreal scrub in North America are accordingly restricted to the subarctic and subalpine regions. Here scrublands usually consist of shrub willows (*Salix* spp.) near timberline and/or stunted trees of the subalpine forest immediately downslope (*Abies, Picea, Pinus*). In the Southwest, such communities are restricted to the southern Rockies and other high mountains between 3,300 and 3,800 m.

131.5 Subalpine Scrub

Scrublands of willows (*Salix* spp.) and other deciduous shrubs, so prevalent in the mesic alpine and subalpine areas to the north, are represented in the arid Southwest only in southern Colorado and northern New Mexico. Upland scrublands in this climatic zone are usually limited to narrow bands of shrubby and prostrate conifers, located at and just below timberline (=krummholz). These "elfinwood" ecotones are limited in area and composed of the same subalpine conifers that form the forests downslope—Engelmann Spruce (*Picea engelmannii*), Bristlecone Pine, (*Pinus aristata*), Limber Pine (*P. flexilis*), and Dwarf Juniper (*Juniperus communis*). Nonetheless, they exhibit a distinct scrub lifeform (Fig. 40). As is the case with Southwestern alpine, subalpine and montane communities, these facultative scrublands can be divided into Rocky Mountain and Sierran-Cascade biomes.

In accordance with the limited development and occurrence of these biomes in the Southwest, few vertebrates are limited to, or well represented in either *Rocky Mountain* or *Sierran-Cascade subalpine scrubland*. The southernmost populations of willow dependant boreal animals such as the Snowshoe Hare (*Lepus americanus*) and White-tailed Ptarmigan (*Lagopus leucurus*) barely reach this area in the alpine regions of the high Sangre de Cristo Mountains of southern Colorado and adjacent New Mexico. More indicative of the conifer scrublands in the subalpine Southwest are nesting populations of the White-crowned Sparrow (*Zonotrichia leucophrys*).

Figure 40. *Subalpine scrubland of "krummholz" on San Francisco Peaks, Coconino National Forest, Arizona, ca. 3,535 m elevation. The Engelmann Spruce (*Picea engelmanni*) and Bristlecone Pine (*Pinus aristata*) are here dwarfed and in shrub form due to reduced available plant moisture from severe cold and wind shear.*

Cold-Temperate Scrublands

These biotic communities have received relatively little recognition and are often considered as seral stages to forest. In other cases they are simply ignored. Nonetheless, some cold-temperate scrublands are distinctive and important biomes that evolved within particular climatic and environmental conditions. Examples of such scrublands are reported for montane areas in Turkey and Chile (Walter, 1973:124, 132), the Rocky and Sierra Nevada Mountains in North America (Shelford, 1963:295), and in other cold-temperate regions (see e.g. Oosting, 1950:258).

Above 1,500 m elevation in the Transverse and Peninsular Ranges, scattered communities representative of *Sierran-Cascade montane scrubland* occur within and adjacent to Sierran-Cascade montane forest. These disturbance-related communities are referred to as montane chaparral (Hanes, 1977) and are generally composed of any of several high elevation species of *Arctostaphylos, Ceanothus, Garrya,* and *Eriodictyon,* and *Quercus wislizenii.* This biome is not differentiated on the color map (Brown and Lowe, 1980), and these communities are discussed in the chapter on Californian coastal chaparral with which they commonly intergrade.

Centered in the montane regions of the Great Basin and on the west slopes of the Rocky Mountains is a *Great Basin montane scrubland.* This biome enters the Southwest from the north where, except in north-central New Mexico, it is almost entirely restricted to mountainous areas north of the 37th parallel.

132.1 Great Basin Montane Scrubland

Great Basin montane scrub is equivalent to "Petran chaparral" (Clements, 1916; Hayward, 1948), "deciduous thicket scrub" (Cooper, 1922), "mountain shrub" (Costello 1954), "Mountain brush" (Vallentine, 1961), "mountain-mahogany-oak scrub" (Küchler, 1964), and "Rocky Mountain bushland" (Shelford, 1963). It is largely a deciduous scrub, although evergreen elements are regularly present and may dominate locally. The physiognomy is that of a thicket of from 1 to 6 m in height which may present a dense homogeneous "chaparral" aspect or be relatively open (Fig. 41).

In the Southwest, Great Basin montane scrub is found in the higher (ca. 2,300-2,400 to over 2,750 m) foothill and mountain regions of Colorado, Utah, Nevada, and in the Sangre de Cristo, Sandia, and Manzano Mountains in New Mexico. It is especially prevalent on mountain slopes around Durango, Colorado, where it is typically positioned on the altitudinal gradient above Great Basin conifer woodland and below subalpine conifer forest. Less commonly (in the Southwest), the lower elevation contact is with sagebrush. The major interaction is with montane conifer forest, whose Ponderosa Pine series it largely replaces on xeric sites.

Considering the high elevations, precipitation is low—mean annual totals range from 380 to 535 mm (Table 8). Moreover, this scanty precipitation is spread throughout the year, with a dry period during the early growing season; the months of May and June are the most moisture deficient. According to Hayward (1948), montane scrub in the Wasatch Mountains is almost bare of snow in winter, indicating that much of the winter precipitation is lost through wind-blow. Low annual precipitation coupled with poor soil development on steep slopes precludes the establishment of montane conifer forest, and it is erroneous to regard this biome as only a "fire climax" community of the montane conifer forest as suggested by Dixon (1935) and others.

Winter temperature minima are commonly below -6° C, and in some years, the growing season is less than 100 days. These cold temperatures are possibly the reason for the replacement of the less cold tolerant Great Basin conifer woodland by montane scrub at higher elevations here and farther north.

In the Southwest, as throughout much of its range, the dominant species in Great Basin montane scrub is a shrub form of the facultative Gambel Oak (*Quercus gambelii* = *Q. gunnisonii, Q. utahensis, Q. vreelandii, Q. novomexicana,* et al.). Other important species include mountain-mahoganies (*Cercocarpus montanus, C. ledifolius*), snowberries (*Symphoricarpos* spp.), serviceberries (*Amelanchier alnifolia, A. utahensis*), Chokecherry (*Prunus virginiana* var. *melanocarpa*), Cliffrose (*Cowania mexicana*), Greenleaf Manzanita (*Arctostaphylos patula*), buckbrushes (*Ceanothus fendleri, C. velutinus*), and New Mexican Locust (*Robinia neomexicana*), any one of which may be locally prevalent in the composition. Other scrub species as Bigtooth Maple (*Acer grandidentatum*) and Bitterbrush (*Purshia tridentata*), although found in the Southwest, are of more importance further north. Sagebrushes (*Artemisia tridentata, A. arbuscula*), and to a lesser extent, rabbitbrushes (*Chrysothamnus viscidiflorus* and others) are almost always well represented. Locally common tree and scrub associates include Aspen (*Populus tremuloides*), hoptrees (*Ptelea* spp.), Hackberry (*Celtis reticulata*), Wildrose (*Rosa woodsii*), Elderberry (*Sambucus cerulea*), currants (*Ribes* spp.), Apache Plume (*Fallugia paradoxa*), barberries (*Berberis* spp.), and Skunkbush Sumac (*Rhus trilobata*). It is not uncommon for individuals of Ponderosa Pines, Douglas-fir or Pinyon to interrupt the brush overstory.

Much of this biotic community has been greatly abused and modified by livestock grazing. As a consequence, grasses are often scarce, and when present, are often the non-native bromes (*Bromus* spp.) and Kentucky Blue-grass (*Poa pratensis*). Native species are those of the Great Basin grassland and montane biomes. Usually the understory and openings are populated by forbs, particularly the less palatable species. Studies in southern Utah by Coles and Pederson (1969) showed the following herbs to be common participants in this biome:

Achillea millefolium	Western Yarrow
Anemone spp.	anemones, windflowers
Antennaria spp.	pussytoes
Aster spp.	asters
Balsamorhiza sagittata	Arrowleaf Balsamroot
Erigeron spp.	fleabanes
Eriogonum spp.	buckwheats
Geranium spp.	geraniums, crane's bills
Lupinus spp.	lupines

Figure 41. *Great Basin montane scrub at Raton Pass between 2,280-2,440 m elevation, near the Colorado-New Mexico state line. Here, as is often the case in this biotic community, the prevalent scrub cover is the deciduous shrub-form Gambel Oak (Quercus gambelii). The scrubland is relatively snow free, the persistent brown leaves of the oak contrasting with the green of the scattered conifers (Pinus ponderosa, P. edulis, Abies concolor). The occasional openings are well clothed with perennial grasses.*

Penstemon spp.	penstemons, beardtongues
Senecio cerra	Butterweed Groundsel
Taraxacum officinale	Dandelion
Thalictrum fendleri	Meadowrue
Thermopsis montana	Golden Pea

Great Basin montane scrub is an important vegetation-type for Rocky Mountain Mule Deer (*Odocoileus hemionus hemionus*); the snow-free hillsides provide winter feeding areas for this animal and for numerous birds (Hayward, 1948). Hayward also reported the following vertebrates to be well represented in this biome in the Wasatch Mountains:

Aphelocoma coerulescens	Scrub Jay
Charina bottae	Rubber Boa
Felis rufus	Bobcat
Hypsiglena torquata	Night Snake
Odocoileus hemionus	Mule Deer
Passerina amoena	Lazuli Bunting

Pipilo chlorurus	Green-tailed Towhee
P. erythrophthalmus	Rufous-sided Towhee
Sceloporus graciosus	Sagebrush Lizard
Spermophilus variegatus	Rock Squirrel
Spizella passerina	Chipping Sparrow
Vermivora celata	Orange-crowned Warbler
Vermivora virginiae	Virginia Warbler
Vireo gilvus	Warbling Vireo

Only the Virginia Warbler and Green-tailed Towhee could be considered characteristic species. This lack of species affinity led Hayward (1948) to conclude reluctantly that Great Basin montane scrub was more an "ecotone" than a biotic community. Although many of the representative plants and animals listed above are also found within montane conifer forest or Great Basin desertscrub, few of them are centered there, suggesting instead that this scrubland is a biotic community of relatively recent derivation.

Table 8. Precipitation data from three stations in the Southwest in and adjacent to Great Basin Montane Scrub.

Station	Elevation (in m)	\multicolumn{13}{c}{Mean monthly precipitation in mm}
		J	F	M	A	M	J	J	A	S	O	N	D	Total
Bryce Canyon N.P., UT 37°39′ 112°10′	2,412	33	31	36	30	22	19	33	61	38	38	27	35	403
Mesa Verde N.P., CO 37°12′ 108°29′	2,155	44	37	40	35	25	18	45	54	33	46	29	47	453
Durango, CO 37°17′ 107°53′	2,012	43	29	37	35	29	22	47	62	40	49	28	51	472

Warm-Temperate Scrublands

Communities of evergreen sclerophyll shrubs have developed around the world, especially where "Mediterranean" type climates prevail. Such climates, in addition to being warm-temperate, are characterized by cool, moist winters and hot, dry summers. Eighty percent or more of the rain falls during the winter, when plant growth is slowed or arrested. These scrublands are variously called "chaparral," "maquis," "garique," and "thamnos," and are found adjacent to, and inland from the Mediterranean coast; in southwest Africa; in Chile; in Australia; and in California, Arizona, and northern Mexico.

The dominant shrubs of these communities represent a wide range of families and species, but all have several important unifying adaptations: a tendency for dense, compact crowns; small, hard, thick, evergreen leaves (sclerophylls); and deep, wide-spreading root systems. They are usually well adapted to fire, and reproduce prolifically from heat-scarified seed that may be stored in the soil for decades. Others sprout vigorously from enlarged root crowns.

Hot, dry summers in areas of dense, leafy compact brush naturally lead to a high fire hazard; and chaparral stands are among the most flammable of vegetation types. Although periodic fires may have profound if temporary effects on both plant and animal life, they are an integral part of the chaparral ecosystem.

In the Southwest, chaparral is represented by Californian (coastal) chaparral and, eastward in Arizona and northern Mexico, by interior chaparral.

Below the chaparral there characteristically occurs a more xeric, short-shrub community commonly referred to as "soft chaparral," sage scrub, dwarf shrub, and in the Middle East, *batha* (Naveh, 1967). In the Southwest, in southern California and Baja California Norte, these communities are represented by *Californian coastalscrub*. Mean annual precipitation in these communities is usually less than 300 mm, and if it were not for such ameliorating coastal factors as moderate maximum temperatures, *velo*[1] and other evapotranspiration saving features, these xerophytic, short-statured coastalscrub communities would be desertscrub. This is the case below interior chaparral, the lower elevations of which are in contact either with Sonoran desertscrub or with semidesert grassland. Because of grazing and fire suppression, this latter biome may take on the appearance and even some of the composition of scrubland (=disclimax semidesert grassland, e.g., Fig. 73).

---

[1] *Velo (= Spanish for veil) is the morning coastal fog or overcast so characteristic of Southern California and other Mediterranean regions.*

133.2 Californian Coastalscrub

Californian coastalscrub is composed mainly of low (0.2-2.5 m), shallow-rooted, often aromatic, shrubs. Many species are drought deciduous, with soft mesomorphic leaves that desiccate and are readily cast as the summer drought progresses. Stem terminals may die back if drought is severe. In northerly, less arid communities, a tendency toward increasing evergreenness is apparent.

The major occurrence of coastalscrub is cismontane, on low hills, foothills and valleys from sea level to the lower edge of the chaparral at ca. 300-600 m (Fig. 42). Much of this former distribution is now cultivated or urbanized. Coastalscrub also occurs on the lower, hotter slopes of the Transverse and Peninsular Ranges and in the inner coastal valleys and foothills (e.g., Perris Valley). There coastalscrub may reach 800 m or higher, occupying a position immediately below the chaparral (Fig. 43). Its intergradation and contact with Californian valley grassland is frequently distinct and marked by a narrow, nonvegetated band (Muller et al., 1964; Halligan, 1974). Southward it merges gradually with the Vizcaíno subdivision of the Sonoran Desert. Its only other major contact with the Sonoran Desert is with the Lower Colorado subdivision in the Banning-Cabazon vicinity where the transformation in San Gorgonio Pass just east of Cabazon is abrupt.

Examination of precipitation data from stations within areas of Californian coastalscrub shows mean annual totals as high as 400 mm in one of the higher intermountain localities to as low as 125 mm near its southern limit, along the coast of Baja California Norte (Table 9). Most coastalscrub receives less than 300 mm per annum, and is significantly drier than the higher chaparral communities. All localities receive 90% or more of their precipitation during the October-April period and, except for lower precipitation and higher summer temperatures, climatographs of coastalscrub stations are similar to those of Californian chaparral (Figs. 44, 45).

Characteristic species include California Sagebrush (*Artemisia californica*), White and Black Sage (*Salvia apiana, S. mellifera*), Whiteleaf Sage (*Salvia leucophylla*), California Buckwheat (*Eriogonum fasciculatum*), California Encelia (*Encelia californica*), Lemonade Sumac (*Rhus integrifolia*), Southern Monkey Flower (*Mimulus longiflorus*), Sawtooth and Damiana Goldenweed (*Haplopappus squarrosus* and *H. venetus*), Our Lord's Candle (*Yucca whipplei*) and Golden Yarrow (*Eriophyllum confertiflorum*) (Munz and Keck, 1949; Horton, 1960). Many herbaceous species are present, particularly after fires. Occasional widely-spaced individuals more characteristic of the higher, moister chaparral may be found—Sugarbush (*Rhus ovata*), California Scrub Oak (*Quercus dumosa*), Hollyleaf Cherry (*Prunus ilicifolia*), and Mountain-mahogany (*Cercocarpus betuloides*), but these add little to the generally short (0.3-1 m) crown cover of the type (Fig. 46).

Southward toward El Rosario (and to a lesser extent eastward), an increasing number of basically desert species become important in the makeup. Especially noticeable are Burbush (*Ambroisia chenopodifolia*), Coffeeberry or Jojoba (*Simmondsia chinensis*), Desert Apricot (*Prunus fremontii*), Maguey (*Agave shawii*), Silver Cholla (*Opuntia echinocarpa*), Beavertail Prickly Pear (*Opuntia basilaris*) and Pitaya Agria (*Stenocereus gummosus*). The passage from coastalscrub to Sonoran desertscrub is a gradual one, but becomes increasingly evident south of Ensenada (Shreve 1936b). This transition region in northern Baja California also contains a number of endemics such as Parry Buckeye (*Aesculus parryi*),

Figure 42. *Californian coastalscrub just north of Dana Point, Orange County, California, ca. 15 m elevation. High bluffs here separate the coastalscrub of sage, buckwheat, sumac (right center), and other species from the coastal strand, minimizing the usual integration between these two California biomes.*

Figure 43. *Californian coastalscrub near Sage, Riverside County, California, ca. 640 m elevation.* Eriogonum fasciculatum, Simmondsia chinensis, Rhus ovata, Salvia apiana, *and a host of herbaceous annuals and perennials in this xeric "interior" fasciation reminiscent of Sonoran desertscrub-semidesert grassland transition.*

Table 9. Precipitation data from 15 stations in the Southwest in or adjacent to Californian Coastalscrub.

Station	Elevation (in m)	J	F	M	A	M	J	J	A	S	O	N	D	Total	Total Oct. thru April	Percent of total
Newport Beach, CA 33°36′ 117°53′	3	53	55	41	33	4	1	0.3	0.8	4	7	39	44	282	272	96
San Diego, CA 32°44′ 117°10′	4	48	38	39	21	4	1	.8	2	3	9	32	44	242	231	95
Chula Vista, CA 32°36′ 117°06′	3	49	54	41	20	5	0.8	.3	3	4	12	22	54	265	252	95
Corona, CA 33°53′ 117°33′	186	60	55	43	27	5	.5	1	2	5	7	38	46	290	276	96
Riverside, CA 33°57′ 117°23′	256	46	44	39	26	5	.8	2	3	5	8	30	43	252	236	94
San Bernardino, CA 34°08′ 117°16′	343	79	73	62	41	11	3	1	3	8	13	49	66	409	383	94
Redlands, CA 34°08′ 117°11′	402	59	52	53	34	9	2	2	4	8	12	39	50	324	299	93
Elsinore, CA 33°40′ 117°20′	392	56	49	42	22	3	.5	.5	3	4	7	36	50	273	262	96
San Jacinto, CA 33°47′ 116°58′	468	51	49	50	30	6	1	3	6	8	13	36	44	297	273	92
Tecate, BCN 32°35′ 116°38′	515	78	28	47	25	10	2	2	3	3	12	16	41	267	247	93
Santo Tomas, BCN 31°32′ 116°25′	150	52	29	35	23	2	—	1	1	1	8	23	38	213	208	98
San Vicente, BCN 30°59′ 116°06′	105	45	26	25	21	2	0.1	1	1	2	6	22	32	183	177	97
San Telmo, BCN 30°59′ 116°06′	100	26	25	13	1	0.3	.3	.3	1	6	6	21	34	134	126	95
El Socorro, BCN 30°20′ 115°49′	10	47	22	13	7	—	—	—	1	0.3	7	15	25	138	136	99
Santa Maria del Mar, BCN 30°26′ 115°54′	30	27	25	14	11	—	.3	—	1	.2	7	8	36	134	138	98

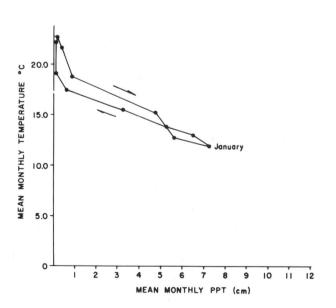

Figure 44. *Monthly precipitation-temperature polygon for Californian coastalscrub (six stations).*

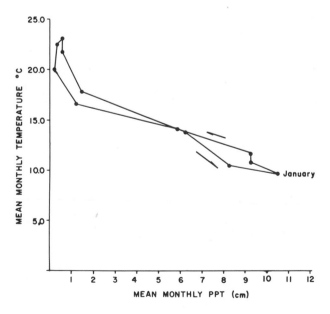

Figure 45. *Monthly precipitation-temperature polygon for Californian chaparral (seven stations).*

Figure 46. *Californian coastalscrub near Julian, San Diego County, California, ca. 825 m elevation. The "patterning" effect with chaparral and woodland is due to variations in available plant moisture, evapotranspiration rates, and possibly fire history. The primary coastalscrub species here is* Eriogonum fasciculatum.

a wild rose (*Rosa minutifolia*), Chaparral Ash (*Fraxinus trifoliata*), Siemprevivos (*Dudleya* spp.) and the Velvet Cactus (*Bergerocactus emoryi*, Fig. 47). The appearance of a buckeye in such an arid region is surprising.

Most species readily sprout after burning, and this, coupled with the general abundance of herbaceous plants, permits rapid recovery after fires. On better sites, complete recovery may take as little as 10 years, but on poor sites recovery may take much longer. After fires at higher elevations, coastalscrub is often a successional stage to chaparral.

The cover in mature stands is generally more open than in the adjacent chaparral, owing to the generally poorer site conditions and lower rainfall in coastalscrub. Ground cover of less than about 50% is common and may not exceed 25% on steep or unstable slopes.

Although Californian coastalscrub can be considered as the ecological center for several mammals and birds—e.g., the Nimble Kangaroo Rat (*Dipodomys agilis*), San Diego Pocket Mouse (*Perognathus fallax*), and California Valley Quail (*Lophortyx californicus*)—it also provides suitable habitats for a number of species adapted to both open chaparral and desertscrub. Also, as an expression of vegetational affinities, some sagebrush species, such as the Sage Sparrow (*Amphispiza belli*) do not differentiate between sagebrush communities of the Great Basin and these of Californian derivation.

Among the vertebrates, reptiles best distinguish this biome as an evolutionary center, although a relatively recent one. Centered here are the: California Side-blotched Lizard (*Uta stansburiana elegans*), San Diego Coast Horned Lizard (*Phrynosoma coronatum blainvillei*), Coastal Whiptail (*Cnemidophorus tigris mundus*), Orange-throated Whiptail (*Cnemidophorus hyperythrus*), Coastal Rosy Boa (*Lichanura trivirgata roseofusca*), California Striped Racer (*Masticophis lateralis lateralis*), Coast Patch-nosed Snake (*Salvadora hexalepis virgultea*), San Diego Gopher Snake (*Pituophis melanolecus annectans*), California Glossy Snake (*Arizona elegans occidentalis*), San Diego Night Snake (*Hypsiglena torquata klauberi*), Southern Pacific Rattlesnake (*Crotalus viridis helleri*), and the Red Diamond Rattlesnake (*Crotalus ruber*). The fact that these taxa are mostly of subspecific rank further attests to the recent differentiation of Californian coastalscrub, contemporary with the Southwest's deserts.

Figure 47. *Californian coastalscrub on Isla Todos Santos, Baja California Norte, ca. 6 m elevation, April, 1979. The profusion of herbaceous vegetation makes the term "soft chaparral" seem especially appropriate here in this southern fasciation after the wet winter of 1978-79. The perennial shrubs, principally* Encelia californica *and* Salvia *spp., are actually rather widely spaced with several "desert" species now inconspicuously present—*Bergerocactus emoryi, Opuntia prolifera, *and* Dudleya, *spp.*

133.1 Californian (Coastal) Chaparral

Chaparral is the main vegetation of southern California and much of northern Baja California. It conspicuously covers extensive mountain, hillside, and foothill landscapes between the Sonoran and Mohave Deserts and the coast. Largely cismontane, it occurs from as high as 2,750 m in the Transverse Ranges down to less than 50 m elevation near the Pacific Ocean (Cooper, 1922). Shrub heights vary from as low as 1 m to more than 3 m on some of the higher slopes, but the physiognomy is remarkably uniform throughout (Fig. 48). Floristic diversity is high, however, with nearly 900 species of vascular plants being associated with this biome (Ornduff, 1974).

Like most chaparral areas around the world, Californian chaparral is found in a "Mediterranean"-type climate, characterized by cool, wet winters, and a hot, dry growing season. Total mean annual precipitation ranges from 300-375 mm to more than 760 mm in the higher montane associations. More than 75% of this precipitation falls from October through April (Table 10), with less than 20% coming in the 6 months when mean monthly temperatures are likely to exceed 15° C (May through October). Much of this rainfall results from a small number of intense polar marine winter storm systems. The periodic 6-month dry period coincides with high temperatures, further accentuating the seasonal drought. As rainfall declines and maximum temperatures begin rising in April or May, the chaparral begins drying, and by July or August is highly flammable. This condition is aggravated by the "Santa Ana winds" and may persist well into fall, sometimes resulting, as in 1977, in uncontrollable fires that sweep over thousands of hectares.

Chaparral is well adapted to seasonal drought and fire. After burning, many chaparral species readily regenerate as sprouts or from heat-scarified seeds. Although the immediate post-fire community may differ markedly from that before burning, at least in herbaceous components, within a few years the original chaparral species reestablish themselves and firmly control the site—until the next fire.

The hard "sclerophyll" leaf is a diagnostic feature of chaparral plants. Leaves tend to be small, hardened (preventing wilting during periods of extreme drought), and evergreen. Leaves of a given year are not cast until the next year's leaves are set; hence the shrubs have some leaves even during protracted drought periods. This physiognomic feature differentiates chaparral from the periodically leafless montane scrub and thornscrub of cold temperate and tropic-subtropic regions.

Chaparral typically occurs on steep mountain slopes but also occupies outwash plains, "bajadas" and other thin-soiled habitats. It is found growing on a variety of geologic parent materials but reaches its best development on deeply fractured and weathered coarse-grained granite and gneiss. Soils may be shallow to deep, are usually well-drained, permitting deep percolation into the regolith. On steeper slopes, erosion rates are commonly too high to permit soil horizon development (Horton and Kraebel, 1955). Nutrient status, at least when compared to woodland or grassland soils, tends to be low.

Because of its importance as watershed cover in high erosion risk areas, Californian chaparral has been the subject of a number of intensive studies. A comprehensive review of major chaparral communities was prepared by Hanes (1977).

Table 10. Precipitation data from five stations in the Southwest within and adjacent to Californian Chaparral.

Station	Elevation (in m)													Total	Total Oct. thru April	Percent of total
		J	F	M	A	M	J	J	A	S	O	N	D			
San Gabriel Dam, CA 34°12′ 117°52′	451	154	132	102	67	10	2	0.5	1	5	23	91	114	702	683	98
Lytle Creek R.S., CA 34°14′ 117°29′	832	189	154	134	83	13	2	2	2	6	20	91	129	825	800	94
Descanso R.S., CA 32°51′ 116°37′	1,067	100	87	106	64	21	3	6	11	8	19	59	90	574	525	91
San Juan de Dios, BCN 32°08′ 116°10′	1,410	51	38	42	38	8	2	23	17	14	12	33	57	335	271	81
Sierra Juarez, BCN 31°45′ 115°49′	1,410	33	35	42	31	5	1	11	26	20	12	40	49	305	242	79

The following vegetative descriptions cover the major chaparral associations in the Southwest in southern California and Baja California Norte.

Chamise Series

This community is a frequent dominant in southern California chaparral, where it forms extensive stands in the mountains of Los Angeles, Orange, San Bernardino, Riverside, and San Diego Counties, and in Baja California Norte. It is characterized by nearly pure Chamise (*Adenostoma fasciculatum*), with minor amounts of *Ceanothus* spp., Bigberry and Eastwood Manzanita (*Arctostaphylos glauca* and *A. glandulosa*), White and Black Sage, California Buckwheat, California Scrub Oak, Sugarbush and Laurel Sumac (*Rhus laurina*). The Chamise chaparral forms dense, matted stands with interwoven branches, making an almost impenetrable stand at maturity. Little or no herbaceous understory occurs in mature or senescent stands, but a moderate postfire flora develops which may persist for several years. Mature stands reach 90% cover in 25 years, but this cover may decline with senescence.

Recovery from fire is somewhat slower in Chamise chaparral than in some other chaparral types, partly because of the generally poor soils on which it grows. Stands become decadent at about 50 to 75 years with little or no replacement. Periodic fires appear essential to maintain this association in a vigorous condition.

Ceanothus Series

Ceanothus may occur as nearly pure stands of a single species, e.g., Hoary Leaf Ceanothus (*C. crassifolius*), or as a codominant with Chamise, Scrub Oak, Toyon (*Heteromeles arbutifolia*), or Sugarbush. Ceanothus usually occurs on wetter sites than Chamise, and is most prevalent on the coastal side of the Peninsular Ranges. It is rarely found above 1,200 m elevation (Horton, 1969).

Ceanothus chaparral regenerates readily from seedlings as well as sprouts after fire, and its early rapid growth rate and high crown density make it a strong early competitor to associated species that commonly regenerate from sprouts. Recovery after fire is more rapid than in Chamise, partly because of the better sites on which these associations occur.

Within a decade of burning, Ceanothus stands approach 50% crown cover, and in 50 years, from 80% to 100%. As a generation of plants mature, the short-lived Ceanothus begins to die out, opening the stand. Overmature 50- to 70-year-old stands may regress to only half the cover value of young, healthy stands, opening the way to replacement by other species. Ceanothus is, therefore, considered a successional chaparral community in southern California.

Manzanita Series

Less extensive than Chamise or *Ceanothus* chaparral, Manzanita (*Arctostaphylos* spp.) is a characteristic, easily identified type of the southern California mountains. It usually occurs at the higher moister elevations and on deeper soils. Consequently, it often occurs below or intermingled with the lower fringe of the montane conifer forest. Some manzanita stands are "pure," but associations found in the Manzanita series include: California Scrub Oak, Chamise, Ceanothus, Birchleaf Mountain-mahogany (*Cercocarpus betuloides*), Chinquapin (*Castanopsis sempervirens*), Coulter Pine (*Pinus coulteri*), and Fremont Silktassel (*Garrya fremontii*).

About half of the dozen or so manzanita species sprout after burning; the rest germinate readily from heat-scarified seeds. Although the series is well adapted to recurrent fires, if the recurrence interval is less than the time required to reach maturity and set seeds, a decline in the non-sprouting species can be expected. Abundant annuals and short-lived perennials develop in the first post-fire rainy period, often turning a fire-blackened landscape into a fantastic flower garden (Sweeney, 1956). A successional stage of a mixture of chaparral and coastalscrub species occurs at lower elevations in these and other chaparral series. Mature stands over 50 years of age may reach 100% cover except on poor sites, where the cover may be less than half that.

Scrub Oak Series

Quercus dumosa is a mesic association unimportant in areas with less than 508 mm precipitation. It occurs mainly on northerly aspects on lower sites but may occupy all aspects at higher elevations. This series is rich in species, and a great variety of large shrubs and woody vines participate. Crowns are from 2 m to 5 m high, and may extend downward almost

Figure 48. *Mixed sclerophyll community in the Santa Ana Mountains, Cleveland National Forest, Orange County, California. Dominant or at least prevalent shrubs in the chaparral at this location (ca. 1,100 m elevation) are California Scrub Oak* (Quercus dumosa), *several species of* Ceanothus, *manzanita* (Arctostaphylos), *and mountain-mahogany* (Cercocarpus). *Note the almost complete ground cover.*

to ground level. A heavy litter is usually present; understory forbs and grasses are sparse, except in scattered openings.

Common associates include Birchleaf Mountain-mahogany, Chaparral Whitethorn (*Ceanothus leucodermis*), Toyon, Holly-leaved Cherry (*Prunus ilicifolia*), Hollyleaf and California Buckthorn (*Rhamnus crocea* and *R. californica*), honeysuckle (*Lonicera* spp.), silktassels (*Garrya* spp.), California Fremontia (*Fremontia californica*), Poison Oak (*Rhus diversiloba*), and a deciduous Two Petal Chaparral Ash (*Fraxinus dipetala*). Scrub Oak, like its major associates, sprouts vigorously from the root crown following fire. Because it usually occurs on better sites, regrowth is rapid. The secondary growth commonly shows a 50% to 60% crown cover at 10 years, reaching a maximum of 80% to 100% cover in 50 years.

A number of other important series of lesser extent have been described within California chaparral, e.g., Red Shanks (*Adenostoma sparsifolium*) chaparral (Hanes, 1977; Fig. 49), Mountain-mahogany chaparral, Live Oak (*Quercus chrysolepis*) chaparral, and mixed chaparral series. These communities share with those already mentioned a more or less short statured, closed physiognomy and dominance by one or more sclerophyll species of *Quercus, Adenostoma, Ceanothus, Cercocarpus, Arctostaphylos, Rhus,* or *Garrya.*

Canyon bottoms and other mesic sites within the chapparal are frequently wooded (Californian evergreen forest or wood-

land or, at its upper limits, Sierran montane conifer forest). The chaparral's lower and more xeric contacts are typically with Californian coastalscrub—less commonly grassland. Frequently, individual, scattered trees of *Pinus, Juniperus,* or *Quercus* punctuate the chaparral and at the higher elevation (1,200 m) there may be much contact with the yellow pines and canyon live oaks of the respective Sierran montane conifer and Californian mixed evergreen forests.

Californian chaparral is the evolutionary center for a large number of vertebrates, many of which are so well adapted to it as to be uncommonly found elsewhere. Foremost of these among the mammals are the Brush Rabbit (*Sylvilagus bachmani*) and California Mouse (*Peromyscus californicus*). Other, more catholic Californian mammals well represented in chaparral are the Mule Deer (*Odocoileus hemionus*), Gray Fox (*Urocyon cinereoargenteus*), Merriam Chipmunk (*Eutamias merriami*), Dusky-footed Woodrat (*Neotoma fuscipes*), Nimble Kangaroo Rat (*Dipodomys agilis*), California Pocket Mouse (*Perognathus californicus*), and Brush Mouse (*Peromyscus boylii*).

Some species of birds are also largely confined to Californian chaparral—Mountain Quail (*Oreotyx pictus*), Wrentit (*Chamaea fasciata*), California Thrasher (*Toxostoma redivivum*), and a few others as the Anna Hummingbird (*Calypte anna*) are best represented there. Still other scrub adapted

Figure 49. *Californian chaparral of Red Shanks* (Adenostoma sparsifolium) *between Santa Rosa and San Jacinto Mountains, San Bernardino National Forest, California, ca. 1,340 m elevation. The light areas on the mountains in background are sites of recent fires.*

species are as equally at home there as in other scrub biomes including *interior chaparral*. Examples of these latter species are the Scrub Jay (*Aphelocoma coerulescens*), Dusky Flycatcher (*Empidonax oberholseri*), Bewick's Wren (*Thryomanes bewickii*), Orange-crowned Warbler (*Vermivora celata*), Bushtit (*Psaltriparus minimus*), Rufous-sided Towhee (*Pipilo erythropthalmus*), Brown Towhee (*P. fuscus*), Lazuli Bunting (*Passerina amoena*), Black-chinned Sparrow (*Spizella atrogularis*), and Rufous-crowned Sparrow (*Aimophila ruficeps*).

Few, if any, reptiles have their distribution restricted to chaparral per se. However, in the Southwest, the San Diego Alligator Lizard (*Gerrhonotus multicarinatus webbi*), Granite Night Lizard (*Xantusia henshawi*), and Striped Racer (*Masticophis lateralis*) are mostly within this biome. Other reptiles to be expected in Californian chaparral habitats in the Southwest are the Western Fence Lizard (*Sceloporus occidentalis*), Sagebrush Lizard (*S. graciosus*), Coast Horned Lizard (*Phrynosoma coronatum*), Coastal Rosy Boa (*Lichanura trivirgata roseofusca*), Western Patch-nosed Snake (*Salvadora hexalepis*), California Glossy Snake (*Arizona elegans occidentalis*), Western Black-headed Snake (*Tantilla planiceps*), California Lyre Snake (*Trimorphodon biscutatus vandenburghi*), Red Diamond Rattlesnake (*Crotalus ruber*), and Western Rattlesnake (*C. viridis helleri*).

133.3 Interior Chaparral

Chaparral is also an important vegetation-type in Arizona and in parts of northern Mexico. In Arizona, interior chaparral discontinuously occupies mid-elevation (1,050-2,000 m) foothill, mountain slope, and canyon habitats from the Virgin Mountains on the extreme northwest border with Nevada, southeastward to the central sub-Mogollon portions of the state, where it is of major occurrence and importance. Southeastward, isolated and disjunct chaparral communities continue into the drier mountains of southeastern Arizona, southern New Mexico (e.g., in the Burro, Florida, Organ, and Guadalupe Mountains), southwest Texas (e.g., in the Glass and Chisos Mountains), and northeastern Chihuahua to the limestone mountains of Coahuila and Neuvo Leon where chaparral is again an important vegetation-type (Muller, 1939, 1947; LeSueur, 1945). The two centers of distribution in the Southwest are mid-elevation watersheds of the Hassayampa, Agua Fria, Verde, and Salt Rivers in central Arizona and the limestone "bufas" and sierras of extreme eastern Chihuahua and western Coahuila.

Recent estimates place chaparral cover in Arizona at about 1.4 million ha (Hibbert et al., 1974). There it occurs mainly between 1,050 and 1,850 m elevation, with stands extending to 2,000-2,150 m and even higher on drier and warmer slopes, as in the Hualapai Mountains and on the south face of the Mazatzal Mountains. Elsewhere, in southern New Mexico, eastern Chihuahua, and Coahuila, interior chaparral is limited to elevations of from ca. 1,700 m to just over 2,450 m. The upper limits normally border Ponderosa Pine or encinal associations, or in the north on poorly drained, fine-textured basaltic soils, pinyon-juniper woodland. Its lower contacts in the west are with Mohave desertscrub, Sonoran desertscrub (Arizona Upland subdivision), or semidesert grassland; eastward from central Arizona chaparral is almost always positioned above semidesert grassland.

Precipitation in interior chaparral varies from 380 to 635 mm per annum, although marginal open chaparral communities may develop on as little as 350 mm (Table 11). Most of this precipitation falls as rain, although in occasional years considerable snow may fall at higher elevations. Although Californian chaparral exists in a Mediterranean climate, precipitation in interior chaparral is distinctly bimodal, with significant but irregular amounts of precipitation falling during the summer monsoon (Fig. 50). Summer rainfall occurs as sharp, high intensity thundershowers, as opposed to winter rainfall which is generally associated with Pacific frontal storms. All Southwestern chaparral communities are characterized by spring drought, the early growing season from April to June being the driest time of the year.

Most chaparral shrubs have dense, compact crowns and small evergreen sclerophyllous leaves. They are deeply rooted, with some species, e.g. Shrub Live Oak (*Quercus turbinella*), having both extensive superficial and tap root systems (Davis and Pase, 1977). Most species sprout readily from an often massive root crown, and quickly regenerate after burning. Notable exceptions to this are Desert Ceanothus and Deerbrush (*Ceanothus greggii* and *C. integerrimus*), and Pointleaf and Pringle Manzanita (*Arctostaphylos pungens* and *A. pringlei*). These species produce prolific seed crops which may be stored in the soil for decades, germinating readily only after fire (Glendening and Pase, 1964; Pase, 1965).

The successional status and nature of interior chaparral has been a matter of debate for several years. The vigor and health of very old chaparral stands and its widespread occur-

Table 11. Precipitation data from 9 stations in the Southwest within and adjacent to Interior Chaparral.

Station	Elevation (in m)	J	F	M	A	M	J	J	A	S	O	N	D	Total	Total Sept. thru April	Percent of total
Bagdad, AZ 8NE 34°39′ 113°05′	1,292	52	39	44	26	8	12	46	82	32	19	37	43	440	292	66
Walnut Grove, AZ 34°18′ 112°33′	1,147	39	35	38	22	6	9	52	85	38	21	26	44	415	263	63
Fossil Springs, AZ 34°25′ 111°34′	1,301	55	43	52	31	14	13	67	81	45	34	39	60	534	359	67
Sunflower, AZ 33°54′ 111°29′	1,134	62	40	57	18	8	8	51	84	56	36	42	58	520	369	71
Sierra Ancha, AZ 33°48′ 110°58′	1,554	72	52	66	29	10	11	68	99	54	43	44	80	628	440	70
Miami, AZ 33°24′ 110°53′	1,085	52	32	45	17	6	7	59	85	39	27	28	61	458	301	66
Salt River, AZ 33°48′ 110°30′	1,100	43	31	37	18	6	13	51	73	25	35	25	36	393	250	64
Cureton, Rch., NM[1] 32°32′ 108°34′	1,585	26	19	20	7	4	12	59	71	39	21	14	25	317	171	54
Sierra Mojada, Coah.[1] 27°17′ 103°42′	1,263	7	9	8	7	23	46	80	67	84	29	9	14	383	167	44

[1]In semidesert grassland, immediately downslope from interior chaparral.

rence under a wide variety of parent materials, exposures, and rainfall regimes strongly suggest that interior chaparral is indeed climax (Carmichael et al., 1978). As in Californian chaparral, high elevation stands marginal to montane forests may be successional. Some old burns inhabited by Pringle Manzanita, Deerbrush, and Fendler Ceanothus (*C. fendleri*) obviously fit this category and in the absence of fire would be replaced by Ponderosa Pine. Grazing by domestic livestock reduces the herbaceous cover in the intershrub spaces but there is little evidence to suggest that grasses and forbs significantly inhibit established chaparral scrub (Pond, 1968). There can be little doubt, however, that the increase in "brush" and other woody species since 1900 is a result of chaparral replacing herbaceous components and the reduction of fire (Leopold, 1924).

Interior chaparral, like Californian chaparral, usually presents a closed or moderately open growth of relatively uniform height between 1 and 2 to 2½ m (Figs. 51, 52). Many of the species present are disjunctive counterparts of those in Californian chaparral—Shrub Live Oak, Sugar Sumac or Mountain Laurel, Hollyleaf Buckthorn, manzanitas, Birchleaf Mountain-mahogany and Hairy Mountain-mahogany (*Cercocarpus breviflorus*), Yellowleaf Silktassel (*Garrya flavescens*), Brickellbush (*Brickellia californica*), Ceanothus and California Fremontia. A number of other species are undoubtedly conspecific.

"Arizona" Chaparral

In Arizona, Shrub Live Oak is the most widespread chaparral species and the common dominant. Sometimes occurring in essentially pure stands, this species is more commonly accompanied by such chaparral shrubs as Birchleaf Mountain-mahogany, Skunkbush Sumac (*Rhus trilobata*), silktassels (*Garrya wrightii, G. flavescens*), and Desert Ceanothus, any of which may locally attain dominance. Less universal but locally important in the composition are Hollyleaf Buckthorn, Cliffrose (*Cowania mexicana*), Desert

Olive (*Forestiera neomexicana*), sophoras (*Sophora* spp.), Arizona Rosewood (*Vauquelinia californica*), the deciduous Lowell Ash (*Fraxinus anomala* var. *lowellii*), and Barberry (*Berberis fremontii*). Manzanita is a frequent associate at higher elevations.

Chaparral is everywhere a drier adapted formation than evergreen sclerophyll woodland. There is, therefore, much integration of interior chaparral with *Madrean evergreen woodland* in the higher, wetter, chaparral reaches, where the presence of one or the other biome is often determined by edaphic controls on plant available moisture. The most xeric Madrean woodlands are usually found on alluvial and volcanic soils; interior chaparral intrudes well into otherwise Madrean woodland areas on limestone substrates. In the rugged mountains immediately below the Mogollon Rim in Arizona, chaparral gradually gives way upslope to taller evergreen oaks (*Quercus emoryi, Q. arizonica*), junipers (*Juniperus deppeana, J. monosperma*), and Ponderosa and Pinyon Pines (*Pinus edulis*) with attendant increase in precipitation. This transition is complex because of the infusion of montane and Great Basin elements in addition to Madrean ones, and has resulted in some workers combining chaparral and Madrean woodland into a broader "chaparral" concept (e.g. Nichol, 1952; Wallmo, 1955; USDA Soil Conservation Service, 1963). This approach, while convenient, fails to consider the major floral and faunal differences and exchanges that exist between these evolutionarily distinct biomes.

In Arizona, chaparral occurs on a variety of parent rock materials, including granites, diabases, gneiss, schist, shales, slates, sandstones, and limestones. The best and most typical development, both in species diversity, structure, and cover, is on the coarser granitic intrusives and on limestone. Soils in chaparral stands tend to be poorly developed; the unstable, coarse soils on steep slopes have little opportunity for horizon development. Organic matter content is ordinarily low, on the order of 0.5% to 2%. Because of the coarse texture, however, infiltration rates are usually high except after severe wildfires when soils may become temporarily non-wettable, a characteristic shared with Californian chaparral soils. Elsewhere in the interior Southwest, chaparral is practically confined to calcereous soils, the porosity of which is presumably important for its presence.

Shrub cover varies but averages about 60% to 70% in mature stands. The drier, rockier, more open sites may contain "thornscrub" elements of Wait-a-bit (*Mimosa biuncifera*) or Catclaw (*Acacia greggii*). Scrub members of the Sonoran desert and semidesert flora are also to be expected and include Coffeeberry or Jojoba (*Simmondsia chinensis*), Crucifixion Thorn (*Canotia holacantha*), Banana Yucca (*Yucca baccata*), agaves (*Agave* spp.), Beargrass or Sacahuista (*Nolina microcarpa*), Sotol (*Dasylirion wheeleri*), Snakeweed (*Gutierrezia sarothrae*), and False-mesquite (*Calliandra eriophylla*). Wright's Buckwheat (*Eriogonum wrightii*) is an important and conspicuous forage plant.

Because of high accessibility and relatively gentle terrain, lower chaparral sites have been heavily grazed, especially between 1880 and 1920, and were, until 1940, the location for a flourishing mohair goat industry. Important range grasses, now largely confined to rocky, protected sites, include Sideoats and Hairy Grama (*Bouteloua curtipendula* and *B. hirsuta*), Cane Bluestem (*Bothriochloa barbinodis*), Plains Lovegrass (*Eragrostis intermedia*), Wolftail (*Lycurus phleoides*), and Spidergrass, Fendler's, and Single Threeawn (*Aristida ternipes, A. fendleriana*, and *A. orcuttiana*). Forbs are not particularly abundant except for a brief period after burns. The more common species include Palmer's, Eaton's, and Toadflax Penstemon (*Penstemon palmeri, P. eatoni*, and *P. linarioides*), Wright's Verbena (*Verbena wrightii*), Few-flowered Goldenrod (*Solidago sparsiflora*), Purple Nightshade (*Solanum xantii*), Hoarhound (*Marrubium vulgare*), White Dalea (*Dalea albiflora*), and Scarlet Starglory (*Ipomoea coccinea*). Patches of bare ground give forth to spring annuals—the introduced Filaree (*Erodium cicutarium*) and Red Brome (*Bromus rubens*) now being particularly prevalent.

At higher elevations and moister sites the chaparral may be dominated by Pringle or Pointleaf Manzanita, or a mixture of sclerophyll shrubs with no single dominant species. These latter associations are high in species diversity, often occupying north slopes and "cove" sites where better soils and more moisture are available. Biomass is high and the variation in species lends a rough, uneven texture to the surface of the community. Most species are crown sprouters, and individual plants may be very old—perhaps hundreds of years—although the above ground portions extend back only to the last fire. Post-fire succession is, therefore, rapid and, except for "new" populations of Desert Ceanothus, species composition is changed little by the natural fires that may sweep the area at intervals as long as from 50 to 100 years. Common codominants are Shrub Live Oak, Pointleaf Manzanita, Desert Ceanothus, Skunkbush Sumac, Yellowleaf Silktassel, Sugar Sumac, Hollyleaf and California Buckthorn (*Rhamnus californica*). The more moist places may even include typical California species as Flannel-bush and Canyon Live Oak (*Quercus chrysolepis*) as well as the more cosmopolitan Chokecherry (*Prunus virginiana*).

An occasional One-seed Juniper, Emory Oak, Pinyon, or other tree is often present and any of these species may form an open, scattered overstory. Arizona Smooth-bark Cypress (*Cupressus glabra*) may occupy north facing slopes, canyons, and canyon bottoms.

Because of the high percentage of crown cover (70+%),

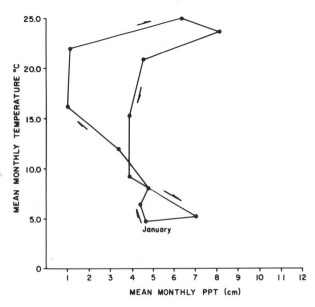

Figure 50. *Monthly precipitation-temperature polygon for Arizona chaparral (six stations).*

forbs and grasses are not abundant except in the scattered interscrub openings, on rocky outcrops, or in the early post-fire succession. The lush post-fire flora includes Red Brome, Red Sprangletop (*Leptochloa filiformis*), Telegraph Plant (*Heterotheca subaxillaris*), Scarlet Starglory, Mexican Morning Glory, Scarlet and Erect Spiderling (*Boerhaavia coccinea* and *B. erecta*), Louisiana Sagebrush (*Artemisia ludoviciana*), and penstemons (*Penstemon* spp.). Perennial grasses, often confined to low shrub density microsites, include Fendler and Single Threeawn, Spidergrass, Green Sprangletop, Cane Bluestem, Arizona Cottontop (*Trichachne californica*), Plains Lovegrass (*Ergrostis intermedia*), Longtongue Mutton Bluegrass (*Poa longiligula*), and Hairy Grama.

Pringle Manzanita grows at the highest chaparral elevations with 635 mm or more of annual rainfall, where it may occur in pure stands or in association with Emory Oak, Yellowleaf Silktassel, Narrowleaf Yerbasanta (*Eriodictyon angustifolium*), and Desert Ceanothus. Both density and canopy coverage are very high and these communities often present an "impenetrable" aspect. Mature Pringle Manzanita may have trunk diameters of 30 cm and individual plants reach a height of 4.5 m. Forest floor weights of up to 18 metric tons per ha have been reported (Glendening and Pase, 1964).

Under the dense canopy of mature and overmature stands, virtually no herbaceous cover can be found, but annuals and short-lived perennials are common after fire. Conspicuous in the brief post-fire flora are Canyon and Beardlip Penstemon (*Penstemon pseudospectabilis, P. barbatus*), and Wright Verbena. Red brome is a prominent annual. Desert Ceanothus and Narrowleaf Yerbasanta, decadent or missing from old stands, often become prominent in the decade or so after burning, but gradually become senescent and disappear long before the longer-lived Pringle Manzanita matures.

In the very old stands, Emory and/or Arizona Oak and occasionally Ponderosa Pine overtops the manzanita and gains ascendancy in the climax. Although Pringle Manzanita is long-lived and is rarely entirely replaced by oaks and pine, it appears unable to reproduce except when its seeds are heat scarified.

Figure 51. *Interior chaparral in Hualapai Mountains, Mohave County, Arizona, ca. 1,980 m elevation. A mixed association of manzanita* (Arctostaphylos), *silk tassel* (Garrya), *Shrub Live Oak* (Quercus turbinella), *mountain-mahogany* (Cercocarpus), *and others. The occasional conifers are Singleleaf Pinyon* (Pinus monophylla).

Pointleaf Manzanita appears to form a seral chaparral community on ridges and steep slopes where higher temperatures and poor soils restrict Pringle Manzanita. Although Pointleaf Manzanita does not sprout from its crown or root, very old shrubs may "ground layer," i.e., the tips of the long outer branches may take root, producing a "fairyring" type of growth. The old plant center dies, leaving a ring of offset shrubs around the old plant center. This center may be devoid of herbaceous cover for decades.

"Coahuilan" Chaparral

Because of their remoteness and poor accessibility, much less is known of chaparral series in Mexico, southern New Mexico, and Texas. Our knowledge of "Coahuilan chaparral" is based on studies by Muller (1939, 1947), LeSueur (1945), and more recently Dick-Peddie and Moir (1970), and on visits to the Sierras Mojada and San Marcos in Coahuila, the Florida and Guadalupe Mountains in New Mexico, and the Chisos Mountains in Texas.

Chaparral associations in these areas are characterized by many of the same genera that characterize chaparral in California and Arizona. A frequent dominant is Coahuila Scrub Oak (*Quercus intricata*) which appears analagous with Shrub Live Oak. Other shrub-scrub live oaks include Vasey Oak (*Q. pungens*), Pringle Oak (*Q. pringlei*), and a number of less known or poorly defined species (e.g., *Q. invaginata, Q. tinkhamii, Q. laceyi, Q. organensis*), as well as facultative Gray (*Q. grisea*), Emory, and Toumey Oaks (*Q. toumeyi*). Some species shared with interior chaparral in Arizona are *Garrya wrightii, Cercocarpus breviflorus, Ceanothus greggii, Fallugia paradoxa, Rhus trilobata,* and *Arctostaphylos pungens*. Other species reported in "Coahuilan" chaparral having equivalents in "Arizonan" chaparral are *Rhus choriophylla* (*R. ovata*), *Garrya obovata* (*G. flavescens*), *Cowania plicata* (*C. mexicana*), *Berberis trifoliata* (*B. fremontii*), *Vauquelinia angustifolia* (*V. californica*), *Fraxinus greggii* (*F. anomala*), and *Fendlera linearis* (*F. rupicola*). Other associates as Madrone (*Arbutus texana, A. arizonica*), and the sages (*Salvia ramosissima, S. roemeriana, S. regla*) are endemics. The nonchaparral associates are also similar to, or closely related to those in Arizona (e.g., *Mimosa biuncifera, Acacia greggii, Ptelea trifoliata, Nolina erumpens, Cupressus arizonica*) so that interior chaparral maintains a recognizable floral continuity from southern Nevada through to the Sierra Madre Oriental.

Although interior chaparral shares a number of plants and animals with Californian chaparral, their separation by

Figure 52. *Interior chaparral. "Coahuilan" fasciation in the Chisos Mountains, Big Bend National Park, Texas. A mixed Texas. A mixed association on limestone of Coahuila Scrub Oak (Quercus pungens), Dwarf Live Oak (Q. intricata), Gray Oak (Q. grisea), Gregg Ash (Fraxinus greggii), Spiny Greasebush (Glossopetalon spinescens), Zexmenia brevifolia, and others. National Park Service photo.*

Sonoran and Mohave desertscrub has resulted in significant differences. Furthermore, the isolated nature of the chaparral communities in northern Mexico, west Texas, and southern New Mexico has left them with a more depauperate chaparral fauna than the larger, more extensive "Arizona" chaparral.

Several small mammals indicative of Californian chaparral are lacking in interior chaparral—Brush Rabbit, Merriam Chipmunk, California Mouse, California Pocket Mouse, and the Dusky-footed Woodrat. Their respective niches in interior chaparral are filled instead by such wide-ranging forms as the *holzneri* subspecies of the Eastern Cottontail (*Sylvilagus floridanus*), the Cliff Chipmunk (*Eutamias dorsalis*), White-footed Mouse (*Peromyscus leucopus*), Rock Mouse (*P. difficilis*), and White-throated Woodrat (*Neotoma albigula*). Other wide-ranging chaparral species as the Mule Deer and Brush Mouse are present in both biomes with distinction only at the subspecific level.

Nesting birds are the more general scrub-adapted species and include the Scrub Jay, Bushtit, Canyon Wren (*Catherpes mexicanus*), Crissal Thrasher (*Toxostoma dorsale*), Rufous-

sided Towhee, Brown Towhee, Rufous-crowned Sparrow, and Black-chinned Sparrow.

Reptile relationships with chaparral are generally ill-defined but subspecies of the Side-blotched Lizard (*Uta stansburiana*), Western Blind Snake (*Leptotyphlops humilis*), Glossy Snake (*Arizona elegans*), Night Snake, Western Black-headed Snake (*Tantilla planiceps*), and Western Rattlesnake (*Crotalus viridis*) are commonly found in both Californian and interior chaparral. The Eastern Fence Lizard (*Sceloporus undulatus*) extends into interior chaparral and is well represented here as is the also widespread Western Fence Lizard (*S. occidentalis*) in Californian chaparral. Other species of narrower distribution encountered in interior chaparral and having Californian counterparts are the Arizona Alligator Lizard (*Gerrhonotus kingi*), Sonora Mountain Kingsnake (*Lampropeltis pyromelana*), Arizona Night Lizard (*Xantusia arizonae*), and the Sonora and Texas Lyre Snakes (*Trimorphodon biscutatus lambda, T. biscutatus vilkinsoni*). The whip-snakes (*Masticophis bilineatus, M. taeniatus*) are interior chaparral equivalents to the Striped Racer.

Tropical-Subtropical Scrublands

Semidesert scrublands dominated by thorny shrubs and small trees characterize much of the world's tropic-subtropic zones. They are known in Australia as mulga, in parts of southern Africa as bush, in parts of South America as chaco-seco, and in Mexico as matorral (=chaparral in Texas).[1] These drought deciduous communities occupy a moisture gradient transitional between desertscrub and woodland or forest, tend toward an irregularly layered overstory between 2 m and 8 m in height, and are typically composed of spinose, microphyllous, and succulent plant life-forms (=thornscrub). Thornscrub species and communities are antagonists with grasses and grasslands (Walter, 1973:75). Scrublands were formerly more confined to the rockier, thinner soils, giving way to subtropic savanna grasslands on level plains and broad river valleys. This partitioning of habitats has been altered, however, with the advent of grazing and fire suppression. Various thornscrub communities have now captured large areas of former grassland.

Thornscrub in North America is a major component of the Caribbean, Tamaulipan, Sinaloan, and San Lucan biotic provinces (Fig. 6). The matorrales of alta espinoso, arborescente, arbocrasicaulescente, and submontano as described and mapped for Mexico by Flores Mata et al. (1971) and the Comisión Técnico Consultiva etc. (1974) are largely thornscrub. In the Southwest, Sinaloan thornscrub covers much of western Sinaloa and southeastern Sonora where it is usually positioned between Sonoran desertscrub and either Sinaloan deciduous forest or Madrean evergreen woodland—in northeast Sonora, semidesert grassland. Tamaulipan and San Lucan thornscrub occur largely outside our boundaries for the Southwest.

---

[1]*Subtropic "brush" communities in southern Texas (Fig. 53) are commonly termed "chaparral" (e.g. Clover, 1937; Tharp, 1939). This terminology is confusing in that chaparral is a Spanish designation more commonly associated with evergreen sclerophyll scrublands existing under warm temperate climate (see e.g. Cronemiller, 1942; Lowe, 1964). The subtropic scrubland of southern Texas is Tamaulipan thornscrub (see Muller, 1947; Blair, 1950, 1952).*

134.3 Sinaloan Thornscrub

Sinaloan thornscrub includes the "Thornforest" of Gentry (1942) and Shreve (1937b, 1951), the "Tropical Thorn Woodland" of Hardy and McDiarmid (1969), and is analogous to the "Tamaulipan Thornscrub" of Muller (1947). We would also include most of Shreve's (1951) "Foothills of Sonora" and other communities of his subdivisions of the Sonoran Desert as *Sinaloan thornscrub* (Brown and Lowe, 1974b; Felger and Lowe, 1976). The reasons for this revision are based on both physiognomic and biotic criteria, and include: (1) the abundance of shrubbery and increased participation of short microphyllous trees which together make up a cover of from 20% to 90% with much of the intervening areas held by perennial forbs and grasses, (2) the fact that a large number of dominant and characteristic plant and animal species have their center of distribution and abundance here, e.g. Tree Ocotillo (*Fouquieria macdougalii*) and Elegant Quail (*Lophortyx douglasii*), (3) the absence or poor representation of numerous characteristic Sonoran desert species (e.g., *Larrea tridentata, Simmondsia chinensis, Cercidium microphyllum,* etc.), and (4) the appearance and often heavy representation of numerous southern tropical forms not found in the Sonoran desertscrub to the north (e.g., *Acacia cymbispina* and *Felis pardalis*).

Sinaloan thornscrub then covers much of southern and southeastern Sonora from near sea level south of parallel 28 to over 900 m in the mountains. Disjunct Sinaloan thornscrub communities occur on Sierra Kunkaak on Isla Tiburón, in the Sierra Santa Rosa, in the Sierra Espinazo Prieto, in the foothills of the Sierra de la Madera, and in other mountains within Shreve's (1951) Plains of Sonora and Central Gulf Coast subdivisions of the Sonoran Desert. Other disjunct thornscrub communities occur at the southwestern edge of semidesert grassland, and some semidesert grassland stands of *Fouquieria splendens* in the Tumacacori, Baboquivari, and Las Guias Mountains in southern Arizona contain *Eysenhardtia orthocarpa, Erythrina flabelliformis, Dodonaea viscosa* and other constituents of Sinoloan thornscrub, and appear to at least approach thornscrub status. Similar stands in the Rincon Mountains, also in southern Arizona, support such additional thornscrub species as *Tecoma stans, Lysiloma watsonii,* and *Acacia millefolia.* Other dense scrub communities occupy mountainous sites and arroyos within Shreve's (1951) "Magdalena" region of the Sonoran Desert and elsewhere in Baja California Sur in which *Lysiloma candida, Bursera odorata, Prosopis palmeri, Opuntia comonduensis, Fouquieria splendens, Stenocereus thurberi, Pachycereus pringlei,* and *Cercidium floridum* ssp. *peninsulare* participate (see e.g. Shreve and Wiggins, 1964:Plate 28).

The basic structure and composition is of drought deciduous, often thorny, pinnate-leaved, multi-trunked trees, and/or shrubs between 2 m and 7½ m in height. Sinaloan thornscrub typically contains as many as 1,600 to 2,000 perennial plants per ha (Gentry, 1942; Fig. 54). Its primary residency is on low hills, bajadas, mesas, and mountain slopes, although it now occupies lowland valleys where it has invaded former savanna grassland. There it still usually presents a more open aspect than on hillsides. Generally, shrubs tend to be more important in the north, trees being more prevalent southward and in the valleys and plains.

Precipitation means normally range from ca. 300-500 mm per year; as low as ca. 230 mm along the coast, to as high as ca. 635 mm (Table 12). Approximately 70% of this rain falls during the July through September growing season. The driest months are April and May, and plant growth is slowed

Table 12. Precipitation data from 11 stations in the Southwest in and adjacent to Sinaloan Thornscrub.

Station	Elevation (in m)	J	F	M	A	M	J	J	A	S	O	N	D	Total	Total July thru Sept.	Percent of total
Arizpe, Son. 30°20' 110°11'	770	21	8	17	1	2	29	165	113	43	13	12	48	472	321	68
Ojo de Agua, Son. 30°04' 109°47'	770	32	9	29	9	0.3	16	129	94	24	20	6	24	392	247	63
Opodepe, Son. 29°56' 110°38'	640	25	10	18	5	.8	13	125	127	60	29	11	20	444	312	70
Moctezuma, Son. 29°48' 109°42'	609	31	16	13	3	2	27	152	119	61	27	6	27	484	332	69
Baviacora, Son. 29°44' 110°10'	560	19	11	7	2	.3	12	93	76	32	10	21	24	307	201	65
Ures, Son. 29°26' 110°24'	390	18	14	11	9	3	17	133	120	59	30	12	36	462	312	68
Mazatan, Son. 29°00' 110°09'	550	29	18	9	2	1	37	150	134	64	26	7	26	503	348	69
Sahuaripa, Son. 29°03' 109°14'	510	30	16	8	2	2	31	102	94	44	15	9	24	377	240	64
Tonichi, Son. 28°36' 109°34'	183	40	18	7	5	2	67	153	157	67	28	10	33	587	377	64
Obregon, Son. 27°30' 109°56'	51	14	4	4	1	.3	7	79	85	59	22	6	21	302	223	74
Navojoa, Son. 27°25' 109°27'	38	19	5	4	2	1	15	87	103	78	30	8	33	385	268	70

or arrested from December to June. The landscape is then thorny and bare and in great contrast to the summer rainy season when it is luxuriant and green (Fig. 55). Nonetheless, many species including the numerous drought deciduous ones bloom in early spring. In fall, a second, shorter period of dryness occurs, and the leaves of some species turn color before being shed. The red foliage of the Brasil (*Haematoxylon brasiletto*) and Torote Papelio (*Bursera* spp.) are then especially conspicuous as are the yellow leaves of *Jatropha cordata* (Shreve, 1951). Freezing temperatures are unexpected and of short duration.

Quantitative data on plant composition in Sinaloan thornscrub are lacking except for studies by Gentry (1942) which were confined to a few areas in the southern extremity of our Southwest. The following shrubs and trees appear to be some of the most common, or at least conspicuous constituents of Sinaloan thornscrub and are from Gentry (1942), Shreve (1951), and field notes. Whether the species is typically encountered as a tree (T), shrub (S), or either, is indicated.

Acacia angustissima (S)	Whiteball Acacia
A. constricta (S)	Whitethorn, Mescat Acacia
A. cymbispina (T)	Espino
A. farnesiana (T)	Sweet Acacia
A. pennatula (T)	Feather Acacia
Aloysia palmeri (S)	Lippia
Ambrosia cordifolia (S)	—
Bursera odorata (T)	Torote, Chutama
Caesalpinia pumila (S)	—
Ceiba acuminata (T)	Pochote
Celtis pallida (S)	Desert Hackberry
Cercidium sonorae (T)	Sonoran Palo Verde
Dodonaea viscosa (S)	Hopbush
Encelia farinosa (S)	Brittlebush
Eysenhardtia orthocarpa (S)	Kidneywood
Fouquieria macdougalii (T)	Tree Ocotillo
Guaiacum coulteri (T)	Guayacan
Haematoxylon brasiletto (S-T)	Brasil
Ipomoea arborescens (T)	Tree Morningglory
Jacquinia pungens (S-T)	San Juan, San Juanito
Jatropha cardiophylla (S)	Limber Bush, Sangre-de-Cristo
Karwinskia humboldtiana (S)	Coyotillo
Lantana velutina (S)	Lantana
Lysiloma divaricata (T)	Mauto
Mimosa laxia (S)	—
Olneya tesota (T)	Ironwood, Palofierro
Piscidia mollis (T)	Palo Blanco
Randia obcordata (S)	Papachillo
Sapium biloculare (S)	Mexican Jumping Bean

Other more or less characteristic shrub and tree species are generally restricted to hillside habitats: e.g., *Randia laevigata* (T), *Acacia willardiana* (T), *Coursetia glandulosa* (S), *Jatropha cordata* (T), *Bursera laxiflora* (T), *Cassia biflora* (S), *Croton fragilis* (S), *Bursera confusa* (T), *Brongniartia alamosana* (T), and *Erythrina flabelliformis* (S-T); while others such as *Pithecellobium sonorae* (T), *P. mexicanum* (T), *Acacia occidentalis* (T), *Prosopis juliflora* (T), *Forchammeria watsoni* (T), *Lycium berlandieri* (S), *Zizyphus obtusifolia* (S), *Condalia spathulata* (S), *Karwinskia parviflora* (T), *Jatropha cinerea* (S), *Atamisquea emarginata* (T), and *Guazuma ulmifolia* (S) favor drainages and moister level habitats.

Cacti are nowhere as prevalent as in some areas of the Sonoran Desert. Two species are particularly conspicuous, however: Organ Pipe Cactus (*Stenocereus thurberi*) and the Hecho (*Pachycereus pecten-aboriginum*). The former is a common participant in the more northern reaches of Sinaloan thornscrub, while the Hecho is increasingly common southward where it typically rises above the upper levels of the thornscrub proper—in contrast to subtropical deciduous forest which contains the Hecho within the tree canopy. Other species of cacti locally common are *Rathbunia alamosensis*, *Opuntia fulgida*, *Opuntia thurberi*, and *Lophocereus schottii*, all of which find their greatest abundance in the plains and valleys in contrast to the leaf-succulents which are largely confined to slopes and ridges. Several agaves (e.g.,

Figure 53. *Tamaulipan thornscrub along lower Rio Grande between Eagle Pass and Laredo, Texas, ca. 260 m elevation. Summer aspect. A landscape of Texas Ranger or Cenizo (***Leucophyllum frutescens***), Mesquite (***Prosopis glandulosa***), acacia (***Acacia romeriana, A. spp.***), and other scrub species. These "brush" and "chaparral" areas have invaded former semidesert grasslands in southern Texas and are now probably best referred to as Tamaulipan thornscrub.*

Figure 54. *Sinaloan thornscrub between Cucurpe and Rayon, Sonora, ca. 500 m elevation. Drought deciduous aspect in March. A large number of shrubs are present and include: Tree Ocotillo (***Fouquieria macdougalii***), Hopbush (***Dodonaea viscosa, Caesalpinia pumila***), Mesquite (***Prosopis juliflora***), and Organ Pipe Cactus (***Stenocereus thurberi***). These and other communities designated by Shreve as the "Foothills of Sonora" subdivision of the Sonoran Desert are Sinaloan thornscrub.*

Figure 55. *Sinaloan thornscrub north of Obregon, Sonora in March of 1981. Because of the mild winter and copious rains most species are in leaf. The principal species are* Fouquieria macdougalii, Olneya tesota, Lantana horrida, Cercidium praecox *and* Opuntia *sp.   Photo by Rich Glinski.*

Agave schotti, A. ocahui) may be well represented locally and the Sotol (*Nolina matapensis*) is found at higher elevations. Yuccas are rare.

In the north, the arroyo and streamside habitats are principally populated by mesquite. Southward this dominance may be increasingly shared by a number of pinnate-leafed and broadleaf evergreen trees and shrubs (Fig. 56). These include the following additional species taken directly from Gentry (1942) and Shreve (1951).

Acacia cymbispina	*Albizzia sinaloensis*
Bumelia occidentalis	*Caesalpinia platyloba*
Cassia emarginata	*Celtis iguanaea*
Cordia sonorae	*Erythea roezlii*
Hymenoclea monogyra	*Jacobinia ovata*
Parthenium stramonium	*Pisonia capitala*
Pithecellobium dulce	*Pithecellobium mexicanum*
Plumeria acutifolia	*Randia echinocarpa*
Sapindus saponaria	*Solanum amazonium*
Stegnosperma ssp.	*Vallesia glabra*
Vitex mollis	*Zizyphus sonorensis*

Here and elsewhere in the thornscrub there are numerous small woody and herbaceous perennials in the understory. These include *Ambrosia ambrosioides, Brickellia coulteri, Brongniartia palmeri, Commicarpus scandens, Antigonon leptopus,* and various species of *Janusia, Ruellia, Salvia, Passiflora, Phaseolus, Talinum, Boerhaavia, Elytraria, Carlowrightia, Ayenia, Desmodium, Turnera,* and *Abutilon.* Certain grasses such as *Bouteloua radicosa* may be locally abundant on the higher and rougher slopes, while several other annual and root-perennial grasses may be seasonally abundant within the more open communities of the plains and valleys.

Sinaloan thornscrub hosts a number of endemic animals as well as a number of more northerly (Sonoran Desert) and southerly (Sinaloan deciduous forest) distribution. The Desert Mule Deer (*Odocoileus hemionus eremicus*) is replaced by the Coues White-tailed Deer (*Odocoileus virginianus couesi*) and the Bobcat (*Felis rufus baileyi*) is sympatric with the Ocelot (*Felis pardalis*).

Mammals not encountered to the north include the Painted Spiny Pocket Mouse (*Liomys pictus*), Coues' Rice Rat (*Oryzomys couesi*), and Sinaloan Pocket Mouse (*Perognathus pernix*). The representative hare is the Antelope or Allen Jackrabbit (*Lepus alleni*). The Elegant Quail and Black-capped Gnatcatcher (*Polioptila nigriceps*) are unique to Sinaloan thornscrub. Other characteristic avian species are also present in Tamaulipan thornscrub, other tropic-subtropic scrublands, or have closely related analogs there. Examples of these include the Ferruginous Owl (*Glaucidium brasilianum*), Sinaloa Wren (*Thryothorus sinaloa*), Wied's Crested Flycatcher (*Myiarchus tyrannulus*), and Beardless Flycatcher (*Camptostoma imberbe*). Still other more or less representative species as the Javelina (*Dicotyles tajacu*), Harris' Hawk (*Parabuteo unicinctus*), Elf Owl (*Micrathene whitneyi*), White-winged Dove (*Zenaida asiatica*), and Gila Woodpecker (*Melanerpes uropygialis*) are also able to inhabit denser habitats within the Arizona upland region of Sonoran desertscrub. Reptiles to be expected in Sinaloan thornscrub include *Urosaurus ornatus* and *Cnemidophorus costatus.* Neither reptiles nor amphibians in Sinaloan thornscrub have been adequately studied, but representative species would probably include *Leptodeira punctata, Masticophis striolatus, Masticophis valida,* and *Crotalus basiliscus.*

Figure 56. *Sinaloan thornscrub near Tezopaco, Sonora, ca. 760 m elevation. A higher elevation community of short-statured trees and scrub adjacent to a drainage bottom where both evergreen (e.g., the palm* **Erythaea roezlii***) and deciduous components participate. Note that the height of the scrub overstory is exceeded by the height of the columnar cactus as compared to the Sinaloan deciduous forest depicted in Figure 39.*

Part 4. Grasslands

David E. Brown

Plains grassland near Sueco Junction, Chihuahua,
at about 1760 m elevation.

Arctic-Boreal Grasslands

Grass and/or herb dominated areas exist under this rigorous climatic regime in the arctic, the subarctic, the subantarctic, and throughout the world's alpine and subalpine areas. These climax grasslands are typically characterized by bunchgrasses as well as forbs and are sometimes referred to as "tussock grassland" (e.g., Walter, 1973).

Arctic-Boreal grasslands in the Southwest are most often referred to as subalpine grassland or mountain meadow grassland (e.g. Moir, 1967; Lowe and Brown, 1973; Turner and Paulsen, 1976). The alpine and subalpine designations are used here because they indicate the relative position of these grasslands on the mountain gradient (the Hudsonian life-zone of Merriam) that in part distinguishes them from the lower montane meadow grasslands existing under the warmer temperate climates of the Canadian and Transition life-zones.

141.4 Alpine and Subalpine Grasslands

These grasslands occupy valleys, slopes, and ridges on usually flat or undulating terrain adjacent to and within subalpine conifer forests. Size of the grassland ranges from a small park-like opening within the forest to extensive landscapes covering several thousand acres (Fig. 57). These grasslands range from the lower edge of the subalpine forest to below and within the alpine tundra; all are at sufficient elevation to experience subalpine conditions. In the Southwest, subalpine grasslands reach their best development between 2500-2600 and 3500 m in the Sangre de Cristo, San Juan, Jemez, and La Plata mountains in Colorado and New Mexico, and in the White Mountains, Lukachukai Mountains, and on the Kaibab Plateau (Buckskin Mountains) in Arizona. Limited areas are also *Rocky Mountain subalpine grassland* in the Sierra Blanca, San Mateo, Magdalena, Cebolleta, Mogollon, Chuska, and other mountains in New Mexico, and in the Pinaleño, Escudilla, and San Francisco mountains in Arizona; there are also some small, high elevation subalpine meadow areas in the Pine Valley Mountains and elsewhere in southern Utah.

Small areas of *Sierran subalpine grassland* are found in the San Gabriel, San Bernardino, and San Jacinto mountains in southern California where they sometimes occur as "snowmelt gullies" (Thorne, 1977). Small meadows also occur in the Sierra San Pedro Mártir in adjacent Baja California Norte. Subalpine grassland communities are absent from the Sierra Madres and other Mexican ranges in the Southwest; they do occupy sizeable areas in the Sierra Madre Oriental (e.g., on Cerro Potosí) and on the high volcanoes of central Mexico. There subalpine and alpine grasslands are represented by "zacatonales" of *Stipa, Muhlenbergia, Calamagrostis,* and *Festuca* (Flores Mata et al., 1971).

Subalpine grassland soils are variable, and often well drained, and yet they possess properties unsuitable for tree growth. Air temperatures are significantly lower and evaporation rates are significantly higher in the grassland than in the adjacent forest but it is not known whether these microclimatic differences are causes or effects. Particularly at the higher elevations, a near timberline situation of short trees often exists at the grassland-forest border, where 2-m trees at the immediate edge may be the same age as 10-m tall trees only 5 m to 6 m back in the forest.

Precipitation averages from as low as 360 mm, as at Eagle Nest in the Sangre de Cristo Mountains, New Mexico, to 500 mm to 1,150 mm annually, most of which falls as snow. The resulting snowpack commonly covers the ground from October to May and may be of considerable depth. Although subzero air temperatures can be expected during winter months, soil temperatures below the snowpack are at or slightly below freezing. Some plants remain green throughout the winter, while others begin growth before the completion of snowmelt (Turner and Paulsen, 1976). The growing season is brief, often less than 100 days, and occasionally interrupted by nighttime frosts.

Well drained sites are commonly dominated, actually or potentially, by perennial bunchgrasses (*Festuca, Agropyron, Stipa, Poa, Muhlenbergia*) with a greater or lesser accompaniment of forbs including species of fleabane or wild-daisy (*Erigeron*), mountain dandelions (*Agoseris, Taraxacum*), cinquefoil (*Potentilla*), larkspur (*Delphinium*), aster (*Aster*), yarrow (*Achillea*), vetch (*Vicia*), clover (*Trifolium*), and many others. These high elevation "prairies" are replaced in the

Figure 57. *Rocky Mountain subalpine grassland and conifer forest mosaic (=parkland) in the White Mountains on the Apache National Forest in Arizona, ca. 2,896 m elevation. The abundance of bunchgrasses (principally* Festuca arizonica*) mixed with forbs is characteristic, and indicates that this is a subalpine grassland in good condition. The small trees invading the hillsides and the "flagged" trees on the ridge crest are Southwestern White Pine* (Pinus strobiformis).

moister "ciénega" sites by communities of sedges (*Carex, Cyperus*) and rushes (*Juncus*), often in combination with a variety of moisture-dependent forbs and grasses, e.g., Mountain Timothy (*Phleum alpinum*) and Tufted Hairgrass (*Deschampsia caespitosa*). Although both of these communities possess numerous distinctive "indicator" species, there is much intermingling with grasses and forbs equally or more representative of either *montane meadow grassland* below or *alpine tundra* above.

Better water supplies—more snow, rain, springs, and streams—favor the higher subalpine grasslands, which also have lower evapotranspiration rates than the montane mountain grasslands below. As a result many alpine and subalpine grasslands in the Southwest still retain much of their original "wet-meadow" characteristics, while many montane meadows are now distinctly drier.

Few subalpine grasslands are in a climax condition because of grazing or, less commonly, fire. Past and present misuse of subalpine grasslands is prevalent but not always readily apparent because of the abundant herbaceous cover usually present and the variations in plant composition potential between sites (Fig. 58). Generally the less palatable forbs and grasses tend to increase on cattle ranges at the expense of the native bunchgrasses, while secondary grasses can be expected to replace the more valuable forbs where sheep are pastured

(Turner and Paulsen, 1976). At the lower elevations and on drier and poorer ranges, shrubs such as *Artemisia tridentata* may now occur abundantly.

Not considered as subalpine grasslands are those grass-forb and shrub areas recently the site of a major forest disturbance and which are obviously in a successional stage toward subalpine conifer forest. Nonetheless, it should be noted that the grassland edge is often in the process of being invaded by small conifers or thickets of Quaking Aspen (*Populus tremuloides*), so that the extent of subalpine grassland appears in many areas to be decreasing.

Of the vertebrates inhabiting subalpine grasslands in the Southwest, mammals are best represented and may include several species that have adapted to the rigorous winters by hibernating or by feeding underground and/or under snow. These include the Marmot (*Marmota flaviventris*), Gray-collared Chipmunk (*Eutamias cinereicollis*), Least Chipmunk (*Eutamias minimus*), Heather Vole (*Phenacomys intermedius*), Meadow Voles (*Microtus montanus, M. longicaudus, M. mexicanus, M. pennsylvanicus*), Golden-mantled Ground Squirrel (*Spermophilus lateralis*), Deer Mouse (*Peromyscus maniculatus*), Jumping Mouse (*Zapus princeps*), Pocket Gophers (*Thomomys talpoides, T. bottae*), several shrews (*Sorex vagrans, S. obscurus, S. cinereus, S. nanus*), and their predators—e.g., the Long-tailed Weasel (*Mustela frenata*), Ermine (*Mustela erminea*),

Figure 58. *Subalpine grassland at and above timberline on the San Juan National Forest, Colorado, ca. 3,566 m elevation. The ground cover is mostly of forbs, the taller bunchgrasses having been cropped or removed by livestock. Photograph by David Cook.*

Badger (*Taxidea taxus*), and Red Fox (*Vulpes vulpes*). The proclivity of the Grizzly Bear (*Ursus arctos*) for open areas such as subalpine grasslands contributed to its early extirpation from the Southwest. Other large mammals such as the Mule Deer (*Odocoileus hemionus*), Wapiti or Elk (*Cervus elaphus*), Bighorn Sheep (*Ovis canadensis*), and Coyote (*Canis latrans*) are more or less migratory to and from these other high elevation sites, as was the now extirpated Gray Wolf (*Canis lupus*). The insectivorous bats (e.g., *Myotis volans, Lasionycteris noctivagans, Lasiurus cinereus*) are also of necessity, summer residents only. In the San Juan and Sangre de Cristo mountains in Colorado and extreme northern New Mexico, the White-tailed Jack-rabbit (*Lepus townsendii*) is characteristically a resident of subalpine grassland.

No species of bird is restricted to or is particularly characteristic of subalpine grassland. Avian inhabitants are those species generally found throughout the higher open landscapes of the Southwest—e.g., Raven (*Corvus corax*), Red-tailed Hawk (*Buteo jamaicensis*), American Kestrel (*Falco sparverius*), Mountain Bluebird (*Sialia currocoides*), Horned Lark (*Eremophila alpestris*), Common Nighthawk (*Chordeles minor*), and Savannah Sparrow (*Passerculus sandwichensis*). Other species found more or less at or near the forest edge include the Yellow-rumped Warbler (*Dendroica coronata*), Robin (*Turdus migratorius*), and Blue Grouse (*Dendragapus obscurus*).

As expected in boreal biotic communities, relatively few reptilian or amphibian species are present. These few include the Short-horned Lizard (*Phrynosoma douglassi*), Wandering Gartersnake (*Thamnophis elegans*), Gophersnake (*Pituophis melanoleucus*), Western Toad (*Bufo boreas*), Chorus Frog (*Pseudacris triseriata*), Leopard Frog (*Rana pipiens*), and Tiger Salamander (*Ambystoma tigrinum*). Populations of these species in the Southwest approach or exceed 3,050 m elevation, where they can be found in these natural openings in the forest.

Cold-Temperate Grasslands

Extensive grasslands recently existed throughout the world's temperate zones under semi-arid and wind-swept conditions, and in other areas unfavorable for tree or scrub growth. These grass-dominated landscapes reach maximum size and development in the Eurasian "steppe" and in the North American "prairie," both of which evolved within a continental-type climatic regime characterized by cold, harsh winters with a substantial part of the limited precipitation falling during a more or less arid growing season (see e.g. Walter, 1973).

In North America the *Plains grassland* (=tall-grass, mid-grass, short-grass prairie, etc.) is by far the most important biotic community in the evolutionary history and development of our continent's grassland biota. This largely mid-summer flowering grassland extends from approximately 55° latitude in the Canadian Provinces of Alberta and Saskatchewan southward to below latitude 30° in Mexico, and once covered most of the American "Midwest" from the Eastern deciduous forest westward to the Rocky Mountains and beyond. More than 70% of the Plains grassland in now under cultivation (Garrison et al., 1977).

Centered in eastern Washington and northeastern Oregon (Columbia Basin) is a grassland region widely referred to as the "Palouse Prairie" (see e.g. Shelford, 1963: 350). This spring-flowering *Great Basin* or *inter-mountain grassland* is restricted to those areas west of the Rockies and east of the Sierra-Cascades that possess favorable soils, climate and grazing history. Much of this grassland has been appropriately described as a shrub-steppe in that pure-grass landscapes without shrubs are limited (see Franklin and Dyrness, 1973). Great Basin grassland merges with Plains grassland over a large transition area adjacent to the Rocky Mountains in Montana, Wyoming, Colorado, New Mexico, and Arizona. Much of this grassland has been converted to cultivated cropland through irrigation, and most of the remainder has experienced a degree of shrub invasion because of grazing and fire suppression.

In addition to Plains and Great Basin grasslands, cold temperate grasslands in North America are represented by other, smaller grass and herb areas in the Pacific Northwest (Oregonian grassland), in the Sierra-Nevada, Rocky Mountains, and Sierra Madres (montane meadow grassland), and also within the Eastern deciduous forest (see e.g. Garrison et al., 1977).

142.4 Montane Meadow Grassland

Montane meadow grasslands are contained largely within the Rocky Mountain, Sierran, and Madrean montane conifer forests. They are not shown on the color map (Brown and Lowe, 1980). These natural openings or parks are generally restricted to flatlands possessing heavy, poorly drained muck soils. Such soils are unsuitable for tree growth, and the forest edge is frequently abrupt, producing a marked "edge effect."

Herbs, forbs, and "weeds" usually outnumber the grasses (Fig. 59). These "flower meadows" of lower elevations are distinct from the higher and sometimes adjacent subalpine grasslands, which in the Southwest are usually above 2,600 m elevation and are commonly dominated by bunchgrasses. Also not to be confused with montane meadow grasslands are extensions of plains grassland into the lower Ponderosa Pine forest on plateaus and mesas between 1,800-2,500 m.

While often not extensive, these attractive landscapes are nonetheless important to the meadow-affiliated animals of the montane conifer forest. These include, in appropriate geographic areas of the Southwest, Elk (*Cervus elaphus*), deer (*Odocoileus hemionus, O. virginianus*), Pocket Gopher (*Thomomys bottae*), meadow voles (*Microtus mexicanus, M. californicus*), Wild Turkey (*Meleagris gallopavo*), the Western Bluebird (*Sialia mexicana*), and the Western Flycatcher (*Empidonax difficilis*). The larger meadow areas may also serve as more or less seasonal habitats for such grassland species as the Pronghorn (*Antilocapra americana*), the American Kestrel (*Falco sparverius*) and other raptors, and the meadowlarks (*Sturnella magna* and *S. neglecta*).

Where a small stream courses through meadow, *Thamnophis elegans* and *Rana pipiens* are usually present and often abundant. Wet meadows that are without stream development, but with a small pond at the lower end, often support populations of *Ambystoma tigrinum* and Chorus Frogs (*Pseudacris triseriata*) as do the subalpine mountain meadows. This is the mountain home of the bright green Arizona Treefrog (*Hyla eximia*) that breeds in water as do the others, but unlike Leopard and Chorus Frogs, lives in the shrubs and trees bordering the montane meadow ponds and streams—where it hides by day and hunts its insect food at night. Thus the amphibian species in montane meadow grasslands are usually found in the lower wetland portions of the meadows where sedges (*Carex, Cyperus*), spike rushes (*Eleocharis*) and rushes (*Juncus*) occur.

Where not heavily grazed, an enormous variety of summer flowering perennial forbs and grasses may be present. These may include some introductions such as Black-eyed Susan (*Rudbeckia hirta*), Mullein (*Verbascum thapsus*), and Kentucky Bluegrass (*Poa pratensis*). Surprisingly many of the same characteristic meadow species are present in montane habitats in the Californias, in the Rocky Mountains, and in the Sierra Madres, and many montane meadow species are conspecific between two or more biotic provinces. These include Bracken Fern (*Pteridium aquilinum*), Corn-lily (*Veratrum californicum*), Monkey Flower (*Mimulus nasutus*), Mountain Brome (*Bromus marginatus*), and Mountain Muhly (*Muhlenbergia montana*). Another shared species is Iris (*Iris missouriensis*) which may dominate completely in heavily grazed meadows. Other common species are regionally restricted as is to be expected within biotic provinces—for example *Blepharoneuron tricholepis* which occurs only in the Rockies.

Figure 59. *Montane meadow grassland of herbaceous forbs within Rocky Mountain montane conifer forest in the Apache-Sitgreaves National Forest, Arizona, ca. 2,590 m elevation. Midsummer aspect. The trees are Douglas-fir (*Pseudotsuga menziesii*) and White Fir (*Abies concolor*) at this site near the upper elevation limit for temperate montane biomes. Photograph by John N. Theobald.*

Other characteristic plants in montane meadows include a large number of usually showy herbs in the genera *Lupinus, Lathyrus, Helianthus, Aster, Penstemon, Senecio, Solidago, Lotus, Astragalus, Vicia, Sphaeralcea, Viola,* and many others.

Because of their initial abundant herbage, meadow grasslands were and still are centers for livestock operations that often subject the meadow to unduly heavy grazing pressures. Such overgrazing commonly results in changes in plant composition from grass and forbs to perennial scrub (*Artemisia, Chrysothamnus, Senecio,* etc.). If drying, trampling, and erosion of the deeper soils continue, the meadows themselves may eventually be replaced by forest.

143.1 Plains and Great Basin Grasslands

These two continental fasciations of the North American prairie both reach their southern terminus in the American Southwest. These grasslands are now much altered but were formerly open, grass-dominated landscapes in which the grasses formed a continuous or nearly uninterrupted cover. Although not uncommonly occurring on virtually any type of terrain, the term Plains grassland is appropriate in that this biotic community is situated largely on high level plains, in valleys, and on intervening and adjacent low hillsides, rises, ridges, and mesas in what is predominantly flat and open country.

Because these grasslands are situated on open and exposed plains, they are subject to high solar radiation and long windy periods, particularly during winter and early spring. In the past, as the summer progressed, these characteristic winds carried lightning set grass fires over many miles when "cured" grasses from prior growing seasons were available as fuel. A natural succession to climax grass-forb associations then took place, as other successions toward climax do today on abandoned farmlands and rangelands. Because most grasslands are grazed, less residual grass is available for fuel, and the incidence of fire is reduced. Natural successions are now usually arrested and instead replaced by fire disclimax associations of shrubs.

Plains grassland enters the Southwest from the northeast in southern Colorado and northwestern Texas, and continues, interrupted, through or to extreme southwestern New Mexico (Animas Valley) to as far west as northwestern Chihuahua (e.g., Valle de Carretas), northeastern Sonora (e.g., vicinity Cananea), and southeastern Arizona (the Sonoita-Elgin and San Rafael valleys), reaching northwestward to west-central Arizona (e.g., the Chino and Williamson valleys and the Coconino Plateau).

Grasslands exhibiting characteristics of a Great Basin (=intermountain grassland) reach this area from the northwest and intergrade with plains grassland over a large transition area which includes southern Utah, northern Arizona, southwestern and south-central Colorado, and northwestern and north-central New Mexico. a rough division between these two grassland fasciations is approximated by the boundary between the Plains and Great Basin biotic provinces shown in Fig. 3. Transitional grassland communities continue as far west in the Southwest as the "Arizona Strip," and near Enterprise in Iron County, Utah.

Plains grasslands in the Southwest are situated above 1,200 m elevation; in southwestern New Mexico, southeastern Arizona, and northeastern Sonora, this grassland is mostly restricted to elevations above 1,500 m and to above 1,700 m in most of Chihuahua. The upper elevation limits are usually at 2,200-2,300 m. The primary grassland contact is woodland (juniper-pinyon or encinal), with some upward extensions into montane conifer forest (Ponderosa Pine series). At its western limits, Plains grassland occasionally comes in contact with interior chaparral.

Precipitation within Plains grassland proper averages between 300 mm and 460 mm per annum, with extremes to as low as 250 mm and as high as 530 mm (Table 13). Much of this precipitation falls during summer thunderstorms; May through August precipitation always averages more than 115 mm and ordinarily is 50% or more of the total (Table 13). In the south the lower elevation contact is with semidesert grassland. There Plains grassland is uninvaded by dry-tropic

Table 13. Precipitation data from 16 stations in the Southwest in Plains Grassland and Great Basin Transition Grassland.

Station	Elevation (in m)	J	F	M	A	M	J	J	A	S	O	N	D	Total	Total May thru Aug.	Percent of total
Bravo, TX 35°39′ 103°00′	1,268	10	12	17	30	56	59	72	66	47	29	11	14	424	254	60
Portales, NM 34°28′ 103°21′	1,222	10	10	13	16	53	64	77	59	49	36	9	14	412	253	61
Fort Sumner, NM 34°28′ 104°15′	1,231	8	8	14	17	29	31	64	59	47	33	9	10	330	183	56
Las Vegas, NM 35°39′ 105°09′	2,093	7	9	12	19	39	42	75	90	42	31	9	13	388	245	63
Albuquerque, NM* 35°03′ 106°37′	1,619	8	10	12	12	13	13	35	34	20	20	7	18	202	96	47
LaJunta, Chih. 28°27′ 107°25′	2,062	2	5	8	3	8	31	143	134	43	25	7	21	427	316	74
Namiquipa, Chih. 29°15′ 107°25′	1,906	11	10	6	8	11	37	86	113	39	29	9	31	390	247	63
Hillsboro, NM 32°56′ 107°34′	1,606	15	9	9	8	7	16	53	54	47	26	10	19	274	130	47
Springerville, AZ 34°08′ 109°17′	2,152	14	9	15	8	8	11	63	74	39	25	9	13	288	156	54
Jeddito, AZ* 35°46′ 110°08′	2,042	24	20	23	19	13	10	37	41	22	25	15	25	274	101	35
Cananea, Son.** 30°59′ 110°18′	1,518	27	34	25	12	8	25	129	124	67	33	23	50	556	285	51
San Rafael Valley, AZ 31°21′ 110°37′	1,745	26	19	22	11	2	13	116	104	42	22	12	30	419	235	56
Chino Valley, AZ 34°45′ 112°27′	1,448	23	20	23	15	7	7	52	61	25	19	16	27	294	126	43
Seligman, AZ 35°19′ 112°53′	1,600	23	18	21	13	6	12	40	57	19	16	14	23	261	115	44
Mount Trumbull, AZ* 36°25′ 113°21′	1,707	21	18	25	18	10	11	43	2	24	23	20	25	240	65	27

* Great Basin Transition
** Adjacent to Plains Grassland in Encinal Woodland

scrub only where the mean annual rainfall exceeds 380 mm or above 1,500-1,700 m elevation where an average of more than 150 days during the year have minimum temperatures below freezing.

Those grasslands possessing characteristics and components of Great Basin affinity tend to be drier (180-300 mm mean annual precipitation) and colder (mean annual growing season 125-200 days) than in locations more representative of Plains grassland. These transitional grassland areas also receive a larger percentage of their precipitation during winter and spring months; May through August rainfall averages less than 127 mm (Table 13). At lower elevations the contact is most frequently with Great Basin desertscrub.

Within Plains grassland, two or three major divisions are commonly recognized—tall grass prairie and mixed and/or short-grass grassland (see Shantz and Zon, 1924; Bruner, 1931; Morris, 1935; Weaver and Clements, 1938; Castetter, 1956; Shelford, 1963, etc.). The presence of tall-grass communities within Plains grassland is determined by available plant moisture and in the Southwest is dependent on edaphic conditions (such as porous loam and sandy soils) and grazing history. Consequently, tall-grass communities in the Southwest are restricted to east of longitude 104° on sandhills on

the Llano Estacado (e.g., near Portales in Roosevelt County, New Mexico), and on some of the high mesas in Colfax County, New Mexico and adjacent Colorado—e.g., on Johnson, Barillo, and Fisher Peak mesas.

Where grazing has not been too severe, the sandhills are dominated by bluestems (Schizachyrium scoparium, Andropogon gerardi, A. gerardi var. paucipilus), frequently in association with Shinnery or Midget Oak (Quercus havardii) and other grasses such as Indian Grass (Sorghastrum nutans), Switchgrass (Panicum virgatum), and Sideoats Grama (Bouteloua curtipendula); see Fig. 60. Heavily grazed "dune fields" possess correspondingly fewer grasses and are more or less populated by Shinnery Oak, Sandsage (Artemisia filifolia), Soapweed (Yucca glauca), and Mesquite (Prosopis glandulosa).

The "caprock" country near Raton and other high elevation areas along the New Mexico-Colorado border may also possess taller grasses more representative of Midwestern locales—again where grazing has not been too intense. In addition to Little Bluestem, the taller grasses here include Western Wheatgrass (Agropyron smithii), Needle and Thread Grass (Stipa comata), Red Three-awn (Aristida longiseta), Galleta (Hilaria jamesii), and Sand Dropseed (Sporobolus cryptandrus) as important species. At the higher elevations,

Figure 60. *Plains grassland tall-grass community on the Bluett State Wildlife Management Area, Roosevelt County, New Mexico, ca. 1,402 m elevation. The dominant tall grasses on these sandhill habitats are Big and Little Bluestem (*Andropogon gerardi, Schizachyrium scoparium*) which completely overshadow an understory of shorter grasses, herbs, Soapweed Yucca (*Yucca glauca*), and Shinnery Oak (*Quercus havardii*).*

Festuca arizonica may be abundant with shrubs more or less prevalent throughout.

These tall-grass communities approximate the present or past distributional limits in the Southwest for the Prairie Chicken (*Tympanuchus pallidicinctus*), Bobwhite (*Colinus virginianus*), and Sharp-tailed Grouse (*Pedicetes phasianellus*). The latter species has now been extirpated (Hubbard, 1970), and those tall-grass areas remaining in good condition can be considered "endangered habitats."

Almost all the Plains grassland in the Southwest is composed of *mixed* or *short-grass communities*. While these communities have been considerably altered by grazing and the results of this practice (fire suppression followed by shrub invasion), much of the grassland remains an uncluttered perennial grass dominated landscape (Figs. 61, 62, 63). The principal grass constituents are perennial sod-forming

species of which Blue Grama (*Bouteloua gracilis*) and/or other gramas (*B. hirsuta, B. chondrosioides, B. eriopoda, B. curtipendula*) are usually important in the make-up. Other important grasses, either locally or generally, include Buffalo-grass (*Buchloë dactyloides*), Indian Rice Grass (*Oryzopsis hymenoides*), Galleta Grass (*Hilaria jamesii*), Prairie Junegrass (*Koeleria cristata*), Plains Lovegrass (*Eragrostis intermedia*), Vine Mesquite Grass (*Panicum obtusum*), Wolftail or Texas Timothy (*Lycurus phleoides*), and Alkali Sacaton (*Sporobolus airoides*).

Shrubs such as Four-wing Saltbush (*Atriplex canescens*), sagebrush (*Artemisia*), Winterfat (*Ceratoides lanata*), wild rose (*Rosa*), cholla (*Opuntia*), Soapweed (*Yucca glauca*), Prairie Sumac (*Rhus copallina* var. *lanceolata*), rabbitbrush (*Chrysothamnus*), and snakeweed (*Gutierrezia*) may be scattered throughout, or because of grazing and/or soils, be noticeably conspicuous. This is particularly true of the snakeweeds (*G.*

Figure 61. *Plains grassland short-grass community near Springer, Colfax County, New Mexico, ca. 1,783 m elevation. A landscape of Blue Grama* (Bouteloua gracilis) *with some Buffalo Grass* (Buchloë dactyloides). *Contrast the sod-cover here with that shown in Figures 62 and 64.*

Figure 62. *Plains grassland short-grass community in the Valle de Carretas, Chihuahua, ca. 1,676 m elevation. The barren "eaten out" appearance is due to heavy, year-long livestock grazing and the presence of Black-tailed Prairie Dogs* (Cynomys ludovicianus arizonensis); *note burrows. Although uninvaded by dry-tropic scrub, the grama grasses and sod are much reduced and the principal ground cover is composed of annual composites and the stalks of less palatable grasses.*

Figure 63. *Plains grassland short-grass community representative of Plains grassland-Great Basin grassland transition, west of Winslow, Coconino County, Arizona at ca. 1,676 m elevation. This grassland is in good condition with representative sod-cover and scrub distribution because of the far distance to water. The grasses are a mixture of both Plains species* (Bouteloua gracilis) *and Great Basin species* (Hilaria jamesii, Oryzopsis hymenoides). *The shrubs are mostly Fourwing Saltbush* (Atriplex canescens) *with an occasional Soapweed Yucca* (Yucca glauca).

sarothrae and closely related "species") which now form sodless disclimax communities over many miles of former grassland, as does rabbitbrush and sagebrush in large areas transitional to and in the Great Basin (Figs. 64, 65).

In this shrubby respect, Plains grassland differs significantly from much of the lower and warmer semidesert grassland, because it tends to be characterized by either invasions of large-shrub monocultures and/or by short-statured semi-shrubs (usually much shorter than one-half meter), most of which are equaled or exceeded by the heights of any surrounding residual grasses. The result is that an overgrazed Plains grassland landscape retains a more or less evenly statured low canopy of plants, common examples of which are snakeweed and grass. In contrast, scrub invasions in the more diverse semidesert grassland often present an aspect of residual grasses and/or semishrubs accompanied by a much taller layer of scrub, e.g. grass and/or Burroweed and Mesquite trees.

It is well-documented that junipers (*Juniperus monosperma, J. scopulorum, J. osteosperma*) have invaded large acreages of all types of grassland in the Southwest during this century, particularly on rocky, thin soil habitats (Fig. 66); (see e.g. Parker, 1945; Arnold et al., 1964).

Under natural conditions, even today, forbs may also equal or even exceed the grasses in abundance. One or several species of primrose (*Oenothera*), bahia (*Bahia*), spiderflower (*Cleome*), four-o'clock (*Mirabilis*), gaura (*Gaura*), mallow (*Sphaeralcea*), aster (*Aster*), scurfpea (*Psoralea*), coneflower

(*Ratibida*), bricklebush (*Brickellia*), and numerous others occur; many that put forth showy flowers may be present, especially during wet periods. Such areas are exceptional, however, and throughout most of the grasslands the most palatable herbs and forbs are in short supply, if not fully eliminated. Today it is usually the tougher, less palatable grasses and "weeds" that are able to achieve aspect dominance; these are often species of composites, e.g. golden-eye (*Viguiera*), groundsel (*Senecio*), thistles and prickly poppies (*Cirsium, Argemone*), and sunflowers (*Helianthus*, etc.).

Assigned to Plains grassland by Benson (1969) are the following cacti: Plains Pricklypear (*Opuntia macrorhiza*), chollas (*O. imbricata, O. arbuscula, O. whipplei*), hedgehogs (*Echinocereus fendleri, E. engelmannii* var. *variegatus*), a pincushion (*Mammillaria wrightii*) and the Grama-grass Cactus (*Pediocactus papyracanthus*). Various morphotypes of Engelmann Pricklypear (*Opuntia phaeacantha*) may be locally common. Club Cholla (*O. clavata*) and a pricklypear (*O. polyacantha*) are characteristic species in the Plains-Great Basin grassland transition, the entrance of which from the east is marked by the appearance of *Hilaria jamesii, Oryzopsis hymenoides*, and *Artemisia tridentata*. There, many typical plains species are lacking or replaced by related forms.

Grassland mammal representatives include the well known Pronghorn (*Antilocapra americana*) and Bison (*Bison bison*); the latter is generally believed to have been confined in historic times to east of the Rio Grande except in Chihuahua where it ranged as far west as the vicinity of Casas Grandes

Figure 64. *Plains grassland disclimax community on the Moenkopi Plateau of the Navajo Indian Reservation, ca. 1,554 m elevation. Sandy habitat now dominated by Snakeweed (*Gutierrezia sarothrae*) with annual weeds (*Eriogonum, Salsola kali*), Soapweed Yucca (*Yucca glauca*), and at the base of dunes, Joint-fir (*Ephedra*). These habitats which formerly supported grassland are now highly susceptible to wind erosion.*

Figure 65. *Great Basin grassland on the "Arizona Strip" in Antelope Valley southwest of Fredonia, Mohave County, Arizona, ca. 1,585 m elevation. Big Sagebrush (*Artemisia tridentata*) is encroaching on the grassland at this deep soil site. The principal grasses are Alkalai Sacaton (*Sporobolus airoides*) and Galleta (*Hilaria jamesii*) with some Blue Grama (*Bouteloua gracilis*) and Indian Ricegrass (*Oryzopsis hymenoides*). Great Basin conifer woodland is on ridge in background.*

Figure 66. *Plains grassland being invaded by junipers—in this case* Juniperus scopulorum *and* J. monosperma *near Chaves Pass, Coconino County, Arizona, at ca. 1,829 m elevation.*

(Wallace, 1883). The list of associated smaller mammals is long, and many, such as the prairie dogs *Cynomys ludovicianus* (Plains) and *C. gunnisoni* (Great Basin)), Thirteen-lined Ground Squirrel (*Spermophilus tridecemlineatus*), Swift Fox (*Vulpes velox*), Plains Pocket Gopher (*Geomys bursarius*), and Plains Harvest Mouse (*Reithrodontomys montanus*), have adapted to spending most of their time underground.

Because the center of the Plains grassland is well outside the boundaries of the Southwest, some of the birds most characteristic of Plains grassland are peripheral as nesting species in this area. This include the Prairie Chicken discussed above, Upland Sandpiper (*Bartramia longicauda*), Mountain Plover (*Charadrius montana*), Lark Bunting (*Calamospiza melanocorys*), Grasshopper Sparrow (*Ammodramus savannarum*), and the Long-billed Curlew (*Numenius americanus*), the nest of which is a grass-lined hollow on the open prairie. Other grassland species such as the meadowlarks, Prairie Falcon (*Falco mexicanus*) and Burrowing Owl (*Athene cunicularia*) may be found throughout these and other open landscapes.

The excavations provided by the burrowing mammals are used, in addition to their owners, by a relatively large snake fauna. Some of the more frequently seen species are the Bull-snake (*Pituophis melanoleucus sayi*), Corn Snake (*Elaphe guttata*), Western Coachwip (*Masticophis flagellum testaceus*), Western Plains Milksnake (*Lampropeltis triangulum celaenops*), and the Prairie Rattlesnake (*Crotalus viridis viridis*) which is a common inhabitant of prairie dog towns. As befits many of their names, other reptiles and amphibians indicative of plains habitats, which have had a long evolutionary history in climax grassland, are the Plains Spadefoot (*Scaphiopus bombifrons*), Great Plains Toad (*Bufo cognatus*), Lesser Earless Lizard (*Holbrookia maculata*), Southern Prairie Lizard (*Sceloporus undulatus consobrinus*), Great Plains Skink (*Eumeces obsoletus*), Prairie-lined Racerunner (*Cnemidophorus sexlineatus viridis*), Western Box Turtle (*Terrapene ornata*), Plains Hognose Snake (*Heterodon nasicus nasicus*), Prairie Ringneck Snake (*Diadophis punctatus arnyi*), Great Plains Ground Snake (*Sonora episcopa episcopa*), and Plains Blackhead Snake (*Tantilla nigriceps*).

Warm-Temperate Grasslands

Warm temperate grasslands are found on six continents and formerly covered sizable areas of southern Eurasia and North America. They also occupy important regions in the southern hemisphere, including the pampas of Argentina in South America, and the South African veldt. Seasonal drought and drying winds are characteristic climatic features, and periodic droughts of some duration can be expected in these semiarid environments. The droughts and mild winter temperatures are important reasons that these grasslands have been greatly altered by people and livestock.

In the Southwest warm temperate grasslands are represented by a *semidesert grassland* with a more or less biseasonal to summer precipitation pattern, and the winter precipitation-summer drought *Californian Valley grassland*. As is the case with other warm-temperate grasslands, the original grass cover in these biomes has been largely replaced. Many semidesert grassland sites have been invaded by woody plants, leaf succulents, and cacti, and their grasses replaced by shrubs. Californian Valley grassland in the Southwest, where it remains at all, is now comprised almost entirely of introduced annuals. Less pronounced but similar displacements have occurred with the native fauna of both warm-temperate biotic communities.

143.1 Semidesert Grassland

Southwestern semidesert grasslands (Figs. 67, 68, 69, 70) were originally described and mapped by Shreve (1917) as desert-grassland transition. Others have described and/or mapped these communities as desert savanna (Shantz and Zon, 1924), mesquite grassland (Brand, 1936; Leopold, 1950), desert plains grassland (Weaver and Clements, 1938; LeSueur, 1945), desert shrub grassland (Darrow, 1944), grassland transition (Muller, 1947), and most frequently as desert grass or desert grassland (Shantz and Zon, 1924; Nichol, 1937; Benson and Darrow, 1944; Buechner, 1950; Castetter, 1956; Humphrey, 1958; Lowe, 1964). The terms scrub-grassland (Brown and Lowe, 1974a, 1974b), or "semidesert grassland" (Little, 1950) is used here because the earlier term "desert grassland," while more euphonious, is less correct. In Mexico, these and other grasslands are known collectively as *pastizals* (Flores Mata et al., 1971). Whatever the terminology, semi-desert grassland is potentially a perennial grass-scrub domi-nated landscape positioned between desertscrub below and evergreen woodland, chaparral, or plains grassland above.

Semidesert grassland adjoins and largely surrounds the Chihuahuan desert, and with the possible exception of some areas as in west central Arizona, it is largely a *Chihuahuan semidesert grassland.* Extensive areas of this grassland occur in the Southwest in Chihuahua, western Coahuila, Trans-Pecos Texas, the southern half of New Mexico, southeast Arizona, and extreme northeastern Sonora. These grasslands extend southward on the Mexican plateau to just northeast of Mexico, D.F., and into the Mexican state of Puebla, covering extensive portions of the states of Durango, Zacatecas, Jalisco, Aguascalientes, and Nuevo Leon. At its western and northeastern edges, the lower elevational limit of semidesert grassland can be expected at approximately 1,100 m. How-ever, the usual lower elevational range in the interior South-west is between 1,100 and 1,400 m, where its contact with Chihuahuan desertscrub is complex and provides alternating landscape mosaics over many miles. Its upper elevational contacts are usually with *Madrean evergreen woodland* or *Plains grassland* (less commonly *Interior chaparral*) and can be expected to occur between 1,500 m and 1,700 m, occasionally higher—to as high as 1,900 m. An exception is the boundary with Plains grassland in and along the Llano Estacado, which is tenuous and with many reversions dependent on edaphic conditions; it may be arbitrarily delineated here at ca. 1,250 m. Within the Chihuahuan Desert it also occupies the numerous *"bolsóns"* (enclosed drainages) as *mogotes* of Tobosa Grass (*Hilaria mutica*) and/or Sacaton (*Sporobolus wrightii*) (See Figs. 71, 72). At its western limits in Mohave, Yavapai, Gila, Pinal, Pima, and Graham counties of Arizona, semi-desert grassland may be encountered above and rarely even within the Sonoran Desert. Its southwestern limits in Sonora are adjacent to and above *Sinaloan thornscrub.*

Most of the semidesert grassland receives an annual average precipitation of between 250 mm and 450 mm. Except for extreme west-central Arizona, over 50% of this total comes during the April-September period when rainfall averages 150 mm or more—rarely as low as 100 mm. Perennial grass production is dependent primarily on the predictability and amount of precipitation during this period (Cable and Martin, 1975).

Precipitation data for 21 representative semidesert grass-land localities are given in Table 14, arranged from east to west on both sides of the international boundary. These data

Table 14. Precipitation data from 21 stations in the Southwest within or adjacent to Semidesert Grassland.

Station	Elevation (in m)	J	F	M	A	M	J	J	A	S	O	N	D	Total	Total April thru Aug.	Percent of total
Lordsburg, NM 32°18′ 108°39′	1,324	18	17	15	6	6	9	43	45	30	21	14	18	242	109	45
Animas, NM 31°57′ 108°49′	1,343	12	16	14	6	7	11	51	54	39	21	13	20	264	128	49
Fronteras, Son. 30°54′ 109°34′	1,136	24	11	13	5	3	13	96	87	30	14	12	21	328	202	62
Pearce, AZ 31°54′ 109°49′	1,347	15	14	13	6	4	13	73	85	29	13	9	20	294	180	61
Fort Huachuca, AZ 31°34′ 110°20′	1,422	19	14	14	5	2	10	114	99	41	19	11	25	372	229	61
Globe, AZ 33°23′ 110°47′	1.082	40	25	36	13	5	7	60	76	33	28	22	48	394	162	41
Willow Springs, AZ 32°43′ 110°52′	1,125	36	28	35	13	4	7	63	78	30	25	24	46	387	165	43
Sasabe, AZ/Son. 31°29′ 111°33′	1,094	31	25	18	7	3	6	98	94	50	14	25	58	428	208	48
Cordes, AZ 34°18′ 112°10′	1,149	31	26	29	17	8	8	45	65	33	22	23	39	344	142	41
Hillside, AZ 34°29′ 112°53′	1,012	35	33	29	14	6	9	46	58	33	22	27	37	349	133	38
Kingman, AZ 35°11′ 114°03′	1,024	26	27	27	20	5	4	20	35	18	16	17	26	239	83	35
Hobbs, NM 32°42′ 103°08′	1,102	14	8	13	25	39	58	48	66	70	41	11	15	405	236	58
Balmorhea Exp. Sta., TX 31°00′ 103°41′	983	19	12	9	20	35	36	42	36	51	36	13	16	325	169	52
Sierra Mojada, Coah. 27°17′ 103°42′	1,263	7	9	8	7	23	46	80	67	84	29	9	14	383	223	58
Roswell, NM 33°18′ 104°32′	1,101	11	12	13	19	33	37	46	37	51	25	10	13	307	172	56
Valentine, TX (10WSW) 30°30′ 104°38′	1,347	17	10	9	6	14	45	55	53	57	29	11	12	318	173	54
Chihuahua, Chih. 28°38′ 106°04′	1,431	4	5	8	8	11	25	80	96	95	37	8	21	394	220	56
Jornada Exp. Sta., NM 32°37′ 106°44′	1,300	12	11	8	5	12	14	44	41	38	24	11	17	238	116	49
San Buenaventura, Chih. 29°50′ 107°30′	1,545	8	17	3	12	9	26	89	108	46	29	9	21	377	244	65
Casas Grandes, Chih. 30°22′ 107°59′	1,487	10	16	7	10	13	21	91	86	53	34	18	22	381	221	58
Ascención, Chih. 31°05′ 107°59′	1,294	7	13	7	6	7	9	48	65	31	24	4	18	240	135	56

show that warm-season rainfall begins earlier (in June, May, or even April) in the eastern locations with winter-spring precipitation predominating toward the west. Unlike Plains grassland, the winters are mild and freezing temperatures occur generally less than 100 days during the year, and always average less than 150. Summers are warm to hot with several days over 100° F usually recorded. Relative humidities are low throughout the year except during storm periods. Winds are especially prevalent during the spring and fall, particularly in the early morning and late afternoon.

Originally the grasses were perennial bunch grasses, the bases of the clumps separated by intervening bare ground. Reproduction of these grasses is principally from seed. In areas of heavy to moderate rainfall, heavy grazing has reduced these bunch grasses and increased low growing sod grasses (e.g. Curly Mesquite Grass, *Hilaria belangeri*). Under low average summer rainfall the shift in grasses has been from bunch grasses to annuals. Where soils are deep, well protected from erosion, and with few shrubs or cacti, perennial grasses may cover extensive stretches of landscape, even today.

Such purely grass landscapes, however, stand in marked contrast to most semidesert grassland cover. There are vast areas where the climax grasses have been much reduced by competition with a wide variety of shrub, tree, and cactus life-forms. As a result, there now are extensive landscapes where shrubs, half-shrubs, cacti, and forbs greatly outnumber, and even completely replace the grasses.

While semidesert grassland is transitional in the sense of

Figure 67. *Semidesert grassland in Sulphur Springs Valley, Arizona, ca. 1,341 m. These yucca-grasslands of Soaptree yucca or Palmilla* (**Yucca elata**) *cover many miles of the high valleys (1,219-1,524 m elevation) and plains in southeastern Arizona, southern New Mexico, southwest Texas, and Chihuahua. The shrubs, partially hidden by the grasses* (**Aristida, Bouteloua, Trichachne**) *are mostly Burroweed* (**Isocoma tenuisecta**), *a frequent and persistent grass competitor and grassland invader.*

Figure 68. *Semidesert grassland 30 miles east of Valentine, Texas, ca. 1,372 m elevation. A mesquite-grassland community in which the "brush" is attaining aspect dominance. In addition to the dominant Mesquite* (**Prosopis glandulosa**), *other encroaching scrub species here are Crucillo* (**Condalia spathulata**), *Palmilla* (**Yucca elata**), *Pricklypear* (**Opuntia phaeacantha**), *and Whitethorn Acacia* (**Acacia constricta**). *Photograph by John S. Phelps.*

Figure 69. *Disjunct semidesert grassland on Picketpost Mountain, Tonto National Forest, Arizona, ca. 1,310 m elevation. Here the scrub species are Golden-flowered Century Plant (***Agave palmeri***), Sotol (***Dasylirion wheeleri***), Burroweed (***Isocoma tenuisecta***), and Mesquite (***Prosopis glandulosa***). Surprisingly, one of the most prevalent grasses at this remote locale in 1977 was an introduced species, Lehmann Lovegrass (***Eragrostis lehmanniana***).*

being positioned geographically between Plains grassland and Chihuahuan desertscrub, and shares some of the floral and faunal constituents of both, it is, nonetheless, a distinctive and separate biome. As such, it is the geographical and evolutionary center for a distinguishable and diverse flora and fauna. These include a variety of summer-active perennial grasses: Black Grama (*Bouteloua eriopoda*), Slender Grama (*Bouteloua filiformis*), Chino Grama (*Bouteloua briseta*), Spruce Top Grama (*Bouteloua chondrosioides*), Bush Muhly or Hoe Grass (*Muhlenbergia porteri*), several species of Three-awn (*Artistida divaricata, A. wrightii, A. purpurea,* and others), Arizona Cottontop (*Trichachne californica*), Curly Mesquite Grass, Slim Tridens (*Tridens muticus*), Pappus Grass (*Pappophorum vaginatum*), Tanglehead Grass (*Heteropogon contortus*), Vine Mesquite Grass (*Panicum obtusum*), and others.

Tobosa Grass is with Black Grama the most diagnostic grass dominant in semidesert grassland (Fig. 72). Black Grama is generally found on gravelly upland sites while Tobosa is most frequently encountered on heavier soils subject to flooding. In some vicinities, particularly at higher elevations, grasses of the Plains grassland—e.g., Blue Grama (*Bouteloua gracilis*), Sideoats Grama (*B. curtipendula*), Hairy Grama (*B. hirsuta*), Buffalo Grass (*Buchloë dactyloides*), Plains Bristlegrass (*Setaria macrostachya*), Plains Lovegrass, Wolftail, and Little Bluestem are mixed with the semidesert grasses and may even dominate local areas. Often only the tougher less palatable grasses are present or abundant, such as Hairy Tridens (*Tridens pilosus*) and Fluffgrass (*T. pulchellus*), Red Three-awn (*Aristida longiseta*), and Burrograss (*Scleropogon brevifolius*). Lehmann Lovegrass (*Eragrostis lehmanniana*), an early "green up" grass introduced from South Africa, now occupies extensive areas in some western portions of the semidesert grassland and appears to be naturally spreading at the expense of more palatable native grasses (Fig. 69).

Forbs and weeds are seasonally abundant with different sets of species growing in spring time—filarees (*Erodium*), lupines (*Lupinus*), buckwheats (*Eriogonum*), and mallows (*Sphaeralcea*)—than in the summer—spiderlings (*Boerhaavia*), white-mats (*Tidestromia*), devils-claws (*Martynia*), and amaranths (*Amaranthus*).

Dry-tropic stem and leaf succulents are particularly well represented and characteristic of semidesert grassland. These include the sotols (*Dasylirion wheeleri, D. leiophyllum*), beargrasses (*Nolina microcarpa, N. texana, N. erumpens*), the agaves (*Agave lechuguilla, A. parviflora, A. schottii, A. scabra, A. parryi,* etc.), and the yuccas (*Yucca torreyi, Y. baccata, Y. rostrata, Y. macrocarpa, Y. carnerosana,* etc.). Palmilla or Soaptree Yucca (*Yucca elata*) is a particularly conspicuous landscape feature over much of the semidesert grassland in the Southwest (Fig. 67).

Other generally or locally important scrub-shrub componenets, several of which may share or assume dominance, are Mesquite (*Prosopis glandulosa, P. juliflora*), One-seed Juniper (*Juniperus monosperma*), Lotebush (*Zizyphus obtusfolia, Condalia spathulata*), Allthorn (*Koeberlinia spinosa*), Mormon or Mexican Tea (*Ephedra trifurca, E. antisyphilitica*), Mimosa (*Mimosa biuncifera, M. dysocarpa*), False Mesquite (*Calliandra eriophylla*), Wright's Lippia (*Aloysia wrightii*), Catclaw Acacia (*Acacia greggii*), Littleleaf Sumac (*Rhus microphylla*), Desert Hackberry (*Celtis pallida*), Javelina-bush (*Condalia ericoides*), Barberry (*Berberis trifoliata*), and Ocotillo (*Fouquieria splendens*) (see e.g., Fig. 68). All of these dry-tropic

Figure 70. *Semidesert grassland in the foothills of the Mazatzal Mountains east of State Highway 87 in the Tonto National Forest, Arizona, ca. 1,250 m elevation. A northwestern fasciation on south-facing hillside habitats possessing such characteristic semidesert grassland species as Wait-a-minute Bush* (Mimosa biuncifera), *Engelmann Pricklypear* (Opuntia phaeacantha), *Sotol* (Dasylirion wheeleri), *Mesquite* (Prosopis velutina), *and because of fire suppression, much One-seed Juniper* (Juniperus monosperma).

species were natural elements in the Chihuahuan semidesert grassland, and still are, albeit now with greater densities.

Semidesert grassland has a naturally high species diversity of dry-tropic shrubby species, most of which tend also to be taller than the shrubs of the Plains grassland. Therefore, the community physiognomy of semidesert grassland—compared to that of the higher-elevated and greater freeze-prone Plains grassland—is one in which the seed stocks of the native grasses are usually much shorter than the heights of their associated shrubby perennials. The frequent result is a grassy landscape broken up by the uneven stature of large, diverse, and well-spaced scrub.

Tarbush (*Flourensia cernua*), Whitethorn (*Acacia neovernicosa*) and Creosotebush (*Larrea tridentata*) are characteristic plants of the Chihuahuan Desert that have invaded extensive areas and continue to increase today; they readily replace the native grasses, as does Mesquite. Numerous common shrubs, including Snakeweed (*Gutierrezia sarothrae*), Burroweed, Jimmyweed, and turpentine bushes (*Isocoma tenuisecta, I. hetero-*

Figure 71. *Semidesert grassland of Western Honey Mesquite* (Prosopis glandulosa *var.* torreyana) *and Sacaton* (Sporobolus wrightii) *on alluvial bottomland adjacent to Willcox Playa, Cochise County Arizona, ca. 1,250 m elevation.*

Figure 72. *Semidesert grassland of Tobosa* (Hilaria mutica) *with Clock-face Prickly pear* (Opuntia chlorotica), *Catclaw* (Acacia greggii), *and Snakeweed* (Gutierrezia spp.) *between Date Creek and Hillside, Yavapai County, Arizona, ca. 975 m elevation. These communities, which have primary residence in Chihuahua, also extend westward in valley bottom habitats to as far west as Kaka and Vekol valleys on the Papago Indian Reservation (see also e.g. Lowe, 1964:44). Photograph taken on February 2, 1950 by R.R. Humphrey.*

phylla, and *Ericameria laricifolia*), several buckwheats (*Eriogonum* spp.), Mariola (*Parthenium incanum*), Desert Zinnia (*Zinnia acerosa*), Threadleaf Grounsel (*Senecio longilobus*), and herbaceous sages such as *Artemisia ludoviciana*, all may be prevalent, scarce, or absent, depending on location, edaphic conditions, and grazing history.

Cacti are important in the structure of many communities. Species centered in, or at least well represented in semidesert grassland include Barrel Cactus or Visnaga (*Ferocactus wislizenii*), Turk's Head (*Echinocactus horizonthalonius*), Cane Cholla (*Opuntia imbricata, O. spinosior*), Desert Christmas Cactus (*Opuntia leptocaulis, O. kleiniae*), the prickly pears (*Opuntia chlorotica, O. phaeacantha, O. violacea* and *O. violacea* var. *macrocentra*), Rainbow Cactus (*Echinocereus pectinatus* var. *rigidissimus*), and other hedgehogs (*Echinocereus*), the pincushions (*Mammillaria wrightii, M. grahami, M. mainiae, M. gummifera*), and *Neolloydia intertexta* and *N. erectocentra*.

In the south, semidesert grassland communities may contain thornscrub species (Fig. 73). In the warmer portions of the grassland in Sonora and southern Arizona these include Kidneywood (*Eysenhardtia orthocarpa*), Little-leaf Lysiloma (*Lysiloma microphylla*), Palo-blanco or Net-leaf Hackberry (*Celtis reticulata*), Gum Bumelia (*Bumelia lanuginosa*), and large shrubs that include Coursetia (*Coursetia glandulosa*), Dodonaea or Hopbush (*Dondonaea viscosa*), Spiny Sageretia (*Sageretia wrightii*), Algodoncillo or Desert Cotton (*Gossypium thurberi*), and Chuparosa (*Anisacanthus thurberi*).

Trees in the semidesert grassland, except for mesquite and One-seed Juniper, are uncommon and usually restricted to drainages where one may encounter in addition to the tree species just listed above, Western Soapberry (*Sapindus saponaria*), Desert Willow (*Chilopsis linearis*), and occasionally one of the lower oaks of the Madrean woodland (*Quercus toumeyi, Q. grisea, Q. emoryi, Q. oblongifolia*, and *Q. chihuahuensis*).

Semidesert grasslands within and adjacent to interior chaparral in west central Arizona (e.g., on Burro, Bozarth, Goodwin, and Aquarius mesas), while formerly populated by perennial grasses of continental origin, are today comprised largely of introduced annuals due to grazing (Fig. 74). Red Brome (*Bromus rubens*) is especially prevalent, although others, such as bristlegrass (*Setaria*) and wild oats (*Avena*), may be locally and seasonally common. Except for steep hillsides and rocky canyons, the only perennial grass often remaining is Tobosa (*Hilaria mutica*), and these grasslands now have the appearance and growth-form composition of California Valley grasslands. There, many of the plant life-forms representative of semidesert grassland regions to the south and east are lacking or poorly represented; Sotol, Palmilla, agaves, and chollas are conspicuously few or absent as are several characteristic semidesert grassland animals. Mesquite is only locally common, and the principal scrub components in these westerly areas are prickly pears (*O. phaeacantha, O. chlorotica*), Catclaw, Dátil, Mimosa, Algerita, juniper, Scrub Oak (*Quercus turbinella*), and Canotia (*Canotia holocantha*). Annual spring forbs both native and introduced are of great importance, and scrub-shrub invasion may be more pronounced than in more typical semidesert grassland locales.

Mammals well represented or to be expected in semidesert grassland are the Black-tailed Jack Rabbit (*Lepus californicus*), Spotted Ground Squirrel (*Spermophilus spilosoma*), Hispid Pocket Mouse (*Perognathus hispidus*), Ord's, Banner-tailed,

Figure 73. *Semidesert grassland dominated by Ocotillo (*Fouquieria splendens*) near Presumido in the Pozo Verde Mountains on the Arizona-Sonora line, ca. 1,158 m elevation. Such thornscrub species as Hopbush (*Dodonaea viscosa*), Kidneywood (*Eysenhardtia orthocarpa*), and Coralbean (*Erythrina flabelliformis*) are present in addition to the Ocotillo, Burroweed, Mesquite, and other "typical" semidesert grassland invaders.*

and Merriam Kangaroo Rats (*Dipodomys ordii, D. spectabilis, D. merriami*), White-footed Mouse (*Peromyscus leucopus*), the cotton rats (*Sigmodon hispidus, S. fluviventer*), Southern Grasshopper Mouse (*Onychomys torridus*), Southern Plains and White-throated Wood Rats (*Neotoma micropus, N. albigula*), Badger (*Taxidea taxus*), and the ubiquitous Coyote (*Canis latrans*).

The variety of birds is also great and a list of the well distributed nesting species would include Swainson's Hawk (*Buteo swainsoni*), Prairie Falcon, Kestrel, Mourning Dove (*Zenaida macroura*), Scaled Quail (*Callipepla squamata*), Roadrunner (*Geococcyx californianus*), Burrowing Owl, Poor-will (*Phalaenoptilus nuttallii*), Ladder-backed Woodpecker (*Picoides scalaris*), Western Kingbird (*Tyrannus verticalis*), Ash-throated Flycatcher (*Myiarchus cinerascens*), Say's Phoebe (*Sayornis saya*), Horned Lark, Barn Swallow (*Hirundo rustica*), White-necked Raven (*Corvus cryptoleucus*), Verdin (*Auriparus flaviceps*), Cactus Wren (*Campylorhynchus brunneicapillus*), Mockingbird (*Mimus polyglottos*), Curve-billed Thrasher (*Toxostoma curvirostre*), Black-tailed Gnatcatcher (*Polioptila*

Figure 74. *Semidesert grassland. A western fasciation on Burro Mesa east of Aquarius Mountains, ca. 1,219 m elevation, Yavapai County, Arizona. Because of past and present grazing, the original perennial grasses are much reduced. Mostly uninvaded by scrub, these grasslands are nonetheless in a disclimax state and are principally populated by winter-spring annuals—Red Brome* (Bromus rubens) *and Yellow Bristlegrass* (Setaria lutescens) *being particularly prevalent this year at this site. The principal remaining perennial grass is Tobosa* (Hilaria mutica); *more characteristic semidesert grassland species, such as Arizona Cottontop, Black Grama, and Sideoats Grama, are confined to rocky hillsides and less accessible areas. The cactus is Clockface Prickly Pear* (Opuntia chlorotica).

melanura), Loggerhead Shrike (*Lanius ludovicianus*), Meadow Lark (*Sturnella magna*—increasingly *S. neglecta*), Brown-headed Cowbird (*Molothrus ater*), Scott's Oriole (*Icterus parisorum*), House Finch (*Carpodacus mexicanus*), Lark Sparrow (*Chondestes grammacus*), and Cassin's Sparrow (*Aimophila cassinii*).

As is to be expected, important animals in semidesert grassland include species from adjacent scrublands and desertlands, e.g., Gambel's Quail (*Lophortyx gambelii*), Mule Deer (*Odocoileus hemionus crooki*), and Black-throated Sparrow (*Amphispiza bilineata*); where dense Ocotillo and/or thornscrub elements are present, such species as Javelina (*Dicotyles tajacu*), and White-tailed Deer occur, as well as a grassland fauna.

There are also a number of endemic taxa centered in the semidesert grassland as Scaled Quail, the Western Yellow Box Turtle (*Terrapene ornata luteola*), Desert-grassland Hognose Snake (*Heterodon nasicus kennerlyi*), Western Hooknose Snake (*Ficimia cana*), the all-female Desert-grassland Whiptail

(*Cnemidophorus uniparens*), the Southwestern Earless Lizard (*Holbrookia texana scitula*), and the Western Green Toad (*Bufo debilis insidior*).

Generally, the grassland and other open landscape-adapted species have fared less well than their scrub-adapted competitors. Antelope, for example, are now totally absent from large areas of their former range in semidesert grassland, whereas Mule Deer and Javelina have extended their range and increased in density in these habitats during this century. This is largely because the reduction and elimination of grasses by livestock has facilitated the invasion of woody and shrubby species by opening the stands of grass and thereby reducing and eventually eliminating the incidence of fire (Humphrey, 1958).

This invasion of semidesert grasslands by scrubby trees and shrubs (brush) since Anglo settlement is well documented (Leopold, 1924; Humphrey, 1958; Castetter, 1956; and Martin, 1975). Mesquite and juniper have invaded large areas of former grassland. Investigations have included before and

Figure 75. *Disclimax semidesert grassland on Tonto National Forest, Tonto Basin, Gila County, Arizona, ca. 1,006 m elevation. Once supporting a community of perennial grasses, these habitats at the lower and western edge of semidesert grassland now host a number of successful rangeland invaders, such as Mesquite (*Prosopis velutina*), Beargrass (*Nolina microcarpa*), Sotol (*Dasylirion wheelerii*), One-seed Juniper (*Juniperus monosperma*), agave (*Agave spp.*), Jimmyweed (*Isocoma heterophylla*), Snakeweed (*Gutierrezia sarothrae*), and Prickly Pear (*Opuntia phaeacantha*). Virtually no grasses remain and the area now has much of the appearance of coastalscrub (=soft chaparral).*

after photographs many of which show striking changes (e.g., Parker and Martin, 1952; Hastings and Turner, 1965b). Differences in the abundance of cacti are also impressive. Less discussed but of equal or greater importance is the disappearance and replacement of the soil-binding perennial grasses by shrubs, both native and introduced. Two especially successful native shrubs, Burroweed and Snakeweed, have now replaced the grass understory over millions of acres and are indicators of former grass areas.

Burroweed and Jimmyweed, which may germinate with either fall or spring precipitation but grow primarily in the spring, occupy extensive areas of former semidesert grassland in Arizona and southwestern New Mexico and are largely concomitant with the range of Gambel's Quail. Snakeweed,

also a "cool-season" germinator, is even more widespread than Burroweed and competes more directly with the grasses (Martin, 1975). These grass competitors have increased westward on the gradient of decreasing summer precipitation (increasing winter precipitation). Much of the former grassland landscape of western New Mexico and Arizona is now a disclimax grassland (=semidesert scrubland) having the appearance of an open short-statured "soft chaparral" (Fig. 75). Eastward, the primary invaders of semidesert grassland tend to be species of the Chihuahuan Desert; much of the former grassland in west Texas, eastern New Mexico, and Chihuahua is now populated by Creosotebush, Tarbush, Acacia, and Mesquite, much of this accomplished during the present century.

143.2 Californian Valley Grassland

The original nature and extent of Californian Valley grassland may never be known. Where not urbanized or under cultivation, it is now an annual grassland much disrupted by more than 200 years of grazing and other disturbances. Today more than 400 alien species account for from 50% to 90% of the vegetative cover (Heady, 1956; McNaughton, 1968). There is much speculation concerning the original vegetation, and while many authors believe the prehistoric vegetation was perennial, historic evidence is meager and the earliest references are to annual prairies (McNaughton, 1968). Because of grazing, much of the cover is herbaceous and largely composed of introduced annual forbs and weeds (Biswell, 1956). The structure and composition of the grassland varies annually and often presents a mosaic of floristic components, depending on the amount of precipitation received, soil habitats, and intensity of use.

Munz and Keck (1949) describe "Valley Grassland" and state that it occurs in California along the coast from San Luis Obispo County south, in the Great Central Valley, in the low, hot valleys of the inner coast Ranges, and in Antelope Valley, ascending to ca. 1,200 m in the Tehachapi Mountains in Kern County, and in eastern San Diego County. Californian Valley grassland is synonymous with Valley Grassland, which occurs southward into Baja California Norte to at least Valle de la Trinidad.

Jensen (1947), Biswell (1956), and Small (1974) map "grass" and "grassland" in California but do not describe or indicate it as occurring south of the Transverse Ranges. Other large scale maps of "California Steppe" (Küchler, 1964) and "Grassland in California" (McNaughton, 1968) include the interior of the Los Angeles basin and plain and the coastal mesas in Los Angeles and Orange counties. Wieslander (1932, 1934a, 1934b, 1934d, 1934e, 1937a, 1937b, 1938, 1940) mapped grasslands in Los Angeles, San Bernardino, Orange, and San Diego counties. Shreve (1937) correctly stated that "some grassland is present in Baja California between the Sierras Juarez and San Pedro Mártir and the Pacific Ocean."

Californian Valley grassland occurs, or occurred, in the Southwest from the Los Angeles Plain or parallel 34° southward in discontinuous coastal and intermountain hills and valleys to Valle de la Trinidad in Baja California Norte. Portions of the low interior valley around Riverside were grassland, as was most of the coastal mesa in Orange County on which Irving Ranch (Irving) is situated. Other remnant areas are found in valleys and on hillsides along the coasts of southern California and Baja California from Orange County to just south of Ensenada. Warner, Ramona, and Coahuila valleys in San Diego County, California, and Valle San Rafael and Valle del Rodeo in Baja California Norte, present higher interior valley examples of Californian Valley grassland.[1]

---

[1]*The composition of certain grasslands within interior chaparral in west central Arizona (e.g., Burro, Bozarth, Goodwin, and Aquarius mesas) today shows affinity with these coastal communities. While classified as semidesert grassland of continental origin, these Arizona communities exist under a precipitation regime where the majority of the mean annual rainfall occurs from October through March. Summer precipitation is significant, however, and the grasses were, and in part remain, a mixture of perennial "continental" grasses and relict perennials of "Mediterranean" origin. Today these grasslands are often extensively populated by introduced annuals (e.g., species of* **Bromus, Avena, Erodium***) and except on steep hillsides the only remaining perennial of consequence is Tobosa* (Hilaria mutica).

Figure 76. *Californian Valley grassland. Disclimax community of annual forbs and grasses in Warner Valley, ca. 914 m elevation, San Diego County, California. Where not destroyed by cultivation or urbanization, the "California Prairies" are, as here, now composed largely of introduced forbs (e.g.,* **Erodium***) and annual grasses (e.g.,* **Bromus***). It is late winter (March) and the green carpet is putting on growth which will be accelerated in the coming months of April and May. Later, with the curing of the vegetation, the landscape will change to golden-brown.*

Figure 77. *A disclimax Californian Valley grassland on the northeast side of Guadalupe Island, 160 miles off the Pacific coast of Baja California Norte. Here as elsewhere the native California grassland flora has given way to aggressive adventive annuals; in this case and year (1979) Wild Oat (***Avena fatua***) is an important participant in the grassland.*

Table 15. Precipitation data from 13 stations within or adjacent to California Valley grassland.

Station	Elevation (in m)	Mean monthly precipitation in mm												Total	Total April thru Sept.	Percent of total
		J	F	M	A	M	J	J	A	S	O	N	D			
Warner Springs, CA 33°17' 116°38'	969	66	80	64	44	6	2	13	26	9	19	32	79	440	100	23
Irvine Ranch, CA 33°44' 117°47'	36	61	73	52	27	6	2	0	1	5	12	31	70	340	42	12
Santa Ana, CA 33°45' 117°52'	35	70	81	60	29	7	1	1	1	5	13	34	81	384	44	11
Pomona, CA 34°04' 117°49'	261	89	94	77	35	7	2	0	2	6	18	38	94	462	52	11
Elsinore, CA[1] 33°40' 117°20'	396	55	71	50	18	3	1	0	3	8	14	24	69	318	34	11
Whittier, CA 33°58' 118°02'	94	69	80	62	28	4	1	—	2	6	12	30	80	375	41	11
Downey, CA 33°56' 118°08'	35	72	81	60	26	3	1	0	1	7	12	32	81	377	39	10
Escondido, CA 33°07' 117°05'	201	71	84	68	33	8	3	0	5	6	22	35	92	426	54	13
Corona, CA[1] 33°52' 117°34'	216	59	72	52	22	18	1	—	2	5	14	28	68	341	47	14
Yorba Linda, CA 33°54' 117°49'	123	70	81	66	28	6	1	—	2	7	15	45	74	395	45	11
Ensenada, BCN 31°53' 116°37'	24	60	54	41	24	6	2	2	2	5	12	24	47	279	41	15
Valle Trinidad BCN 31°20' 115°47'	899	26	34	24	13	1	1	13	25	13	14	25	40	229	66	29
El Alamo, BCN 31°36' 116°06'	1,149	38	26	32	23	3	0	11	16	9	6	26	39	230	62	27

[1]Mapped as coastalscrub; at edge of Californian Valley grassland.

Californian Valley grassland occurs from ca. 30 m near the coast to over 1,100 m in the interior valleys. Its usual contact near the coast and on lower elevation hillsides is with coastalcrub. At higher elevations in the interior valleys it is adjacent to coastal (hard) chaparral on hillsides or, with decreasing frequency southward, encinal woodlands. Where perennial or near-perennial streams drain the valleys, a narrow band of riparian deciduous forest or woodland may be present.

Climate is warm-temperate Mediterranean, characterized by mild, moderately wet winters and warm to hot summer drought. The growing season is from 7 to 11 months with 205 to 325 frost free days (Munz and Keck, 1949). Summer temperatures frequently exceed 41°C (Biswell, 1956) and, while winter temperatures rarely drop below -4°C (Bakker, 1971), winter frosts may be heavy. Precipitation can be expected to begin in late fall (October-November) and end in April or early May with two-thirds or more falling in the December-March period.

While a total mean annual precipitation of 150 mm to 500 mm is reported for grassland in California (Munz and Keck, 1949; Biswell, 1956; Small, 1974), data for Californian Valley grasslands south of the Transverse Ranges indicate an average annual precipitation of 230 mm to 460 mm. Also, late summer precipitation may contribute more to the totals in the higher interior valleys than previously reported. Mean annual and monthly precipitation for 13 representative stations for which a sufficient number of years of records are available, and which are approximate to or in grassland in southern California and Baja California Norte, are given in Table 15.

Vegetation, particularly under grazing, is herbaceous and greens soon after the first rains (Fig. 76). Growth in early fall may be considerable if conditions are favorable (Biswell, 1956). With the advent of first freezing temperatures, usually in December, growth is arrested. While the landscape remains green, growth is slow until the advent of warmer weather in early spring. Rapid development then begins, and if the winter rains were bountiful the observer may be greeted by a spectacular display of annual flowers. This may be especially so in the so-called "vernal pools" which are hardpan depressions or sinks within the grassland mosaic (Holland and Jain, 1977). Both flowering forbs and grasses mature and dry by late April or early May with some of the perennials remaining green slightly longer. From then until the commencement of the fall rains the landscape is a golden brown broken by the green of an occasional evergreen oak—Coast Live Oak (*Quercus agrifolia*) or Interior Live Oak (*Q. engelmannii*). Cacti, and stem and leaf succulents are essentially lacking.

As previously noted, there is disagreement regarding the prehistoric composition of Californian valley grassland. Clements (1934), Munz and Keck (1949, 1950), Twisselman (1956), Bakker (1971), Ornduff (1974), Heady (1977), and others, give the opinion that the original grasslands were probably dominated by perennial bunchgrasses, particularly species of *Stipa*, *Poa*, and *Festuca*. Investigations cited by Biswell (1956), White (1967), and McNaughton (1968) seem to support the opinion that much of the grassland in California may have been composed of native annual grasses before the advent of civilized man and his livestock. White (1967) investigated an area protected from grazing for 27 years and found that while the composition of annuals was significantly different (71% being native) from adjacent grazed areas (20% native), the standing crop of the perennial Purple Needlegrass (*Stipa pulchra*) was no more abundant on the reservation than in comparable areas subject to moderate grazing. He concluded that the climax vegetation was annual grassland, at least under present environmental conditions, and that grazing favored the replacement of native annual species by exotic annuals.

Because Purple Needle-grass tends to become established after burning, this perennial species may have occupied extensive areas before the advent of fire suppression and the removal of potential fuel by grazing (Biswell, 1956). The absence of large numbers of large, grass-consuming herbivores before domestic livestock were introduced (Tule Elk, *Cervus elaphus nannodes*, were not present south of the San Joaquin Valley), and periodic burning by the aborigines certainly would have permitted this possibility. Biswell (1956) reports that grazing also favored an increase in rodent populations to the detriment of perennial native forbs such as Blue Dick (*Dichelostemma pulchellum*). For a particularly good treatment and discussion of Californian Valley grassland, see Bakker's (1971) discussion of "California's Kansas."

Examination of adobes from known-age structures, including missions, have provided valuable insights into the chronology of alien arrivals (Burcham, 1957). Most of the early adventives were the most valuable from a forage standpoint and include Wild Oats, Bur-clover, and Filaree. The most recent introductions, such as Halogeton and Goat Grass, tend to be noxious. The most successful invaders of California grasslands came from Spain and include species of *Avena*, *Bromus*, *Erodium*, and *Medicago*. An excellent discussion of the progressive invasion of open, unstable (annual) grasslands is given by Naveh (1967).

Table 16 gives a short list of the most frequent or characteristic annual and perennial plants of Californian Valley grassland and is from Twisselman (1956), Biswell (1956), Heady (1956), and to a lesser extent from Small (1974). Intensive field investigation of grassland south of the Transverse Ranges are apparently lacking so that the listing in Table 1 is based on collections from elsewhere in southern and central California. Alien species are indicated by an asterisk; most of these are annuals.

As is the case with the native grassland itself, almost all of the larger influent vertebrates have been destroyed and replaced. These include the California Grizzly Bear, Pronghorn Antelope, and California Condor (*Gymnogyps cali-*

Table 16. Some common grasses and forbs of Californian Valley grassland. Introduced taxa are indicated by asterisk.

Annual Grasses and Forbs	
Bromus mollis	*Soft Chess
Bromus rubens	*Red Brome
Bromus rigidus	*Ripgut Grass
Avena barbata	*Slender Oat
Avena fatua	*Wild Oat
Hordeum spp.	*foxtails
Festuca confusa	Hairy-leaved Fescue
Festuca dertonensis	*Sixweeks Fescue
Festuca megalura	Foxtail Fescue
Lolium multiflorum	*Italian Ryegrass
Medicago hispida	*Bur Clover
Eschscholtzia californica	California Poppy
Brassica spp.	*wild mustards
Lepidium spp.	peppergrasses
Orthocarpus spp.	owlclovers
Nemophila menziesii	Baby Blue-eyes
Erodium cicutarium	*Red-stem Filaree
Erodium botrys	*Broadleaf Filaree
Trifolium spp.	annual clovers
Lotus americanus	Spanish Clover
Lupinus bicolor	Ground Lupine
Hemizonia spp.	tarweeds
Layia platyglossa	Tidytips
Eriogonum spp.	buckwheats
Astragalus spp.	locoweeds
Malva parviflora	*Cheeseweed

Perennial Grasses and Forbs	
Stipa pulchra	Purple Needlegrass
Stipa speciosa	Desert Needlegrass
Elymus triticoides	Creeping Wildrye
Melica californica	Melic Grass
Poa scabrella	Pine Bluegrass
Ranunculus californicus	California Buttercup
Solidago californica	California Goldenrod
Brodiaea capitata	Blue Dick
Calochortus spp.	mariposa lilies

fornianus). The list of vertebrates associated with Californian Valley grassland is relatively short and none of these species is restricted in its distribution to this biotic community, the depauperate grassland fauna being continental in origin. Only two animals could be considered endemic to California grasslands, the Tule Elk and the Yellow Billed Magpie (*Pica nuttalli*), neither of which extends south of the Transverse Ranges.

Because few examples of Californian Valley grassland are in public ownership, it will continue to deteriorate and shrink in extent. Representative areas south of the Transverse Ranges are particularly limited and, with continued rapid urbanization, these examples may be restricted to islands and the interior valleys of Baja California in the forseeable future (Fig. 77).

Tropical-Subtropical Grasslands

As the equatorial region is approached, an increasingly warmer winter coupled with an increasingly longer and wetter growing season, favors the establishment of forest so that a gradual transition from scrub and grassland to equatorial forest takes place. In the subtropics grasslands are restricted, and are only well developed on deep loam or sandy soils receiving between 100 and 400 mm of precipitation. These grasslands often have the physiognomy of savanna grasslands—grass-dominated landscapes with scattered woody plants.[1]

Perennial grasses and woody species are competitors. On shallow stony soils, thornscrub predominates. Or, if precipitation is inadequate for thornscrub communities, desertscrub predominates. Only on the finer soil types is an equilibrium reached between the grasses and scrub. Edaphic factors are important in determining which will dominate, with grasses occupying the valley bottoms and plains (=llanos), and scrub on the hillsides.

Grasses have an intensive root system in contrast to the extensive system of thorny woody plants. With the removal of grasses, such as by excessive numbers of grazing animals, transpirational water loss above the soil surface almost stops and more water is retained in the soil for advantageous use by woody plants (commonly legumes). Scrub or brush encroachment then begins, and unless the grasses are allowed to recover (as was the case when the principal large herbivores, if any, were migratory or nomadic species of big game) and provide fuel for either lightning or human-caused fires, scrub invasion intensifies at the expense of the grassland.

Domestic livestock have greatly changed these grasslands and accelerated their conversion to scrublands over wide areas of Africa, Australia, North, Central, and South America. In the North American Southwest a subtropical Sonoran savanna grassland climax occurs in Sonora, Mexico, and in limited areas in extreme southern Arizona. These grasslands have now been greatly altered if not entirely destroyed.

---

[1]The term savanna or sabana is here defined biogeographically. Those areas where the crowns of trees cover 15% or more of the ground are woodlands (savanna-woodland). Those areas consisting of a mosaic of grasslands and smaller or larger stands of shrubs and trees are parklands; parklands are composed of two or more ecologically distinct plant communities (Walter, 1973).

144.3 Sonoran Savanna Grassland

Savanna grasslands, while never extensive in the Southwest, once occupied favorable sites within Shreve's (1951) Plains of Sonora and Foothills of Sonora subdivisions of the Sonoran Desert (Fig. 78). Certain areas within the Altar and Santa Cruz valleys in Arizona were, or approached Sonoran savanna grassland, as did suitable areas in extreme north-central Sonora (Fig. 79). These subtropical fire-climax grasslands were encountered between 90 m and 1,000 m elevation on level plains and at intervals along the larger river valleys on deep, fine textured soils (Brown, 1884; Stephens, 1885; Ligon, 1942, 1952; Brown and Ellis, 1977).

J.T. Wright, who collected the grassland-affiliated Masked Bobwhites at Rancho Noria de Pesqueira, Sonora, in 1931, told Tomlinson (1972) "that the country at that time consisted of wide, grass-covered valleys with certain grasses reaching over the heads of the native white-tailed deer." Brand (1936) mapped large areas in central and eastern Sonora as *Sonoran mesquite-grassland,* and Shreve (1951) in a description of certain areas in the central portion of the *Plains of Sonora* states that "Grasses often form as much as 75% of this cover, which gives an aspect of abundant verdure after the period of summer rain." Because of grazing by livestock these grasslands were greatly altered or destroyed largely by 1900 in Arizona, and after 1940 in Sonora (Tomlinson, 1972; Brown and Ellis, 1977). Whatever the past situation, these habitats with few exceptions are now perhaps more properly classified as Sonoran desertscrub or Sinaloan thornscrub (Fig. 80). Exceptions are small areas not shown on the color map. While they are small in geographic extent, these few relicts are of great biotic and historic interest.

Past accounts and recent investigation of those "grassland llanos" that are considered to be the least altered indicate that the principal grass species were summer-active root perennials (Fig. 81). Of these grasses, the principal species encountered today are Rothrock Grama (*Bouteloua rothrockii*) and three-awns (*Aristida hamulosa, A. wrighti, A. ternipes, A. californica,* and others). Other characteristic species are the subtropic gramas *Bouteloua aristidoides, B. radicosa, B. filiformis, B. parryi,* and *B. barbata,* the False Grama (*Cathestecum erectum*), Tanglehead Grass (*Heteropogon contortus*), and windmill grasses (*Chloris* spp.). Sod-forming or other perennial grasses of warm temperate semidesert grassland when present are usually restricted to favored sites along drainages and on north-facing slopes; e.g., Curly-mesquite Grass (*Hilaria belangeri*), Vine Mesquite Grass (*Panicum obtusum*), and Side-oats Grama (*Bouteloua curtipendula*). Herbaceous shrubs were important in the grassland composition and include ragweeds (*Ambrosia*), purslane (*Portulaca*), several spurges (*Euphorbia*), spiderling (*Boerhaavia*), Janusia (*Janusia gracilis*), species of *Isomeris,* Croton (*Croton sonorae*), and a pigweed (*Amaranthus palmeri*). Most of the dry-tropic scrub species prevalent and characteristic of semidesert grassland such as Palmilla, Sotol, agaves, Burroweed, and Snakeweed are lacking.

Trees and scrub components, while always present, vary in composition and density from site to site. In the southern and eastern portions of Sonoran savanna grassland, an enormous variety of tropic-subtropic thorny shrubs and trees may be present. At the northern limits, Mesquite (*Prosopis velutina*) is often the primary or exclusive treeform constituent. Other important tree or tall shrub species are Ironwood (*Olneya tesota*), paloverdes (*Cercidium microphyllum, C. floridum, C. praecox*), and locally the Retama (*Parkinsonia aculeata*), Jito

Figure 78. *Sonoran savanna grassland within Shreve's "Plains of Sonora" subdivision of the Sonoran Desert, Rancho Carrizo, Sonora, ca. 610 m elevation. Summer aspect, 1968. The principal grasses are Rothrock Grama (*Bouteloua rothrockii*) and threeawn (*Aristida*); the trees are paloverdes (*Cercidium microphyllum, C. floridum*). The shrub in the immediate center foreground is an acacia (*Acacia angustissima*). This area, which was habitat for the Masked Bobwhite (*Colinus virginianus ridgwayi*), has since been increasingly invaded by scrub. Photograph by R.E. Tomlinson.*

(*Forchammeria watsoni*), *Guaiacum coulteri, Atamisquea emarginata*, and acacias (*Acacia angustissima, A. farnesiana*, and others). Depending on location and site, the following may be important scrub-shrub species: Tomatillo (*Lycium brevipes*), *Caesalpinia pumila, Croton sonorae*, Desert Hackberry (*Celtis pallida*), Kidneywood (*Eysenhardtia orthocarpa*), *Coursetia glandulosa*, Tree Ocotillo (*Fouquieria macdougalii*), Limberbush (*J. cardiophylla*) and other species of *Jatropha*, and species of *Cassia*. Along the drainages and flood plains, trees and shrubs may form dense thickets with numerous tangles of vines.

The larger cacti, while present, are not prevalent, and are usually one of four species of cholla (*Opuntia thurberi, O.*

arbuscula, O. fulgida, O. leptocaulis), Sina (*Rathbunia alamosensis*), Senita (*Lophocereus schottii*), Pitahaya or Organ Pipe (*Stenocereus thurberi*), less frequently Saguaro (*Carnegiea gigantea*).

The monthly and seasonal precipitation for eight locations in, near, and just outside of Sonoran savanna grassland is summarized in Table 17. Mean annual precipitation ranges from 275 mm to 525 mm—more commonly between 300 and 500 mm. The greater percentage of this precipitation falls during the July through September period and will average over 150 mm.

Freezes, while to be expected any winter, are not of long duration and rarely drop much below -4° C. Killing frosts are

Figure 79. *Sonoran savanna grassland in northern Altar Valley, Pima County, Arizona, ca. 945 m elevation. October, 1970. This community and others, now virtually destroyed in Altar and Santa Cruz Valleys, Arizona, were the northernmost representatives of Sonoran savanna grassland which had primary residence in north-central Sonora. The shrubs mixed with three-awn grasses and forbs are mostly Burroweed—a temperate species not characteristic of the subtropic Sonoran savanna grassland. The overstory trees however, are such distinctly Sonoran species as Ironwood, Blue Paloverde, and Velvet Mesquite.*

Table 17. Precipitation data from 7 stations in the Southwest within and directly adjacent to former Sonoran Savanna Grassland.

Station	Elevation (in m)	J	F	M	A	M	J	J	A	S	O	N	D	Total	Total July thru Sept.	Percent of total
Punta de Agua, Son. 28°25′ 110°25′	220	26	9	5	6	—	13.7	107	124	59	43	7	30	431	290	67
Suaqui, Son. 29°11′ 109°41′	250	31	16	11	5	3	27	163	141	53	27	12	30	519	356	69
La Colorado, Son. 29°48′ 110°35′	390	22	15	5	1	1	14	106	112	36	10	9	26	357	255	71
Mazatan, Son. 29°00′ 110°09′	550	28	18	9	2	0	37	150	—	134	26	7	26	502	348	69
Rancho Carizzo,[1] Son. 30°03′ 111°15′	732	16	7	8	2	—	7	96	121	43	14	8	21	344	260	76
Bacoachi, Son. 30°30′ 109°58′	1,050	7	19	13	5	3	19	132	112	34	24	16	29	436	279	64
Tumacacori, AZ 31°34′ 111°03′	996	21	15	17	6	3	10	101	92	34	18	15	30	361	226	63

[1]Data from Tomlinson, (1972).

Figure 80. *Former Sonoran savanna grassland converting to a Sonoran desertscrub association on Palo Alto Ranch, Altar Valley, Arizona, ca. 975 m elevation. This site, which is similar to the one shown in Figure 79, was known to be "open prairie" in 1885. The overstory trees are mesquite and Foothill Paloverde; the understory, Burroweed and Pigweed* (Amaranthus palmeri). *The grasses are annual gramas* (Bouteloua), *Poverty Three-awn* (Artistida divaricata), *and Rothrock Grama* (Bouteloua rothrockii).

Figure 81. *Former Sonoran savanna grassland near Benjamin Hill, Sonora, ca. 671 m elevation. December, 1968. A dense grassland composed mostly of subtropic three-awns and Rothrock Grama then extended to the wooded drainages.*

Figure 82. *Sonoran savanna grassland in the process of being transformed to Sinaloan thornscrub near Mazatán, Sonora, ca. 457 m elevation. This photograph, taken in 1964, shows the effects of heavy grazing on a former savanna grassland; numerous young sprouts of* **Acacia** *are already becoming established and with the lack of fuel and absence of fire, these and successive generations of scrub have joined the overstory (principally Mesquite and Sonoran Paloverde) to form dense thornscrub. Photograph by Steve Gallizioli.*

therefore infrequent.

From November to the abrupt onset of the summer rains the following year, the landscape becomes increasingly desolate and bare. Cattle have eaten the last vestiges of grass and except for dried and leafless scrub, groundcover is almost nonexistent (Fig. 82). By late June temperatures in the late afternoon commonly exceed 38°C and the humidity gradually rises. Usually the summer rains begin during the first half of July and last through September; if summer moisture is adequate, growth is rapid and continues through October. Grasses and forbs spring up and develop rapidly, trees and shrubs develop leaves and several species commence blooming—those that haven't bloomed in spring. The amount of herbaceous growth is determined by the generosity of the rains and little or no growth may occur in some years.

The transformation of these grasslands to thornscrub and desertscrub has resulted in the displacement of an interesting and unique subtropic grassland fauna including a number of invertebrates. While these changes have benefitted some scrub-adapted animals, e.g., the Javelina and Antelope Jackrabbit (*Lepus alleni*), other subtropic grassland affiliated vertebrates appear to have been negatively affected, e.g., the Caracara (*Caracara cheriway*) and White-tailed Hawk (*Buteo albicaudatus*). The numbers and distribution of three endemic animals, the Masked Bobwhite Quail (*Colinus virginianus ridgwayi*), the Rufous-winged Sparrow (*Aimophila carpalis*), and the Sonoran Green Toad (*Bufo retiformis*) are greatly reduced, and the Masked Bobwhite is now faced with extinction (Brown and Ellis, 1977).

Part 5. Desertlands

Mohave desertscrub dominated by joshua trees
*(*Yucca brevifolia*) and creosote bush (*Larrea triden-*
tata) between Wickenburg and Wikieup, Arizona.

Cold-Temperate Desertlands

Often simply called "cold deserts," these desertlands are chiefly found in vast rain-shadowed interiors of the northern hemisphere. In the Turkestan, Takla Makan, and Gobi deserts in Eurasia, and in the Great Basin Desert in North America, there are great similarities in plant life forms and even genera—low, widely spaced hemispherical shrubs of *Artemisia, Atriplex, Ceratoides,* and scattered bunchgrasses (*Agropyron, Stipa*). This is a function of a recent, shared evolutionary history and the fact that these deserts all experience cold, harsh winters, low precipitation scattered more or less throughout the year, and great extremes in both daily and seasonal temperatures. Further, they share large, enclosed basins of saline soils dominated by halophytes in communities of low species diversity. The sole biotic representative of desertlands in this climatic zone in North America—*Great Basin Desertscrub*—finds its southernmost limits in the Southwest.

152.1 Great Basin Desertscrub

The most northerly of the four North American deserts, the Great Basin Desert, evolved from both cold-temperate and warm-temperate vegetation. Its affinities with cold-temperate progenitors sets its apart from the other three deserts, which have almost exclusive ties with warm-temperate and tropical-subtropical archetypes. Major plant dominants having cold-temperate affinities are sagebrushes (*Artemisia*), saltbushes (*Atriplex*), and Winterfat (*Ceratoides lanata*). These distinctly cold-temperate dominants are joined in varying degrees by species having evolutionary ties with warmer climates. Included here are species of rabbitbrush (*Chrysothamnus*) blackbrush (*Coleogyne*), hopsage (*Grayia*), and horsebrush (*Tetradymia*) (Axelrod, 1950).

Rzedowski (1973) and MacMahon (1979) show by using coefficients of generic similarity that the affinities between the cold Great Basin desertscrub biome and the warm Sonoran and Chihuahuan desertscrub biomes are weaker than the ties between the last two deserts and the Argentine "Monte." That the intercontinental floristic links are stronger than the intracontinental ones illustrates that there is strong resistance for the evolutionary process at the generic level to transcend temperature boundaries.

The major series within this biome, both in the Southwest and farther north, are those dominated by various species of sagebrush (*Artemisia*) or by Shadscale (*Atriplex confertifolia*). In this region these two series may be joined by several of lesser extent including Blackbrush (*Coleogyne ramosissima*), Winterfat (*Ceratoides lanata*), Greasewood (*Sarcobatus vermiculatus*), or rabbitbrush (*Chrysothamnus*).

These principal scrub species are much-branched, nonsprouting, aromatic, semishrubs with soft wood and evergreen leaves. These shrubs are mostly without spines. There are few cacti—either in numbers of individuals or species. Those present tend to be of short stature or prostrate and include a few chollas (*Opuntia whipplei, O. pulchella*), prickly pears (*Opuntia polyacantha, O. gracilis, O. erinacea*), and hedgehog cacti (*Echinocereus triglochidiatus* var. *melanacanthus, E. fendleri* var. *fendleri*). Small cacti (*Pediocactus, Sclerocactus*) and *Echinocactus polycephalus* var. *xeranthemoides* occur in the more southern locales.

Species diversity is characteristically low in all major communities of this biome with a dominant shrub occurring to the virtual exclusion of other woody species. Another feature setting this desert apart from others of the region is the absence of characteristic desert plants in minor waterways; nor is there a fringe of more closely spaced upland plants along these habitats of slightly more favorable moisture conditions (Shreve, 1942). There are, however, both cosmopolitan and characteristic plants along flood plains of the larger waterways: included here are Greasewood (*Sarcobatus vermiculatus*), Four-wing Saltbush (*Atriplex canescens*), and New Mexican Forestiera (*Forestiera neomexicana*). The introduced Russian Olive (*Elaeagnus angustifolia*), and in the warmer regions Saltcedar (*Tamarix chinensis*), may be present along wetland stream channels.

As with the other deserts of North America, there is evidence that the Great Basin Desert evolved comparatively recently, perhaps only 5,000 to 12,000 years ago (Butler, 1976; Stutz, 1978). The evidence for this comes from a variety of sources. Of particular significance to the present discussion is the observation that two major genera of Great Basin desertscrub, *Artemisia* (section *Tridentatae*) and *Atriplex*, are

Table 18. Precipitation data from 14 stations in the Southwest within and directly adjacent to Great Basin Desertscrub.

Station Lat./Long.	Elevation (m)	J	F	M	A	M	J	J	A	S	O	N	D	Total
Bluff, UT 37°17' 109°33'	1,315	15	13	14	13	9	7	17	25	19	25	13	19	192
Cortez, CO 37°21' 108°34'	1,883	26	21	26	27	24	14	29	6	29	39	21	31	293
Alamosa, CO 37°27' 105°52'	2,297	6	7	9	16	16	13	30	29	18	18	6	9	176
Taos, NM 36°23' 105°36'	2,124	20	17	23	24	25	23	44	52	27	28	18	19	321
Aztec, NM 36°50' 108°00'	1,719	19	16	18	18	15	12	22	33	23	30	14	23	242
Farmington, NM 36°45' 108°10'	1,644	14	12	13	15	12	12	20	30	22	30	12	17	205
Fredonia, AZ 36°57' 112°32'	1,425	26	22	24	17	11	7	17	31	17	21	20	25	238
Page, AZ 36°56' 111°27'	1,302	7	11	17	8	11	5	9	18	17	17	11	15	146
Kayenta, AZ 36°44' 110°16'	1,725	14	13	13	10	9	8	28	34	15	26	11	13	194
Tuba City, AZ 36°08' 111°15'	1,504	11	9	15	9	9	6	17	25	16	17	10	14	157
Winslow, AZ 35°01' 110°44'	1,492	11	10	12	8	7	9	31	38	19	16	10	15	187
Ganado, AZ 35°43' 109°34'	1,932	17	16	20	17	10	11	40	46	28	32	16	23	277
Sanders, AZ 35°13' 109°20'	1,779	19	19	23	14	4	12	41	47	21	19	18	26	263
Saint Johns, AZ 34°30' 109°22'	1,747	15	14	20	11	9	11	52	53	31	26	10	18	270

extremely plastic groups, and there is strong evidence of a continuing fast tempo of evolution in both (Beetle, 1960; Stutz, 1978). Recent work on *Atriplex* suggests that the genus is "awesomely genetically rich" (Stutz, 1978). Perennial members of this genus are all possessors of the "Kranz" type of leaf anatomy, an anatomical "flag" denoting the C_4 photosynthetic pathway, a physiological condition adapting plants to hot, bright desert conditions. Much evidence points to northeastern Mexico as a likely evolutionary center from which most perennial North American *Atriplex* species have arisen (Stutz, 1978). That a species with a physiological adaptation for existence under hot, arid conditions does so well in the winter-cold Great Basin desertscrub, may be explained by the preadaptive tolerance to low temperatures that arose incidentally as the new taxa in Mexico evolved mechanisms to withstand the special conditions imposed by the gypsiferous soils so prominent in the area. The ability to accumulate salts as a means of surviving on highly saline soils may have provided the preadaptive ability to withstand low temperatures. The accumulated salts could act both as an osmotic adjustment for accommodating physiologic and climatologic drought and as an "antifreeze" in colder climates.

Great Basin desertscrub is located mostly north of the 36th parallel although the biome is represented south of that line along the Little Colorado River drainage in Arizona and New Mexico, where, in some places, it is referred to as the Painted Desert (e.g. Benson, 1969). It occupies roughly 59,570 km² at elevations mostly between 1,200 m and 2,200 m (sometimes even higher to 2,600 m). Most of the desert receives less than

250 mm of precipitation per year. Mean monthly precipitation shows a strong winter dominated pattern on the west, with a gradual shift eastward toward a stronger summer influence with wet and dry seasons less distinct than in the other deserts (Table 18).

Part of the area of summer rainfall dominance has approximately the same amount of winter precipitation as the drier parts of the Sonoran Desert (Tables 18 and 23). Northeastern Arizona, for example, receives monthly amounts for December, January, and February in excess of 50 mm only 1 year out of 25 (Sellers and Hill, 1974). These monthly amounts, based on values averaged for the entire area regardless of elevation, would be even less were only low-lying desert localities considered.

The low winter season precipitation is compensated for by relatively low temperatures. Maximum daily values may remain below freezing during many days of the three coldest months—December, January, and February. Minimum daily temperatures probably reach freezing or lower most nights during December and January, although nighttime freezing can occur in all but the warmest months of summer. Summer temperatures are moderately high. For much of the area, the spring season of increasing air temperatures is also a period of gradual decrease in available moisture supplies.

Sagebrush Series

The total area of the West now dominated by sagebrush has been estimated at 1,093,690 km² (Beetle, 1960). This estimate includes all biotic communities in which *Artemisia* species in

Figure 83. *Big Sagebrush* (Artemisia tridentata *var.* tridentata) *community on Fishtail Mesa, Grand Canyon National Park, Coconino County, Arizona, ca. 1,860 m elevation. This remote, inaccessible site has never been grazed by livestock (Jameson et al. 1962). The open growth of essentially one woody dominant and the lack of significant amounts of perennial grass is typical of much of this community in our area even in the absence of livestock grazing. The grasses present include such "cold season" species as Desert Needlegrass* (Stipa speciosa), *Indian Ricegrass* (Oryzopsis hymenoides) *and Longtongue Mutton Bluegrass* (Poa longiligula).

section Tridentatae are important members, and thus includes considerably more than Great Basin desertscrub. This large area is occupied by 18 closely related taxa of *Artemisia*, "each defining in its own way a different ecological area" (Beetle, 1960).

Sagebrush-dominated communities within the Southwest have as dominants mainly three species of the 18 taxa noted above: Big Sagebrush (*Artemisia tridentata* var. *tridentata*), Bigelow Sagebrush (*A. bigelovii*), and Black Sagebrush (*A. arbuscula* ssp. *nova*), although at least three other species occur here to a limited extent (Beetle, 1960). *Artemisia arbuscula* ssp. *nova* is much more restricted in occurrence in our area than are the other two species. *A. bigelovii* has a range that is generally more southerly and easterly than that of *A. tridentata* and within our region is the most widespread of the three species, occurring along the valleys and canyons of the Colorado, Little Colorado, and the San Juan rivers, and in the upper reaches of the Rio Grande. The range of *A. tridentata* frequently overlaps that of *A. bigelovii* but the former is missing from virtually all of the Little Colorado River drainage (Beetle, 1960).

Studies specifically outlining the ecologic roles of the different sagebrush types are lacking for our area but in general *A. tridentata* var. *tridentata* is found from 1,500 to 2,150 m on a variety of deep soil sites. *Artemisia arbuscula* ssp. *nova* is found on shallow soils from 1,500 m to 2,500 m elevation. *A. bigelovii* is found in canyons, gravelly draws and on dry flats at 1,800 m or lower (Beetle, 1960; Munz, 1974). In all cases these species extend beyond the limits of Great Basin desertscrub, occurring as subdominants in other communities such as Great Basin conifer woodland and Great Basin grassland.

Sagebrush communities are regarded by many as steppe or shrub steppe because of the usual importance of grasses (Young et al., 1976; Daubenmire, 1970). In the Columbia River area and elsewhere, grasses, if not eliminated by grazing, are important understory elements in distinctly shrub-steppe communities. Increasingly to the south, however, sagebrush may grow to the virtual exclusion of grasses even in areas that have never been grazed by domestic livestock (Vale, 1975; Jameson et al., 1962, Fig. 83) and, unlike

Figure 84. *Bigelow Sagebrush* (Artemisia bigelovii) *community on Boysag Point, Grand Canyon National Park, Coconino County, Arizona, ca. 1,675 m elevation. Boysag is a small promontory accessible only across a narrow, broken approach. The area received light sheep grazing intermittently between 1920 and 1943 but has not been grazed by livestock since. The introduced* Bromus rubens *(midground) was probably carried here during the grazing period and continues to reproduce locally even though the area has been free of grazing for 36 years (date of photograph, June 1979). Grass cover is roughly 8% (Schmutz et al., 1967) and is provided by such species as Black Grama* (Bouteloua eriopoda), *needlegrasses* (Stipa), *muttongrasses* (Poa), *and Hairy Tridens* (Tridens pilosus). Opuntia erinacea *and* Agave utahensis *are prominent succulent members of this desertscrub community. Photograph by R.M. Turner.*

Table 19. Percentage of plant cover (by line-intercept method) for Fishtail Mesa (from Jameson et al. 1962) and Boysag Point (from Schmutz et al. 1967).

	Fishtail Mesa (altitude 1,860 m)	Boysag Point (altitude 1,675 m)
Shrubs		
Artemisia tridentata var. *tridentata* and *A. bigelovii*	24.4	8.8
Opuntia (prickly-pear) species	0.6	1.1
Opuntia (cholla) species	—	0.0
Ephedra viridis	0.7	0.8
Gutierrezia sarothrae	1.1	0.3
Ceratoides lanata	0.4	0.3
Yucca baccata	0.0	0.1
Atriplex canescens	0.1	0.1
Mammillaria species	—	0.2
Fallugia paradoxa	0.0	0.1
Coleogyne ramosissima	0.8	0.0
Eriogonum species	—	0.7
Polygala rusbyi	—	0.0
Chrysothamnus greenei	0.0	1.0
Agave utahensis	0.0	0.3
Grasses		
Bouteloua eriopoda	0.0	2.3
Bouteloua gracilis	—	0.2
Hilaria jamesii	0.0	1.8
Poa species	—	1.2
Stipa species	—	2.6
Sporobolus cryptandrus	0.0	0.5
Oryzopsis hymenoides	—	—
Erioneuron pilosum	0.0	—
Sitanion hystrix	0.0	—
Poa secunda	—	0.0
Forbs	—	0.3

— = less than 0.1 percent.

related vegetation to the north, sagebrush communities lacking a significant graminoid component are not necessarily in a grazing disclimax (Young et al., 1976). The near absence of grasses has been attributed to strictly climatic controls— i.e., the timing and amount of precipitation—the paucity of grasses being correlated with low annual precipitation, falling predominantly in the winter (Christensen, 1959; Cronquist et al., 1972). Therefore, sagebrush is recognized as occurring in both a Great Basin grassland and in a Great Basin desertscrub.

Plant species found on two sagebrush sites with little or no livestock grazing are shown in Table 19. One of the sites, Fishtail Mesa (Jameson et al., 1962), is an ungrazed island plateau in northwestern Arizona (Fig. 83) and the other site, Boysag Point (Schmutz et al., 1967; Fig. 84), is a relatively ungrazed promontory of the Kaibab Plateau, 12 miles to the southwest. The presence of such grasses as gramas (*Bouteloua*), Galleta (*Hilaria jamesii*), and Desert Needlegrass (*Stipa speciosa*) mark these stands as a southern fasciation of the Great Basin desertscrub (Young et al., 1977). These communities provide valuable references against which to judge grazing effects. It is noteworthy that the stands differ greatly in both the amount of sagebrush and the amount of perennial grasses present. At Boysag Point, the total cover of perennial grasses (8.6%) was as great as the sagebrush cover (8.8%); at Fishtail Mesa the perennial grass coverage was only 0.12%, that of the sagebrush, 24.39%.

Seral communities within the sagebrush series are heavily influenced by one or two forces,—grazing and fire. Sagebrush foliage is not readily eaten by domestic or native ruminants and contains oils that inhibit microbial activity in rumens (Nagy et al., 1964). Avoidance of sagebrush results in reduction of the more palatable grasses and forbs and an increase in sagebrush. Since approximately 1900, introduced annuals have become increasingly conspicuous throughout this biome and must be considered prime forces in arresting succession in many areas. These annuals, largely from Eurasia, filled a near void when introduced, beginning about 1900. Before

Figure 85. *Shadscale (Atriplex confertifolia) community in the basin of the Little Colorado River on the Navajo Indian Reservation near Cameron, Arizona, ca. 1,280 m elevation. This is another example of shrub dominance by a single low shrub species in this cold-arid biotic community.*

that time, domestic livestock had greatly reduced the herbaceous component of this community. Under conditions of heavy grazing, native annuals were unable to exist and alien species, once introduced, quickly occupied the abandoned niche (Young et al., 1972, 1976). These new species include Cheatgrass Brome (*Bromus tectorum*), Russian thistles (*Salsola paulsenii* and *S. iberica*), Filaree (*Erodium cicutarium*), and Tumble Mustard (*Sisymbrium altissimum*). All originated in Eurasia and have accompanied domestic livestock herds here and throughout semi-arid parts of the world. Once these aliens have taken their place in the disclimax community, they relinquish this position to native species slowly or not at all, even if grazing pressures are eliminated (Fig. 84). Cheatgrass Brome has spread to become a common codominant of seral stands of sagebrush and other communities in this biome and, indeed, in many stands it may be considered a disclimax dominant.

Although the history of fire in the Great Basin Desert has not been specifically studied in this area, it has been much studied to the north, and findings there should be applicable elsewhere. Fire has been considered a minor force in presettlement Great Basin Desert areas (Galbraith and Anderson,

1971), yet during the last 50 years it has become a potent force in shaping communities in this biome. Where Cheatgrass Brome has become a significant constituent of sagebrush stands, the incidence of fire is greatly increased. Because sagebrush does not sprout after burning, it is slow to reoccupy a site after fire. The highly flammable Cheatgrass thus plays a double role in altering the stands of sagebrush it has come to occupy: under grazing pressure it increases in density, effectively closing the ecosystem to native perennial grasses; the increased fuel it creates when dry promotes more frequent fires not withstood by sagebrush. The two-phase system of disturbances has placed much of the sagebrush zone in a stage of arrested succession (=Cheatgrass Brome fire disclimax).

After fire, the role of the nonsprouting sagebrush may eventually be taken by sprouting species of such genera as *Chrysothamnus*, *Tetradymia*, and *Gutierrezia*. These more fire adapted plants arose from neotropical ancestors in the Madro-Tertiary geoflora; the sagebrush they replace is from the Arcto-Tertiary geoflora. The seral sequence, progressing from plants in the fire disclimax with tropical or warm temperate affinities to cold temperate climax species, is often

Table 20. Presence of woody and succulent plants at 9 widely distributed shadscale stands in southern Nevada and southern California (from Billings, 1949).

Species	Number of Stands
Atriplex confertifolia	9
Artemisia spinescens	7
Ephedra nevadensis	5
Forsellesia spinescens	5
Grayia spinosa	3
Lycium andersonii	3
Yucca brevifolia	3
Ceratoides lanata	2
Tetradymia glabrata	2
Ambrosia dumosa	2
Ephreda funerea	2
Lycium sp.	2
Sarcobatus baileyi	1
Lycium cooperi	1
Dalea polyadenia	1
Hymenoclea fasciculata	1
Tetradymia spinosa	1
Kochia americana	1
Opuntia bigelovii	1
Echinocactus polycephalus	1
Eriogonum umbellatum	1
Lepidium fremontii	1
Yucca schidigera	1

repeated in Great Basin communities (Young et al., 1976). A similar example of seral shifts in composition from warm to cold adapted species was noted by Schmutz et al. (1967): "cool season" versus "warm season" grass cover occurred at a ratio of 44 to 56 in a relatively ungrazed sagebrush community (Boysag Point) compared with a ratio of 20 to 80 in a nearby heavily grazed sagebrush stand. "This strange union of contrasting evolutionary lines in one seral sequence...leaves open to question the evolutionary equilibrium of the pristine flora" (Young et al., 1976; see also Fig. 88).

Shadscale Series

Shadscale (Atriplex confertifolia) is a wide ranging saltbush found from Chihuahua and the panhandle of Texas (Correll and Johnston, 1970) northward to northeastern Montana and eastern North Dakota (Branson et al., 1967) and from southern California to northeastern Oregon. Except for its absence in Washington, it occurs in all states in which extensive stands of sagebrush occur, and it is only in and adjacent to the Great Basin Desert that it is found as a wide ranging dominant.

The largest continuous area of Shadscale dominance is in southeastern California and Nevada, extending from west-central Nevada (north of our area) to west of Death Valley, California (Billings, 1949; Cronquist et al., 1972). Billings (1949) considered Shadscale vegetation in this part of its range to be transitional between sagebrush and the Creosotebush series of the Mohave desertscrub biome. He noted that on some mountain ranges these same zones appear on a smaller scale with Shadscale occupying a belt intermediate between Creosotebush below and sagebrush above. Eastward, extensive areas of Shadscale occur within Great Basin desertscrub along drainages of Colorado River tributaries such as the Little Colorado, San Juan, and Green Rivers in Arizona, Utah, New Mexico, and Colorado (Figs. 85, 86). In the last two

states it also occurs in the upper Rio Grande basin.

In southern Nevada and southeastern California, the Shadscale series is found from 990 m to 1,775 m (Billings, 1949). Eastward and northward this vegetation is found mainly between 1,220 m and 1,525 m.

The general appearance of this community is one of open starkness with the dominant woody plants attaining heights of only 0.3 m to 0.6 m (Figs. 85, 86). Stands of this plant in Tooele Valley, Utah, were aptly described by Kearney et al. (1914): "No other vegetation in this valley gives the impression of being so nearly conquered by the environment. Even the few species which grow on the salt flats have the appearance of finding their habitat more congenial."

Few quantitative studies of Shadscale communities have been made in the Southwest. In one such study, Billings (1949) tallied all woody and succulent plants found at nine widely distributed stations in the western transition to Mohave Desert. These are listed in Table 20.

In two quantitative studies in the Great Basin Desert north of this area in Colorado (Branson et al., 1967) and in Utah (West and Ibrahim, 1968), the only shrub besides Shadscale found on plots examined by Billings (1949) was Bud Sagebrush (Artemisia spinescens). In addition to the above species, other important shrubs found throughout the Shadscale series in Great Basin desertscrub include:

Shadscale (Atriplex confertifolia)
Gardner Saltbush (Atriplex gardneri)
Nuttall Saltbush (Atriplex nuttallii)
Greene Rabbitbrush (Chrysothamnus greenei)
Rubber Rabbitbrush (Chrysothamnus nauseosus)
Broom Snakeweed (Gutierrezia sarothrae)
Black Greasewood (Sarcobatus vermiculatus)
Seep Weed (Suaeda fruticosa)

Although widely scattered, perennial grasses are commonly found in the Shadscale community; few species are represented and include mainly Galleta (Hilaria jamesii), Indian Rice Grass (Oryzopsis hymenoides), Bottlebrush Squirreltail (Sitanion hystrix), Desert Needlegrass (Stipa speciosa), and Alkali Sacaton (Sporobolus airoides),—again showing the Great Basin affinity of this biome.

Much of the early work on this community was done in habitats where subsoil accumulations of salts were common (Clements, 1920; Kearney et al., 1914; Shantz and Piemeisel, 1940). As a result, an early consensus developed that Shadscale usually indicated a saline soil,—a belief that has been shown to be incorrect by more recent studies (Billings, 1949; West and Ibrahim, 1968; Branson et al., 1976). Although often occurring in valleys, the plant is frequently also found on upland sites; for example, on pediment remnants in southeastern Utah (West and Ibrahim, 1968), on rolling hills in northern Arizona, and on rocky slopes in southeastern California (Vasek and Barbour, 1977).

Soils within the Shadscale series are often covered with desert erosion pavement. The rocks of this covering are commonly dark in color and contrast sharply with the lighter soil beneath (Billings, 1951).

Shadscale grows where the precipitation may be considerably lower than that of adjacent sagebrush. In Nevada and eastern California, annual precipitation ranges from 78.2 mm to 144.3 mm for the series (Billings, 1949). At six stations in the adjacent sagebrush vegetation, precipitation ranged from 196.3 mm to 240.3 mm with an average for the series of 223.3

Figure 86. *Mixed shrub community of Great Basin Desertscrub, Valley of the Gods, San Juan County, Utah, ca. 1,400 m elevation. Blackbrush* (Coleogyne ramosissima), *the dominant foreground plant, gives way downslope to Shadscale* (Atriplex confertifolia). *The dark narrow column of plants along the watercourse is composed of Greasewood* (Sarcobatus vermiculatus). *Additional plant species here include* Ephedra torreyana, Gutierrezia *sp.,* Hilaria jamesii, *and* Bromus tectorum. *Photograph by R.M. Turner.*

mm. As an example of rainfall conditions farther to the east, Ibrahim et al. (1972) reported an average annual rainfall of 156.2 mm for a large Shadscale area in southeastern Utah.

That Shadscale is characteristic of the driest Great Basin desertscrub region is shown by Figure 85. A large low rainfall (150-200 mm) "sink" (including Greenriver, Utah; Shiprock, New Mexico; Lees Ferry and Winslow, Arizona) circumscribes much of the Shadscale land in the north-central part of the North American Southwest. Farther west, the vast rainshadow in the lee of the Sierra Nevada is an area of low precipitation also largely dominated by Shadscale. On a broad scale, because low precipitation is correlated with low altitude, Shadscale generally occurs below sagebrush.

Where the Shadscale community occurs in a Mohave desertscrub-Great Basin desertscrub transition zone, as for example at the Nevada Test Site (Beatley, 1975), Shadscale occupies the floors of enclosed basins and Creosotebush occurs next above at the higher positions in the basin, in a reversal of the usual zonation sequence. The lower tempera-

tures of the bottom lands were inferred to have excluded Creosotebush from these sites.

Like the Sagebrush series, the Shadscale series has been widely used for grazing livestock. Certain associated shrubs such as *Artemisia spinescens* and *Ceratoides lanata* are palatable as are most of the perennial grasses. It is used mostly as a winter range for sheep, but cattle are grazed year-long where water is available. Cheatgrass Brome is found in those parts of this series with highest rainfall, but this exotic species apparently requires more moisture than is available throughout the drier sections (Billings, 1949).

Blackbrush Series

In addition to sagebrush- and Shadscale-dominated areas, some southern parts of the Great Basin desertscrub biome are dominated by Blackbrush (Fig. 86). Very little has been written about this community, which occurs primarily in southern Nevada, southeastern California, northcentral Arizona, and southeastern Utah. Beatley (1976) considered it as

Figure 87. *Blackbrush* (Coleogyne ramosissima) *community near Indian Gardens, Grand Canyon National Park, Coconino County, Arizona, growing on soil derived from Cambrian Bright Angel Shale, ca. 1,125 m elevation. Beginnings of the Tonto Platform can be seen on the right. This broad flat shelf is a prominent feature of the Grand Canyon landscape and is formed because the Bright Angel Shale is highly susceptible to erosion. The Tonto Platform, straddling the Mohave Desert-Great Basin Desert "tension line" is almost uniformly dominated by Blackbrush. The foreground area is watered by a spring and supports a riparian scrubland of Common Reed* (Phragmites communis) *and Coyote Willow* (Salix exigua). *Photograph by D.T. MacDougal, 1905.*

belonging to the Mohave desertscrub biome but recognized it as ecotonal to Great Basin and Mohave desertscrub communities. Cronquist et al. (1972), however, noted that this community is best developed along the valleys of the Colorado and lower Green Rivers in southeastern Utah, an area clearly within Great Basin desertscrub. It is found in extensive stands on the Tonto Platform in the Grand Canyon (Fig. 87). Hackman (1973), Hunt el al. (1953), and Everitt (1970) showed this vegetation covering extensive areas in the San Rafael, Fremont, and Dirty Devil River drainages in southeastern Utah. Its presence in both Mohave desertscrub and Great Basin desertscrub mosaics is a condition similar to that of other communities (e.g., Creosotebush-White Bursage and Mesquite) which transcend biome boundaries but whose associates differ when passing from one biome to another.

Bowns and West (1976) have provided the most compre-

hensive discussion of ecological and morphological information about Blackbrush. They note that fire effectively destroys this species and the plant does not aggressively reoccupy these sites. The post-burn communities are variable, although *Bromus rubens* and *B. tectorum* are common pioneer species (Beatley 1966, 1976). The replacement shrubs are usually undesirable as forage for domestic livestock (Bowns and West, 1976). Perennial grasses are commonly prevalent in unburned stands and Bradley (1964) considered this series an important vegetation type in the Mohave Desert for Desert Bighorn Sheep (*Ovis canadensis nelsoni*) because of the occurrence of these grasses (e.g., *Oryzopsis, Hilaria*).

Beatley (1976) concluded that temperature conditions alone do not separate this vegetation from adjacent Creosotebush communities. Rainfall, however, did seem to be an important controlling factor with Blackbrush occupying sites

Figure 88. *Sand Sagebrush* (Artemisia filifolia) *stand in contact with Mohave Desertscrub species such as Creosotebush* (Larrea tridentata, *large plant at left), 2 miles east of Washington, Utah, ca. 900 m elevation. In 1914 (upper photo), Sand Sagebrush was the clear dominant. By 1979 (lower photo), Creosotebush was the dominant, raising the question of what environmental conditions had changed to cause the transfer of dominance from a plant with cold-temperate affinities to one with sub-tropical affinities. Upper photograph by H.L. Shantz; lower by R.M. Turner.*

Figure 89. *Saltbush series of Greasewood (*Sarcobatus vermiculatus*), on Monte Vista National Wildlife Refuge in San Luis Valley, Colorado, ca. 2,320 m elevation. Low precipitation, saline soils, and cold nighttime temperatures extend Great Basin desertscrub eastward to the high, rain-shadowed basin of the upper Rio Grande. Greasewood is one of the few dark green shrubs among the mostly gray shrubs of this desert biome. It and other salt tolerant plants, such as* **Allenrolfea**, **Atriplex**, *and* **Suaeda**, *range over a wide amplitude of environments from cold-temperate to subtropic, their occurrence primarily dependent upon soil salinity and occasional flooding.*

where rainfall exceeds the needs of many Mohave desertscrub species. In cold drainage basins, Blackbrush is found only on upper bajadas and is excluded from the basin floors under the influence of cold air accumulation. Low temperatures probably control its upper limits on bajadas and northward where it abuts Shadscale or sagebrush communities.

In southern Nevada, crown coverage reaches values of 45-51% in nearly pure stands of Blackbrush, the highest coverage values of any desertscrub communities of the region, and is probably indicative of the relatively high rainfall of the sites it dominates (Beatley, 1976). The physiognomy of this community is uniform as is typical of many Great Basin desertscrub series. Widely scattered individuals of the following species may be present (Cronquist et al., 1972):

Sand Sagebrush (*Artemisia filifolia*)
Parry Sagebrush (*Artemisia parryi*)
Big Sagebrush (*Artemisia tridentata* var. *tridentata*)
Shadscale (*Atriplex confertifolia*)

Rough Joint-fir (*Ephedra nevadensis*)
Torrey Joint-fir (*Ephedra torreyana*)
Bush Buckwheat (*Eriogonum fasciculatum*)
Three-leaved Snakeweed (*Gutierrezia microcephala*)
Diamond Cholla (*Opuntia ramosissima*)
Stenopsis (*Stenopsis linearifolius*)

Other Series

Within Great Basin desertscrub are several additional plant communities imbedded within the matrix of the more widespread Sagebrush, Shadscale, and Blackbrush series. The occurrence of these minor communities appears largely dependent upon edaphic conditions rather than being the result of particular climatic conditions. The dominant species of these minor communities are all members of two key Great Basin desertscrub plant families — Compositae and Chenopodiaceae.

Sand Sagebrush (*Artemisia filifolia*) is a widespread species occurring from Nevada to Chihuahua and north as far as Wyoming and Nebraska. It is a plant of sandy soils, often occupying dunes or areas of deep, loose sand. It is found and is locally dominant in regions representative of Great Basin desertscrub, Mohave desertscrub (Fig. 88), Plains grassland, and semidesert grassland.

Common occupants of valley bottoms and playa margins are stands representing a saltbush series. The dominant species of this series is often Greasewood (*Sarcobatus vermiculatus*) (Fig. 89), a plant able to tolerate high salt content (Branson et al., 1976) but it is not an infallible indicator of high salinity; its occurrence may be dependent solely upon associated high soil moisture content (Fautin, 1946; Branson et al., 1976).

Plants that locally grow with Greasewood or are themselves dominants are species of *Atriplex*, particularly Fourwing Saltbush (*Atriplex canescens*), Fivehook Bassia (*Bassia hyssopifolia*), Inland Saltgrass (*Distichlis spicata* var. *stricta*), Common Russian Thistle (*Salsola kali*), and seepweeds (*Suaeda* spp).

Winterfat (*Ceratoides lanata*) is another species that has broad ecological amplitude, occurring mainly in the Shadscale region but extending into the arid limits of sagebrush and even the mesic limits of the Creosotebush series (Mohave desertscrub) (Cronquist et al., 1972). Germination of its seeds has been shown to have broad genetically defined responses to temperature and salinity, helping to explain the broad ecologic amplitude of the plant (Workman and West, 1967).

A distinct fauna is centered in Great Basin desertscrub. Several distinctive mammalian representatives follow the sagebrush communities of this biome into the boundaries for the Southwest: Townsend's Ground Squirrel (*Spermophilus townsendi*), Dark Kangaroo Mouse (*Microdipodops megacephalus*), and Sagebrush Vole (*Lagurus curtatus*). Others as the Pallid Kangaroo Mouse (*Microdipodops pallidus*) and Chisel-toothed Kangaroo Rat (*Dipodomys microps*) appear to favor the *Atriplex* and other desertscrub series. Still others such as the Townsend Pocket Gopher (*Thomomys townsendi*), Belding Ground Squirrel (*Citellus beldingi*), and Pygmy Rabbit (*Sylvilagus idahoensis*) do not reach the area. The Merriam's Shrew (*Sorex merriami*), Great Basin Pocket Mouse (*Perognathus parvus*), Ord's Kangaroo Rat (*Dipodomys ordii*), and the Montane Vole (*Microtus montanus*) are centered in, but not restricted to, this biome and are best represented at the desert's higher altitudes. A number of mammals, e.g., the Coyote (*Canis latrans*) and Black-tailed Jackrabbit (*Lepus californicus*), are more or less influent and found throughout this and other western biomes.

Unlike the cold temperate deserts of Eurasia with its onagers, camel, and gazelles, large ungulates are generally poorly represented in the Great Basin Desert. The Pronghorn (*Antilocapra americana*) occurs largely as an incursionary species from adjacent or former grassland. The Desert Bighorn (*Ovis canadensis nelsoni*) is now generally lacking in this biome—probably because the cold winters encouraged a history of domestic sheep grazing in these habitats with their attendant introduction of diseases transmittable to the resistance-poor native sheep.

As some of their names imply, several birds are characteristic of sagebrush communities—Sage Thrasher (*Oreoscoptes montanus*), Sage Sparrow (*Amphispiza belli*), and Sage Grouse (*Centrocercus urophasianus*). This last species, while well adapted and restricted to Big Sagebrush communities, also requires insect-rich wet meadows or interspersion of grassland for nesting and rearing young. The widely introduced Chukar Partridge (*Alectoris chukar*) has been successfully established largely in rocky precipitous habitats within Great Basin desertscrub. There it feeds almost' entirely on the invading winter-spring growing annuals of its Eurasian homelands—*Salsola*, *Erodium*, and *Bromus tectorum*.

Because of its long, cold winters, reptiles are not as well represented in the Great Basin Desert as in the warmer biomes. Some of the more common representative species such as the Sagebrush Lizard (*Sceloporus graciosus*) and the Great Basin Spadefoot Toad (*Scaphiopus intermontanus*) occur throughout other Great Basin biomes whereas others present, as the Leopard Lizard (*Crotaphytus wislizenii*), Collared Lizard (*C. collaris*), and Northern Side-blotched Lizard (*Uta stansburiana*), are found in several other biomes as well. A number of subspecies are indicative of Great Basin desertscrub and a recent history of evolutionary separation. These include the Northern Desert Horned Lizard (*Phrynosoma platyrhinos platyrhinos*), Great Basin and Northern Whiptails (*Cnemidophorus tigris tigris* and *C. tigris septentrionalis*), Great Basin and Northern Plateau Fence Lizards (*Sceloporus occidentalis biseriatus* and *S. undulatus elongatus*), Great Basin Gopher Snake (*Pituophis melanoleucus deserticola*), Wandering Garter Snake (*Thamnophis elegans vagrans*), and the Great Basin and Hopi Rattlesnakes (*Crotalus viridis luteosus, C. viridus nuntius*).

Warm-Temperate Desertlands

All or parts of the Sahara, Arabian, Iranian, Indian, Atacamian, and North American deserts are warm-temperate. Those warm-temperate extensions of the North American Desert are best known as the Mohave and Chihuahuan deserts. The former is the smallest of the subdivisions of the North American Desert and is characterized by predominantly winter rainfall. The Chihuahuan Desert is the larger and receives most of its scanty precipitation during summer months. Almost all the Mohave Desert is located within the Southwest mapped area, whereas much of the Chihuahuan Desert lies largely outside these boundaries, which do not include all of the North American Southwest.

153.1 Mohave Desertscrub

The Mohave desertscrub biome, the smallest of the four North American desertland biomes, is intermediate between Great Basin desertscrub and Sonoran desertscrub. It intervenes between these two biomes, both spatially and floristically. Its fauna, although also shared, is more representative of the Lower Colorado subdivision of the Sonoran Desert. Main plant dominants of this biome include series of Creosotebush (*Larrea tridentata*), All-scale (*Atriplex polycarpa*), Brittlebush (*Encelia farinosa*), Desert Holly (*Atriplex hymenelytra*), and White Burrobrush (*Hymenoclea salsola*), all of which occur as dominants in Sonoran (and, in the case of *Larrea*, Chihuahuan) desertscrub series. Shadscale (*Atriplex confertifolia*) and Blackbrush (*Coleogyne ramosissima*) may dominate either Mohave desertscrub or Great Basin desertscrub biomes. The arboreal leaf succulent Joshua Tree (*Yucca brevifolia*) provides one of the best examples of the few Mohave endemics. Of these species, all but the *Atriplex* have close evolutionary ties with warm temperate and tropic-subtropic vegetation.

Until recently, hypotheses for desert evolution were formulated from meager fossil evidence in combination with inferences based upon cliseral relations between the deserts and adjacent communities, and by the presence of relics within or adjacent to desert areas (Clements, 1936; Cain, 1944; Axelrod, 1950). Refinement of these early views came with the use of fossil pollen extracted from arid region sediments (Martin and Mehringer, 1965).

The history of North American desert development was known in only the broadest terms. Earlier deserts, some believed, occupied far less area than today and were thought to have been restricted to a refugium near the mouth of the Colorado River during the full glacial, 20,000 years ago (Martin and Mehringer, 1965).

Through the new technique of fossil packrat (*Neotoma*) midden analysis (Wells and Jorgensen, 1964), a rather clear picture is emerging of shifts in dominance of the major desert species. Although, as surmised earlier, the area of extreme aridity was once probably much smaller, the plants characteristic of present deserts were sorted into assemblages no longer common or even known (Van Devender and Spaulding, 1979). Thus, in the Ajo Mountains, Arizona (Lat. 32° 07', Long. 112° 42' W; altitude 975 m) in an area of Sonoran desertscrub today, there occurred together during the period 13,000 to 20,000 years before the present such species as Big Sagebrush (*Artemisia tridentata*, now typical of Great Basin desertscrub), Joshua Tree (*Yucca brevifolia*, now typical of Mohave desertscrub), and Shrub Live Oak (*Quercus turbinella*, now typical of interior chaparral). These findings and others (Van Devender and Spaulding, 1979) suggest that a broadscale sorting of southwestern vegetation occurred and that recent conditions are far more stressful than those during the period of late Wisconsin climate. The desertscrub biomes of today have arisen during relatively recent times, through a sorting of species from preexisting more generalized vegetation.

Recognition of the Mohave Desert as a distinct desert apparently arose when contrasting it with the adjacent Colorado Desert (Shreve's Lower Colorado subdivision of Sonoran Desert) of extreme southeastern California (Parish, 1930; Munz, 1935). When viewed within this limited provincial scope, elevation of Mohave desertscrub to biome status paralleling that of the other desertscrub biomes within

Figure 90. *Mohave desertscrub east of Spring Valley Mountains, Clark County, Nevada, ca. 1,220 m elevation. A Creosotebush* (Larrea tridentata)-*Bursage* (Ambrosia dumosa) *series with yuccas present* (Yucca schidigera, Y. baccata, Y. brevifolia). *These open communities, with yuccas either present or absent, characterize many square miles of gravel bajadas and outwash plains in the Mohave Desert of southeastern California, southern Nevada, extreme southwestern Utah and northwestern Arizona.*

California is understandable. However, when viewed in the larger context of the North American Southwest, its recognition as a separate biome is less easily justified.

Separation of the Mohave along its boundary with Sonoran desertscrub, with which it is most closely allied, is commonly blurred because distinct coincidental breaks in indicator species' ranges are lacking. Furthermore, there are areas within Sonoran desertscrub with important Mohave desertscrub species. For example, in northern Baja California, the characteristic Mohave Yucca (*Yucca schidigera*) is at its southern limit (Hastings et al., 1972). A number of other species that occur widely in both the Mohave Desert and the Sonoran Desert are also found here (e.g., *Eriogonum fasciculatum. Encelia farinosa*). This Vizcaíno Sonoran Desert vegetation is rich in low rosette-forming leaf succulents (*Agave, Yucca, Dudleya*), a growth form that is also important in many parts of the Mohave Desert (Fig. 90). Because of long established usage and its warm temperate climate, we continue recognition of the Mohave desertscrub as a distinct entity at the biome level, realizing that a strong case can be made for inclusion of its biota within the Sonoran desertscrub biome.

The Mohave desertscrub biome is contained in the northwestern part of our area, occupying perhaps 1/16 of the North American Southwest. On the north and west its boundaries are sharply defined; here it abuts upon Great Basin desertscrub, Great Basin conifer woodland, and Californian coastal

chaparral vegetation. Along its northeastern boundary in southwestern Utah, its margin is also usually sharply defined as a result of the sharp elevational gradients found there, and Mohave desertscrub gives way to Great Basin desertscrub at elevations of about 900-1,050 m as *Larrea* wanes and *Artemisia* or its associates assume dominance (Fig. 88). Southward, this easier distinction becomes blurred and one encounters anomalous combinations of "indicator" species from Mohave desertscrub, Sonoran desertscrub, Great Basin desertscrub, and Great Basin conifer woodland. For example, Mohave desertscrub in the vicinity of the Cerbat Mountains in Mohave County, Arizona is an assemblage which includes species that are mainly from Mohave desertscrub (*Yucca schidigera*), mainly from Great Basin desertscrub (*Coleogyne ramosissima, Tetradymia canescens, Ephedra torreyana*), a species that ranges widely through Mohave, Sonoran, and Chihuahuan desertscrub (*Larrea tridentata*), and a species that is predominantly of the Sonoran and Chihuahuan desertscrub (*Acacia greggii*) (Shreve, 1942c: Plate 4).

To the south, near the Hualapai Mountains, there occurs a different but equally anomalous combination including *Yucca brevifolia* (typical of Mohave desertscrub), Saguaro (*Carnegiea gigantea*, typical of Sonoran desertscrub), and California Juniper (*Juniperus californica*, typical of Great Basin conifer woodland) (Shreve, 1942c: Plate 3).

Generally, however, the Mohave species are best represented on the level (colder) plains; the Sonoran species favor

Figure 91. *Mohave Desert-Sonoran Desert transition at Mopah Springs in the Turtle Mountains, San Bernardino County, California, ca. 745 m elevation. The springs occur in the foreground arroyo and support five mature palms (***Washingtonia filifera***) and many seedlings. In addition, halophytic species such as ***Atriplex*** spp. and ***Distichlis spicata*** var. stricta occur in the moist soils that are heavily charged with minerals. The surrounding vegetation is Mohave desertscrub (Creosotebush-Bursage series); the fan palm and the ***Cercidium floridum*** along the drainage (midground center) are Sonoran desertscrub species here at the northern margin of their ranges (Brown et al., 1976; Hastings et al., 1972). The dominants on the slopes are ***Larrea tridentata***, ***Ambrosia dumosa***, and ***Yucca schidigera***. Photograph by R.M. Turner.*

the bajadas and hills. To the south and west from the Hualapai Mountains, where Mohave desertscrub contacts first the Arizona Upland and then the Lower Colorado Valley subdivisions of the Sonoran desertscrub biome, boundaries separating the two are increasingly difficult to discern. This is especially so on both sides of the Colorado River near Needles, California, and southwesterly to the Eagle Mountains (Fig. 91). Along this entire interface, definition of the boundary separating these two desertscrub biomes is largely arbitrary "regardless of whether one relies on physiographic, climatic, or biotic criteria" (Johnson, 1976). Westward from the Little San Bernardino Mountains, the Mohave boundary extends beyond the Sonoran Desert to its contacts with Californian coastal chaparral and Great Basin conifer woodland communities in the San Bernardino, San Gabriel, Tehachapi, and El Paso Mountains.

Species that serve to separate Sonoran from Mohave desertscrub include such widespread Sonoran species as Ironwood (*Olneya tesota*), Blue Palo Verde (*Cercidium floridum*), and Chuparosa (*Justicia californica*). Sonoran desertscrub species that there may be used to separate the two desert areas but elsewhere make slight incursions into Mohave desertscrub are American Trixis (*Trixis californica*) and Teddy

Bear Cholla (*Opuntia bigelovii*). Additional Sonoran plants are Bitter Condalia (*Condalia globosa*), Emory Dalea (*Psorothamnus emoryi*), Smoketree (*P. spinosa*), Longleaf Ephedra (*Ephedra trifurca*), Crucifixion Thorn (*Canotia holacantha*), Western Honey Mesquite (*Prosopis glandulosa* var. *torreyana*), and Jojoba (*Simmondsia chinensis*). In eastern California the northern limits of these species coincide roughly with a shift to Mohave taxa (Hastings et al., 1972; Johnson, 1976). Mohave taxa of indicator significance include Spiny Menodora (*Menodora spinescens*), sages (*Salvia funerea, S. mohavensis*), Desert Senna (*Cassia armata*, chiefly Mohavean), Mohave Dalea (*Psorothamnus arborescens*, chiefly on granitic and volcanic soils), Fremont Dalea (*P. fremontii*, on calcareous soils), Goldenhead (*Acamptopappus shockleyi*), Scalebroom (*Lepidospartum latisquamum*), and *Ephedra funerea* (Benson, 1957; Hastings et al., 1972).

Cacti are well represented in Mohave desertscrub. Besides a number of more widely distributed species, there are several indigenous taxa centered in this biome: a variety of Engelmann Hedgehog (*Echinocereus engelmannii* var. *chrysocentrus*), Silver Cholla (*Opuntia echinocarpa*), Mohave Prickly Pear (*Opuntia erinacea*), Beavertail Cactus (*O. basilaris*), Many-headed Barrel Cactus (*Echinocactus polycephalus*), and

Table 21. Data for 17 weather stations in the Mohave desertscrub biome. Data for each station include latitude and longitude (values rounded to closest 0.1°C), altitude (meters), average annual precipitation (mm), and abbreviated name as used in climographs of Fig. 92. Data from Ruffner (1978), U.S. Environmental Data Service (1959-1979) and Laboratory of Climatology (1975).

Station name	Abbreviated name	Latitude	Longitude	Altitude (m)	Average annual precipitation (mm)
Barstow, CA	BAR	34.9	117.0	658	104.9
Beaverdam, CA	BEDA	36.9	113.9	570	168.7
Boulder City, NV	BOCI	36.0	114.9	769	130.8
Daggett, CA	DAG	34.9	116.8	585	92.4
Davis Dam, CA	DADA	35.2	114.6	201	106.4
Death Valley, CA	DEVA	36.5	116.9	−58	46.2
Desert NWRCC, NV	DES	36.4	115.4	890	103.7
Iron Mtn., CA	IRO	34.1	115.1	281	67.8
Las Vegas, NV	VEG	36.1	115.2	658	95.6
Phantom Ranch, AZ	PHRA	36.1	112.1	783	213.1
Pierce Ferry, AZ	PIFE	35.9	114.1	1,177	252.8
Randsburg, CA	RAN	35.4	117.6	1,088	143.5
St. George, UT	STGE	37.1	113.6	841	192.0
Searchlight, NV	SEA	35.5	114.9	1,079	168.2
Trona, CA	TRO	35.8	117.4	516	93.8
Twenty Nine Palms, CA	TWE	34.1	116.0	601	96.2
Victorville, CA	VIC	34.5	117.3	871	119.9
Subdivision Average					129.2

Neolloydia johnsonii. A few others, mostly of subspecific rank—Buckhorn Cholla (*Opuntia acanthocarpa* var. *acanthocarpa*), *Opuntia whipplei* var. *multigeniculata*, Parish Cholla (*Opuntia stanleyi* var. *parishii*), and *Coryphantha vivipara* var. *deserti*—are restricted to it.

The Mohave Desert is especially rich in ephemeral plants, many of which are endemic. Of the approximately 250 Mohave taxa with this life form, perhaps 80-90 are endemic to that biome (Shreve and Wiggins, 1964). These short-lived plants may be placed in two phenologic groups: those that are attuned to winter conditions and those that germinate and grow in response to summer conditions. As might be expected, most of the Mohave species are winter annuals. In southern Nevada, Beatley (1974b) found only seven plant species in the summer annual flora and these are unknown from most undisturbed communities below roughly 1,200 m elevation. Johnson (1976) reported for the entire Mohave Desert only 17 summer annuals and none are endemic. These usually germinate in response to August or September rain and remain physiologically active only until the occurrence of subfreezing temperatures, a period usually measured in terms of only a few weeks.

In southern Nevada, Beatley (1974b) found that mass seed germination of winter annuals is triggered by critical rains between late September and early December. The critical rainfall, arriving in one storm, should exceed 25 mm to produce abundant germination. A heavy rain in late September is ineffective unless temperatures have shifted by then from summer to autumn regimes. At the other extreme of the critical fall season, a similar rain in early December will be ineffective for certain members of the ephemeral flora if winter temperatures have already commenced. Conditions prevailing at opposite ends of this autumn period are different enough to promote growth in two rather distinctive ephemeral plant assemblages: following early December rain, germination in the families Cruciferae and Boraginaceae is high compared to that in other families; late September or early October rain, in contrast, promotes germination espe-

cially in species in the families Polemoniaceae, Hydrophyllaceae, Polygonaceae, Leguminosae, and Onagraceae.

A strong correlation has been shown between the characteristic season of activity of the summer and winter annual species and the photosynthetic carbon fixation pathway they possess. Those with a three-carbon compound as the first organic product of photosynthesis are called C_3 plants and those with a four-carbon compound as that product, C_4 plants. Plants with the C_4 trait are peculiarly well adapted to hot-dry conditions in that optimal rates of photosynthesis occur at high temperatures and light intensities compared to C_3 plants and the former also possess high net productivity rates and high relative water-use efficiencies (Johnson, 1976). A total of 66 summer and 64 winter annuals were examined from the Mohave and Sonoran Desert regions (Mulroy and Rundel 1977). The winter annuals were all C_3 plants except for three species (*Aristida adscensionis* and two species of *Euphorbia*) that are active during both seasons. The vast majority of summer annuals possess the C_4 pathway of carbon fixation, a condition suiting them to the high temperature and light conditions of that season. The few Mohave Desert summer annuals are truly distinctive in this regard in that they are all C_4 plants (see also Johnson, 1976). In the adjacent Sonoran Desert, areas with reliable summer precipitation (e.g., southern Baja California and Sonora) have a significant number of C_3 summer annuals along with a preponderance of C_4 ephemeral species. The proportion of C_3 summer annuals increases with increasing quantity and reliability of summer rainfall (Mulroy and Rundel, 1977).

Mulroy and Rundel (1977) also found that the growth form of the two groups of annuals was distinctive. The majority of winter annuals begins growth as rosettes. This growth habit occurs during the coolest part of the year in the Mohave Desert when nighttime temperatures often reach freezing and daytime temperatures may not exceed 10° C. Later in the season as ground temperatures increase, the rosette leaves die or lift away from the soil and the photosynthetic activity is often transferred to cauline leaves on newly produced upright

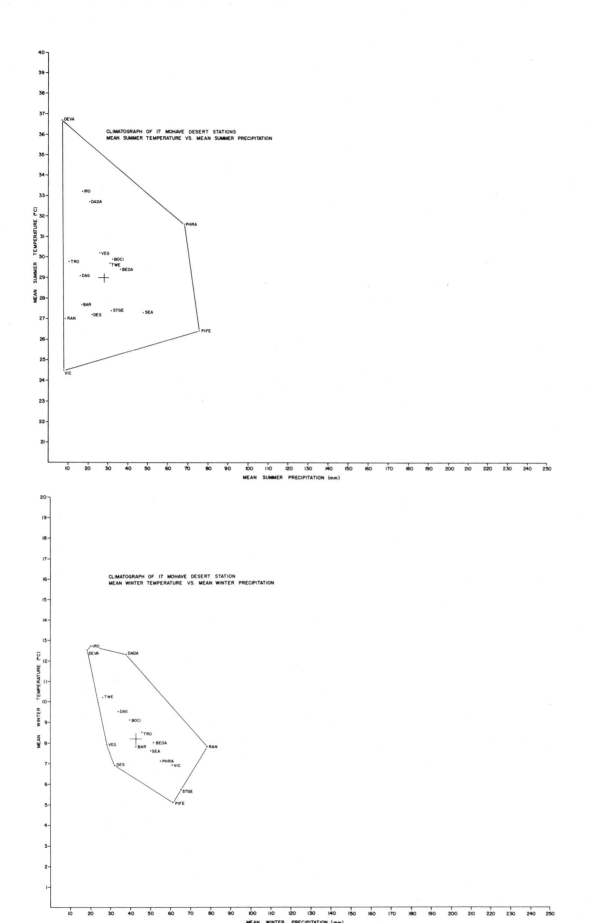

Figure 92. *Climographs for the Mohave desertscrub biome. The polygons circumscribe scattered points representing either summer or winter temperature and rainfall conditions at the weather stations indicated by abbreviated names. The corresponding full names of the stations are given in Table 21.*

stems. Examples of plants with this habit are Chia (*Salvia columbariae*), Filaree (*Erodium cicutarium*), Large-flowered Stork's Bill (*E. texanum*), and Mohave Desert Star (*Monoptilon bellioides*).

The elevation range of the Mohave desertscrub biome is broader than that of the other desertscrub biomes. Sub-sea level altitudes to the north in Death Valley reach -146m. Upper elevational limits reach the 1,370 m contour along the biome's northern limit in Nevada and California and 1,675 m on some mountains. As the southern border of the desert is approached, elevational limits descend to values below 300 m in some areas. There is only a small part of the desert below this level, however, with roughly three fourths of the biome lying between 610 and 1,220 m giving use to the general term "high desert" (Shreve, 1942a). Dry lakes are a common landscape feature.

Precipitation is scanty with annual values ranging from 46 mm at Death Valley to 253 mm at Pierce Ferry, Arizona. Both stations are poor examples of Mohave desertscrub: the Death Valley record is anomalously low and Pierce Ferry is at the edge of desertland. Annual precipitation for most of the Mohave lies between 65 mm and 190 mm (Table 21). Similar low values occur in other deserts of our region but the strong winter precipitation dominance, the relatively low winter temperature and the high summer temperature combine to set the unique conditions of this desert (Fig. 92).

A sampling of January average daily minimum temperatures for Mohave desertscrub stations shows values of -0.1°C at Lancaster, California; -0.7°C at Palmdale, California; and 5.3°C at Death Valley. Values for the same month from nearby Sonoran desertscrub stations range from 4.7°C at Blythe to 4.4°C at El Centro and 3.5°C at Brawley. A sampling of average daily maximum values for July, the hottest month, at Mohave stations shows Death Valley, 46.2°C; Lancaster, 35.2°C; Palmdale, 36.3°C. A sampling of Sonoran Desert stations shows Imperial, 41.5°C; Blythe 42.6°C; and Brawley, 41.9°C. By using as an index of continentality the difference between the average temperature of summer and winter, the Mohave desertscrub is more continental than any of the Sonoran desertscrub subdivisions (Table 23).

Vasek and Barbour (1977) listed five major series within the Mohave desertscrub biome of California. These are Creosotebush scrub, Saltbush scrub, Shadscale scrub, Blackbrush scrub, and Joshua Tree woodland. All but the last occur in some form in either the Sonoran desertscrub biome or the Great Basin desertscrub biome, emphasizing the transitional nature of this desert.

Beatley (1976) listed as typical bajada communities of central southern Nevada those dominated by Creosotebush. She listed an additional set of communities of the prevailingly limestone and dolomitic mountains that are dominated by Shadscale.

In broad outline, the Mohave desertscrub biome can be viewed as described by Vasek and Barbour (1977), with each series possessing more or less distinctive Mohave associates. Our treatment follows the series they recognize.

Creosotebush Series

Creosotebush (*Larrea tridentata*) is a wide-ranging dominant of the three warm deserts in the North American Southwest. It is also one of approximately 50 species or species-pairs that occurs in both northern and southern hemisphere American warm deserts (Solbrig, 1972). The North American subspecies occurs as three chromosome races over its range, with rather sharp differentiation among populations in the three warm deserts: diploid (n=13), Chihuahuan Desert; tetraploid (n=26), Sonoran Desert; hexaploid (n=39), Mohave Desert (Yang, 1970; Barbour, 1969). This sequence suggests that the Mohave Desert population has been derived most recently, raising provocative questions of timing if the hypothesis (Wells, 1977; Wells and Hunziker, 1976) is correct that *Larrea* was absent from the Chihuahuan Desert, its North American center of dispersal, until post-pluvial times (approximately 9,000 years before present).

Larrea occurs prevailingly on the bajadas and well-drained sandy flats, its dominance often asserted more by its tall stature in these two-layered communities than by its density. Shreve (1942a) described as the most characteristic association of the Mohave regions, that association formed by *Larrea* and White Bursage (*Ambrosia dumosa*). Over most of the Mohave region the usual codominants are, besides *Ambrosia dumosa*, Anderson Thornbush (*Lycium andersonii*), Spiny Hopsage (*Grayia spinosa*), Paper Bag Bush (*Salazaria mexicana*), and Shadscale (*Atriplex confertifolia*).

Larrea-dominated communities are mostly below about 1,220 m elevation although they are found more than 300 m higher on south-facing slopes along the northern edge of the plant's range. In the northern part of the Mohave, it is replaced at elevations above the 1,220 m by communities transitional to Great Basin desertscrub: e.g., the Blackbrush and Shadscale series.

Beatley (1976) lists as common shrub associates in the *Larrea tridentata-Ambrosia dumosa* association the following: *Psorothamnus fremontii, Krameria parvifolia, Ephedra nevadensis, Ceratoides lanata,* and, less commonly, *Acamptopappus shockleyi* or *Menodora spinescens*. This association is best developed on deep, loose, sandy soils lacking surface pavement.

Another widespread variant of the Creosotebush series is the community dominated by *Larrea* and *Atriplex confertifolia*. This association predominates on calcareous soils derived from limestone-dolomite mountains and hills and is a major association of the Mohave Desert of south central Nevada (Beatley, 1976). The soil normally has a well developed surface pavement and a shallowly-placed caliche layer. This layer restricts root penetration making the habitat edaphically more arid than that of deeper soils. Because the mountains from which these soils are derived are found at the lower elevations of the region, this vegetation is already growing under the low rainfall associated with these lowlands. Other plants often occurring with the two dominants are *Psorothamnus fremontii, Krameria parvifolia, Ephedra* species, *Acamptopappus shockleyi, Lycium pallidum, Yucca schidigera, Y. baccata,* and *Y. brevifolia*. This is the only community on which Diamond Cholla (*Opuntia ramosissima*) occurs. It is often joined by the more widespread Beavertail Cactus (*Opuntia basilaris*).

Beatley (1976) recognized one other *Larrea*-dominated association in which *Lycium andersonii* and *Grayia spinosa* are codominants. The soils of sites occupied by this assemblage commonly have a more or less poorly developed surface pavement, with hardpan mostly missing from near the surface. The soil matrix is sandy loam with large rock fragments scattered throughout.

In the northern Mohave Desert (Death Valley), Hunt (1966)

Figure 93. *Mohave desertscrub, Valley of Fire area, Clark County, Nevada, ca. 610 m elevation. A common Mohave Desert landscape of mountain, outwash plains, and bajadas dominated by Creosotebush* (Larrea tridentata), *Bursage* (Ambrosia dumosa), *and Bladder-sage* (Salazaria mexicana).

described the dominant communities on the gravel fans noting a tendency for overlapping ranges. Creosotebush is the dominant unifying component of these communities, other species joining Creosotebush with change in elevation or latitude. Toward the lower parts of the fans to the north, *Atriplex hymenelytra* is codominant, occurring in pure stands as the fan base is reached. Southward another saltbush, *Atriplex polycarpa*, occupies an analogous position forming pure stands at the fan base and merging upslope with *Larrea tridentata*. On the upper third of the fans to the north, White Bursage (*Ambrosia dumosa*) occurs with *Larrea*; to the south this position is taken by Brittlebush (*Encelia farinosa*). *Larrea* also occurs along many of the washes with or without *Atriplex hymenelytra* and *A. polycarpa*. Along washes at the base of some fans, White Burrobrush (*Hymenoclea salsola*) occurs as a series in nearly pure stands.

Shadscale Series

This series, dominated by Shadscale (*Atriplex confertifolia*), has been discussed in detail in the chapter on the Great Basin Desert. This transitional community, like the *Larrea-Ambrosia* and *Atriplex polycarpa* associations, is not confined

to only one biome. Its main center lies in the Great Basin desertscrub and its occurrence in the Mohave Desert is in areas transitional to the two deserts, where the mosaic of communities and their associates shifts from one biome to the other as conditions related to soils or relief exceed the ecologic amplitude of the indicator species. Indeed, Beatley (1975) has expressed the point of view that such *Atriplex* communities may constitute a broad zone of xerophytic vegetation transitional between the Mohave Desert and the Sagebrush series of the Great Basin desertscrub and "that *Atriplex* is perhaps the shrub species of greatest ecological amplitude (and ecotoypic variation) in both deserts. Tolerant of most extremes in temperature and rainfall and a wide range in edaphic conditions, including salt content, the species occurs as a dominant where the tolerances of other potential dominants have been exceeded with regard to one or more of the climatic and/or edaphic variables" (Beatley, 1975: 69). Farther south from this zone of transition, Shadscale is a dominant plant of the exceedingly arid habitats of steep hills and mountain slopes—these generally having rocky calcareous soils. The community is also found in valleys in contact with *Larrea* on the slopes above (Beatley, 1976: Fig. 11). This

Figure 94. *Mohave desertscrub saltbush community on Pahranagat National Wildlife Refuge, Lincoln County, Nevada, ca. 1,050 m elevation. The principal plants are species of saltbush (*Atriplex*) and other alkaline tolerant species.*

inversion of the usual zonal sequence between cold-adapted and warm-adapted species results from the ability of the cold-adapted *Atriplex* to better withstand low winter temperature extremes in the depression of enclosed basins than can the warm-adapted *Larrea* (Beatley, 1975, 1976).

Saltbush Series

The communities here are characterized by one or more species of *Atriplex*, such as All-scale (*A. polycarpa*), Shadscale (*A. confertifolia*), Four-wing Saltbush (*A. canescens*), and Desert Holly (*A. hymenelytra*), commonly in combination with other halophytic chenopods such as Pickleweed (*Allenrolfea occidentalis*), alkali weeds (*Nitrophila* spp.), glassworts (*Salicornia* spp.), seep weeds (*Suaeda* spp.), and Greasewood (*Sarcobatus vermiculatus*). This group of communities, while subject to flooding on occasion, occupies habitats of extreme aridity and may be divided into two groups: those occupying sites that are extremely arid by reason of topography, climate, or soil physical properties and those that are extremely arid because of soil chemical properties. The former, or xerophytic phase, occurs on hills, in basins, or around the margins of playas. The other group, or halophytic phase, is found on

saline or alkali soils on playas, in sinks, or near seeps where the available ground water is heavily charged with minerals.

These communities are well represented in the adjacent Great Basin and Sonoran desertscrub biomes as well as in Mohave desertscrub. Among the most widespread is that dominated by *Atriplex polycarpa* which in the Mohave Desert is especially prevalent around Rabbit, Lucerne, and Soggy dry lakes (Vasek and Barbour, 1977) and in Death Valley (Hunt, 1966). This community is also widespread to the south and east in the Sonoran desertscrub biome, occurring as the dominant vegetation along much of the Gila River Valley in Arizona (Aldous and Shantz, 1924; Shantz and Piemeisel, 1924; Turner, 1974) and as a prominent community in the Coachella Valley, California (Shantz and Piemeisel, 1924; Küchler, 1977).

Soils supporting stands of *Atriplex polycarpa* generally contain few salts and, although the water table may be as shallow as 5 m (Hunt, 1966), permanent water is mostly at much greater depth. The soil of this community is often highly productive if cultivated, resulting in the destruction of many stands supporting this species, as at Lucerne Dry Lake. The vegetation may be closed but is often very open, sup-

Figure 95. *Mohave desertscrub of Blackbrush* (Coleogyne ramosissima) *and Banana Yucca* (Yucca baccata) *immediately west of Grand Wash Cliffs, Mohave County, Arizona, ca. 760 m elevation. Blackbrush associations characterize many northern Mohave desertscrub communities transitional to Great Basin desertscrub. The cliffs are occupied by an open pinyon-juniper association dominated by* Pinus monophylla.

porting only about 124 plants per hectare (Hunt, 1966). The xerophytic phase vegetation of mixed saltbush species is generally situated between the halophytic phase below and Creosotebush communities above.

Halophytic saltbush associations (Fig. 94) occur at topographically defined sites such as playas and sinks and may be found in close association with Mohavian marshland if surface water is present. The vegetation is dominated by succulent chenopods such as *Bassia hyssopifolia, Allenrolfea occidentalis, Nitrophila occidentalis, Salicornia subterminalis, Suaeda* spp., and *Sarcobatus vermiculatus.* Other chenopods, especially in the genus *Atriplex*, may be present. Saltgrass, *Distichlis spicata* var. *stricta,* is a common associate. Many of these species occur together elsewhere in moist, saline habitats throughout much of arid western North America, and the associations they constitute may be viewed as intrazonal because of their wide occurrence.

Blackbrush Series

Associations dominated by Blackbrush (*Coleogyne ramosis-*

sima) (Fig. 95) are widespread across the southern limits of the Great Basin desertscrub biome and the northern limits of the Mohave desertscrub biome. In the latter part of its range, Blackbrush is often viewed in the role of a community transitional between Mohave and Great Basin desertscrub biomes and in that context it has been discussed in detail under Great Basin Desertscrub.

Joshua Tree Series

Yucca brevifolia is the best known (and most photographed) Mohave Desert endemic; yet, in few places can it be used to characterize this biome because of the plant's rather limited occurrence there. It is found along almost the entire periphery of the Mohave Desert where this biome grades upslope into cooler moister vegetation (Benson and Darrow, 1954) and thus may at times be associated with species from Great Basin desertscrub and Great Basin conifer woodland. It is absent from much of the southeastern margin of the Mohave where contact is made at lower elevations with Sonoran desertscrub. Joshua Tree does, however, make contact with the Sonoran

Figure 96. *A diverse Joshua Tree "woodland" community near Dolan Springs, Mohave County, Arizona, ca. 975 m elevation. Joshua Tree* (Yucca brevifolia) *dominates the association containing as subdominants and understory species: Buckhorn Cholla* (Opuntia acanthocarpa), *Club Cholla* (Opuntia clavata), *Diamond Cholla* (Opuntia ramosissima), *Beavertail Cactus* (Opuntia basilaris), *Banana Yucca* (Yucca baccata), *Goldenhead* (Acamptopappus sphaerocephalus), *and Galleta* (Hilaria rigida).

Desert in west-central Arizona in the area north of Wickenburg (Lowe, 1964, Fig. 14). This Mohave Desert enclave predictably occurs near the upper elevational limits of the two deserts. Because of the varied contacts made with other biomes, *Yucca brevifolia* may be in codominance with *Larrea, Coleogyne, Juniperus,* or *Carnegiea* at various places in its near-circular range around the margin of the Mohave Desert. Because of the area's higher elevation and more generous precipitation, understory species are often diverse—some such as *Acamptopappus sphaerocephalus* are endemic.

Although the "woodland" is dominated by the bizarre Joshua Tree, the tree's dominance is more visual than real. The understory of various shrubs and herbaceous plants may contain several species of greater importance than the *Yucca* (Fig. 96). Until recently, quantitative studies of Joshua Tree communities were almost nonexistent. Vasek and Barbour (1977) presented data for a dense Joshua Tree community north of Cima, San Bernardino County, California. *Yucca brevifolia* had an importance value (IV) of 4.36 (based upon density, cover, and frequency) and ranked ninth out of 22

plants present. The shrubby Big Galleta Grass (*Hilaria rigida*), was the clear dominant (IV=96.9), with *Haplopappus cooperi* (IV=49.1), *Salazaria mexicana* (IV=26.2), *Ephedra nevadensis* (IV=12.2), *Hymenoclea salsola* (IV=12.7), *Muhlenbergia porteri* (IV=12.6), and *Yucca baccata* (IV=13.4), sharing roles of somewhat lesser dominance. Rowlands (1978) has recently made a thorough study of vegetation in which Joshua Tree grows and concludes that "Joshua Trees are not restricted to any desertscrub or xeric woodland community; they contribute very little, in a quantitative vegetational sense, to total stand composition wherever they occur, and can be considered dominants only in terms of stature. The Joshua Tree woodland as a community type is a non-entity" (Rowlands, 1978: 171). He refers to Joshua Tree as a ubiquit that is found in many different plant communities.

Joshua Trees normally grow on sandy, loamy, or fine gravelly soils where runoff is probably minimal, suggesting the species has relatively high moisture requirements (Fig. 97). Went (1957) published data suggesting that after *Yucca brevifolia* has reached a certain age the plant requires an

Figure 97. *Mohave desertscrub near Searchlight, Nevada, ca. 1,145 m elevation. This community appears dominated by yuccas (*Yucca brevifolia, Y. baccata*) but these arborescent monocots are probably exceeded in biomass by the shrubby grass,* Hilaria rigida, *and by* Coleogyne ramosissima.

annual cool-season exposure to low temperatures for optimal growth later during the warm season. Although ideal temperature combinations were not determined, this work may help to explain Joshua Tree's restriction to higher, presumably cooler, sites at the Mohave Desert periphery.

Other Series

Minor associations dominated by *Artemisia, Salazaria, Acamptopappus,* and even cacti occur within Mohave desertscrub. These series are always of small extent when compared to *Larrea, Atriplex,* and *Coleogyne.* However, even the drainageways along the arroyos fail to present much variety in either species composition or stature. All in all, the Mohave Desert, or "high desert" as it is called in southeastern California, exhibits general low species composition within basin and range topography as compared with its progenitor, the Sonoran Desert.

The conclusion that the Mohave is a transitional desert differentiated at the subspecific level was derived from analysis of distributional data for amphibians, reptiles, birds, and mammals. Among the more instructive is an analysis of Dice's (1943) Mohavian Biotic Province. Data for mammalian distribution (Hall and Kelson, 1959) reveal that there are no modern species or genera of mammals that are essentially or wholly restricted to the Mohave Desert; none are Mohavian taxa.

As befits the sparse vegetation of this desert, large mammals are few and are represented principally by Desert Bighorn Sheep (*Ovis canadensis nelsoni*) and Coyote (*Canis latrans*). Mule Deer (*Odocoileus hemionus*) and Pronghorn (*Antilocapra americana*) occurred only at the desert's edge. Smaller, less wide-ranging mammals abound, however. Rodents are among the most sedentary of mammals and, among the rodent species endemic to the North American Desert, a few on the Mohave are clearly differentiated at the subspecific level. Even there, the data show a relatively low percentage for Mohavian differentiation and endemicity compared to that for the larger Great Basin Desert northward and the larger Sonoran Desert southward.

For example, in the Mohave Desert near Las Vegas, Nevada, the vegetation is dominated by *Larrea tridentata* and associated shrubby desert perennials including *Ambrosia dumosa, Ceratoides lanata, Ephedra funerea, Hilaria rigida, Yucca schidigera, Krameria parvifolia, Thamnosma montana, Psorothamnus fremontii,* and *Sphaeralcea ambigua.* In an excellent study of desert animals in this Mohave desertscrub community, Bradley and Mauer (1973) trapped 11 species of native rodents over a 32-month period. Eight of the rodents in residence are especially characteristic mammals in Mohave Creosotebush communities in Nevada and adjacent California, Arizona, and Utah; the other three species (*Peromyscus maniculatus, P. truei, Thomomys umbrinus*) have their widest distributions outside of desert environments. The characteristic eight are listed as follows in decreasing order of their relative abundance.

Dipodomys merriami	Merriam's Kangaroo Rat
Perognathus longimembris	Little Pocket Mouse
Ammospermophilus leucurus	Whitetailed Antelope Squirrel
Neotoma lepida	Desert Woodrat
Onychomys torridus	Southern Grasshopper Mouse
Perognathus formosus	Longtailed Pocket Mouse
Peromyscus eremicus	Cactus Mouse
Peromyscus crinitus	Canyon Mouse

From the taxonomic and distributional data for each of these rodent species in western North America (see Hall and Kelson, 1959) we find that the systematic data are typical for faunal taxa in the Mohave Desert. There are 22 subspecies (in a total of 117 represented by the 8 species above) that occur in the Mohave. However, only 7 of these in the total of 117 subspecies are either wholly or essentially Mohavian (6%).

The avifauna and the herpetofauna are similar to the mammal fauna in the relatively low level of taxonomic differentiation exhibited within Mohave desertscrub. Of the birds, only LeConte's Thrasher (*Toxostoma lecontei*) could be considered as centered there, and it also occurs in the open Lower Colorado and Vizcaíno subdivisions of the Sonoran Desert. Otherwise, the relative paucity of bird species is represented by Sonoran and Great Basin taxa—particularly the former. Even the Desert Night Lizard (*Xantusia vigilis*), which occurs so characteristically and abundantly in association with *Yucca brevifolia* in the Mohave Desert—and is correctly regarded as one of the most representative vertebrate residents there—also occurs in parts of the Sonoran Desert and the Chihuahuan Desert and is represented by one or more subspecies in each of the three desert regions. Similarly, other of the many lizards represented in the Mohave, are subspecies or populations of: Banded Gecko (*Coleonyx variegatus*), Chuckwalla (*Sauromalus obesus*), Desert

Iguana (*Dipsosaurus dorsalis*), Zebratail Lizard (*Callisaurus draconoides*), Mohave Fringe-toed Lizard (*Uma scoparia*), Leopard Lizard (*Crotaphytus wislizenii*), Collared Lizard (*Crotaphytus collaris*), Banded Gila Monster (*Heloderma suspectum cinctum*), Yellowback Spiny Lizard (*Sceloporus magister uniformis*), Western Brush Lizard (*Urosaurus graciosus graciosus*), Desert Side-blotched Lizard (*Uta stansburiana stejnegeri*), Regal Horned Lizard (*Phrynosoma solare*), Southern Desert Horned Lizard (*P. platyrhinos calidiarum*), Western Whiptail (*Cnemidophorus tigris tigris*), and Desert Tortoise (*Gopherus agassizi*).

Snakes are almost as numerous and include the Western Leafnose Snake (*Phyllorhynchus decurtatus perkinsi*), Western Blind Snake (*Leptotyphlops humilis*), Desert Rosy Boa (*Lichanura trivirigata gracia*), Coachwhip (*Masticophis flagellum*), Striped Whipsnake (*M. taeniatus*), Mohave Patchnose Snake (*Salvadora hexalepis mojavensis*), Great Basin Gopher Snake (*Pituophis melanoleucus deserticola*), Desert Glossy Snake (*Arizona elegans eburnata*), California Kingsnake (*Lampropeltis getulus californiae*), Western Longnose Snake (*Rhinocheilus lecontei lecontei*), Western Ground Snake (*Sonora semiannulata*), Mohave Shovelnose Snake (*Chionactis occipitalis occipitalis*), Nevada Shovelnose Snake (*C. occipitalis talpina*), Desert Night Snake (*Hypsiglena torquata deserticola*), Sonoran Lyre Snake (*Trimorphodon biscutatus lambda*), Sidewinder (*Crotalus cerastes*), Mohave Rattlesnake (*C. scutulatus*), and Speckled Rattlesnake (*C. mitchelli*).

The continuing controversy over the introduced Wild Burro (*Equus asinus*), which has been destroying desert habitat and widely eliminating native Desert Bighorn Sheep, centers on the Mohave Desert in southern Nevada and adjacent California and Arizona, including Death Valley National Monument and Grand Canyon National Park (see Transactions Desert Bighorn Council, yearly since 1957). Bradley (1964, 1965) provided a classification and field survey of the major plant communities at 350 stations in the Desert Game Range (Desert National Wildlife Range) and discussed their relationship to Desert Bighorn Sheep distribution.

Bradley's (1965) analysis revealed that Blackbrush (*Coleogyne ramosissima*) communities, with their associated grass, *Hilaria rigida,* on the upper bajadas at elevations of 1,280 m to 1,830 m are areas of heavy utilization and preferred habitat for *Ovis canadensis nelsoni* in southern Nevada.

153.2 Chihuahuan Desertscrub

First delineated and described by Forrest Shreve (1942a, 1951), the Chihuahuan Desert is the largest of the three Creosotebush dominated deserts in North America. It covers large tracts of eastern Chihuahua, western Coahuila, San Luis Potosí, southern Nuevo Leon, northeast Zacatecas, eastern Durango, southwest Texas, and southern New Mexico, as well as smaller but equally distinctive areas in southeast Arizona and northeast Sonora. It is also one of the least known biomes. This incongruous fact has been somewhat alleviated lately by the recent compendium of available knowledge on the Chihuahuan Desert's living resources edited by Wauer and Riskind (1977). The contributing authors to this and other recent works now allow a reasonable discussion of the region's geography, climate, flora, and fauna.

The Chihuahuan Desert boundaries used here are based on floristic criteria and agree well with those of Henrickson and Straw (1976) and the climatological based outline of Schmidt (1979) (Fig. 98). Although the Desert's eastern and southern regions are outside the delineated boundaries, these outlying areas are within the scope of the Southwest and will be treated as such. Indeed, the actual delineation of the Chihuahuan Desert is a matter for some discussion (Shreve, 1942a; Morafka, 1977; Schmidt, 1979). Its boundaries are further complicated by the desert's recent dynamic expansion into former semidesert grassland—much of which has happened not only within the memory of living people, but continues to be readily observed today (Castetter, 1956; Hastings and Turner, 1965b; York and Dick-Peddie, 1969).

The heart and center of the Chihuahuan Desert is the arid highland plains and basins of north-central Mexico, between the Sierra Madre Occidental and the Sierra Madre Oriental and their outliers. More than 80% of this desertscrub resides on limestone, and the gray gravel plains of this substratum are the characteristic landscape feature over hundreds of kilometers. Although there are more than 1,000 species of plants endemic to the Chihuahuan Desert (Johnston, 1977), many of these are local in distribution and restricted to one or several portions of the desert. As a result the vegetation maintains a recognizable homogeneity in its dominants from west to east and north to center—much more so than in the Sonoran Desert. A greater appearance of large succulents (e.g., *Ferocactus pringlei, Echinocactus platyacanthus*), takes place southward, below the Rio Nazas. This region has simply been crossed by the author, the southeastern portions not having been visited. Because the literature is incomplete on this subject, little can be said about the nature and extent of this change in the physiognomy.

Basin and range landscapes prevail throughout, and the desert occupies a vast expanse of rain shadowed basins, outwash plains, low hills, and bajadas. Many of the mountains are large anticlines, some of which rise 2,000 m or more from the barrial of the valley floor and capped in chaparral, pines, oak, and even fir (Fig. 104). After rains, these barren limestone edifices may take on a green appearance with the new growth of the appropriately named Resurrection Plant, *Selaginella lepidophylla.* Drainages are gash-like, without perennial streams; bajadas are steep and often of meager development. Except for local situations along the upper Gila, San Pedro, and Bavispe watersheds in southeast Arizona and in northeast Sonora, the principal external drainage is via the Rio Grande (Bravo) and its major tributaries, the Conchos,

Table 22. Precipitation data for 19 stations in the Southwest within and adjacent to Chihuahuan Desertscrub.

Station Lat./Long.	Elevation (in m)	J	F	M	A	M	J	J	A	S	O	N	D	Total Ppn (mm)	Total May through Sept. Ppn	$\dfrac{\text{Total Annual Ppn}}{\text{May through Sept. Ppn}}$ x 100
Socorro, NM 34°05′ 106°53′	1,398	8	9	9	10	11	14	35	37	26	22	6	15	201	122	61
Elephant Butte Dam, NM 33°09′ 107°11′	1,395	9	7	6	7	7	15	40	49	31	18	6	13	207	141	68
White Sands Nat. Monument, NM 32°47′ 106°11′	1,218	12	8	8	7	6	15	39	35	30	18	7	13	197	125	63
Jornada Exp. Range, NM 32°37′ 106°44′	1,303	12	9	8	5	5	11	45	43	33	22	9	14	215	137	64
Carlsbad, NM 32°25′ 104°14′	951	11	8	13	12	38	37	41	45	41	37	9	10	303	202	67
San Simon, AZ 32°22′ 109°08′	1,100	21	13	13	6	3	9	43	55	20	16	10	15	224	131	58
N Lazy H Ranch (Pantano, AZ) 32°07′ 110°41′	930	27	21	22	10	4	6	66	65	31	18	17	29	314	171	54
Hachita, NM 31°56′ 108°20′	1,370	16	12	12	6	2	9	55	55	31	19	8	15	240	152	64
El Paso, TX 31°48′ 106°24′	1,194	10	11	10	6	8	15	39	28	29	20	8	13	197	120	61
Fairbank, AZ 31°21′ 109°32′	1,173	18	12	11	5	3	12	87	75	30	13	9	18	293	206	71
Tombstone, AZ 31°42′ 110°03′	1,405	20	14	15	7	3	11	96	83	31	16	9	19	324	224	69
Pecos, TX 31°25′ 103°30′	796	11	8	8	15	27	27	34	23	35	24	10	9	232	146	63
Douglas, AZ 31°21′ 109°32′	1,211	17	17	13	6	5	13	83	73	30	19	16	20	311	203	65
Candelaria, TX 39°09′ 104°41′	876	11	7	6	7	17	27	47	50	38	27	8	10	253	178	70
Ojinaga, Chih. 29°34′ 104°25′	849	3	6	4	3	9	12	27	45	41	32	14	12	209	134	64
El Mulato, Chih. 29°23′ 104°10′	789	8	9	4	11	13	22	26	35	34	14	7	13	196	130	66
Panther Jct., TX 29°19′ 103°13′	1,140	11	12	9	10	38	40	54	44	61	42	14	13	348	237	68
Meoqui, Chih. 28°16′ 105°29′	1,162	17	3	4	2	12	26	71	64	40	23	13	20	294	213	72
Camargo, Chih. 27°42′ 105°10′	1,663	13	4	4	5	8	25	80	59	54	31	14	21	318	228	71

Nazas, and Pecos. Otherwise, most of the region has no outlet to the sea, the drainages ending in closed basins or bolsones. As in the Mohave and Great Basin Deserts, the smaller, ephemeral watershed drainages are without distinctive vegetative development.

Hot in summer, the Chihuahuan Desert can be cold in winter. Summer temperatures over 40°C are common and a low of -30°C has been recorded at Ahumada, Chihuahua (Schmidt, 1975). Freezing temperatures can be expected in its northwestern reaches on more than 100 nights a year, and some days do not get above freezing. Even at the desert's lowest elevations along the Rio Grande, and in the south, the climate is distinctly warm temperate with only an expected 200-250 days between the last 0°C temperature in March and the first frost of autumn.

Precipitation means range from a low of ca. 200 mm per annum to more than 300 mm (Table 22). Those stations receiving more than this are usually at the edge of the desert or are areas that historically were semidesert grassland. These rainfall amounts are relatively generous for desertscrub. However, most of the precipitation falls during summer thunderstorms when evapotranspiration rates are high, thereby losing much of its effectiveness—well over half of the precipitation is received during the months of May through September. Variability is also high, and the area that received 500 mm during one summer may get only 50 mm the next. Although Chihuahuan desertscrub is characterized by distinctly summer rainfall regime, low intensity precipi-

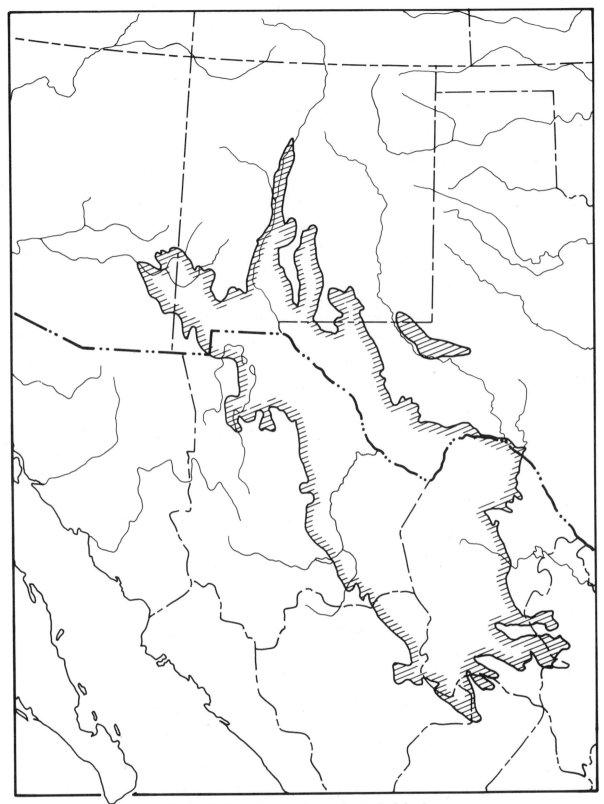

Figure 98. *Generalized outline of the "real Chihuahuan Desert" according to Schmidt (1979). Based on mean temperature and precipitation data, the general outline and extent of Schmidt's climatic 'desert' agrees well with our delineation of Chihuahuan desertscrub based on floristic criteria.*

Figure 99. *Ecotype portraits of* **Larrea tridentata** *characteristic of three divisions of desertscrub. Upper left: Mohave ecotype. Lower left: Sonoran ecotype. Upper Right: Chihuahuan ecotype. Plants with the "typical" Chihuahuan growth form are shorter, have sparser foliage and straighter stems, and are more open at the base than their Sonoran counterparts. The latter have a fuller, less angular appearance. The Mohave Desert ecotype is shorter and bushier (more "shrub-like") than the Sonoran Desert form. Each ecotype also has a characteristic number of chromosomes, — n=13 in the Chihuahuan Desert, n=26 in the Sonoran Desert, and n=39 in the Mohave Desert (Yang, 1970). Photographs by R.M. Turner.*

tation associated with Pacific frontal storms occurs with some regularity. It is, therefore, an oversimplification to consider the Chihuahuan Desert as being without meaningful winter precipitation. This is especially true at the north, where winter precipitation is consistent enough to support several taxa of spring annuals (Burgess, 1980).

The Chihuahuan Desert is a "high" desert. Even its lowest elevations near Langtry, Texas along the Rio Grande are at ca. 400 m. Altitudes gradually increase up the Rio Grande and Rio Conchos to as high as from 1,400 m to 1,600 m. All of the interior basin locales are higher than 700 m, and most are well over 1,000 m. The upper limits in the Southwest are commonly between 1,400 m and 1,500 m (somewhat higher in the south to over 2,000 m) where the desert intergrades with and is finally replaced by semi-desert grassland.

Large expanses of outwash plains, low hills and valleys are characterized by the diploid form of *Larrea tridentata* (Yang,

1970; Fig. 99). Here this phenotypically recognizable form of Creosotebush often shares or temporarily relinquishes dominance with Tarbush (*Flourensia cernua*) or Whitethorn Acacia (*Acacia neovernicosa*) (Fig. 100). Except for the occasional appearance of Ocotillos (*Fouquieria splendens*) and an individual Allthorn (*Koeberlinia spinosa*) or clump of mesquite, these three shrub species are the dominants over hundreds of miles of "plains communities" Chihuahuan Desert. The only other desertscrub communities of widespread occurrence on these valley locales are saltbushes on fine grained soils and open stands of shrub Mesquite (*Prosopis glandulosa* var. *torreyana*) on sandy, wind eroded hummocks (Fig. 101).

Major understory associates on the "plains" are Mariola (*Parthenium incanum*), Guayule (*P. argentatum*), Goldeneye (*Viguiera stenoloba*), *Coldenia canescens*, desert zinnias (*Zinnia acerosa*, *Z. grandiflora*), dogweeds (*Dyssodia* spp.), cacti, and locally, to the east, Jatropha (*Jatropha dioica*). At higher

Figure 100. *"Plains" community of* Larrea tridentata *and* Flourensia cernua *near Fairbanks, Arizona, ca. 1,250 m elevation.* Larrea tridentata *is the most prevalent plant followed by* Flourensia cernua *and lesser numbers of* Acacia neovernicosa, Fouquieria splendens, Koeberlinia spinosa, Parthenium incanum *and small cacti. Such landscapes, with alternating dominants, characterize millions of hectares of calcareous plains, low hills and bajadas in the Chihuahuan Desert in Arizona, New Mexico, Texas, Chihuahua and Coahuila.*

elevations and in areas having more relief, species of the more diverse "succulent-scrub" community increasingly make their appearance — the economically valuable Candelilla (*Euphorbia antisyphilitica*), used for making wax, a bromeliad (*Hechtia scariosa*), Lecheguilla (*Agave lechuguilla*), Ocotillo, dogweeds (*Dyssodia pentachaeta, D. acerosa*), Ratany (*Krameria parvifolia* var. *glandulosa*), condalias (*Condalia* spp.), and many, many more.

Upslope, above the shrubby plains, outcrops, arroyo, bajada, and foothill habitats are populated by an increasingly rich assemblage of scrub in which leaf succulents (*Agave, Hechtia*) and stem succulents (*Yucca, Dasylirion*) are well represented (Fig. 102). One of the most prevalent of these is the low growing Shindagger or Lecheguilla (*Agave lechuguilla*) which covers thousands of acres of ridges, slopes, and benches. Other succulents usually present as local dominants or representatives include a number of yuccas (*Yucca elata, Y. rostrata, Y. thompsoniana, Y. filifera, Y. carnerosana, Y. torreyi,* etc.), sotols

(*Dasylirion leiophyllum, D. wheeleri*), agaves (*Agave scabra, A. falcata, A. neomexicana, A. parryi, A. striata,* etc.), and nolinas (*Nolina microcarpa, N. erumpens, N. texana*).

A great variety of large woody shrubs and cacti are also present in these "succulent-scrub upland" communities and include such species as Ocotillo, Coldenia (*Coldenia greggii*), Catclaw (*Acacia greggii*), cenizos (*Leucophyllum minus, L. frutescens*), condalias (*Condalia* spp.), *Zizyphus obtusifolia*, Lippia or Oreganillo (*Aloysia wrightii*), and Little Leaf Sumac (*Rhus microphylla*). Another species, Sandpaperbush (*Mortonia scabrella*) may be locally abundant in southeast Arizona and dominate limestone hills to the exclusion of most other species (Fig. 103). Other species dominate locally elsewhere.

These succulent-scrub communities grade into semi-desert grassland at their upper limits. Consequently, their herbaceous components are varied and abundant as compared to the lower and more depauperate plains communities. Clumps of species of *Bouteloua* and other summer growing grasses may therefore

Figure 101. *Mesquite scrub community south of Deming, New Mexico, ca. 1,280 m elevation. Classification of these hummock forming Mesquite communities as disclimax semidesert grassland or Chihuahuan desertscrub is moot.*

be common along with a myriad of soft-stemmed shrubs and forbs. An occasional juniper (*Juniperus monosperma, J. pinchottii*), as well as any of several chaparral species may also be encountered at the desert's upper limits.

The usual upper and lower elevation contacts with Chihuahuan desertscrub are with semidesert grassland—mixed bunch grasses and scrub above, Tobosa (*Hilaria mutica*) swales or mogotes below (Fig. 104). In the east as near the junction of the Rio Grande and Devil's River, and near Puerto Sacramento east of the Cuatro Ciénegas Basin, the lower elevational contacts are (or were) with Tamaulipan thornscrub. Much of this region of contact is now under cultivation. Occasionally Chihuahuan desertscrub comes in contact directly with Madrean evergreen woodland or interior chaparral as on the northeast slopes of the Chiricahua Mountains where it butts up against *Quercus grisea* woodland. Another unusual intermingling is the gradual replacement of Chihuahuan desertscrub by Sonoran desertscrub along the Bavispe River in Sonora and along the San Pedro River in southeast Arizona (Fig. 105).

The areas drained by the Rio Grande, Gila, and Yaqui rivers and their tributaries are less than one-third of the desert's total area (Henrickson, 1977). Most of the drainage is internal, ending in enclosed basins with no access to the sea. These bolsón depressions receive infrequent runoff and are therefore more mesic than the surrounding desertscrub plains. Accordingly, the bottom of even the smallest basins is often an island of semi-desert grassland clothed in *Hilaria mutica* or sacatons (*Sporobolus wrightii, S. airoides*) with associated semidesert grassland species of scrub, e.g. Mesquite (*Prosopis glandulosa*), Mormon Tea (*Ephedra trifurca*), or Palmilla (*Yucca elata*). In the center of the larger basins, these grasslands may encircle a saline marsh, or, as is more likely, a playa—an intermittently flooded wetland almost devoid of vegetation.

These barren playas were once Pluvial lakes, and on their downwind side, dune fields have often developed as a result of wind carrying the finer dried deposits. These "sand dunes" are of quartz in the volcanic areas or, as is most often the case in this calcereous region, are composed of gypsum (hydrous calcium sulfate). Some of the larger and better known examples are the "White Sands" in New Mexico, the dunes west of Cuatro Ciénegas, Coahuila, and the large dune fields

Figure 102. *"Foothill" or mixed succulent-scrub community on slopes of Sierra San Marcos, Coahuila, ca. 825 m elevation. A diverse scrub, the composition at this locale includes* Yucca macrocarpa, Dasylirion leiophyllum, Agave lecheguilla, Hechtia *spp.,* Opuntia leptocaulis, Euphorbia antisyphilitica, Leucophyllum frutescens, Dyssodia pentachaeta, Fouquieria splendens, *and* Larrea tridentata.

Figure 103. *"Mattoral" of Sandpaperbush (*Mortonia scabrella*) and other Chihuahuan desertscrub species (*Larrea tridentata, Fouquieria splendens, Nolina microcarpa, Yucca baccata, *and in immediate right foreground,* Rhus microphylla*) on limestone in the Tombstone Hills, Cochise County, Arizona, ca. 1,425 m elevation. These relatively dense and local desertscrub communities are restricted to calcareous soils at the Chihuahuan Desert's upper limits in southeastern Arizona, where annual precipitation approaches or exceeds 300 mm.*

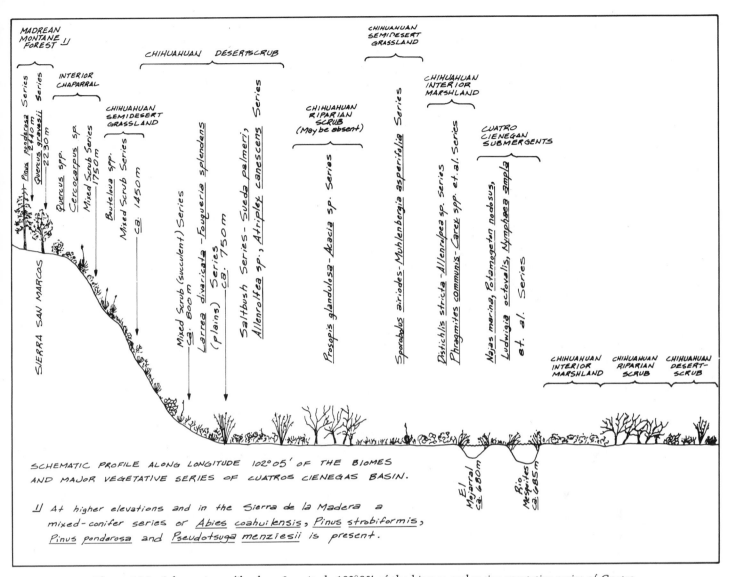

Figure 104. *Schematic profile along Longitude 102°05′ of the biomes and major vegetative series of Cuatro Ciénegas Basin.*

near Samalayuca, Chihuahua (Fig. 106). These dunes are populated by open assemblages of *Artemisia filifolia, Yucca elata, Prosopis glandulosa, Ephedra trifurca, Tiquilia hispidissima, Atriplex canescens, Varilla mexicana,* and *Hymenoclea monogyra* as well as a number of gypsophilous endemics, more or less local in distribution (Powell and Turner, 1977). Grasses such as *Sporobolus airoides, S. flexuosus, S. contractus, S. giganteus,* and *Oryzopsis hymenoides* may be important associates depending on the stability of the dunes and on grazing history. Annuals (e.g., *Croton, Helianthus, Euphorbia, Tidestromia*) are seasonally abundant.

Saline depressions not clothed in grasses (e.g., *Distichlis spicata* var. *stricta*) or marsh plants take on an open desertscrub appearance. These low communities occupy extensive acreages in and adjacent to playas and are dominated by a number of halophytes of general distribution—species of *Suaeda, Allenrolfea,* and *Atriplex.*

Cacti are only locally dominant and not often conspicuous, at least not to the extent that they are in the Arizona Upland and Gulf Coast subdivisions of the Sonoran Desert. The Chihuahuan Desert is essentially a shrub dominated biome in which leaf and stem succulents sometimes heavily participate. Nonetheless, small cacti are well represented and a diligent search along a rocky canyon, bajada, or other suitable environment will usually result in the discovery of a surprising number of species over even a small area. It should not go unnoticed, then, that there are almost as many species of cacti in the Chihuahuan Desert in Texas as in the rest of the United States combined. A definitive work on the cacti of the entire region is much needed.

The only prominent cacti in the "Southwest" portion of the Chihuahuan Desert are local populations of *Opuntia bradtiana* in Coahuila, the Cane Cholla (*Opuntia imbricata*), and prickly pears (*Opuntia violacea* var. *macrocentra, O. phaeacantha* var. *major, O. p.* var. *discata*). Otherwise the cacti are mostly low growing and often clumped or prostrate. These include a number of widespread and endemic forms of Turk's heads and hedgehogs (e.g., *Echinocactus horizonthalonius, E. texensis, Ancistrocactus uncinatus, Ferocactus hamatacanthus*), claret cups, rainbows, and strawberries (*Echinocereus triglo-*

Figure 105. *Passing from Chihuahuan desertscrub to Sonoran desertscrub north of Colonia Morelos, Sonora, ca. 1,017 m elevation. Of the dominants shown,* Acacia neovernicosa *is an indicator of Chihuahuan desertscrub and* Opuntia bigelovii *is a Sonoran desertscrub species. The shorter life span of the cholla and the area's recent semidesert grassland history probably make the Chihuahuan desertscrub designation more appropriate at this locale; Sonoran desertscrub is entered farther to the south with the advent of* Stenocereus thurberi *and a host of other Sonoran Desert species prior to entry into Sinaloan thornscrub.*

chidiatus, *E. pectinatus* var. *rigidissimus, e.p.* var. *neomexicanus, E. chloranthus, E. enneacanthus* var. *stramineus*), hedgehog coryphanthas (e.g.,*Coryphantha strobiliformis, C. scheeri* var. *valida, C. echinus, C. macromeris, C. pottsii, C. vivipara, C. ramulosa*), pincushions and nipple cactus (*Mammillaria gummifera* var. *meiacantha, M. pottsii, M. gummifera* var. *applanata*), a fish-hook cactus (*Ancistrocactus scheerii*), Texas Cactus (*Neolloydia intertexta*), Button Cactus (*Epithelantha micromeris*), Texas Pride (*Thelocactus bicolor*), and several low statured or prostrate chollas (*Opuntia leptocaulis, O. kleiniae, O. schottii, O. tunicata*). Several others, such as the Night Blooming Cereus (*Peniocereus greggii*), and the infamous Peyote (*Lophophora williamsii*), while widespread, are often local and can be difficult to locate. A truly bizarre species is the Living Rock or Star Cactus (*Ariocarpus fissuratus*) which may go virtually unnoticed until after a rain when a greatly increased moisture content expands its size.

Because of its recent origin, few warm-blooded vertebrates are centered in, or restricted to Chihuahuan desertscrub. A few mammals such as the Desert Pocket Gopher (*Geomys arenarius*), Yellow-faced Pocket Gopher (*Pappogeomys castanops*), Nelson's Kangaroo Rat (*Dipodomys nelsoni*), Nelson's Pocket Mouse (*Perognathus nelsoni*), Southern Grasshopper Mouse (*Onychomys torridus*), Goldman's Woodrat (*Neotoma goldmani*), Texas Antelope Squirrel (*Ammospermophilis interpres*), and Desert Pocket Mouse (*Perognathus penicillatus*) are resident there, but mostly the desert represents a southeastern extension for more general desert-adapted species, e.g., Desert Shrew (*Notiosorex crawfordi*), Desert Mule Deer (*Odocoileus hemionus crooki*), Desert Bighorn Sheep (*Ovis canadensis mexicana*), Merriam's Kangaroo Rat (*Dipodomys merriami*), and Desert Cottontail (*Sylvilagus auduboni*). This is true of the birds also; only the Scaled Quail (*Callipepla squamata*) and White-necked Raven (*Corvus cryptoleucus*)

Figure 106. *Dunes near Estación Samalayuca, Chihuahua, ca. 1,340 m elevation. Principal plants are* Artemisia filifolia *(shown);* Prosopis glandulosa, Atriplex canescens *and* Ephedra *spp.*

can be considered as "characteristic species," and both of these exceed the range of desertscrub. The Scaled Quail, while periodically one of the desert's most abundant birds, is equally or more at home in semi-desert grassland, and the desert's avifauna is largely a mixture of "grassland" species and desert birds of wider distribution—Mourning Dove, Roadrunner (*Geococcyx californianus*), Lesser Nighthawk (*Chordeiles acutipennis*), Scott's Oriole (*Icterus parisorum*), Cactus Wren (*Campylorhynchus brunneicapillus*), Curve-billed Thrasher (*Toxostoma curvirostre*), and Black-throated Sparrow (*Amphispiza bilineata*).

A Chihuahuan Desert herpetofauna is more apparent. Lizards characteristic of, or at least centered in, the Chihuahuan Desert, according to Morafka (1977), include Texas Banded Gecko (*Coleonyx brevis*); Reticulated Gecko (*C. reticulatus*); Greater Earless Lizard (*Cophosaurus texanus*); Round-tail Horned Lizard (*Phrynosoma modestum*); the spiny lizards, *Sceloporus cautus, S. maculosus, S. merriami, S. ornatus, S. poinsetti,* and *S. magister bimaculosus;* the Fringe-footed Lizard (*Uma exsul*) and the Little Striped and Marbled

Whiptails (*Cnemidophorus inornatus, C. tigris marmoratus*). Of special interest are two other whiptails (*C. neomexicanus* and *C. tesselatus*), which occur as all female parthenogenic clones in select disturbed habitats within the Chihuahuan Desert (see e.g. Wright and Lowe, 1968). Representative snakes include the Trans-Pecos Ratsnake (*Elaphe subocularis*), Western Hooknose Snake (*Gyalopion canum*), Texas Black-headed Snake (*Tantilla atriceps*), and whipsnakes (*Masticophis taeniatus* and *M. flagellum lineatus*). The Mohave Rattlesnake (*Crotalus scutulatus*) is, along with the Western Diamondback (*C. atrox*), the commonly encountered rattler, both occurring here as well as in the Mohave Desert and Sonoran Desert.

A number of species such as the amazing Bolson Tortoise (*Gopherus flavomarginatus*), while contained within the boundaries of the Chihuahuan Desert, are in fact relict grassland species. These and other grassland animals, e.g., the Pronghorn (*Antilocapra americana*) face deteriorating habitats throughout most of the Chihuahuan Desert region, and several can be expected to disappear in the foreseeable future.

Tropical-Subtropical Desertlands

Many of the world's desert regions are included in this climatic zone—the Kalahari, Namib, Eastern and Western Sahels, Arabian, Central Australian, Atacama, Patagonian—and in North America, the Sonoran Desert. These deserts all lie on or near the Tropic of Cancer or Tropic of Capricorn, and are the result of prevailing onshore winds that have either dropped their moisture elsewhere or have had their tropic air cooled by cold ocean currents inhibiting rainfall inland. Accordingly, the most arid portions of these deserts are mostly near the sea, and their flora and fauna tend to be drought-adapted species derived from more humid adjacent tropic biomes.

The Sonoran Desert is unique compared with the others in that its eastern and western extremities exist under different dominant precipitation regimes. In the west, on the Pacific side of the Baja California peninsula, and to a lesser extent northward, scanty winter-spring rainfall from exhausted Pacific frontal storms prevails; summer precipitation is erratic and unusual. At its southeastern edge, along the coast of Sonora, summer convection storms of tropical origin predominate. No part of the Sonoran Desert is entirely free of either of these events, however, and its unifying theme is that of an unreliable and uneven biseasonal rainfall pattern, separated by periods of spring and fall drought. Freezing temperatures are of short duration.

154.1 Sonoran Desertscrub

This large, arid region is centered at the head of the Gulf of California and takes in the western half of the state of Sonora, Mexico, as well as large areas in southeastern California, southwestern Arizona, and the Baja California peninsula. The desert spans 12 degrees of latitude from 23° to 35° north. Although many of its distinguishing characters were known earlier (see e.g. Parish, 1930), it remained for that greatest of North American desert ecologists, Forrest Shreve (1942a, 1951), to describe its unifying features and standardize the appropriate designation of Sonoran Desert.

Shreve recognized seven subdivisions in this desert. Of these, his Foothills of Sonora and parts of his Magdalena Region as well as parts of other of his subdivisions, are considered Sinaloan thornscrub here and are discussed in another section. His remaining five subdivisions (Lower Colorado River Valley,[1] Arizona Upland, Plains of Sonora, Vizcaíno, and Central Gulf Coast) are discussed in this section. Only three of these, the Lower Colorado River Valley subdivision, the Central Gulf Coast, and parts of the Vizcaino, subdivisions would be recognized as desertscrub by the majority of biogeographers. The other two are so dominated by arborescent plants of low stature that by some criteria they could be classed as depauperate thornscrub communities.

Sonoran desertscrub merges in the east with semidesert grassland, or more unusually, Chihuahuan desertscrub; to the south the vegetation gradually increases in density and stature to become Sinaloan thornscrub. In the north, the transition is commonly to chaparral. In southeastern California, this desert is commonly called the "Colorado Desert," and includes the area between the Colorado River and the Coast Ranges south of the Little San Bernardino Mountains and the Mohave Desert. Most of Baja California is Sonoran desertscrub, except for the "Californian" northwest and in the south, where, as in Sonora, it grades into thornscrub. Almost all of southwestern Arizona below 1,050 m is in this biome. As we define the North American Southwest, only that portion of the Sonoran Desert in Baja California south of latitude 27° N is excluded.

The flora of the Sonoran Desert is clearly derived from subtropical elements and its affinities are in large part to the south. The present distribution of desertscrub is recent, with the desert as we understand it undeveloped until 8,000-9,000 years before the present (Axelrod, 1979a, 1979b). Much of the northern area was occupied by woodland until the early Holocene when species of *Juniperus* and *Quercus*, and *Artemisia tridentata*, *Pinus edulis*, and *Pinus monophylla* shifted out of the region as drought conditions developed (Van Devender, 1977; Van Devender and Spaulding, 1979; Wells and Hunziker, 1976). This Holocene shift from woodland to desert was synchronous in the Chihuahuan, Mohave, and Sonoran deserts, and apparently occurred in response to a northward migration of the Aleutian low and winter storm track which resulted in drastically reduced winter rainfall. The plants which moved in to fill these abandoned niches include genera that even today are in great flux. Woody members of such widespread and important genera as *Atriplex* and *Ambrosia* show evidence of having rapidly evolved new species as arid areas were opened to them (Stutz, 1978; Raven et al., 1968). The now widespread Creosotebush (*Larrea tridentata*) presumably underwent autoploidy to produce separate chromosome races in the three deserts (Yang, 1961, 1970; Yang and Lowe, 1968; Wells and Hunziker, 1976).

---

[1]*Shreve (1951) applied the term Lower Colorado Valley to this subdivision.*

Table 23. Continentality of Mohave desertscrub and five Sonoran desertscrub subdivisions as determined by the difference between the average temperature of summer (June, July, August) and winter (December, January, February). Data from Hastings and Humphrey (1969a, 1969b) and Sellers and Hill (1974).

Desert biome region	Average summer temperature –average winter temperature (°C)	Number of stations
Mohave	20.2	17
Lower Colorado River Valley	18.2	64
Arizona Upland	17.7	38
Plains of Sonora	13.6	8
Central Gulf Coast	13.0	25
Vizcaíno	8.6	27

The bimodal rainfall pattern of the Sonoran Desert allows for a greater structural diversity than in the Great Basin, Mohave, and Chihuahuan Deserts. The Sonoran Desert differs markedly from the other North American desert biomes, which are dominated by low shrubs, in its arboreal elements and its truly large cacti and succulent constituents. Even in its most arid parts, the Sonoran Desert exhibits tree, tall shrub and succulent life-forms along drainages and in other favored habitats. These provide for distinctive landscapes—some of which can only be termed bizarre.

Shreve (1951) provisionally characterized Sonoran Desert climate by using temperature data from 8 stations (3 each from Arizona and Sonora and 2 from Baja California) and precipitation data from a total of 40 stations (4 from California, 17 Arizona, 16 Sonora, and 3 Baja California). Since Shreve's initial analysis, dense networks of weather stations have been established on both sides of the International Border and several useful summaries of the resulting data have appeared (Hastings, 1964a, 1964b; Hastings and Humphrey, 1969a, 1969b; Sellers and Hill, 1974). It is now possible to define more precisely the climate of this region.

In the present analysis, data are used from 162 weather stations located within the Sonoran Desert. The stations are listed in the appropriate sections of this chapter. Separate tables also give latitude, longitude, altitude, average annual precipitation, and abbreviated station names used in presenting the data (Figs. 107-111).

Characterizing desert climate using average seasonal precipitation and temperature values is simple but revealing. Plotting temperature against precipitation at each station for the three coldest months (December, January, February) and for three months of the hot season (June, July, August), provides a series of fairly tightly clustered dots, the patterns of which may be used to characterize the subdivision limits of these two seasonal climatic characteristics. Although there is considerable overlap between climographs of some subdivisions, the relative position of the several polygons when plotted at the same scale provides a ready insight into the most likely climatic control for each region (Figs. 107-112). Similar climographs could no doubt also be constructed for the spring and fall seasons, providing additional information. Also it is recognized that minimum temperatures and median rainfall values might be more meaningful than mean data in this and other biomes.

As a measure of continentality, average winter temperatures

were subtracted from average summer temperatures. The differences, averaged for all stations of each desert region, were then used as simple descriptors of the degree of continentality (Table 23).

A map of average annual rainfall throughout the Sonoran Desert region is shown in Figure 113.

Considering the recent and synchronous evolution of North America's deserts, it is not surprising that many Sonoran Desert animals are species found throughout the Southwest's warmer and drier regions. Among the mammals are several bats including the California Leaf-nosed Bat (*Macrotus californicus*), Coyote (*Canis latrans*), Ring-tailed Cat (*Bassariscus astutus*), Black-tailed Jack-rabbit (*Lepus californicus*), Desert Cottontail (*Sylvilagus auduboni*), Merriam's Kangaroo Rat (*Dipodomys merriami*), White-throated Woodrat (*Neotoma albigula*), and Desert Pocket Mouse (*Perognathus penicillatus*). Others, as the Desert Kangaroo Rat (*Dipodomys deserti*), are generally limited to the Sonoran and Mohave Deserts, and a few such as Bailey's Pocket Mouse (*Perognathus baileyi*) and Round-tailed Ground Squirrel (*Spermophilus tereticaudus*) are restricted to the Sonoran Desert region. Still others, (e.g., Arizona Cactus Mouse, *Peromyscus eremicus eremicus*), have adapted at the subspecific level to one or more Sonoran desertscrub communities.

The recent arrival of Sonoran desertscrub is also apparent in its bird inhabitants. Again, some of the more frequently encountered and widely occurring species are found generally throughout the Southwest's desertscrub and coastalscrub—Roadrunner (*Geococcyx californianus*), Mourning Dove (*Zenaida macroura*), Lesser Nighthawk (*Chordeiles acutipennis*), Verdin (*Auriparus flaviceps*), Cactus Wren (*Campylorhynchus brunneicapillus*), Black-tailed Gnatcatcher (*Polioptila melanura*), Phainopepla (*Phainopepla nitens*), and Black-throated Sparrow (*Amphispiza bilineata*). As with the mammals, a number of others are restricted to one or more parts of the Sonoran Desert and to adjacent biomes, and many Sinaloan thornscrub species find suitable habitats in the denser Sonoran desertscrub and adjacent riparian scrub. Several birds are centered in the Sonoran Desert: Gambel's Quail (*Lophortyx gambelii*), Costa's Hummingbird (*Calypte costae*), Gilded Flicker (*Colaptes auratus*), and Gila Woodpecker (*Melanerpes uropygialis*).

A somewhat similar case can be made for Sonoran desertscrub reptiles. Although there are several species restricted to certain locales within particular subdivisions of the Sonoran Desert, others occur widely in appropriate habitats in several Sonoran desertscrub subdivisions as well as in Mohave desertscrub. These include such distinctive animals as the Desert Tortoise (*Gopherus agassizi*), Chuckwalla (*Sauromalus obesus*), Desert Iguana (*Dipsosaurus dorsalis*), Spotted Leaf-nosed Snake (*Phyllorhynchus decurtatus*), Rosy Boa (*Lichanura trivirgata*), and Western Shovelnose Snake (*Chionactis occipitalis*). Some, such as Zebra-tailed Lizard (*Callisaurus draconoides*), also extend into Great Basin desertscrub and others, such as the Mohave Rattlesnake (*Crotalus scutulatus*), occur in Chihuahuan desertscrub. Others e.g., Banded Gecko (*Coleonyx variegatus*), Desert Spiny Lizard (*Sceloporus magister*), Patch-nosed Snake (*Salvadora hexalepis*), Glossy Snake (*Arizona elegans*), Western Ground Snake (*Sonora semiannulata*), and Western Diamondback (*Crotalus atrox*), can be found throughout our warm deserts and drier areas, but have Sonoran Desert subspecies.

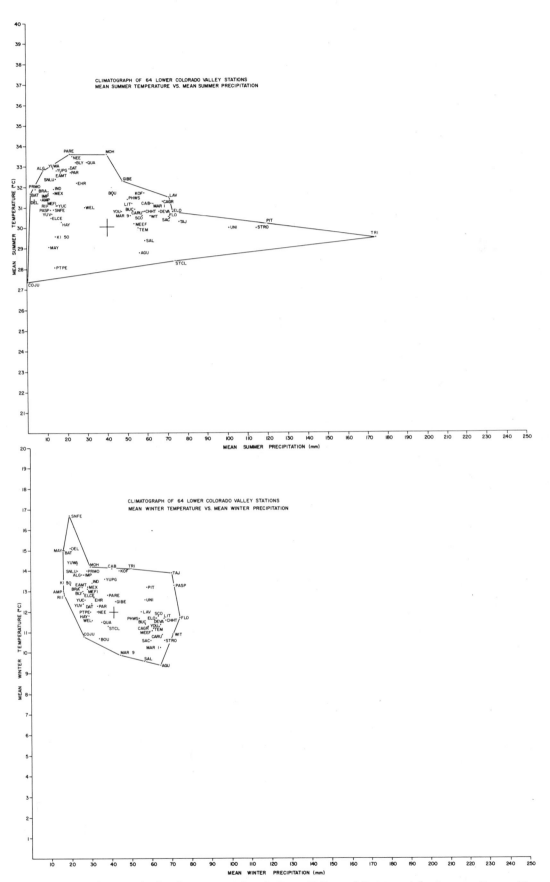

Figure 107. *Climographs for the Lower Colorado River Valley subdivision of the Sonoran Desert. Names of the stations and other data are given in Table 24. See Figure 92.*

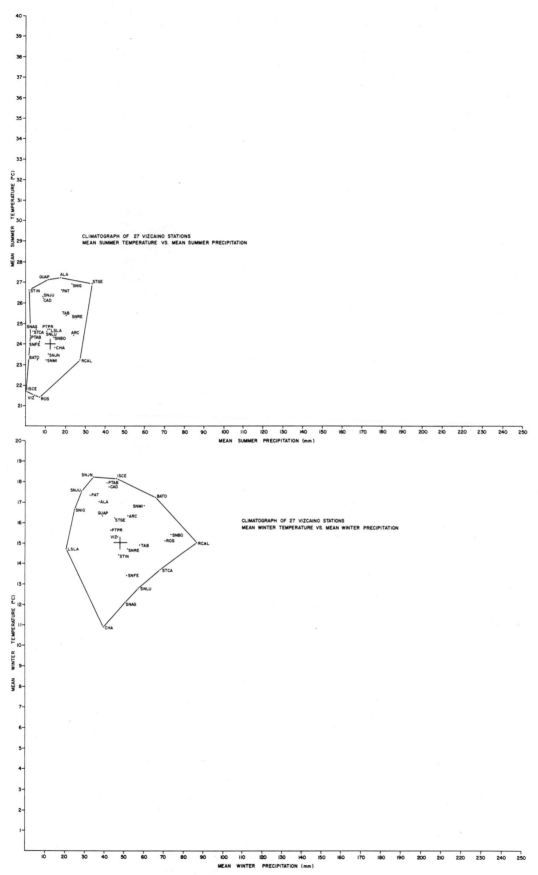

Figure 108. *Climographs for the Vizcaino subdivision of the Sonoran Desert. Names of the stations and other data are given in Table 28. See Figure 92.*

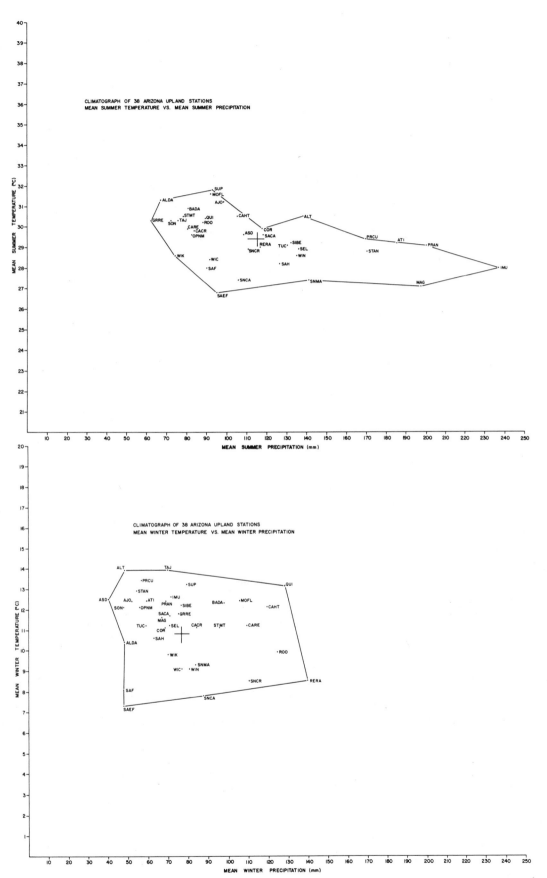

Figure 109. *Climographs for the Arizona Upland subdivision of the Sonoran Desert. Names of the stations and other data are given in Table 27. See Figure 92.*

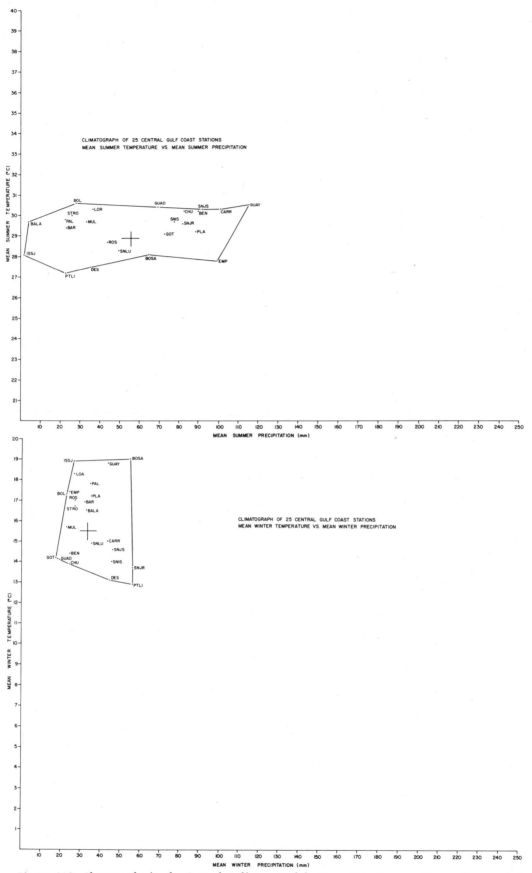

Figure 110. *Climographs for the Central Gulf Coast subdivision of the Sonoran Desert. Names of the stations and other data are given in Table 30. See Figure 92.*

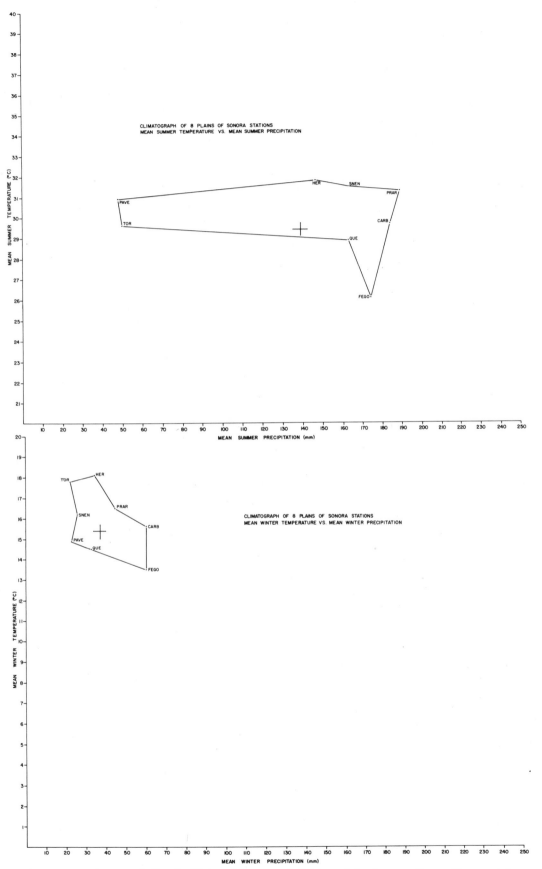

Figure 111. *Climographs for the Plains of Sonora subdivision of the Sonoran Desert. Names of the stations and other data are given in Table 31. See Figure 92.*

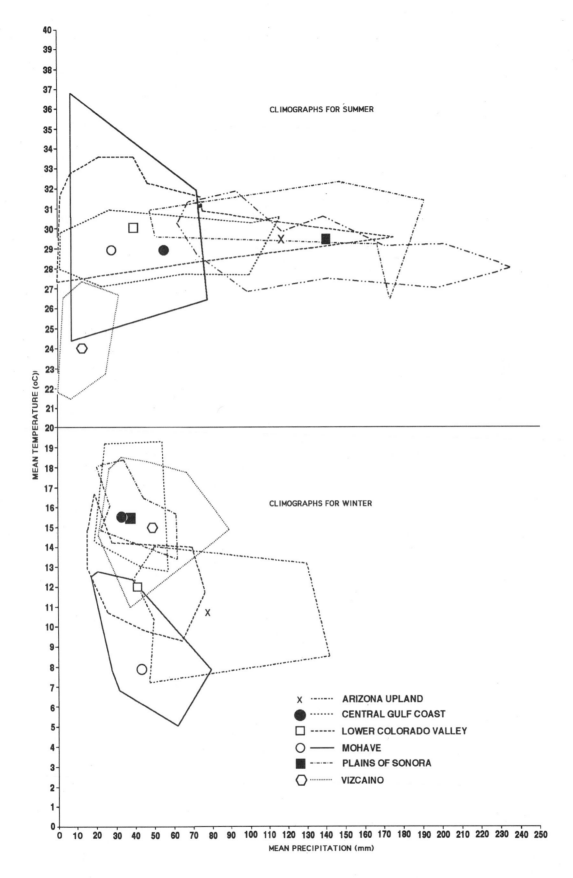

Figure 112. *Comparison of seasonal climographs for the Mohave desertscrub biome with those for the five Sonoran desertscrub subdivisions. See Figures 92 and 107-111. The short crossed lines associated with each climograph represent the means of the station means for temperature and rainfall. The intersections of these average lines provide rough measures for comparing the seasonal climate in the various desert areas.*

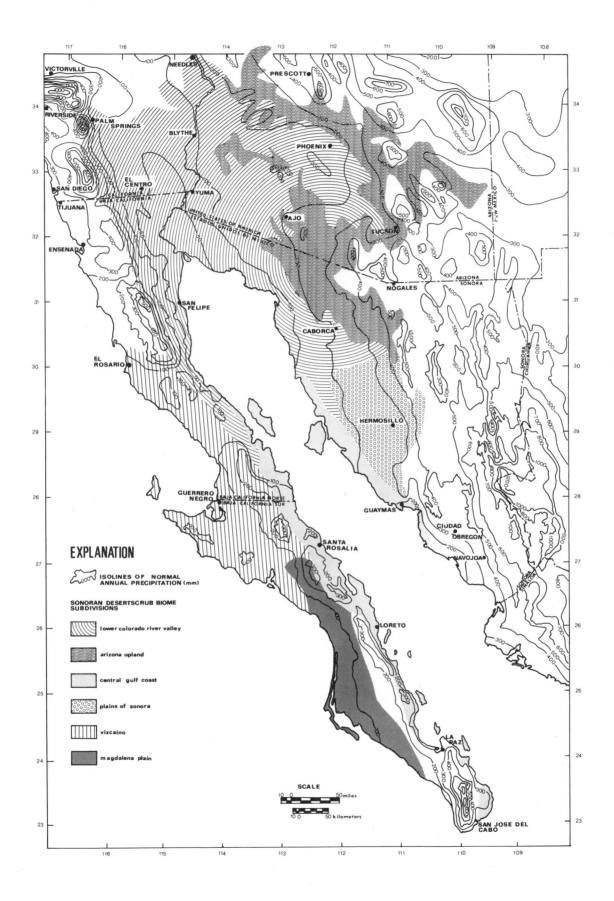

Figure 113. *Map of the outline of the Sonoran Desert biome showing average annual rainfall amounts. Subdivision boundaries are from Shreve (1951) as modified by Brown and Lowe (1980).*

Lower Colorado River Valley Subdivision

Also termed the microphyllous desert by Shreve, the Lower Colorado River Valley is the largest and most arid subdivision of the Sonoran Desert. It extends southward from its center around the head of the Sea of Cortez (Gulf of California) to about 28° 30' north in Baja California and to about 29° 15' north in Sonora. This subdivision plays a central role in the Sonoran Desert because it alone is in contact with all that desert's subdivisions as well as with the Mohave Desert and with Californian coastalscrub. Although many miles of shoreline fall within its boundaries, it is the most "continental" of the subdivisions (Table 23). Its lengthy coastal position is broken in the south on both sides of the Gulf by the Central Gulf Coast subdivision, probably in response to increased winter temperature minima (Figs. 107-111).

The tenuous border with the Mohave Desert is northward of the Salton Basin, and extends northeasterly from Joshua Tree National Monument to about 35° N along the Colorado River near Needles, California. This Mohave Desert-Sonoran Desert transition appears to be determined by winter temperatures with the average winter values at Mohave Desert stations almost 4° C colder than for Lower Colorado River Valley stations (Fig. 112). To the west, in California and northern Baja California, the contact is largely with Californian coastal chaparral, the change there resulting from cooler temperatures and greater winter precipitation in the chaparral. Farther south, its western edge abuts the Vizcaíno subdivision from which it is sharply separated by the Vizcaíno's lower summer temperature values (Fig. 112). To the east, this subdivision contacts the Arizona Upland subdivision, from which it is differentiated by the Arizona Upland's higher summer rainfall and winter temperature minima, and in the southeast the Plains of Sonora, from which it differs both in lower average summer rainfall (Lower Colorado River Valley 40 mm versus 139 mm in the Plains of Sonora) and in lower average winter temperatures (12.0° C in the lower Colorado River Valley and 15.4° C in the Plains of Sonora) (Figs. 107-112). Figure 107 gives winter and summer climatic data in the form of climographs for 64 Lower Colorado River Valley stations. Table 24 gives certain geographical data, average annual precipitation, and abbreviated names used in climographs (Fig. 107).

Because of the combination of high temperature and low precipitation, this subdivision is the driest of the Sonoran Desert subdivisions (Fig. 113). Plant growth is typically both open and simple, reflecting the intense competition existing between plants for the scarce water resource.

As noted by Shreve (1951:51-52), drainageways in the Lower Colorado River Valley subdivision assume two forms. In the drier parts of the desert, if relief is low, the channels conveying the infrequent flows are connected into a network of shallow rills that fail to provide through flow. The goal of the heavily laden water in these anastomosing systems is normally a playa or a plain. The other drainage pattern is dendritic in form, and occurs in areas of greater rainfall or relief. As small channels upslope converge, they form a definite course that ultimately, through repeated convergences, carries runoff to some regional drainage. The minor runnels of both drainage patterns are lined by small trees and shrubs not characteristic of the drier interfluves. The drainageway vegetation is irregularly scattered in the reticulate type and, because the indistinct runnels may be numerous, the illusion is presented of trees and shrubs forming a homogeneous community over the entire bajada. In contrast, the plants of dendritic drainageways grow in rows along the margins of the water courses, clearly set apart from the intervening vegetation of the interfluves. The species of plants are the same, but the drainage pattern has a significant influence on the physiognomy. Species commonly found along larger drainageways include small trees, all of which have a high proportion of their chlorophyll in or beneath the bark of stems and are aphyllous or microphyllous (Johnson, 1976): Western Honey Mesquite (*Prosopis glandulosa* var. *torreyana*), Ironwood (*Olneya tesota*), Blue Palo Verde (*Cercidium floridum*), and Smoketree (*Psorothamnus spinosa*). These species (except *P. spinosa*) are also found outside the washes and might be considered facultative wash species. Other plants are almost wholly obligate in these wash habitats and include Desert Willow (*Chilopsis linearis*), Chuparosa (*Justicia californica*), Desert Honeysuckle (*Anisacanthus thurberi*), and Canyon Ragweed (*Ambrosia ambrosioides*). These occur as mere shrubs (or they are absent) in minor water courses where more common occupants are Catclaw Acacia (*Acacia greggii*), Burrobrush (*Hymenoclea salsola* var. *pentalepis*), Anderson Thornbush (*Lycium andersonii*), and Desert Broom (*Baccharis sarothroides*).

In the more arid parts of this subdivision are extensive areas where the soil is covered by a single layer of tightly packed pebbles, most of which are coated with a dark varnish. Development of this "desert pavement" is either from *in situ* weathering or transport of weathered pebble-size fragments from nearby slopes (Evanari et al., 1971). Such areas are commonly devoid of perennial plants, supporting instead a sparse seasonal cover of ephemerals such as Wooly Plantain (*Plantago insularis*) and *Chorizanthe rigida*, and occur as low flat ridges separated by runnels (Fig. 114). The runnels often support dense growths of such dominant Sonoran desertscrub perennials as Creosotebush (*Larrea tridentata*), White Bursage (*Ambrosia dumosa*), Ocotillo (*Fouquieria splendens*), Brittlebush (*Encelia farinosa*), Foothill Palo Verde (*Cercidium microphyllum*), Saguaro (*Carnegiea gigantea*), and Ironwood (*Olneya tesota*) (Fig. 115). *Psorothamnus spinosa* is a Lower Colorado River Valley endemic that is restricted largely to washes and these runnels. In western Arizona, the absence of plants on the interfluves has been attributed to the presence of large amounts of exchangeable sodium in the layer of vesicular soil lying immediately beneath the pebbles (Musick, 1975). These soils tend to have low infiltration rates, hence high runoff characteristics. Plants growing on the interfluves are less well watered than those on nearby soils lacking desert pavement while the runnels receive more water than falls directly.

Sand substrates are also commonplace in this part of the desert, and dunes are characteristic features of some areas. Although the "Algodones" west of Yuma are the largest and most famous, other examples occur west of the Mohawk Mountains, west of Laguna Salada, on the Cactus Plain near Parker, and elsewhere. Otherwise, the general habitats consist of rugged mountains, some half buried in their own alluvium, their outwash plains and bajadas, and valleys of alluvial fill.

Table 24. Data for 64 weather stations in the Lower Colorado River subdivision. Data for each station include latitude and longitude (values rounded to closest 0.1°), altitude (m), average annual precipitation (mm), and abbreviated name used in climographs of Fig. 107. Data from Hastings and Humphrey (1969a, 1969b), Sellers and Hill (1974) and U.S. Environmental Data Service (1959-1979).

Station name	Abbreviated name	Latitude	Longitude	Altitude (m)	Average annual precipitation (mm)
Aguila, AZ	AGU	33.9	113.2	661	190.3
Ampac, BCN	AMP	32.6	115.4	5	39.3
Bataques, BCN	BAT	32.5	115.1	5	30.6
Blythe, FAA, CA	BLY	33.6	114.7	119	80.5
Bouse, AZ	BOU	33.9	114.0	283	128.5
Brawley 2S, CA	BRA	32.9	115.5	−30	54.6
Buckeye, AZ	BUC	33.4	112.6	265	179.8
Caborca, Son.	CAB	30.7	112.2	292	175.2
Casa Grande, AZ	CAGR	32.9	111.7	427	206.2
Casa Grande Ruins, National Monument, AZ	CARU	33.0	111.5	433	208.8
Chandler Heights, AZ	CHHT	33.2	111.7	434	218.9
Colonia Juarez, Son.	COJU	32.3	115.1	20	42.0
Dateland, CA	DAT	32.8	113.5	136	77.0
Deer Valley, AZ	DEVA	33.6	112.1	383	205.3
Delta, BCN	DEL	32.4	115.2	20	35.4
Eagle Mtn., CA	EAMT	33.8	115.4	297	70.8
Ehrenberg, AZ	EHR	33.6	114.5	98	89.6
El Centro, CA	ELCE	32.8	115.6	−9	58.5
El Mayor, BCN	MAY	32.1	115.8	20	49.1
El Riito, BCN	RII	32.2	115.0	5	40.9
Eloy, AZ	ELO	32.7	111.5	475	214.6
Florence, AZ	FLO	33.0	111.4	458	241.8
Gila Bend, AZ	GIBE	32.9	112.7	224	146.4
Hayfield P, CA	HAY	33.7	115.6	418	74.9
Imperial, CA	IMP	32.9	115.6	−19	59.8
Indio USDG, CA	IND	33.7	116.2	8	76.3
Kilómetro 50, Son.	KI50	32.3	115.1	12	48.8
Kofa Mts., AZ	KOF	38.3	113.9	541	156.2
La Unión, Son.	UNI	30.2	112.3	390	194.4
Laveen, AZ	LAV	33.3	112.1	339	193.0
Litchfield Park, AZ	LIT	33.5	112.4	314	192.0
Los Algodones, Son.	ALG	32.7	114.7	35	49.4
Maricopa 9SSW, AZ	MAR9	32.9	112.1	427	159.3
Maricopa 1, AZ	MAR1	33.1	112.0	354	186.7
Mecca FIR, CA	MEFI	33.6	116.1	−55	70.1
Mesa Exp. Farm, AZ	MEEF	33.4	111.9	375	191.10
Mexicali, BCN	MEX	32.6	115.4	4	69.1
Mohawk, AZ	MOH	32.7	113.8	164	112.4
Needles, FAA, CA	NEE	34.8	114.6	278	111.3
Palm Springs, CA	PASP	33.8	116.5	129	135.4
Parker, AZ	PAR	34.2	114.3	130	97.1
Parker Reservoir, CA	PARE	34.3	114.2	225	110.5
Phoenix, AZ	PHWS	33.4	112.0	339	171.2
Pitiquito, Son.	PIT	30.7	112.1	330	245.4
Presa Morelos, Son.	PRMO	32.7	114.7	35	65.9
Puerto Peñasco, Son.	PTPE	31.3	113.5	4	85.6
Quartzsite, AZ	QUA	33.7	114.2	271	111.1
Sacaton, AZ	SAC	33.1	111.7	391	212.6
Salome, AZ	SAL	33.8	113.6	579	180.1
San Felipe, BCN	SNFE	31.0	114.8	10	57.6
San Luis, Son.	SNLU	32.5	114.8	27	54.7
Sta. Clara, BCN	STCL	31.3	115.2	540	203.3
Sta. Rosa School, AZ	STRO	32.3	112.1	561	255.4
Scottsdale, AZ	SCO	33.5	111.9	374	204.8
Tajitos, Son.	TAJ	30.9	112.4	340	214.0
Tempe, AZ	TEM	33.4	111.9	351	198.8
Trincheras, Son.	TRI	30.4	111.5	680	287.3
Wellton, AZ	WEL	32.7	114.1	79	96.8
Wittmann, AZ	WIT	33.8	112.5	518	216.6
Yuma Citrus Sta., AZ	YUC	32.6	114.6	58	70.4
Yuma Proving Ground, AZ	YUPG	32.8	114.4	98	89.5
Yuma Valley, AZ	YUV	32.7	114.7	37	65.8
Yuma WSO, AZ	YUWA	32.7	114.6	59	68.7
Youngtown, AZ	YOU	33.6	112.3	346	191.6
Subdivision Average					131.2

Figure 114. *Area of desert pavement in the Lower Colorado River Valley subdivision east of Cibola, Arizona. Perennial species of the Creosotebush-White Bursage series occur in the runnels where runoff from the interfluves promotes more favorable moisture conditions. Altitude 75 m.*

Figure 115. *Lower Colorado River Valley subdivision vegetation in MacDougal Crater, Sierra Pinacate region, Sonora, Mexico. Rainfall here is about 125 mm. The plants are arranged in lines along small and large runnels on the crater floor. The minor rills receive extra water from the barren interfluves that are covered with desert pavement. The major drainageways receive runoff from that source and from the crater walls. The small light colored shrubs are* **Encelia farinosa** *and* **Ambrosia dumosa.** *The larger dark shrubs are* **Larrea tridentata** *and the small widely scattered trees are* **Cercidium microphyllum, C. floridum,** *and hybrids between these. The coarse grass, forming large dense clumps, is* **Hilaria rigida.** *The columnar cactus is* **Carnegiea gigantea.** *Foreground plants (lower left) include* **Fouquieria splendens, Jatropha cuneata, Encelia farinosa,** *and* **Cercidium microphyllum.** *The slope in the background is dominated by the Brittlebush series. Altitude 275 m. Photograph by R.M. Turner.*

Figure 116. *The Creosotebush-White Bursage series that exemplifies the Lower Colorado River Valley subdivision. Here these species grow on sandy alluvium near the Gila River around Arlington, Arizona. Altitude 245 m.*

Creosotebush-White Bursage Series

These two plants, either together or alone compose the most widespread and important community of the Lower Colorado River Valley subdivision, occurring over many thousands of hectares in the broad valleys common to the region (Fig. 116). Both species normally decrease in importance upslope on the bajadas; White Bursage (*Ambrosia dumosa*) barely extends above the broad valley floors, whereas Creosotebush (*Larrea tridentata*) continues to hold a position on the upper-most bajadas and even continues on into the mountains (Figs. 116, 117). This community may occur even on steep slopes, especially on hills derived from young volcanic rocks. "These two plants are either dominant or abundant under differences of substratum which would support dissimilar vegetation under more favorable moisture conditions. This is due to the remarkable physiological constitution of these shrubs, and also to the simplicity of the flora, which has given them little competition" (Shreve 1951:49). This community is similar to the Creosotebush-White Bursage community that occurs within the Mohave Desert, except that in the Sonoran Desert *Larrea* has the tetraploid chromosome complement versus hexaploid in the Mohave (Yang, 1970; Barbour, 1969).

Competition between the two dominants of this series was recently studied by Fonteyn and Mahall (1978). Although made in the Mohave Desert, their study probably has application here also. They found that, at times of low water

availability, interspecific competition is more intense than intraspecific competition. Another finding of their study was that *Larrea* is regularly spaced and *Ambrosia* is clumped. The authors speculated that, at one time, *Larrea* too had been clumped, but intense competition thinned the stand, resulting in wide spacing with minimized competition. That *Ambrosia* remains clumped may result from growth activity occurring only when water is not a limiting factor. In the Mohave and western Sonoran Deserts, as rainfall increases, the density of *Larrea* increases (Woodell et al., 1969), although the latter plants remain evenly spaced. Toward the eastern limits of the Lower Colorado River Valley subdivision in Avra Valley, Arizona, rainfall is near the maximum for the series. *Larrea* plants are clumped about mounds and the mounds themselves are contagiously distributed on a larger scale (Wright, 1970). At the moister limits of the series, *Larrea* escapes the extremes of aridity that induce stand thinning and regular spacing farther west. Burk (1977) discussed the various hypotheses used to explain spacing of *Larrea* and concluded that the relative importance of several factors at a given site determines the occurrence pattern. These include water availability, soil depth, mode of reproduction, and root system symmetry.

Marks (1950) described the vegetation-soil characteristics of the Lower Colorado River Valley east and west of Yuma, Arizona. With emphasis on plants as indicators of soil arability, he described a progression of communities occupying sandy loams to coarse clean sand. On the sandy loams,

Figure 117. *Sonoran desertscrub near the Eagle Tail Mountains west of Phoenix, Arizona. This area is ecotonal between the Creosotebush-White Bursage series of the Lower Colorado River Valley subdivision and the Paloverde-cacti series of the Arizona Upland subdivision. In this view the plants of the latter community, such as* **Carnegiea gigantea, Cercidium microphyllum,** *and* **Olneya tesota,** *have slightly higher water requirements and occur in depressions in the landscape where runoff water provides improved moisture relations.* **Larrea,** *here without* **Ambrosia dumosa,** *is the dominant in all positions of the terrain. Photograph by R.M. Turner. Altitude 450 m.*

Larrea and *Ambrosia* are dominants. As the sand fraction increases, Big Galleta (*Hilaria rigida*) appears, followed by Indigo Bush (*Psorothamnus schottii*). The importance of *Larrea* is reduced as sand increases; *Ambrosia* withstands sand somewhat better. On the sandiest soils, *Ambrosia* disappears and Longleaf Ephedra (*Ephedra trifurca*) and *Eriogonum deserticola* dominate. His delineation provides a basis for identifying several additional series that might be recognized for the region.

In a few areas, as in the San Cristobal Valley in southwestern Arizona, *Ambrosia dumosa* occurs as a dominant without *Larrea* (Fig. 118). The community is found on a sand sheet apparently formed during the Pleistocene by aeolian material carried up from the nearby Gila River (Martin, 1979).

Saltbush Series

This series is a community of gently sloping lands and valleys. Much of this series is now under cultivation, but if areas formerly supporting this community are included, it ranks second to *Larrea* in area occupied. It was the most widespread community in the Gila Valley, Arizona (Fig. 119), and the Coachella Valley, California, before cultivation (Shantz and Piemeisel, 1924). Formerly, and even today, these lands are subject to flooding, however infrequently. In general, the soil occupied by this community is finer than that of the *Larrea-Ambrosia* community and the water retention capacity is therefore greater. This feature decreases the penetration of water on lands occupied by *Atriplex*. Soils supporting the *Atriplex* community are therefore generally more saline than *Larrea* habitats but, except where the *Atriplex* is stunted signalling extremes in soil chemical and physical properties, salt content is not too great for farming purposes. This community, too, is widespread in the Mohave Desert.

Figure 118. Ambrosia dumosa *occurs without* **Larrea** *on certain sandy soils as in the San Cristobal Valley near Dateland, Arizona. Coppice dunes in the background dominated by* **Prosopis glandulosa** *var.* torreyana. *Altitude 130 m.*

Figure 119. *Saltbush series south of Phoenix, Arizona. Few unaltered stands of this community remain; most have become irrigated cropland and the rest have been disturbed by livestock with the resulting establishment of such exotic annual species as Mediterranean grass* (Schismus arabicus *and* S. barbatus), *Red Brome* (Bromus rubens), *and London Rocket* (Sisymbrium irio). *Photograph by R.M. Turner. Altitude 345 m.*

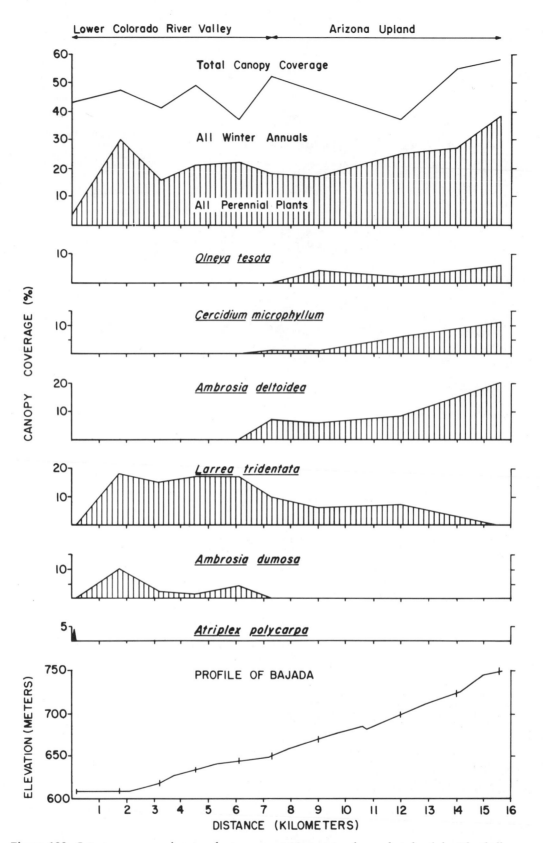

Figure 120. *Canopy coverage of major plant species at 10 stations along a bajada of the Silverbell Mountains, Arizona. Tic marks on the profile of the bajada (bottom) indicate the location of the 10 stations. The lower five stations support* **Atriplex polycarpa** *or* **Ambrosia dumosa** *and* **Larrea tridentata***, plants characteristic of the Lower Colorado River Valley subdivision. The upper five stations on the bajada fall within the Arizona Upland subdivision and support such characteristic species as* **Ambrosia deltoidea***,* **Cercidium microphyllum***, and* **Olneya tesota***. Total canopy coverage, comprising values from both perennial and winter annual species, is shown at the top of the diagram. Note that* **Larrea** *is most abundant at lowland stations also supporting* **Ambrosia dumosa***. Upslope,* **Larrea** *is still present but of diminished importance.*

Ecotones between areas of saltbush and the Creosotebush-Bursage series are sharp in some places but more commonly gradual. In some stands of this series, *Atriplex polycarpa* is the only woody species; elsewhere, such plants as *Atriplex canescens, Lycium,* and *Prosopis* are common associates. Shantz and Piemeisel (1924) listed additional species for the Coachella and Gila River valleys. In the Coachella Valley, they note the common or frequent occurrence of Screwbean Mesquite (*Prosopis pubescens*), Goldenbush (*Isocoma acradenia*), and the more unusual occurrence of species such as Arrow-weed (*Tessaria sericea*), Quail Brush (*Atriplex lentiformis*), and *Larrea tridentata*. In the Gila Valley, Fremont Thornbush (*Lycium fremontii*), Narrow-leaved Wingscale (*Atriplex canescens* ssp. *linearis*), and Cilindrillo (*Lycium berlandieri*) are important additions to the community with Corona de Cristo (*Castela emoryi*), Coulter Globe Mallow (*Sphaeralcea coulteri*), Fendler Globe Mallow (*S. fendleri*), and Jimmy Weed (*Isocoma heterophylla*) less common or rare. In the Avra Valley, west of Tucson, Arizona, where this series reaches its eastern limit, additions to this community include Burrow-weed (*Isocoma tenuisecta*), *Larrea tridentata, Lycium* spp., *Ambrosia dumosa,* and *A. deltoidea* (Table 25). In one stand in this valley, on sandy soil with a hard pan at a depth of about 20 cm, crown coverage of perennial species is only 3.5% (Table 25, Fig. 120). Yet, during the spring period of maximum ephemeral growth, coverage of ephemerals alone may amount to 39% (Table 26, Fig. 120).

When destroyed by fire or cultivation, stands of this vegetation in the Coachella Valley are first occupied by annuals and biennials, then by *Isocoma acradenia,* followed by the return of *Atriplex polycarpa* (Shantz and Piemeisel, 1924). In recent work in Avra Valley, Arizona, Karpiscak and Grosz (1979) and Karpiscak (1980) provided more detailed information about succession in this and other Lower Colorado River Valley subdivision series. Although distinctions between *Atriplex-* and *Larrea-*dominated communities are not made, their study shows a rather regular successional sequence of mainly exotic annuals. Russian Thistle (*Salsola kali*) quickly follows retirement of agricultural fields. This species is replaced in 2 to 3 years by several species of mustards, which in turn are succeeded by a dense growth of introduced annual grasses accompanied by other non-native species including Filaree (*Erodium cicutarium*) and Prickly Lettuce (*Lactuca serriola*) and a few native plants such as *Sphaeralcea.* Ruderal shrubs such as *Baccharis sarothroides* and *Isocoma* usually become established before *Larrea* or *Atriplex* reoccupy the site.

This *Salsola kali-*mustards-grass sequence is similar to that reported for the sagebrush series of Great Basin desertscrub (Piemeisel, 1938, 1951). The mustards that enter the sequence in the two deserts are the same or belong to closely related species pairs [Great Basin desertscrub: Tumble Mustard (*Sisymbrium altissimum*) and Tansy Mustard (*Descurainia pinnata*); Sonoran desertscrub: Yellow Rocket (*Sisymbrium irio*), *Descurainia pinnata* and *D. sophia*]. Cheatgrass Brome (*Bromus tectorum*) is the dominant seral grass of the Great Basin desertscrub whereas in Sonoran desertscrub the role is taken by Arabian Grass (*Schismus arabicus*), Mediterranean Grass (*S. barbatus*), and Red Brome (*Bromus rubens*).

Other Series

In addition to the two preceding series, which are by far the most important, a number of additional communities occur

Table 25. Canopy coverage (percent) of perennial plants at three stations in Avra Valley west of Tucson, Arizona (1962).

Species	Lower Colorado River Valley		Arizona Upland
	Atriplex polycarpa series (alt. 610 m)	Larrea-Ambrosia series (alt. 610 m)	Cercidium-Carnegiea series (alt. 715 m)
Ambrosia deltoidea	present		10.4
Ambrosia dumosa	present	10.5	
Atriplex polycarpa	3.5		
Carnegiea gigantea			0.009*
Cercidium microphyllum			2.8
Echinocereus fendleri			0.0014*
Ferocactus wislizenii			0.0007*
Isocoma tenuisectis	present		
Jatropha cardiophylla			present
Krameria grayi			0.1
Larrea tridentata	present	18.0	9.2
Lycium species	present		
Mammillaria microcarpa			0.0014*
Olneya tesota			3.5
Opuntia fulgida			2.1
O. versicolor			0.2
Perezia nana		present	
Prosopis velutina		1.5	
Psilostrophe cooperi		0.5	
Zinnia pumila		present	
Total coverage	3.5	30.5	27.3

*Plants/m²

Table 26. Canopy coverage (percent) of winter annuals at three stations in Avra Valley near Tucson, Arizona. Values are averages taken from thirty 0.1 m² plots at each site (spring 1962).

Species	Lower Colorado River Valley		Arizona Upland
	Atriplex polycarpa series	Larrea tridentata-Ambrosia dumosa series	Cercidium-Carnegiea series
Amsinckia intermedia		0.33	
Aristida sp.			0.17
Astragalus nuttallianus	0.17	0.17	
Chaenactis stevioides		2.17	12.58
Calycoseris wrightii		0.08	
Chorizanthe brevicornu			0.50
Chorizanthe rigida			1.58
Cryptantha spp.	4.50	0.08	0.17
Eriastrum diffusum	2.92		
Eriophyllum lanosum	7.00		0.50
Erodium cicutarium		0.17	0.08
Erodium texanum		0.17	0.17
Festuca octoflora	2.42		
Lepidium lasiocarpum	4.00	2.83	
Lesquerella gordonii	0.08	4.67	
Linaria texana	0.67		
Loeflingia squarrosa	3.67		
Lotus humistratus	2.58		
Lotus tomentellus	0.50	0.08	
Nemacladus glanduliferus		0.92	
Oenothera spp.		0.08	
Pectocarya recurvata	3.58	0.17	12.67
Phacelia distans		0.33	
Phlox spp.	1.67		
Plantago insularis		3.75	
Plantago purshii	2.00		
Schismus arabicus		0.83	
Thelypodium lasiophyllum	0.58	0.17	
Tillaea erecta	2.75		
Total coverage	39.09	17.00	28.42

Figure 121. *Brittlebush series on coastal slope overlooking the Sea of Cortez. This area south of San Felipe, Baja California del Norte, is among the most arid on this continent. The open, simple community is found on rocky basaltic soils. Photograph by J.R. Hastings. Altitude 75 m.*

with *Encelia farinosa, Prosopis glandulosa* var. *torreyana, Hilaria rigida,* and *Frankenia palmeri* as dominants.

Consociations of *Encelia* occur in open sparse stands on volcanic rocks near the head of the Sea of Cortez (Fig. 121). These desert pavement soils are probably characterized by low rates of infiltration in this area of severely deficient rainfall. In the vicinity of Yuma, Arizona, some gravelly bajadas were reported as dominated by *Encelia farinosa* and *Atriplex hymenelytra* (Marks, 1950).

Creosotebush-Big Galleta Series

Shreve estimated that nearly one-seventh of this subdivision is occupied by sandy plains or dunes. These sandy areas are generally found in the lowest and hottest reaches of the desert in the vicinity of the Colorado River and the Sea of Cortez. There a series dominated by *Larrea* and *Hilaria rigida* is found in places on the Yuma and San Luís Mesas (Marks, 1950). *Hilaria rigida,* unlike most grasses, has a woody structure and elevated renewal buds. Because the renewal buds are not protected from climatic extremes by a mantle of soil, its growth form is similar to that of shrubs. This grass is the main stabilizer over extensive areas of sand dunes that occur in the subdivision (Fig. 122) and in some Mohave Desert communities (Vasek and Barbour, 1977).

Frankenia Series

This series is found not only in the Lower Colorado River Valley subdivision, which it enters along coastal beaches and low terraces about the head of the Sea of Cortez, but also in similar habitats in the Central Gulf Coast and Vizcaino subdivisions. The species diversity is exceedingly low there with *Frankenia* almost the sole perennial (Fig. 123).

Mixed Scrub Series

Along washes and similar places are more diverse communities within the overall Creosotebush-White Bursage series. There one can find open to fairly dense assemblages in which Blue Palo Verde (*Cercidium floridum*), Ironwood (*Olneya tesota*), Desert Lavender (*Hyptis emoryi*), *Psorothamnus schottii,* Jojoba (*Simmondsia chinensis*), and other typically Sonoran species may participate. These associations differ from structurally similar ones in the Arizona Upland subdivision by the poorer representation or absence of Little-leaf Palo Verde (*Cercidium microphyllum*) and Velvet Mesquite (*Prosopis velutina*) and by a more open ground cover in which Triangle-leaf Bursage (*Ambrosia deltoidea*), Bush Muhly (*Muhlenbergia porteri*), and the several species of *Opuntia* associated with the Arizona Upland subdivision are conspicuously lacking.

A few minor series are formed by the local occurrence of

Figure 122. *Big Galleta series within the Lower Colorado River Valley subdivision south of Casa Grande, Arizona. Although technically a grass, the dominant* **Hilaria rigida** *is shrub-like. It is semi-woody with renewal buds on the stems, as in shrubs, instead of below the soil surface, as in most grasses. In that sense and because of the bare soil separating the individual plants, this series might best be regarded as desertscrub. Altitude 455 m.*

Figure 123. *Coastal flat, dominated by* **Frankenia palmeri**, *37 km south of San Felipe, Baja California del Norte. This essentially monotypic community is common along both shores of the upper Sea of Cortez and along beaches of the Pacific Ocean in Baja California. It is found in three Sonoran Desertscrub subdivisions: Lower Colorado River Valley, Central Gulf Coast, and Vizcaino. Photograph by J.R. Hastings. Altitude 5 m.*

such species as Western Honey Mesquite (*Prosopis glandulosa* var. *torreyana*) on dunes (Fig. 118), Teddy Bear Cholla (*Opuntia bigelovii*), Desert Agave (*Agave deserti*), and Ocotillo (*Fouquieria splendens*). These situations become increasingly common as the edge of the subdivision is approached and contacts with the constituents of other subdivisions and biomes increase. A typical contact between any two is commonly marked by a broad ecotone in which there is a sharp increase in floristic complexity as the relatively simple composition of the Lower Colorado River Valley subdivision gives way to more diverse associations containing species of other subdivisions or biomes.

The following cacti are well represented or are largely confined to this subdivision (Benson, 1969): *Opuntia wigginsii*, Silver Cholla (*O. echinocarpa*), Diamond Cholla (*O. ramosissima*), Beavertail (*O. basilaris*), Teddy Bear Cholla (*O. bigelovii*), Kunze Cholla (*O. stanlyi* var. *kunzei*), *Mammillaria tetrancistra, Neoevansia striata*, Nightblooming Cereus (*Peniocereus greggii*), Engelmann Hedgehog (*Echinocereus engelmannii*), and Compass Barrel Cactus (*Ferocactus acanthodes*).

Two ungulates have surprisingly adapted to this arid region. The Lower Colorado subdivision's rugged mountains are now the metropolis for the Desert Bighorn Sheep (*Ovis canadensis nelsoni*), and some of its sandy plains and desert pavement are the habitat for the Sonoran Pronghorn (*Antilocapra americana sonoriensis*). Otherwise, large mammals, except for the introduced Burro (*Equus asinus*) and ubiquitous Coyote (*Canis latrans*), are unusual. Most of its mammals have adapted to high diurnal temperatures by spending much of the day underground or aestivating. Consequently, the subdivision's sandy plains may host large populations of burrowing rodents, one of which at least, the Round-tailed Ground Squirrel (*Spermophilus tereticaudus*), is characteristic of the subdivision. Others of somewhat more widespread occurrence are the Kit Fox (*Vulpes macrotis*), White-tailed Antelope Squirrel (*Ammospermophilus leucurus*), Desert Pocket Mouse (*Perognathus penicillatus*), Long-tailed Pocket Mouse (*Perognathus formosus*), and Desert and Merriam Kangaroo Rats.

This is the poorest subdivision of the Sonoran Desert for birds. Its open, sparsely vegetated habitats simply do not support the more diverse avifauna associated with the structurally taller and denser habitats of the other subdivisions. Its avian inhabitants are largely lesser numbers of arid-adapted desert species. Its only diagnostic bird is LeConte's Thrasher (*Toxostoma lecontei*), which is shared with the Vizcaino subdivision and with Mohave desertscrub.

The sandy plains and dunes of the Lower Colorado River Valley subdivision have resulted in a number of unique sand-adapted lizards and snakes, some of which are restricted to this subdivision, whereas others are found only in similar habitats in Mohave desertscrub. Outstanding examples of these are provided by the fringe-toed lizards (*Uma inornata, U. notata*), Flat-tailed Horned Lizard (*Phrynosoma m'calli*), Banded Sand Snake (*Chilomeniscus cinctus*), and the notorious Sidewinder (*Crotalus cèrastes*). Rocky outcrops, bajadas, talus slopes, washes, and gravel plains each support a varied and often different herpetofauna—Chuckwalla, Desert Spiny Lizard, Brush Lizard (*Urosaurus graciosus*), Southern Desert Horned Lizard (*Phrynosoma platyrhinos calidiarum*), Western Whiptail (*Cnemidophorus tigris tigris*), Desert Glossy Snake (*Arizona elegans eburnata*), and many, many more, making this a most productive region for herpetology.

Arizona Upland Subdivision

Variously referred to as the Arizona Desert, Paloverde-Cacti Desert, and *Cercidium-Opuntia* Desert (all from Shreve, 1951), this subdivision includes some of the most famous and picturesque portions of the Sonoran Desert. It forms a narrow curving border at the northeast edge of the Sonoran Desert from the Buckskin Mountains, Arizona (35° north, 14° west) southeastward to northeast of Phoenix, Arizona (parallel 34° north), and south to about Altar, Sonora (parallel 30° 15' north). Most, perhaps more than 90%, of this region is on slopes, broken ground, and multi-dissected sloping plains— hence the name Arizona Upland. Truly spectacular, it is the best watered and least desert-like desertscrub in North America.

The vegetation most often takes on the appearance of a scrubland or low woodland of leguminous trees with intervening spaces held by one to several open layers of shrubs and perennial succulents. Many of these trees are the same species that to the west, in the more arid Lower Colorado River Valley subdivision, are confined to runnels and washes. Included among them are Blue Palo Verde (*Cercidium floridum*), Ironwood (*Olneya tesota*), mesquites (*Prosopis* spp.), and Cat-claw Acacia (*Acacia greggii*). In the Arizona Upland, Foothill Palo Verde (*Cercidium microphyllum*) is the characteristic palo verde except at the desert's upper, northern, and eastern limits where it may be accompanied or replaced by Crucifixion Thorn (*Canotia holacantha*). The arboreal nature of this desert subdivision is often strikingly similar to Sinaloan thornscrub, and indeed, the vegetation could be considered a depauperate northern form of that tropic-subtropic biome.

So important are cacti in this subdivision that Shreve termed it the crassicaulescent or stem succulent desert, in reference to the important role that may be played by this group. The following cacti are largely confined to, or best represented in, this subdivision (Benson 1969): *Opuntia acanthocarpa* var. *major*, Thornber Buckhorn Cholla (*O. acanthocarpa* var. *thornberi*), Cane Cholla (*O. spinosior*), Staghorn Cholla (*O. versicolor*), Chain Fruit Cholla (*O. fulgida*), Teddy Bear Cholla (*O. bigelovii*), Desert Christmas Cactus (*O. leptocaulis*), Pencil Cholla (*O. arbuscula*), Klein Cholla (*O. kleiniae*), Devil's Club Ground Cholla (*Opuntia stanlyi*), *O. stanlyi* var. *peeblesiana, O. phaeacantha*, Saguaro (*Carnegiea gigantea*), Organ Pipe (*Stenocereus thurberi*), Senita (*Lophocereus schottii*), Night-blooming Cereus (*Peniocereus greggii* var. *transmontanus*), *Echinocereus fasciculatus*, Fishhook Pincushion (*Mammillaria microcarpa*), Thornber Pincushion (*M. thornberi*), *M. grahami* var. *oliviae*, Fish-hook Barrel Cactus (*Ferocactus wislizenii*), Compass Barrel Cactus (*F. acanthodes*), *Echinocactus horizonthalonius* var. *nicholii*, and *Neolloydia erectocentra* var. *acunensis*.

Many of the woody plants are spiny; those that are not generally possess aromatic terpenes or other chemical principles that serve to discourage use by herbivores.

The lower contact of this subdivision is with the Lower Colorado River Valley subdivision along its western edge and in valleys. The lower elevation contact may be as low as ca. 300 m (e.g. near Ajo, Arizona, and Sonoita, Sonora) or as high as 650 m (e.g. in Avra Valley near Tucson, Arizona). Upslope this region of desertscrub extends from ca. 900 m to above 1,000 m in elevation to merge at its upper limits with colder

Figure 124. *Vegetation representing the Paloverde-cacti series of the Arizona Upland subdivision north of the Tortolita Mountains, Arizona. The trees are* Cercidium microphyllum. *Cacti are conspicuous members of this subdivision and are here represented by* Carnegiea gigantea, Opuntia phaeacantha *var.* discata, *and* O. fulgida. Larrea tridentata *and* Fouquieria splendens *are common associates. The low abundant shrub is* Ambrosia deltoidea. *Altitude 810 m.*

and wetter interior chaparral or semidesert grassland. In the southeast, still greater summer precipitation marks its transition to thornscrub proper.

Average annual precipitation for stations in this subdivision lies mainly between 200 mm (Sonoita, Sonora) and 425 mm (Reno Ranger Station, Arizona) (Table 27). Summer (June, July, August) rainfall accounts for 30% to 60% of the annual total with smaller proportions to the north and larger to the south. Average summer precipitation at the southernmost stations of Imuris, Presa la Angostura, Magdalena, and Atil is 237 mm, 200 mm, 197 mm, and 186 mm, respectively; the lowest values of this season are at localities near the northern and western margins (e.g., Granite Reef, 62 mm; Alamo Dam, 67 mm; Sonoita, 72 mm; Wikieup, 75 mm) (Fig. 108). Winter rainfall ranges from 10% to 40% of the annual total with the larger proportions to the northwest. The three warmest stations in the winter (Altar, Tajitos, Presa Cuauhtemoc) all border upon the Lower Colorado River Valley subdivision in Sonora, and the three coldest lie along the Gila River Valley toward the northeastern limit of the subdivision near its transition to the Chihuahuan Desert (Fig. 108). The most pronounced precipitation difference separating this subdivision from the Mohave Desert and the other four subdivisions is probably winter rainfall amount: the average for all its stations is 75 mm; that for stations in the other desert areas lies between 30 mm and 50 mm (Fig. 112).

Paloverde-Cacti-Mixed Scrub Series

The most extensive series of this subdivision is dominated by a leguminous tree, *Cercidium microphyllum,* with the columnar cactus, *Carnegiea gigantea,* commonly reaching through the upper strata (Fig. 124). This community is best developed away from valley floors on bajadas and mountain sides. *Olneya tesota,* another tree legume, often plays a secondary role to *Cercidium microphyllum.* The two species occupy closely similar ranges within the Sonoran Desert (Hastings et al., 1972) with *Cercidium* occurring in this community with the greater fidelity. *Olneya* is more frost sensitive than *Cercidium* and is excluded from lower slopes adjacent to cold valley floors and from north slopes that support *Cercidium.*

The ecotone between this and the Creosotebush-White Bursage series is a common feature along the margins of the many valleys of the region. As the gently sloping valley floor

Table 27. Data for 38 weather stations in the Arizona Upland subdivision. Data for each station include latitude and longitude (values rounded to closest 0.1°), altitude (m), average annual precipitation (mm), and abbreviated name as used in climographs of Figure 109. Data from Hastings and Humphrey (1969b) and Sellers and Hill (1974).

Station name	Abbreviated name	Latitude	Longitude	Altitude (m)	Average annual precipitation (mm)
Ajo, AZ	AJO	32.4	112.9	537	227.4
Alamo Dam, AZ	ALDA	34.3	113.5	451	189.7
Altar, Son.	ALT	30.7	111.8	397	266.2
Arizona-Sonora Desert Museum, AZ	ASD	32.2	111.2	858	244.4
Atil, Son.	ATI	30.8	111.6	575	334.3
Bartlett Dam, AZ	BADA	33.8	111.6	503	300.7
Carefree, AZ	CARE	33.8	111.9	771	314.7
Castle Hot Springs, AZ	CAHT	34.0	112.4	640	382.7
Cave Creek, AZ	CACR	33.8	111.9	646	282.2
Cortaro 3SW, AZ	COR	32.3	111.1	692	281.2
Granite Reef Dam, AZ	GRRE	33.5	111.7	404	228.8
Imuris, Son.	IMU	30.8	110.9	340	407.1
Magdalena, Son.	MAG	30.6	111.0	760	397.7
Mormon Flat Dam, AZ	MOFL	33.5	111.4	827	328.4
Organ Pipe National Monument, AZ	OPNM	31.9	112.8	512	232.9
Presa Cuauhtemoc, Son.	PRCU	30.9	111.5	590	305.9
Presa la Angostura, Son.	PRAN	30.4	109.4	800	366.5
Quitovac, Son.	QUI	31.5	112.7	350	280.1
Reno Ranger Station, AZ	RERA	33.9	111.3	738	425.4
Roosevelt, AZ	ROO	33.7	111.1	672	359.4
Sabino Canyon, AZ	SACA	32.3	110.8	805	300.9
Safford, AZ	SAF	32.8	109.7	884	214.1
Safford Experimental Farm, AZ	SAEF	32.8	109.7	900	215.1
Sahuarita, AZ	SAH	32.0	111.0	820	275.6
San Carlos, AZ	SNCA	33.3	110.4	806	301.8
San Carlos Reservoir, AZ	SNCR	33.2	110.5	772	359.4
San Manuel, AZ	SNMA	32.6	110.6	1,085	331.0
Sta. Ana, Son.	STAN	30.5	111.1	686	313.5
Sells, AZ	SEL	31.9	111.9	733	305.2
Silver Bell, AZ	SIBE	32.4	111.5	835	312.5
Sonoita, Son.	SON	31.9	112.9	393	188.0
Superstition Mountain, AZ	SUP	33.4	111.4	594	267.8
Stewart Mountain Dam, AZ	STMT	33.6	111.5	434	290.5
Tajitos, Son.	TAJ	30.9	112.4	340	214.0
Tucson Weather Station (Airport), AZ	TUC	32.1	110.9	787	278.8
Wickenburg, AZ	WIC	34.0	112.7	639	273.5
Wikieup, AZ	WIK	34.7	113.6	647	238.8
Winkleman, AZ	WIN	32.9	110.7	632	327.3
Subdivision Average					293.8

makes contact with the base of the adjacent bajada, *Ambrosia deltoidea* is commonly found growing along minor water courses where moisture conditions are more favorable. As greater elevation is attained, the ecologically complementary *Ambrosia dumosa* wanes and *A. deltoidea* is found on all aspects of the terrain. *Cercidium microphyllum, Olneya tesota,* and *Carnegiea gigantea* become increasingly prominent and are largely responsible for the arborescent physiognomy of the vegetation. *Larrea* remains but is mostly relegated to an understory role. Many additional species are encountered that contribute to the diverse and complex nature of the community. These may include Whitethorn Acacia (*Acacia constricta*), Limber Bush (*Jatropha cardiophylla*), Ocotillo (*Fouquieria splendens*), Jojoba (*Simmondsia chinensis*), Little-leaved Ratany (*Krameria parvifolia*), Desert Hackberry (*Celtis pallida*), Fairy Feather Duster (*Calliandra eriophylla*), Bush

Buckwheat (*Eriogonum fasciculatum*), Desert Zinnia (*Zinnia acerosa*), Cilindrillo (*Lycium berlandieri*), and various cacti such as Desert Christmas Cactus (*Opuntia leptocaulis*), *O. phaeacantha* var. *major*, Engelmann Prickly Pear (*O. p.* var. *discata*), Fish-hook Pincushion (*Mammillaria microcarpa*), and Fendler Hedgehog (*Echinocereus fendleri*). Locally any or several of these can assume numerical superiority. Coverage data for a stand of this series at an elevation of 735 m in the Silverbell Mountains near the Avra Valley, Arizona, are given in Tables 25 and 26.

Jojoba-Mixed Scrub Series

One of a number of Sonoran desertscrub plants that may attain local dominance, *Simmondsia chinensis* or Jojoba, is an evergreen shrub with thick, bluish-green elliptic leaves and fruits resembling acorns. It is especially valuable as a forage

Figure 125. *Creosotebush-Crucifixion-thorn series near Montezuma Well, Arizona. Canotia is the low tree with the unkempt appearance on the slope opposite the camera. Altitude 975 m.*

plant for game and domestic stock and has recently gained notice as a water-soluble oil source. This plant attains maximum abundance in the Arizona Upland subdivision where annual precipitation ranges from 300-450 mm (Gentry, 1958), more than 100 mm of which occurs during the winter months. Consequently, this local series is best developed at the desert's upper limits and in transition areas between Sonoran desertscrub and interior chaparral (Brown, 1978). Because its distribution is almost wholly within the Sonoran Desert (Hastings et al., 1972), we include it as a Sonoran desertscrub series regardless of its "chaparral-like" physiognomy.

Creosotebush—Crucifixion-thorn Series

This series is also found at upper elevation ecotones. These communities are especially common on limestone substrates at the northern and eastern edges of the Sonoran Desert (e.g., near Camp Verde and San Carlos Reservoir, Arizona). *Canotia holocantha* or Crucifixion-Thorn is a low, spiny, leafless tree frequently encountered on hillsides and slopes from roughly Safford to Kingman, Arizona (Fig. 125). It often grows with *Larrea*, but may extend upslope beyond the limits of the desert proper where it mixes with chaparral species (e.g., *Berberis trifoliata*) in semidesert grassland. Other Arizona Upland localities, some of which possess only *Larrea* and a few sparse Sonoran Desert plants, occur on marginal, cold alluvium sites, as in the Gila Valley near Safford and east of Tucson, Arizona. Plants more typical of Chihuahuan desertscrub, such as *Acacia neovernicosa* and *Parthenium incanum*, are often present and this series shares many transitional features with that warm temperate biome and with the Mohave Desert.

Because of the structural density and adequate winter precipitation, some habitats in the Arizona Upland subdivision support moderate densities of Desert Mule Deer (*Odocoileus hemionous crooki*) and Javelina (*Dicotyles tajacu*). Numerous smaller mammals are also at home there, including the California Leaf-nosed Bat (*Macrotus californicus*), California Myotis (*Myotis californicus*), Black-tailed Jackrabbit, Desert Cottontail, Arizona Pocket Mouse (*Perognathus amplus*), Bailey's Pocket Mouse (*P. baileyi*), Cactus Mouse (*Peromyscus eremicus*), White-throated Wood Rat (*Neotoma albigula*), Gray Fox (*Urocyon cinereoargenteus*), and the endemic Harris Antelope Squirrel (*Ammospermophilus harrisii*).

The paloverde-cacti-mixed scrub series is particularly noted for its rich birdlife. Many of its best known inhabitants —the Harris' Hawk (*Parabuteo unicinctus*), White-winged Dove (*Zenaida macroura*), Inca Dove (*Scardiafella inca*), Elf Owl (*Micrathene whitneyi*), Wied's Crested Flycatcher (*Myiarchus tyrannulus*), and Pyrrhuloxia (*Cardinalis sinuatus*)—are in fact thornscrub species extending northward in suitable habitats. Even the "cactus" woodpeckers—*Melanerpes uropygialis, Colaptes chrysoides,* and *Picoides scalaris*—are in fact quite widespread and not nearly as dependent on the saguaro as is popularly thought. Other "desert" birds (e.g., Curve-billed Thrasher (*Toxostoma curvirostre*), and Cactus Wren) also find these communities acceptable, further enriching the avifauna.

In addition to having a generous complement of Sonoran and other desert reptiles, this subdivision is also the distribution center for a number of lizard species and snakes more limited in range. These include the Regal Horned Lizard (*Phrynosoma solare*), Western Whiptail (*Cnemidophorus tigris gracilis*), Gila Monster (*Heloderma suspectum*—especially the reticulated form), Arizona Glossy Snake (*Arizona elegans noctivaga*), Arizona Coral Snake (*Micruroides euryxanthus*), and Tiger Rattlesnake (*Crotalus tigris*).

Vizcaíno Subdivision

This desert region lies wholly within the Baja California peninsula, where it extends southward from just north of the 30th meridian to 26° 15' north latitude. Its southern boundary is thus a full degree south of the limit of the area considered in this publication (Shreve, 1951). It extends from the Pacific eastward to the crest of the drainage divide separating the east and west slopes of the peninsula. Contact on the north is a gradual merging with Californian coastalscrub and chaparral (Shreve, 1936), on the east with the Lower Colorado River Valley and Central Gulf Coast subdivisions of the Sonoran Desert, and on the south with what Shreve referred to as the Magdalena Region, some of which has ties with the Sinaloan thornscrub of mainland Mexico.

The physiography of the northern portion of the region is marked by a coastal plain interrupted by many low hills, few of which exceed 600 m elevation. In the vicinity of Miller's Landing and southward, the plain becomes broader and is almost wholly without relief in its extent from the Pacific shore to the mountains forming the crest of the peninsular divide—the so-called "Vizcaíno Desert" proper (Nelson, 1922; Humphrey, 1974). These mountain backdrops lie much closer to the Gulf than the Pacific and generally are low with few non-desert areas extending high enough to exceed the desert's upper boundary at roughly 1,700 m. The larger areas above this elevation are in the Sierra San Borja, Sierra Calmallí, and the Sierra Calamajué. Roughly 75% of the terrain lies below 500 m with the broadest expanses of low

relief occurring within the Vizcaíno Cape and its eastern extensions, the Llano de Berrendo and Desierto de Vizcaíno. Few mountains break the monotony of this plain between the Pacific shore and the foothills of the mountainous peninsular backbone. The most important of these are the Sierra Pintada and the Sierra Santa Clara, which form a single low range, and the Sierra Placeras, extending for about 160 km parallel to the coast along the southwestern side of the Vizcaíno Cape.

That this area is exposed on the west to the Pacific, and insulated on the east from the Gulf by a mountainous ridge, explains in large part the temperature regime of the subdivision (Fig. 109). Mean summer temperatures are 5°-6° C lower than for any of the other Sonoran Desert subdivisions (Fig. 112), a result of the cooling effect of the Pacific waters.

Baja California climate has received attention from several investigators (Markham, 1972; Pyke, 1972; Hastings, 1964a; Hastings and Humphrey, 1969a; Garcia and Mosiño, 1968), as have vegetation-climate relationships (Hastings and Turner, 1965a). Because of these contributions, a rather detailed description of the Vizcaíno subdivision's climate is now possible.

Using the means of station annual means as a measure of aridity, the Vizcaíno subdivision appears at first to be the most arid of the Sonoran Desert subdivisions with a value of only 99 mm (Table 28). This position of preeminent dryness was assigned to the Lower Colorado River Valley subdivision by Hastings and Turner (1965a), using areally weighted averages and dealing solely with the Baja California peninsula. Shreve (1951) believed the driest subdivision to be the Central

Table 28. Data for 27 weather stations in the Vizcaíno subdivision. Data for each station include latitude and longitude (values rounded to closest 0.1°), altitude (m), average annual precipitation (mm), and abbreviated name as used in climographs of Figure 109. Data from Hastings and Humphrey (1969a).

Station name	Abbreviated name	Latitude	Longitude	Altitude (m)	Average annual precipitation (mm)
Bahía Tortugas, BCS	BATO	27.7	114.9	5	92.5
Cadejé, BCS	CAD	26.4	112.5	70	84.6
Chapala, BCN	CHA	29.5	114.5	580	90.5
El Alamo, BCN	ALA	27.1	112.9	125	77.5
El Arco, BCN	ARC	28.0	113.4	300	140.3
El Tablon, BCS	TAB	27.6	113.4	80	119.3
Guadalupe, BCS	GUAP	27.3	113.4	120	76.0
Isla Cedros, BCN	ISCE	28.1	115.2	10	70.8
Los Lagunas, BCS	LSLA	27.5	113.6	30	57.1
Patrocinio, BCS	PAT	26.8	112.8	400	75.0
Punta Abreojos, BCS	PTAB	26.7	113.6	15	57.8
Punta Prieta, BCN	PTPR	29.0	114.2	200	90.5
Rancho Alegre, BCN	RCAL	28.3	113.9	500	193.8
Rosarito, BCN	ROS	28.6	114.1	190	128.7
San Augustín, BCN	SNAG	29.9	115.0	580	99.9
San Borja, BCN	SNBO	28.8	113.9	375	134.9
San Fernando, BCN	SNFE	30.0	115.2	539	87.8
San Ignacio, BCS	SNIG	27.3	112.9	105	86.0
San Juan, BCS	SNJU	27.2	113.1	125	48.8
San Juanico, BCS	SNJN	26.3	112.4	12	64.2
San Juis, BCN	SNLU	29.7	114.7	500	104.3
San Miguel, BCN	SNMI	28.6	113.9	200	134.5
San Regis, BCN	SNRE	28.6	113.9	300	116.7
Santa Catarina (sur), BCN	STCA	29.7	115.2	450	121.1
Santa Gertrudis, BCN	STGE	28.1	118.1	550	150.6
Santa Inés, BCN	STIN	29.7	114.7	500	87.8
Vizcaíno, BCS	VIZ	28.0	114.1	10	81.7
Subdivision Average					99.0

Figure 126. *Maguey-Boojum series approximately 30 km east of El Rosario, Baja California del Norte. The low shrubs are* Eriogonum fasciculatum, Ambrosia camphorata, *and* A. chenopodifolia. *The cactus in the foreground is* Opuntia prolifera. *Here, near the northern border of the desert, this community is best developed on ridge tops and south-facing slopes. Adjacent north-facing slopes support a scattering of chaparral species. Photograph by J.R. Hastings. Altitude 395 m.*

Gulf Coast based upon but few rainfall stations and the appearance of the vegetation. Maps by Garcia and Mosiño (1968) and by the Instituto de Geografía (1970) appear to confirm the conclusion of Hastings and Turner (1965a) that the Lower Colorado Valley subdivision is indeed the driest. Regardless of which subdivision is visited by the least rainfall, the Vizcaíno is less dry than it appears when judged by rainfall alone because of its exposure to the cool California current which brings both lower temperatures and fog to this coastal desert.

By reason of its mid-peninsular location, the Vizcaíno subdivision is subjected to the interplay between all the precipitation regimes under which the peninsula falls. Although meant to apply to the entire peninsula, the following description also applies to this centrally located subdivision: "[It] lies at the southern edge of the system of

winter cyclones that are imbedded in the westerlies and that play so important a part in weather farther north; it lies at the western edge of the monsoons originating in the Gulf of Mexico and dominating northern Mexico in summer; it lies at the northeastern edge of the complex of tropical storms and hurricanes that range across the eastern North Pacific in autumn; it lies at the western limit of fall activity in the easterlies" (Hastings and Turner, 1965a: 210).

Although visited weakly by moisture from four sources, the Vizcaíno is under the dominant control of two—winter cyclonic storms and fall huricanes. Of these two rainy seasons, winter ranks higher at all but 3 of the 27 stations listed in Table 28. Even so, this winter season of greatest rainfall is hardly one of bounteous moisture—the greatest average rainfall at any station for the 3-month period is 86.6 mm (Rancho Alegre) and the areally weighted seasonal

Figure 127. *White Bursage-Agave series near Agua Dulce, Baja California del Norte. This community is predominantly of low plants with a scattering of individuals of 1.0-1.5 m height in such species as* Yucca schidigera *and* Larrea tridentata. *Taller plants of* Fouquieria splendens *and* F. columnaris *are widely scattered throughout. Certain similarities with Mohave desertscrub are apparent and found in* Yucca schidigera, Larrea tridentata, *and* Ceratoides lanata. *Photograph by J.R. Hastings. Altitude 640 m.*

average for the entire subdivision is only 56 mm (Hastings and Turner, 1965a).

Spring is the most arid of the two remaining seasons, ranking fourth at 20 of the 27 stations, although areally weighted averages (Hastings and Turner, 1965a) show summer to be hardly more moist than spring (11 mm versus 10 mm).

A poorly quantified climatic feature of great importance is the fog which develops during the night and remains until midmorning of the following day during the two; driest seasons of spring and summer. During periods of fog, water condenses on plant surfaces and often drips from leaf tips and spines, moistening spots of soil below. This fog, even if acting only to reduce transpiration for a few hours each morning, plays a large role in ameliorating the effect of the low spring-summer rainfall of the subdivision. Markham (1972) noted that fog visits the Pacific slope to 1,000 m altitude and Shreve (1951) stated that the morning fog extends inland 5 to 6 km.

The long separation of the Baja peninsula from the mainland coast of Sonora (4-10 million years ago, according to Larson et al., 1968), has resulted in the absence from the Vizcaíno subdivision of a number of characteristic Sonoran Desert vertebrates. Included among these are *Heloderma suspectum, Gopherus agassizi, Sauromalus obesus,* Gambel Quail, Curve-billed Thrasher, and Javelina. This absence is somewhat compensated for by such Vizcaíno inhabitants as the Small-scaled Lizard (*Urosaurus microscutatus*), Orange-throated and Coastal Whiptails (*Cnemidophorus hyperythrus, C. tigris multiscutatus*), Red Diamond Rattlesnake (*Crotalus ruber*), California Quail (*Lophortyx californicus*), and Vizcaíno Desert Kangaroo Rat (*Dipodomys peninsularis peninsularis*) which, although having closely related Sonoran forms, are

derived out of Californian coastalscrub species. The Vizcaíno also shares with other Sonoran desertscrub subdivisions such typical arid region animals as *Dipsosaurus dorsalis, Callisaurus draconoides, Coleonyx variegatus, Chilomeniscus cinctus,* Harris' Hawk, White-winged Dove (*Zenaida asiatica*), Cactus Wren, Bendire's Thrasher (*Toxostoma bendirei*), and Gila and Gilded Woodpeckers.

The Vizcaíno subdivision, unlike most other parts of the Sonoran Desert, is dominated by fleshy-leaved plants as *Agave, Yucca,* and *Dudleya* and was labeled the Sarcophyllous Desert by Shreve (1951) in recognition of this abundant growth form. In addition, this subdivision is the home of the fleshly-stemmed Boojum Tree (*Fouquieria columnaris*) and Torote Blanco (*Pachycormus discolor*). Cardón (*Pachycereus pringlei*), a giant columnar cactus that may well be the largest cactus in the world, takes the place here that in the Arizona Upland is occupied by *Carnegiea gigantea.*

Shreve (1951) also used the name Agave-Franseria [=Ambrosia] Region for the Vizcaíno subdivision in recognition of both the abundance and diversity of these two genera. The Vizcaíno is home for several species of *Ambrosia* (Hastings et al. 1972), among which the following occur extensively in several series: San Diego Bursage (*Ambrosia chenopodifolia*), Magdalena Bursage (*A. magdalenae*), *A. camphorata,* White Bursage (*A. dumosa*), Bryant Bursage (*A. bryantii*), *A. divaricata,* and Canyon Ragweed (*A. ambrosioides*). The Vizcaíno is also the center of distribution of several species of *Agave* including Blue Agave (*A. cerulata*), Maguey (*A. shawii*), and *A. avellanidens* (Gentry, 1978) and these leaf succulents are commonly the dominant members of the communities in which they occur. Grasses, except for annuals, are almost entirely lacking from most of this area (Humphrey, 1974).

Figure 128. *Bursage-cholla series near Aguajito, Baja California del Norte. Leaf succulents such as* **Agave** *and* **Yucca** *are absent. The cacti* **Opuntia cholla,** **O. prolifera,** *and* **Ferocactus acanthodes** *are conspicuous. Tall plants such as* **Fouquieria columnaris** *and* **Pachycereus pringlei** *are rare. Photograph by J.R. Hastings. Altitude 425 m.*

Figure 129. *Ragged-leaf Goldeneye-Boojum series near La Virgen, Baja California del Norte. This community is common on granitic soils in the Vizcaino Subdivision. The large cactus is* **Pachycereus pringlei.** *The low shrub stratum is dominated by* **Viguiera laciniata** *but also includes* **Ephedra, Ambrosia chenopodifolia, A. dumosa, Larrea tridentata, Atriplex polycarpa,** *and* **Acalypha californica.** *Photograph by J.R. Hastings. Altitude 610 m.*

The central Vizcaíno is a region of range overlap of *Fouquieria splendens* and *F. diguetii*. The former extends into desert areas to the north and the latter ranges southward (Hastings et al., 1972). Aschmann (1959) suggests that these distribution patterns are a response to rainfall regimes: *F. splendens* flowers following winter rains, *F. diguetii* flowers mostly following summer rains.

The Vizcaíno-chaparral ecotone was examined by Shreve (1936), who concluded that chaparral species extend farther southward into the desert than desert species extend northward into the chaparral. He listed desert species that are almost wholly confined to the desert—(*Cercidium microphyllum, Fouquieria splendens, Yucca schidigera*); those that range northward somewhat more widely—(*Myrtillocactus cochal, Euphorbia misera, Ambrosia chenopodifolia*); chaparral species that are almost wholly confined to chaparral—(*Rhus integrifolia, Ceanothus tomentosus, Quercus dumosa*); and chaparral species that extend as far as 250 km southward beyond the transition—(*Malosma laurina, Isomeris arborea, Ribes tortuosum*). Shreve concluded that the requirements for growth of desert plants were more exacting than those for chaparral species and that freedom from freezing temperatures of long duration, exposure to a high percentage of sunshine and well-drained soils were conditions required by most desert species but not provided by most habitats in the chaparral. Conversely, for chaparral species to enter the desert required only that they find habitats with relatively moist soils.

Some of the more common plant communities occurring in this region are the following: Maguey-Boojum series, White Bursage-Agave series, Bursage-Cholla series, Ragged-leaf Goldeneye-Boojum series, Agave-Boojum series, Frankenia-Ocotillo-Datilillo series, and a Saltbush series.

Maguey-Boojum Series

Near the northern limits of the Vizcaíno region and always within a few kilometers of the Pacific shore occurs a community dominated by Maguey (*Agave shawii*) and Boojum Tree (*Fouquieria columnaris*) (Fig. 126). Low shrubs found here include *Eriogonum fasciculatum, Ambrosia camphorata*, and *A. chenopodifolia. Opuntia prolifera* and *Echinocereus maritimus* are common cacti. The soils are derived mostly from volcanic rock and have high clay content and poor permeability.

White Bursage-Agave Series

A community dominated by *Ambrosia dumosa* and *Agave cerulata* also is found in the northern Vizcaíno region (Fig. 127). It is a low community with only widely scattered tall plants of such species as *Fouquieria columnaris, F. splendens, Pachycereus pringlei, Pachycormus discolor*, and *Yucca schidigera*. The dominants are clearly *Ambrosia dumosa* and *Agave cerulata* although *Encelia californica* and *Ambrosia camphorata* are frequent members. *Eriogonum fasciculatum* and *Ceratoides lanata*, two species with wide distribution to the north, are prominent low shrubs at many sites. *Larrea tridentata* is an uncommon but consistent member. Other species commonly seen here are the shrubs *Simmondsia chinensis, Viguiera laciniata, Krameria parvifolia, K. grayi*, and such cacti as *Lophocereus schottii, Opuntia molesta, Ferocactus* species, and *Opuntia ciribe*. Humphrey (1974) listed quantitative data for a stand near Agua Dulce, Baja California del Norte.

Table 29. Coverage and density data for a stand of the Ragged-leaf Goldeneye – Boojum series near Cataviñá, Baja California del Norte and for a stand of the Agave – Boojum series near Punta Prieta, Baja California del Norte. Data at each site by line intercept from 30 m lines. Density determined by the method of Strong (1966).

	Cataviñá (alt. 52 m)		Punta Prieta (alt. 380 m)	
	Coverage (%)	Density (Plants/ha)	Coverage (%)	Density (Plants/ha)
Abutilon	0.00	0.0	0.02	9.3
Acacia greggii	0.34	16.2	0.00	0.0
Acalypha californica	0.10	150.7	0.00	0.0
Agave cerulata	0.33	156.0	5.66	1580.3
Ambrosia chenopodifolia	3.04	972.0	0.69	182.3
Ambrosia dumosa	0.25	115.0	0.00	0.0
Ambrosia magdalenae	0.00	0.0	3.55	382.4
Atriplex polycarpa	0.09	5.7	1.24	232.7
Bebbia juncea	0.24	63.0	0.00	0.0
Dyssodia porophylloides	0.23	345.2	0.00	0.0
Encelia californica	0.66	203.9	0.00	0.0
Encelia farinosa	0.00	0.0	0.54	448.5
Eriogonum fasciculatum	0.93	2863.8	0.00	0.0
Euphorbia misera	0.00	0.0	1.41	151.2
Euphorbia tomentulosa	0.08	67.1	0.00	0.0
Fagonia californica	0.08	227.9	0.74	1805.5
Fouquieria columnaris	1.26	154.2	1.09	62.8
Fouquieria splendens	present	214.6	0.45	12.7
Hosackia glabra	0.06	18.7	0.00	0.0
Krameria grayi	0.79	195.6	0.00	0.0
Larrea tridentata	1.14	107.7	0.97	44.9
Lophocereus schottii	0.36	27.3	0.06	23.1
Lycium californicum	0.00	0.0	2.22	130.0
Lyrocarpa coulteri	0.31	311.3	0.00	0.0
Mammillaria	present	429.2	0.02	429.2
Mirabilis laevis	0.15	117.7	0.09	118.2
Opuntia acanthocarpa	0.04	66.8	0.00	0.0
Opuntia cholla	0.00	0.0	0.42	78.8
Opuntia echinocarpa	0.04	39.9	0.00	0.0
Opuntia molesta	0.07	14.3	0.00	0.0
Opuntia tesajo	0.00	0.0	0.12	64.5
Pachycereus pringlei	0.06	21.4	0.23	3.3
Pachycormus discolor	0.68	21.7	0.00	0.0
Pedilanthus macrocarpus	0.00	0.0	0.61	157.7
Prosopis juliflora	0.39	7.2	1.83	382.9
Simmondsia chinensis	0.51	106.9	0.13	7.0
Solanum hindsianum	0.00	0.0	0.33	173.3
Sphaeralcea sp.	present	18.7	0.01	76.2
Stenocereus gummosus	0.00	0.0	3.33	365.6
Stillingia linearifolia	0.12	66.0	0.00	0.0
Trixis californica	0.00	0.0	0.20	156.2
Unknown composite	0.03	14.3	0.00	0.0
Viguiera deltoidea	0.24	17.3	0.00	0.0
Viguiera laciniata	8.97	2446.8	1.84	183.5
Yucca valida	0.00	0.0	0.04	5.5
Rocks (>159 mm across one dimension)	13.45		0.08	
Total plant coverage	21.61		27.84	
Total plants/hectare		9594.1		7267.6
Additional species in area				
Asclepias subulatus	Present			
Bursera microphylla	Present			
Euphorbia sp.			Present	
Euphorbia xanti			Present	
Hibiscus denudatus			Present	
Lycium berlandieri	Present			
Pachycormus discolor			Present	
Viscainoa geniculata			Present	

Figure 130. *The Agave-Boojum series, one of the most diverse communities in the Vizcaino region, is seen north of Punta Prieta, Baja California Norte. Tall species such as* Pachycormus discolor *(thick-stemmed tree of midground),* Pachycereus pringlei, Fouquieria columnaris, *and* Yucca valida *appear to dominate although* Agave cerulata *is clearly the dominant if measured by canopy coverage. The ground cover shrubs found here include* Viguiera laciniata, Ambrosia chenopodifolia, *and* A. magdalenae. *Photograph by J.R. Hastings. Altitude 425 m.*

Bursage-Cholla Series

In the northern Vizcaíno region on volcanic soils with especially high clay content is found a series of *Ambrosia camphorata* and cylindropuntias (Fig. 128). *Ambrosia camphorata* and, to a lesser extent, *A. chenopodifolia* form a broken low cover with a scattering of cylindropuntias such as *Opuntia cholla, O. prolifera,* and *O. molesta. Ferocactus acanthodes* is conspicuous in many places. Leaf succulents are strikingly absent.

Ragged-leaf Goldeneye-Boojum Series

Granitic soils of the northern Vizcaíno region support a community dominated by Ragged-leaf Goldeneye (*Viguiera laciniata*) and *Fouquieria columnaris,* this characterized by a greater representation of tall plants than any of the above series (Fig. 129). In addition to *Fouquieria,* the tall species here include *Pachycereus pringlei* and *Pachycormus discolor.* Coverage data, based upon 30-m lines for a representative locality in this series near Cataviña, are shown in Table 29. *Larrea tridentata* is commonly an important member of this community.

Agave-Boojum Series

On basaltic soils, toward the center of the Vizcaíno region, in the vicinity of Punta Prieta, is a rich community dominated by *Agave cerulata* and Boojum Tree (*Fouquieria columnaris*). Plants of various growth forms growing to 4 or 5 m tall occur so abundantly as to give an open woodland appearance in some areas (Fig. 130). With the exception of *Fouquieria splendens,* the tall plant dominants are fleshly leaved (*Yucca valida*), fleshly stemmed (*Pachycormus discolor, Fouquieria columnaris*), or succulent (*Pachycereus pringlei, Lophocereus schottii*). *Agave cerulata* is the dominant low species with lesser biomass contributed by *Ambrosia magdalenae, A. chenopodifolia, Viguiera laciniata,* and *Stenocereus gummosus.* This agave, probably the most abundant in Baja California, is largely confined to the Vizcaíno. Gentry (1978) noted that this plant occurs between latitude 30°N and 27°N and that this is "a wild species whose population is probably exceeded only by *Agave lechuguilla* of the northern desert region of Mexico." Coverage data from 30-m lines in an area 19 km north of Punta Prieta are given in Table 29.

Figure 131. *A stand of* Frankenia palmeri, Fouquieria diguetii, *and* Yucca valida *southeast of Miller's Landing, Baja California del Norte. This site is about 8 km from the coast. Fog, which occurs here most nights during the warm season, is just lifting in this early morning view. The top of the vegetation canopy is at about 2 m although occasional individuals of* Yucca valida *may extend to twice that height. The shrubs defining the upper canopy are largely* Fouquieria diguetii, Jatropha cinerea, *and* Yucca valida. *Most of these support a dense epiphytic growth of* Tillandsia recurvata *and lichens. The low shrubs are* Frankenia palmeri, Atriplex julacea, *and* A. polycarpa. *Photograph by J.R. Hastings. Altitude 23 m.*

Figure 132. Fouquieria diguetii *covered with lichens southeast of Miller's Landing, Baja California del Norte. Epiphytes are common near the Pacific coast in the Vizcaino subdivision. Another epiphyte,* Tillandsia, *can be seen at upper right. Photograph by J.R. Hastings. Altitude 230 m.*

Figure 133. *Saltbush series 13 km northeast of Guerrero Negro, Baja California del Norte. Dominant plants here are* **Atriplex julacea** *and* **Frankenia palmeri**. *Few plants are over 0.3 m high in this open windswept community. Photograph by J.R. Hastings. Altitude 30 m.*

Frankenia-Ocotillo-Datilillo Series

Southward on the coastal strip toward Guerrero Negro, arborescent dominants are lost and the vegetation assumes a monotonous aspect where the tall dominants, only 2-4 m high, are mainly Datilillo (*Yucca valida*), *Fouquieria diguetii*, and *Jatropha cinerea*. There is a distinct layer of low shrubs to about 0.3 m high comprising *Frankenia palmeri*, *Atriplex julacea*, and *A. polycarpa*. The cacti *Stenocereus gummosus* and *Opuntia cholla* are common members of the community. This coastal area is much visited by fog (Fig. 131), and Ballmoss (*Tillandsia recurvata*) and lichens hang densely from many of the shrubs or cover the soil surface (Fig. 132). Ice Plant (*Mesembryanthemum crystallinum*), an introduced annual, also covers the soil surface in many areas.

Saltbush Series

Extensive coastal flats in approximately the southern half of this region are dominated by a low community of *Frankenia palmeri* and species of *Atriplex*, especially *A. julacea* (Fig. 133). The agaves that appear in similar near-coastal positions to the north are missing and no shrubs or trees break the monotony of flat panoramas. The introduced *Mesembryanthemum crystallinum* and various lichens often densely cover much of the soil surface. The annuals *Dyssodia anthemidifolia*, Wooly Plantain (*Plantago insularis*), three-awns (*Aristida* spp.), Sand Verbena (*Abronia villosa*), and *Coreocarpus parthenioides* are conspicuous following significant rainfall (Brown and Webb, 1979).

Central Gulf Coast Subdivision

The Central Gulf Coast subdivision consists of two disjunct units on either side of the Sea of Cortez, separated by an expanse of water 100 km wide or more. This subdivision was also referred to as the Sarcocaulescent Desert by Shreve (1951) to emphasize the common dominance of stem succulents. There is a general absence of a low shrub layer that elsewhere in the Sonoran Desert commonly comprises such genera as *Ambrosia, Viguiera,* and less commonly *Eriogonum* and *Encelia farinosa.* Elephant Tree (*Bursera microphylla*), *B. hindsiana,* Sangre de Drago (*Jatropha cuneata*), and *J. cinerea* are the common perennials here with *Larrea tridentata* only locally found in pure stands.

In Sonora this subdivision extends from about latitude 30° 30'N southward to about 27° 45'N and is confined to a coastal strip no more than 40 km wide. Most of the two largest islands in the Gulf, Tiburón and Angel de la Guarda, are included. The extent of this vegetation on the Baja California peninsula is more southerly than its mainland counterpart, commencing just north of the 29th parallel and extending south beyond the southern limit of our area.

The vegetation occurs predominantly on coastal plains, broken in Sonora by such mountains as the Sierra el Aguaje north of Guaymas and the Sierra Seri and Sierra Bacha near Puerto Libertad. In Baja California at the extreme south of the map area, the narrow coastal strip is bounded to the west by the Sierra de la Giganta and the impressive volcanic cone, Tres Virgenes. Northward the subdivision forms a narrow coastal strip terminating opposite the north end of Isla Ángel de la Guarda where it merges into the *Larrea* plains of the Lower Colorado River Valley region.

In discussing the climate of the Central Gulf Coast subdivision, it is convenient to consider the entire region, although it lies partly outside the area included within the North American Southwest. Average annual precipitation ranges from less than 100 mm to roughly 300 mm, with the smaller amounts usually occurring at the north (Fig. 110). The weather station with the lowest average annual rainfall (71.3 mm) is Bahía de los Angeles, a northern station lying near the southern boundary of the Lower Colorado River Valley subdivision in Baja California (Table 30). The station with greatest annual rainfall is the most southerly peninsular station, Boca del Salado at latitude 23°N. Nonetheless, regression analysis of latitude versus annual rainfall shows a weak relationship (correlation coefficient = -0.24).

When precipitation values are examined by season, those for summer show the greatest geographic effect. Multiple regression analysis shows a strong tendency for values to increase from west to east and a high but less significant tendency for north to south increases. Thus, a general statement would be a tendency for summer precipitation to increase from west north-west toward east south-east. In the early fall, there is also a geographic effect with a relatively strong north to south increase combined with a less strong west to east increase. Elevations in this subdivision are rather unimportant predictors of rainfall at any season.

For the peninsular portion of this subdivision, Hastings and Turner (1965a) showed that summer rainfall is the most reliable of that falling during the four seasons, although the largest mean amount falls during the autumn.

Judging from climographs for summer, this subdivision has close ties with the Lower Colorado River Valley subdivision; winter climographs show close ties with the Plains of Sonora

Table 30. Data for 24 weather stations in the Central Gulf Coast subdivision. Data for each station include latitude and longitude (values rounded to closest 0.1°), altitude (m), average annual precipitation (mm), and abbreviated name as used in climographs of Figure 110. Data from Hastings and Humphrey (1969a, 1969b).

Station name	Abbreviated name	Latitude	Longitude	Altitude (m)	Average annual precipitation (mm)
Bahía de Los Angeles, BCN	BALA	28.9	113.5	5	71.3
Boca del Salado, BCS	BOSA	23.2	109.4	10	268.7
El Barril, BCN	BAR	28.3	112.9	100	132.8
El Bénjamin, Son.	BEN	28.6	111.5	40	192.6
El Boleo, BCS	BOL	27.3	112.3	30	108.1
El Carrizal, Son.	CARR	28.9	111.7	25	193.9
La Desemboque, Son.	DES	29.5	112.4	10	111.1
La Palma (norte), BCS	PAL	27.6	112.7	110	106.0
El Rosarito, BCS	ROS	26.4	111.6	125	116.8
Empalme, Son.	EMP	28.0	110.8	2	178.3
Guadalupe, Son.	GUAD	28.6	111.4	44	138.9
Guaymas, Son.	GUAY	27.9	110.9	8	232.6
Isla San José, BCS	ISSJ	24.9	110.6	10	91.8
La Chupaclilla, Son.	CHU	28.7	111.5	60	145.1
La Gotera, Son.	GOT	28.6	111.6	36	175.3
Loreto, BCN	LOR	26.0	111.3	3	141.9
Los Planes, BCS	PLA	24.0	109.9	60	212.1
Mulegé, BCS	MUL	26.9	112.0	15	107.4
Puerto Libertad, Son.	PTLI	29.9	112.7	5	120.5
San Isidro, Son.	SNIS	28.8	111.7	36	169.7
San Jorge, Son.	SNJR	28.7	111.6	35	200.0
San José, Son.	SNJS	28.8	111.7	36	194.3
San Lucas, BCS	SNLU	27.2	112.2	5	152.7
Santa Rosalía, BCN	STRO	27.3	112.3	26	108.2
Subdivision Average					152.9

Figure 134. *This valley at the north end of the Sierra Bacha, near Puerto Libertad, Sonora, supports the Torchwood-Cardón series. The large woody plants include* Bursera microphylla, B. hindsiana, *and* Cercidium microphyllum. *Smaller shrubs are represented by* Solanum hindsianum, Ambrosia dumosa, Encelia farinosa, Jatropha cinerea, J. cuneata, Hyptis emoryi, *and* Justicia californica. *The conspicuous cacti are* Pachycereus pringlei, Lophocereus schottii, *and* Opuntia bigelovii. *On the north-facing slope opposite the camera station is* Fouquieria columnaris, *occurring here as a population isolated from the main body of its range in the Vizcaino subdivision. Photograph by H.L. Shantz in 1932. Altitude 75 m.*

and Vizcaíno subdivisions (Fig. 112). During the summer the difference in average temperature throughout the subdivision is slight, ranging from about 27°C at Puerto Libertad, Sonora, to 30.5°C at El Boleo, Baja California del Sur (Fig. 110). The range in average winter temperatures is greater, lying between 13°C (Puerto Libertad, Sonora) and 19°C (Boca del Salado, Baja California del Sur) (Fig. 110).

The index of continentality for the Central Gulf Coast is 13.0 (Table 23) placing it near the center of the range of values for the five Sonoran Desert subdivisions. That it is so strongly continental, in spite of its maritime location, is the result of the inland character of the Sea of Cortez upon which this subdivision abuts.

The main floral elements of this subdivision, whether on the coast of Sonora or Baja California, are widely spaced shrubs of *Jatropha, Euphorbia, Fouquieria, Larrea tridentata,* and small trees of *Cercidium, Olneya tesota,* and *Bursera.* There is no low shrub synusia comparable to that of the Arizona Upland where *Ambrosia deltoidea* forms a dense small shrub layer or to that of the Vizcaíno where species of *Ambrosia* and *Viguiera* occupy this position. The general aspect in the Central Gulf Coast communities is best characterized by heterogeneity of composition and openness of stand.

The mainland section of the Central Gulf Coast is of great interest to phytogeographers. From the vicinity of Guaymas northward to Puerto Libertad, occurs a large number of plants with highly localized occurrences. These may be found nowhere else on the mainland but occur widely in Baja California. Among these are Boojum Tree (*Fouquieria columnaris,* Fig. 134), *F. diguetii,* Palo Blanco (*Lysiloma candida,* Fig. 135), *Pithecellobium confine,* Palmer Fig (*Ficus palmeri*), *Viguiera laciniata, Ambrosia magdalenae, A. camphorata, A. chenopodifolia, A. divaricata, Bourreria sonorae, Stenocereus gummosus, Cordia brevispicata, Desmanthus fruticosus,* Magdalena Spurge (*Euphorbia magdalenae*), *E. tomentulosa, E. xantii, Ruellia peninsularis,* and *Viscainoa geniculata* (Johnston, 1924; Gentry, 1949; Hastings et al., 1972).

The local colony of Boojum or Cirio (*Fouquieria columnaris*) in the Sierra Bacha between Puerto Libertad and Desemboque, Sonora has been examined closely (Humphrey, 1974) with the conclusion that this species probably reached here by long

Figure 135. *Riparian thornscrub vegetation at the mouth of a canyon in the Sierra el Aguaje within the Central Gulf Coast subdivision of the Sonoran Desert near Bahia San Pedro, Sonora. The white-stemmed tree is* **Lysiloma candida,** *a plant known only from two areas: Baja California, where it occurs widely, and here, where it is found along a few kilometers of canyon bottoms. Photograph by R.M. Turner. Altitude 25 m.*

Figure 136. *View of the Torchwood-Cardón series, Sierra Bacha, near Puerto Libertad, Sonora. This is a close-up view of the same valley shown in figure 134. Photograph by R.M. Turner. Altitude 60 m.*

distance dispersal of the winged seeds. The presence of many other species having mainly peninsular ranges and with disseminules of various forms, some not at all amenable to wind dispersion, makes this hypothesis appear forced. Although long distance dispersal might explain the presence of some of these many disjunctions, the local occurrences might also be survivors from a former wider range. Fossil packrat middens from the region about the head of the Gulf might yield important evidence bearing on this question.

Whatever the method by which *Fouquieria columnaris* travelled between the Sonoran coast and its peninsular distribution center, it has become established on the mainland in the one locale with summer climate most like that of its peninsular home, the Vizcaíno subdivision (Fig. 112). Desemboque and Puerto Libertad possess average summer rainfall and summer temperature values only slightly above those of the Vizcaíno subdivision (Figs. 109, 110). Because *Fouquieria columnaris* in the Sierra Bacha occurs mostly on north-facing slopes (Humphrey, 1974), the temperature experienced there would be even lower than at the nearby weather stations, placing the growing sites well within the limits of the Vizcaíno as shown by the climographs. The explanation for the plant's presence, attributed to high humidity by Humphrey and Marx (1980), is probably related to the relatively low summer temperature of the area. The anomalous temperatures are possibly a consequence of local upwelling in nearby gulf waters.

The following series form the predominant vegetation of this subdivision: the Torchwood-Cardón series; the Ocotillo-Limberbush-Creosotebush series; the Frankenia series; and the Cactus-Mesquite-Saltbush series.

Torchwood-Cardón Series

On deep soils of granitic origin on both sides of the Sea of Cortez occurs a tall community, 3-4 m high, dominated by small trees of Torchwood (*Bursera hindsiana, B. microphylla*) and Palo Verde (*Cercidium microphyllum*), large shrubs (*Fouquieria splendens*), and the truly tall columnar Cardón (*Pachycereus pringlei*). Shrubs 2 m high or less, while not universal, include *Solanum hindsianum, Ambrosia dumosa, Encelia farinosa, Jatropha cinerea,* and *J. cuneata. Opuntia bigelovii* is often prominent. As with other series of this subdivision, there is much open ground between the perennial plants (Figs. 136, 137). The community is best represented in broad valley bottoms and on fans adjacent to mountain fronts where supplemental water from nearby slopes provides moisture in addition to rainfall.

Ocotillo-Limberbush-Creosotebush Series

This series is confined to the northern portion of the Central Gulf Coast within the range of *Fouquieria splendens*. Southward from about Bahía de los Angeles, Baja California del Norte, and in the vicinity of Guaymas, Sonora, this species wanes and is replaced by the more southerly *Fouquieria*

Figure 137. *Torchwood-Cardón series growing in the lava flow at the base of the volcano Tres Virgenes in Baja California del Norte. The dominant small tree is* **Bursera microphylla** *and the tall cactus is* Pachycereus pringlei. *Additional important species are* **Encelia farinosa** *(foreground),* Cercidium microphyllum, Fouquieria diguetii, Jatropha cuneata, J. cinerea, Solanum hindsianum, *and* Stenocereus thurberi. *Photograph by J.R. Hastings. (From Shreve 1964, Plate 27). Altitude 400 m.*

Figure 138. *Ocotillo-Limberbush-Creosotebush series near Bahia de los Angeles, Baja California del Norte.* Fouquieria splendens *is the most conspicuous tall shrub and is joined by approximately 25 species of woody and succulent perennials including* Larrea tridentata, Jatropha cuneata, J. cinerea, Opuntia acanthocarpa, Simmondsia chinensis, *and* Encelia farinosa. Fouquieria diguetii *and* Pachycereus pringlei *grow nearby but are not seen in this view. Photograph by J.R. Hastings. Altitude 20 m.*

Figure 139. *On marine deposits near the shore of Puerto Libertad, Sonora, is found the Frankenia series. Frankenia palmeri and Atriplex polycarpa are almost the sole occupants of the interfluves. In the runnels are* Encelia farinosa, Jatropha cuneata, Euphorbia misera, *and* Prosopis glandulosa *var.* torreyana. *Photograph by H. L. Shantz. (From Hastings and Turner, 1965b). Altitude 15 m.*

diguetii (Hastings et al., 1972). The series is common in the course outwash on bajadas at the base of the coastal mountains, and occurs on both sides of the Sea of Cortez (Fig. 138). East of Puerto Libertad, Sonora, at an altitude of roughly 120 m, the three dominants, *Fouquieria splendens*, Sangre de Drago (*Jatropha cuneata*), and *Larrea tridentata*, are accompanied by *Encelia farinosa* var. *phenicodonta* as a subdominant, with lesser amounts of *Ferocactus acanthodes* and *Opuntia arbuscula*. Along runnels occur *Ambrosia deltoidea*, *Opuntia bigelovii*, *Pachycereus pringlei*, and *Lophocereus schottii*.

Frankenia Series

This community was described in the Lower Colorado River Valley section. In the vicinity of Puerto Libertad, Sonora, on a nearly level coastal plain, *Frankenia* is the clear dominant and *Atriplex polycarpa* the subdominant. *Euphorbia misera* and *Jatropha cuneata* are virtually the only other shrubs. Cacti here include *Opuntia fulgida* and *Ferocactus acanthodes*. Shallow runnels support *Lycium californicum*, *Prosopis glandulosa* var. *torreyana*, *Larrea tridentata*, and *Fouquieria splendens* (Fig. 139).

Cactus-Mesquite-Saltbush Series

This series is located on the level coastal plain between Empalme and Potám, Sonora. The community is striking because it includes five species of columnar cacti: Hecho (*Pachycereus pecten-aboriginum*), Saguaro (*Carnegiea gigantea*), Senita (*Lophocereus schottii*), Organ Pipe (*Stenocereus thurberi*), and Sina (*S. alamosensis*). It is floristically diverse and includes several genera represented by two or three species (*Atriplex: A. canescens* and *A. polycarpa*; *Bursera: B. microphylla*, *B. odorata* and *B. laxiflora*; *Jatropha: J. cardiophylla* and *J. cinerea*; *Lycium: L. brevipes* and *L. berlandieri*). Western Honey Mesquite (*Prosopis glandulosa* var. *torreyana*) is the dominant plant and with *Cercidium praecox* and the *Bursera* species forms a conspicuous small tree layer. Desert Seep Weed (*Suaeda torreyana* var. *ramosissima*) and the two species of *Atriplex* are the dominant shrubs. In addition there occur *Citharexylum flabellifolium*, Sonoran Caper (*Atamisquea emarginata*), Candelilla (*Pedilanthus macrocarpus*), and Papache (*Randia thurberi*).

No vertebrates are endemic to the Central Gulf Coast subdivision; the mammalian, avian, and reptilian inhabitants are species of more general distribution in the Sonoran Desert and elsewhere. Although some less mobile forms as the Banded Sand Snake (*Chilomeniscus cinctus*) are found on both sides of the Gulf, many others are restricted to one side or the other. Apparently, the long separation of the peninsula from the mainland has had a greater effect on animals than on plants. The latter have maintained a remarkable similarity in species composition on both sides of the Gulf throughout the subdivision.

Plains of Sonora Subdivision

Although the Plains of Sonora is the smallest of Shreve's (1951) Sonoran Desert subdivisions, the actual area occupied by desertscrub is even smaller than the area outlined. The region is at the upper limits of rainfall for desertscrub, and almost all of the subdivision's hills and mountains are better described as thornscrub (= Sinaloan thornscrub). In addition, the numerous small drainages are riparian scrub or woodland, and much of the area's extensive level plains were formerly clothed by summer-active root perennial grasses and are discussed as Sonoran savanna grassland. What remains is desertscrub, consisting of open stands of low branching trees interspersed with an open vegetal cover of shrubs, short-lived herbaceous plants, and bare ground.

In terms of winter rainfall and temperature, the Plains of Sonora is closely allied to the Central Gulf Coast and Vizcaíno subdivisions. In summer its climatic affinities are more similar to the Arizona Upland. Thus, it is characterized by warm, dry winters and moderately warm and relatively wet summers (Fig. 111). There is a large range in summer precipitation values from approximately 190 mm at Presa Alvarado Rodriquez and Carbó, Sonora, to 50 mm at Palo Verde and Torres, Sonora.

The index of continentality of this subdivision lies between the Central Gulf Coast and Arizona Upland (Table 23), occupying an intermediate position among the Southwest desert regions. Its vegetative participants show a marked increase in tropical species from north to south, while maintaining a relatively uniform structure throughout. Unlike other subdivisions of Sonoran desertscrub, the vegetation is dominated by trees, shrubs and forbs rather than a combination of trees, shrubs, and cacti. All of this subdivision is confined to the state of Sonora, except for depauperate phases of Plains of Sonora communities that are present on a few level habitats within the Arizona Upland subdivision. These small enclaves lack cacti and are dominated by *Olneya*, *Prosopis* and *Cercidium*.

As its name implies, most of the Plains of Sonora subdivision is on level valley fill. Mountains, hills, bajadas, and other upland habitats take up less than 15% of the region. A reddish clay soil predominates. Drainages are mostly small and rarely incised. Elevations range from just below 100 m in the west and south to above 800 m in the northeast.

Its northwestern boundary with the Lower Colorado River Valley subdivision is determined by the decrease in summer precipitation northward and westward as the gradually descending plain approaches the Magdalena River. The beginning of the Plains of Sonora is marked by the appearance of trees, a general reduction in *Larrea*, and, often, an increase in herbaceous cover. To the west, the transition to the Central Gulf Coast region is poorly defined but occurs between 100 and 400 m elevation where the mean annual precipitation decreases to less than 200 mm. In the northeast and in the mountains, an increase in mean annual precipitation to more than 300 mm results in the formation of thornscrub or, in the more xeric mountain locales, Arizona Upland communities of Sonoran desertscrub. The gradual increase in precipitation to the south and east, coupled with greater intervals between freezes, results in an uneven increase in density of vegetation toward thornscrub.

The subdivision's mean annual precipitation, based upon the averages of eight stations, is 238.5 mm (Table 31) with more than two thirds falling from July through September. Freezing temperatures are of short duration and, although night frosts may occur in most winters in the north, they are infrequent to the south.

The dominant landscape physiognomy is an open stand of leguminous, sun-loving trees (Fig. 140), and were it not for the barren intervening ground, one could call the Plains of Sonora a woodland. These tree participants, which are mostly from 4-5 m to 8-10 m in height, are the unifying vegetative feature of the region. The most prevalent species is the fire tolerant *Olneya tesota*, often with good representation of *Cercidium microphyllum*, and *Prosopis velutina*, the first and last of which may dominate in size and numbers. Other important trees are *Cercidium praecox*, *Cercidium floridum*, Retama (*Parkinsonia aculeata*), and *Atamisquea emarginata*. The last has the appearance of a small oak. Another unusual tree is the sclerophyllous evergreen Jito (*Forchammeria watsoni*) which, although localized, helps characterize this part of Sonora (Fig. 141). Examples with shorter stature are from the adjacent Sinaloan thornscrub and include *Jatropha cordata*, *Bursera laxiflora*, the red-orange flowered Tree Ocotillo (*Fouquieria macdougalii*), and the bright blue-flowered Guayacan (*Guaiacum coulteri*). When in flower, the last two can give the desert a far different aspect than the yellow and white flowers most commonly seen during the spring season. Sonoran Palo Verde (*Cercidium sonorae*), a name applied to hybrids between *C. praecox* and *C. microphyllum* (Carter, 1974), is commonly found in this subdivision because the region lies in a major area of range overlap of the parent species. Still other trees, such as Tree Morning-glory (*Ipomoea arborescens*) and Pochote (*Ceiba acuminata*), are indicative of short tree forest and are

Table 31. Data for 8 weather stations in the Plains of Sonora subdivision. Data for each station include latitude and longitude (values rounded to closest 0.1), altitude (m), average annual precipitation (mm), and abbreviated name as used in climographs of Figure 111. Data from Hastings and Humphrey (1969b).

Station name	Abbreviated name	Latitude	Longitude	Altitude (m)	Average annual precipitation (mm)
Carbó, Son.	CARB	29.7	110.9	464	309.3
Felix Gomez, Son.	FEGO	29.8	111.5	675	293.9
Hermosillo, Son.	HER	29.1	111.0	211	242.3
Palo Verde, Son.	PAVE	29.1	111.4	60	125.7
Presa A. Rodriguez, Son.	PRAR	29.1	110.9	211	303.4
Querobabi, Son.	QUE	30.0	111.0	650	261.3
San Enrique, Son.	SNEN	29.0	111.4	90	227.6
Torres, Son.	TOR	28.8	110.8	275	144.5
Subdivision Average					238.5

Figure 140. *View of the Brittlebush-Ironwood series of the Plains of Sonora subdivision north of Hermosillo, Sonora. The upper canopy of this vegetation is dominated by* Olneya tesota, *but is also characterized by* Fouquieria macdougalii, Bursera confusa, Cercidium praecox, C. microphyllum, *and* Prosopis. *Three species of large cacti are common:* Lophocereus schottii, Stenocereus thurberi, *and S. alamosensis.* Encelia farinosa, *almost leafless in this view, is the main low shrub. Shrubs of taller stature here include* Mimosa laxiflora, Guaiacum coulteri, Phaulothamnus spinescens, *and* Randia thurberi. *Photograph by J.R. Hastings. Altitude 395 m.*

mostly confined to drainages.

Understory species are mostly irregularly spaced and shrubbery is more dispersed than in other Sonoran Desert subdivisions. An exception is brittlebush, or *Encelia,* which reaches its maximum development here and, with *Olneya tesota,* forms the region's most prevalent series and provides the alternate term—Olneya-Encelia region (Shreve, 1951). *Larrea,* except in the northwest, is local. Other species of frequent but rarely monotypic occurrence include:

Whitethorn Acacia (*Acacia constricta*)
Caesalpinia (*Caesalpinia pumila*)
Fairy Feather Duster (*Calliandra eriophylla*)
Desert Hackberry (*Celtis pallida*)

Cordia (*Cordia parvifolia*)
Sonoran Croton (*Croton sonorae*)
Coursetia (*Coursetia glandulosa*)
Kidneywood (*Eysenhardtia orthocarpa*)
Jacobinia (*Jacobinia ovata*)
Hierba de Cristo (*Lantana horrida*)
Tomatillo (*Lycium brevipes*)
Mexican Jumping Bean (*Sapium biloculare*)
Yellow Trumpet Bush (*Tecoma stans*)
Graythorn (*Zizyphus obtusifolius*)

Cacti, while common, are less prevalent than in all other Sonoran Desert subdivisions but the Lower Colorado River Valley subdivision. The most conspicuous and widespread

Figure 141. *Jito (Forchammeria watsoni) (upper left) in the Plains of Sonora west of Ures, Sonora. The appearance of a large evergreen tree in such an arid region is somewhat a surprise. Altitude 375 m.*

species are: Sina (*Rathbunia alamosensis*), Organ Pipe (*Stenocereus thurberi*), Senita (*Lophocereus schottii*), and four chollas (*Opuntia arbuscula, O. fulgida* var. *mammillata, O. leptocaulis, O. thurberi*). Indeed the region's plant diversity is low when compared to such adjacent tropic-subtropic biomes as Sinaloan thornscrub and the Arizona Upland and Central Gulf Coast subdivisions of Sonoran desertscrub.

Because its arboreal dominants and open ground cover are found elsewhere in the Sonoran Desert, and because of its transitional nature toward Sinaloan thornscrub, the animal inhabitants of the Plains of Sonora show affinities to both of these biomes. No known vertebrate is restricted to any of its

series. Instead its fauna are the tree-requiring (e.g., Harris Hawk, White-winged Dove, and *Urosaurus ornatus*) and burrowing species e.g., Kit Fox, Burrowing Owl (*Athene cunicularia*), *Gopherus agassizi, Heloderma suspectum,* and *Crotalus cerastes* found in other Sonoran Desert communities. It is also here that southern thornscrub species such as the Elegant Quail and Beaded Lizard (*Heloderma horridum*) reach their northern limits. A very few species such as the Sonoran Green Toad (*Bufo retiformis*) are centered here; it and such grass-scrub species as the Masked Bobwhite and *Terrapene ornata*, which find their western limits here, attest to the region's former and recent grassland history.

Figure 142. *Magdalena subdivision community near San Javier, Baja California del Sur. The absence of* Fouquieria columnaris *and* Pachycormus discolor, *the presence of relatively short* Pachycereus pringlei, *and the low frequency of* **Agave** *and* Yucca *set this subdivision apart from the adjacent Vizcaino subdivision to the north. Species growing at this site include* **Lysiloma candida, Larrea tridentata, Jatropha cuneata, Cercidium microphyllum, Pachycereus pringlei, Stenocereus thurberi,** *and* **Opuntia cholla.** *Photograph by J.R. Hastings. Alt. 425 m.*

Magdalena Region

This southernmost subdivision of Shreve's Sonoran Desert barely crosses our southern boundary for the Southwest on hills and slopes near San Ignacio in Baja California del Sur. As do other of Shreve's subdivisions, the moister mountainous habitats of the Magdalena region support vegetation that is too dense to be classified as desertscrub (Fig. 142). Most of the "desert" area in this subdivision occupies a plain with rainfall generally less than 200 mm or, near the Pacific shore, even less than 100 mm. The vegetation here is unmistakably desertscrub. Southward toward the tip of the Baja California peninsula, the arid coastal strip narrows and the vegetation becomes increasingly dense and tall in response to the more favorable moisture conditions. Here, and along the eastern drainage divide between the Pacific Ocean and the Sea of Cortez, the desertscrub increases in density, forming a transitional community between desertscrub and forest. Toward the "cape," it forms a biome analogous to Sinaloan thornscrub and Tamaulipan thornscrub, both of which also merge into tropic-subtropic forest with increased precipitation.

Therefore at least some Magdalena communities may be better described as northern extensions of a San Lucan deciduous scrub. This descriptive terminology is based on Dice's (1943) terminology for the biome's biogeographic center and the fact that most of the plants shed their leaves during the winter-spring dry season. Avoidance of the term thornscrub is in deference to Shreve (1937) and later Humphrey (1974) who objected to the designation "thorn forest" for the short tree forest (= San Lucan deciduous forest) that characterizes the Cape region farther south; thorny acacia-type trees being minor constituents in this biome. This is true also of the shorter deciduous scrub which shares many of the same species as the forest, although its cacti (*Opuntia cholla, Stenocereus gummosus, Lophocereus schottii,* and *Stenocereus thurberi*) are certainly spiny and a number of its non-tree constituents are thorny (e.g., *Fouquieria diguetii*).

Whatever the designation, these landscapes lie almost entirely outside the area under present consideration and need not be discussed further here. Those extending their travels and investigations on the Baja California peninsula south of parallel 27° 30′ N are urged to consult Nelson's (1922) comprehensive, and still valuable, work on the biota of Baja California, as well as Shreve's (1937, 1951) classical treatments on the region's vegetation.

Part 6. Wetlands

W. L. Minckley
Department of Zoology
Arizona State University

and David E. Brown
Arizona Game and Fish Department

*Cold temperate marshland about 1800 m elevation
along Silver Creek, Navajo County, Arizona.*

Figure 143. *Integration of aquatic and riparian communities – only emergent aquatic plants (marshlands) are absent, and they occurred a few hundred meters from this site. Submergent (mostly the alga* Cladophora glomerata) *and strand (barren) habitats, lined by riparian scrub* (Baccharis salicifolia), *subtropical deciduous woodland* (Prosopis velutina), *and warm temperate riparian broadleafed forest* (Populus fremontii, Salix *spp., etc). Bonita Creek, Graham County, Arizona. Elevation ca. 960 m; photograph by Stuart G. Fisher, June 1979.*

Southwestern Wetlands

Wetlands are periodically, seasonally, or continuously submerged landscapes populated by species and/or life forms differing from immediately adjacent biotas. They are maintained by and depend upon circumstances more mesic than those provided by local precipitation. Such conditions occur in or adjacent to drainage-ways and their floodplains (riparian zones), on poorly drained lands, along seacoasts, and in and near other hydric and aquatic situations, i.e., springs and their outflows, ponds, margins of lakes, etc. The various wetland and riparian communities may be represented as forest, woodland or scrubland, marshland or strand, or be composed largely or entirely of submergent vegetation (Fig. 143).

Although wetland formations may be remarkably distinct, they are also often highly integrated, or occur as intermittent stands within other communities (Fig. 144). In riparian habitats that pass through many biomes, high elevation species often extend downslope into grassland or desert within canyons that lead cooler and moister air downward (Lowe 1964; Fig. 145). Also, formations encountered are often

successional as a result of periodic disturbance from flood scouring, inundation, desiccation, grazing by animals, or other factors. The communities and associations are dynamic, subject to frequent change, displacement, replacement, and succession (Campbell and Green, 1968; Everitt, 1968; Johnson et al., 1976; Reichenbacher, 1980).

Although formation-classes of wetland habitats may frequently be complex, the number of dominants often is surprisingly few, especially of aquatic plants where monospecific stands (consociations) are remarkably recurrent in time and space (Gessner, 1955 et seq.). Growth of aquatic and semiaquatic plants is rapid, and succession to at least a quasi-climax may occur in a period of weeks or months.

Aquatic, riparian and other wetland biotic communities of the American Southwest have rarely been differentiated or shown on maps. They tend to be small relative to other communities, but possess an importance and biological interest totally disproportionate to their limited geographic occurrence.

Figure 144. *Integration of aquatic and riparian communities— cold temperate marshland* (Scirpus *spp.*), *cold temperate woodland (Red Willow,* Salix laevigata), *submergent (open water) and at left, mostly out of the picture, scrubland communities. Meadow Valley Wash north of Prescott, Yavapai County, Arizona. Elevation ca. 1,800 m.*

Physical Environments of Southwestern Wetlands

The Southwest's major rivers (Fig. 146) have been altered by dams for many years, their flows diverted and changed, and their once-perennial lower reaches de-watered (Deacon and Minckley, 1974). In fact, the Southwest's largest river, the Colorado, with the exception of a few important reaches (Figs. 147, 148), has been reduced to a series of impoundments connected by canals. These regulated streams have much of their former nutrient loads trapped by reservoirs, to the detriment of downflow systems (Paulson and Baker, 1980). Further, discharge, temperature, and sedimentation regimes are now unsuitable for a native aquatic and semi-aquatic biota adapted to a seasonally turbid and variably aggrading and degrading, warm, and vernal-flooded system (Minckley, 1979). Their highly specialized, endemic fish faunas have been largely destroyed and replaced by non-native species (see e.g. Miller, 1961; Minckley and Deacon, 1968; Moyle and Nichols, 1973). Some lesser rivers are as yet unregulated, however, and these along with many smaller streams at intermediate elevations still support native aquatic faunas variously influenced by introduced forms.

Prior to 1880, alluvial plains of river bottoms at lower elevations (<1,000 m) were wetter and less well drained than at present. Streams were commonly characterized by boggy margins, marshy sloughs, and backwaters (Fig. 149), which were of great annoyance to early travelers and a health hazard to personnel at 19th Century military posts (Hastings, 1959). Stream channels were typically shallow and braided, with deeper water in meanders and oxbows, and where Beaver (*Castor canadensis*) activity was prevalent (Davis, 1982). The well known but poorly understood cycle of arroyo cutting (Fig. 150) that began in the 1880's and 1890's (Bryan, 1925a, 1928; Cooke and Reeves, 1976; Leopold, 1976), coupled with deliberate river channelization, streamflow impoundment and diversion, and mining of ground water, have caused these riverine and adjacent spring-fed marshlands (*ciénegas*) to become an almost extinct Southwestern landscape feature.

Natural lakes and ponds, other than those associated with local fluviatile action, are rare in the Southwest. This results largely from a lack of recent glaciation, general aridity, high evaporation and siltation rates, and the steep gradients of much of the topography. With the exception of some high-elevation glacial lakes in southwestern Colorado, most smaller natural lakes (*lagunas*) are sinkholes resulting from solution and gradual erosion of soluble, underlying rocks (Cole, 1963, 1968). Subsidence basins resulting from underflow or degassing of lava are locally present, and the resulting calderas may hold transitory or permanent waters. Larger lakes are (or were in the past) mostly tectonic in origin, occupying closed basins (*bolsones*) or occurring as a result of structural uplift that impounded ancient waterways. Most are presently more or less marshy and seasonal.

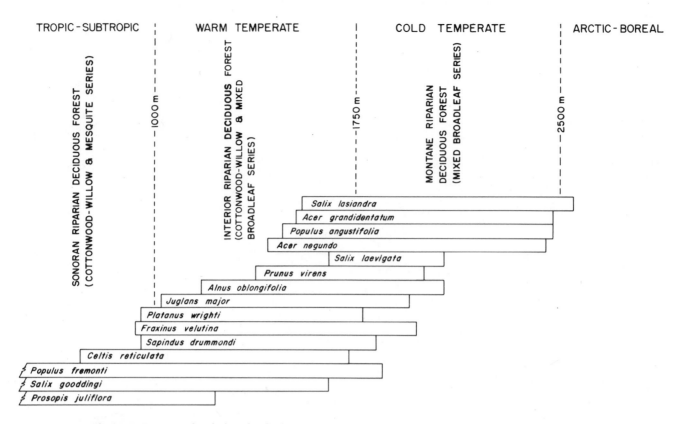

Figure 145. *Generalized altitudinal, climatic and community distribution for some deciduous riparian trees in the sub-Mogollon Rim region of central Arizona.*

Most large, natural lentic habitats in the Southwest are endorheic, fed by ephemeral inflows and by definition lacking surface outflows. Water passes out of such systems by evaporation and/or downward percolation if the lake is perched above the groundwater table. Such lakes often are saline, and although they are heavily populated by diverse invertebrates and, therefore, attractive to waterfowl, shore-birds and wading birds, provide little habitat for fishes. Normally dry, closed basin lakes are called *playas*. Although commonly considered characteristic landscapes of the Chihuahuan Biotic Province, playas are found in all four North American deserts, and throughout the Southwest. Two of the largest of these—the Salton Sea in California and Laguna Salada in Baja California del Norte—have been, or are in the process of being, transformed into permanent bodies of water through diversion of Colorado River water.[1] Some smaller closed basin lakes at higher elevations, with attendant increased precipitation and reduced evaporation, provide relatively dependable aquatic habitats (e.g., Mormon Lake in Arizona, elevation ca. 2,150 m; Lake Elsinore in California, ca. 1,000 m). A few depressions, such as Laguna Prieta in north-western Sonora, that intersect groundwater tables thus remain relatively permanent (Fig. 151).[2]

Rockpools (*tinajas*) in arid mountain ranges have long been known as oases critical for survival of ancient peoples (Taylor, 1962) and for later desert travelers (Bryan, 1925b). They also support local and often distinctive populations of plants, invertebrates, and vertebrates (the last ranging from amphibians through large mammals such as Bighorn Sheep). These small habitats, typically scoured from bedrock by boulders powered by infrequent flash floods, qualify as special wetlands within the Southwest's myriad of systems (Fig. 152). They are so poorly known biologically that we scarcely treat them further, but seepages downslope from such places may support diverse and special riparian communities—dense stands of cottonwood and willow may be present at some locales, or groves of palms in places like the Kofa Mountains of Arizona.

Far more numerous than natural freshwater habitats are artificial reservoirs (*presas*), farm ponds (*estanques*) and the innumerable cattle tanks (*represos* or *charcos*) of varying degrees of permanence. These, and pumped ponds, have created a scattering of aquatic communities in arid parts of the Southwest that were otherwise devoid of surface water. Such impoundments created lentic habitats in a region formerly dominated by streams.

Extensive canal systems provide linear, flowing systems across desertlands, often interconnecting formerly isolated bodies of water to allow dispersal of organisms across vast reaches formerly denied to them, and also providing a transportation system for terrestrial organisms and their propagules. Lining of canals with concrete in the past few years has drastically reduced these substitute riparian habitats by suppressing seepage, and their steep-walled channels often become a death trap for desert animals.

Springs, some within bolsones, have provided aquatic habitats for a large percentage of the unique "desert" fishes of

---

[1] *For an informative account of the filling of the Salton Sea see Sykes (1937),* **The Colorado Delta**.
[2] *Recent photographs (1979) of Laguna Prieta show it to be almost dry, probably as a result of normal annual fluctuation but possibly also the result of new water-well fields in the vicinity.*

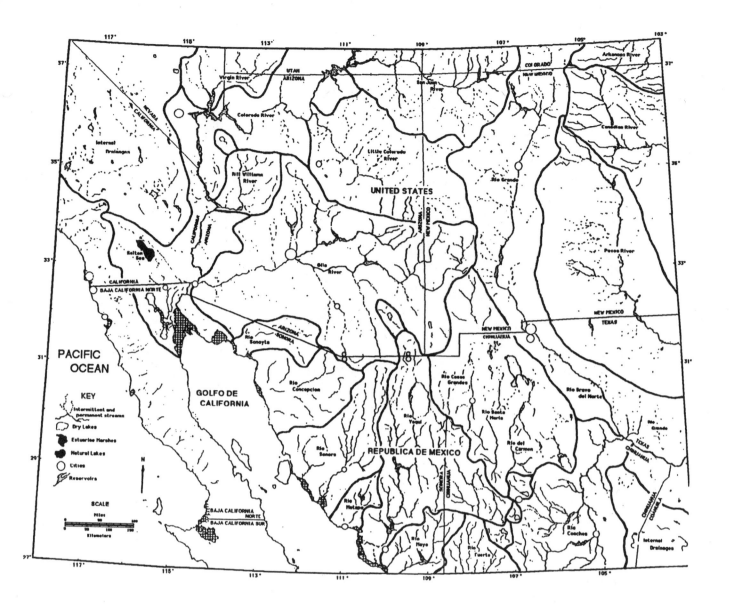

Figure 146. *Semi-diagramatic sketch map of major river systems of the American Southwest.*

the region. These habitats vary from a few square meters of surface water to several hectares and have provided permanent refugia for plants, invertebrates and vertebrates that date from long ago. Biotic elements in such places are often far from their expected natural ranges. Water mining has proven exceptionally damaging to these relict wetlands, an example of which is the total and recent destruction of surface discharge in massive limestone springs of western Texas as a direct result of pumping for agriculture (Brune, 1975).

Coastline beaches of Baja California, southern California and Sonora, with their adjacent marine environments, provide the richest and most extensive, as well as most spectacular, wetland and aquatic habitats in the American Southwest. In addition to strand vegetation on rocky cliffs, rubble beaches, mud and sand, there are tidal marshes now greatly reduced but once occupying most estuaries (*esteros*). Of great importance, but outside the scope of this discussion, are offshore submergent communities, as yet unclassified and only partially known. There forests of Kelp (*Macrocystis pyrifera*), meadows of Eelgrass (*Zostera marina*), and many less structurally distinct underseascapes, have a flora and fauna known only to a relative few. Life histories and ecological

relationships of even many of the larger and more important species are poorly understood and open to discovery, e.g., the recent observations of Felger et al. (1976) on over-wintering of the Green Sea-turtle (*Chelonia mydas*) in the Sea of Cortez. This is surprising considering the economic, scientific and recreational importance of these marine resources both in the United States and Mexico.

Estuarine, riparian and other wetland communities are continuing to be rapidly destroyed by reductions of streamflows and water tables. The impact of nutrient entrapment in Colorado River reservoirs, and the near cessation of freshwater inflow to the upper Sea of Cortez from upstream uses (Thomson et al., 1979), or the antithesis of enrichment of coastal waters by sewage and other wastes off southern California, have had profound impacts on the region's marine communities that have been scarcely assessed. Further, increasing demands placed upon all the arid Southwest's water resources makes the future outlook for remaining communities tenuous at best. Within a single generation, unique riparian and marshland ecosystems, varying in size from tiny springfed marshes to the recently extensive backwaters of the Colorado River and its delta, have been lost.

Figure 147. *Aerial view of the Colorado River mainstream between Parker, Yuma County, Arizona, and Blythe, Riverside County, California, a reach yet to be severely modified by channelization, levees, and dredging. Floodplain vegetation where undisturbed by agriculture consists of subtropical deciduous woodland* (Prosopis velutina, P. pubescens, Tamarix chinensis, *etc.*) *and a complex scrubland of diverse genera* (Atriplex, Baccharis, Pluchea, Tessaria, Tamarix, *etc.*). *Riparian deciduous forests of Cottonwood* (Populus fremontii) *and Willow* (Salix gooddingii) *persist as small stands not visible in the photograph, and marshlands* (Scirpus californicus, S. americanus, Typha domingensis, *and many others) line the submergent communities of the river. Elevation ca. 210 m; photograph by Robert D. Ohmart, 1976.*

Figure 148. *Colorado River mainstream in Grand Canyon National Park, Coconino County, Arizona. Although the river now is controlled by Glen Canyon Dam, closed in the early 1960's, a border of riparian scrub (principally* Prosopis velutina) *above the zone of scour by flooding in years past remains evident. Elevation ca. 500 m; photograph by W.L. Minckley, August 1978.*

Figure 149. *Interior view of mature ciénega habitat at Atascosa, Sonora, México. Submergent communities were principally of algae, with emergent* Polygonum *spp.,* **Salix gooddingii** *and marginal* S. exigua *and sedges (*Scirpus *spp.). Elevation ca. 1,200 m; photograph by Gary Meffe, July 1980.*

Figure 150. *Cut banks and dead mesquites along Santa Cruz River, near San Xavier, Pima County, Arizona. Once one of the finest mesquite bosques in the Southwest, ground water pumping has now virtually destroyed this interesting community.*

Figure 151. *Laguna Prieta, a permanent lake in the Gran Desierto region of the Sonoran Desert, Sonora Mexico. Sierra Del Rosario is in left center background. A surprise is the presence here of a natural lake, albeit saline, fed by springs, and of a high water table in such an arid region. Elevation ca. 50 m; photograph by Peter Kresan.*

Problems with Classification of Aquatic Habitats

Because lakes and large bodies of quiet water were unusual and typically ephemeral in the recent Southwest, the region's unique aquatic biota became largely stream adapted, or was physically restricted to remnant springs and seeps. Thus, although the evolution of aquatic organisms has not been completely independent of terrestrial changes in the Southwest, relationships of such differentiation to terrestrial biomes are often obscure. Numbers of fish species present in a particular drainage basin, for example, depends more on geologic, climatic, and evolutionary histories, and size and complexity of the watershed, than on elevation or terrestrial vegetation.

The ecology of aquatic systems, while differing substantially from terrestrial habitats, is not entirely independent of climate. Streams originating in Arctic-Alpine areas flow downslope through other biomes, and some aquatic species occupy a number of defined (or definable) zones of terrestrial landscape. Physiological constraints obviously restrict such animals as trouts to cool waters at higher altitudes. Other fish groups of tropical or marine affinities (e.g., cyprinodontids and poeciliids) are similarly restricted to lower, warmer places. A vast majority of native fishes present in the Southwest, however, consists of the minnow and sucker families (Miller, 1959), an array of diversity that precludes the luxury of generalizations. Exceptions to these rules also occur when special habitats such as thermal springs are at high elevations, or deep, cool lakes are present in deserts. Also, special conditions such as downcanyon drainage of cool, montane air may allow cold-water fishes, other animals, and riparian plants to penetrate far into other biomes. Some species are thus included in discussions of characteristic biotic elements of more than one biome—far more in this section than is the case elsewhere in the text.

Figure 152. *Permanent rockpool (tinaja) in the Hieroglyphic Mountains, Maricopa County, Arizona, a habitat too erosive to support other than algae and other transitory plants, but extremely important as a water supply to numerous desert animals. Elevation ca. 680 m.*

Table 32. Distribution of native, freshwater fishes in drainage basins of the American Southwest.

Species	Plains[1]	Rio Grande basin				Guzman Basin[5]	Rio Yaqui	Coastal drainages		Death Valley System	Colorado River basin			
		Pecos River[2]	Upper[3]	Lower[4]	Rio Conchos			Baja California[6]	Pacific Coast[7]		Lower[9]	Gila River	Middle[10]	Upper[11]
Petromyzontidae (Lampreys)														
Lampetra tridentata (Pacific Lamprey)	-	-	-	-	-	-	-	-	X	-	-	-	-	-
Acipenseridae (Sturgeons)														
Scaphirhynchus platorhynchus (Shovelnosed Sturgeon)	X[a]	-	X	X	-	-	-	-	-	-	-	-	-	-
Lepisosteidae (gars)														
Lepisosteus oculatus (Spotted Gar)	X[a]	X	X	X	-	-	-	-	-	-	-	-	-	-
L. osseus (Longnose Gar)	X[a]	X	X	X	X	-	-	-	-	-	-	-	-	-
Clupeidae (shads)														
Dorosoma cepedianum (Gizzard Shad)	X	X	X	X	X	-	-	-	-	-	-	-	-	-
D. smithi (Pacific Shad)	-	-	-	-	-	-	X	-	-	-	-	-	-	-
Elopidae (Tenpounders, Tarpons)														
Elops affinis (Machete)	-	-	-	-	-	-	X	-	-	-	X	X	-	-
Salmonidae (Salmon, Trouts, Whitefishes)														
Prosopium williamsoni (Mountain Whitefish)	-	-	-	-	-	-	-	-	-	-	-	-	-	X[c]
Salmo apache (Arizona Trout)	-	-	-	-	-	-	-	-	-	-	-	X	X	-
S. gairdneri (Rainbow Trout)	-	-	-	-	-	-	-	X	X	-	-	-	-	-
S. gilae (Gila Trout)	-	-	-	-	-	-	-	-	-	-	-	X	-	-
Salmo sp. (Yaqui Trout)	-	-	-	-	-	X	X	-	-	-	-	-	-	-
Characidae (Tetras)														
Astyanax mexicanus (Mexican Tetra)	-	X	X	X	X	-	-	-	-	-	-	-	-	-
Cyprinidae (Carps, Minnows)														
Agosia chrysogaster (Longfin Dace)	-	-	-	-	-	-	X	X	-	-	-	X	-	-
Campostoma anomalum (Plains Stoneroller)	X	-	-	X	-	-	-	-	-	-	-	-	-	-
C. ornatum (Mexican Stoneroller)	-	-	-	X	X	X	X	X	-	-	-	-	-	-
Codoma ornata (Ornate Minnow)	-	-	-	-	X	-	X	-	-	-	-	-	-	-
Dionda diaboli (Devil's River Minnow)	-	-	-	X[b]	-	-	-	-	-	-	-	-	-	-
D. episcopa (Roundnosed Minnow)	-	X	X	X	X	-	-	-	-	-	-	-	-	-
Gila bicolor snyderi (Owens Tui Chub)	-	-	-	-	-	-	-	-	-	X	-	-	-	-
G. bicolor mohavensis (Mohave Chub)[13]	-	-	-	-	-	-	-	-	-	X	-	-	-	-
G. cypha (Humpback Chub)	-	-	-	-	-	-	-	-	-	-	X	-	X	X
G. ditaenia (Sonora Chub)	-	-	-	-	-	-	-	X	-	-	-	-	-	-
G. elegans (Bonytail Chub)	-	-	-	-	-	-	-	-	-	-	X	X	X	X
G. intermedia (Gila Chub)	-	-	-	-	-	-	-	-	-	-	-	X	-	-
G. nigrescens (Guzman Chub)	-	-	-	-	-	X	-	-	-	-	-	-	-	-
G. orcutti (Arroyo Chub)	-	-	-	-	-	-	-	-	X	-	-	-	-	-

Species	Plains[1]	Pecos River[2]	Upper[3]	Lower[4]	Rio Conchos	Guzman Basin[5]	Rio Yaqui	Baja California[6]	Pacific Coast[7]	Death Valley System	Lower[9]	Gila River	Middle[10]	Upper[11]
		Rio Grande basin						Coastal drainages			Colorado River basin			
G. pandora (Rio Grande Chub)	-	X	X	-	-	-	-	-	-	-	-	-	-	-
G. pulchra (Mesa del Norte Chub)	-	-	-	-	X	-	X	-	-	-	-	-	-	-
G. purpurea (Yaqui Chub)	-	-	-	-	-	-	X	X	-	-	-	-	-	-
G. robusta (Roundtail Chub)	-	-	-	-	-	-	X	-	-	-	X	X	X	X
Hybognathus nuchalis (Silvery Minnow)	X	X	X	X	X	-	-	-	-	-	-	-	-	-
H. placitus (Plains Minnow)	X	X	X	X	-	-	-	-	-	-	-	-	-	-
Hybopsis aestivalis (Speckled Chub)	X	X	X	X	X	-	-	-	-	-	-	-	-	-
Lepidomeda albivalis (White River Spinedace)	-	-	-	-	-	-	-	-	-	-	-	-	X	-
L. altivelis (Pahranagat Spinedace)	-	-	-	-	-	-	-	-	-	-	-	-	X	-
L. mollispinis (Middle Colorado Spinedace)	-	-	-	-	-	-	-	-	-	-	-	-	X	-
L. vittata (Little Colorado Spinedace)	-	-	-	-	-	-	-	-	-	-	-	-	X	-
Meda fulgida (Spikedace)	-	-	-	-	-	-	-	-	-	-	-	X	-	-
Notropis amabilis (Texas Shiner)	-	X	X	X	X	-	-	-	-	-	-	-	-	-
N. braytoni (Tamaulipan Shiner)	-	X	X	X	X	-	-	-	-	-	-	-	-	-
N. buchanani (Ghost Shiner)	X[a]	-	-	X	-	-	-	-	-	-	-	-	-	-
N. chihuahua (Chihuahua Shiner)	-	-	-	X	X	-	-	-	-	-	-	-	-	-
N. formosus (Beautiful Shiner)	-	-	-	-	-	X	X	-	-	-	-	-	-	-
N. girardi (Arkansas River Shiner)	X[a]	-	-	-	-	-	-	-	-	-	-	-	-	-
N. jemezanus (Rio Grande Shiner)	-	X	X	X	X	-	-	-	-	-	-	-	-	-
N. lutrensis (Red Shiner)	X	X	X	X	X	-	-	-	-	-	-	-	-	-
N. panarcys (Conchos Shiner)	-	-	-	-	X	-	-	-	-	-	-	-	-	-
N. proserpinus (Prosperine Shiner)	-	X	X	X[b]	-	-	-	-	-	-	-	-	-	-
N. rutilus (Salado-San Juan Shiner)	-	-	-	X[b]	-	-	-	-	-	-	-	-	-	-
N. shumardi (Silverband Shiner)	X[a]	-	-	-	-	-	-	-	-	-	-	-	-	-
N. simus (Bluntnose Shiner)	-	-	X	X	-	-	-	-	-	-	-	-	-	-
N. stramineus (Sand Shiner)	X	X	-	X	-	-	-	-	-	-	-	-	-	-
N. venustus (Spottail Shiner)	-	-	-	X[b]	-	-	-	-	-	-	-	-	-	-
Phenacobius mirabilis (Suckermouth Minnow)	X	-	-	-	-	-	-	-	-	-	-	-	-	-
Phoxinus erythrogaster (Red-bellied Dace)	X[a]	-	-	-	-	-	-	-	-	-	-	-	-	-
Pimephales promelas (Flathead Minnow)	X	X	X	X	X	X	-	-	-	-	-	-	-	-
Plagopterus argentissimus (Woundfin)	-	-	-	-	-	-	-	-	-	-	X	X	X	-

Species	Plains[1]	Rio Grande basin				Guzman Basin[5]	Rio Yaqui	Coastal drainages		Death Valley System	Colorado River basin			
		Pecos River[2]	Upper[3]	Lower[4]	Rio Conchos			Baja California[6]	Pacific Coast[7]		Lower[9]	Gila River	Middle[10]	Upper[11]
Platygobio gracilis (Flathead Chub)	X	X	X	-	-	-	-	-	-	-	-	-	-	-
Ptychocheilus lucius (Colorado Squawfish)	-	-	-	-	-	-	-	-	-	-	X	X	X	X
Rhinichthys cataractae (Longnose Dace)	X	X	X	X	X	-	-	-	-	-	-	-	-	-
R. osculus (Speckled Dace)	-	-	-	-	-	-	-	-	X	X	-	X	X	X
Semotilus atromaculatus (Creek Chub)	X	X	-	-	-	-	-	-	-	-	-	-	-	-
Tiaroga cobitis (Loach Minnow)	-	-	-	-	-	-	-	-	-	-	-	X	-	-
Catostomidae (Suckers)														
Carpiodes carpio (River Carpsucker)	X	X	X	X	X	-	-	-	-	-	-	-	-	-
Castostomus bernardini (Yaqui Sucker)	-	-	-	-	-	-	X	-	-	-	-	-	-	-
C. commersoni (White Sucker)	X	-	-	-	-	-	-	-	-	-	-	-	-	-
C. conchos (Conchos Sucker)	-	-	-	-	X	-	-	-	-	-	-	-	-	-
C. fumeiventris (Owens Sucker)	-	-	-	-	-	-	-	-	-	X	-	-	-	-
C. insignis (Sonora Sucker)	-	-	-	-	-	-	-	-	-	-	-	X	-	-
C. latipinnis (Flannelmouth Sucker)	-	-	-	-	-	-	-	-	-	-	X	X	X	X
Catostomus sp. (A) (Little Colorado Sucker)	-	-	-	-	-	-	-	-	-	-	-	-	X	-
Catostomus sp. (B)	-	-	-	-	-	X	X	-	-	-	-	-	-	-
Catostomus sp. (C)	-	-	-	-	-	-	X	-	-	-	-	-	-	-
C. wigginsi (Opata Sucker)	-	-	-	-	-	-	X	X	-	-	-	-	-	-
Cycleptus elongatus (Blue Sucker)	X[a]	X	X	X	X	-	-	-	-	-	-	-	-	-
Ictiobus bubalus (Smallmouth Buffalofish)	X[a]	X	-	X	-	-	-	-	-	-	-	-	-	-
I. niger (Black Buffalofish)	X[a]	X	X	X	X	-	-	-	-	-	-	-	-	-
Moxostoma austrinum (Mexican Redhorse)	-	-	X	-	X	-	-	-	-	-	-	-	-	-
M. congestum (Gray Redhorse)	-	X	X	X	-	-	-	-	-	-	-	-	-	-
Pantosteus clarki (Desert Mountain-sucker)	-	-	-	-	-	-	-	-	-	-	X	X	X	-
P. discobolus (Bluehead Mountain-sucker)	-	-	-	-	-	-	-	-	-	-	-	-	X	X
P. plebeius (Sierra Madre Mountain-sucker)	-	X	X	-	X	X	X	-	-	-	-	-	-	-
P. santaanae (Santa Anna Mountain-sucker)	-	-	-	-	-	-	-	-	X	-	-	-	-	-
Xyrauchen texanus (Razorback Sucker)	-	-	-	-	-	-	-	-	-	-	X	X	X	X
Ictaluridae (North American Freshwater Catfishes)														
Ictalurus furcatus (Blue Catfish)	X[a]	X	X	X	X	-	-	-	-	-	-	-	-	-
I. lupus (Headwater Catfish)	-	X	X	X	X	-	-	-	-	-	-	-	-	-
I. melas (Black Bullhead)	X	-	-	-	-	-	-	-	-	-	-	-	-	-
I. natalis (Yellow Bullhead)	X[a]	-	-	-	-	-	-	-	-	-	-	-	-	-
I. "pricei" (Yaqui Catfish)[14]	-	-	-	-	X	-	X	-	-	-	-	-	-	-
I. punctatus (Channel Catfish)	X	-	-	X	-	-	-	-	-	-	-	-	-	-

Species	Plains[1]	Rio Grande basin				Guzman Basin[5]	Rio Yaqui	Coastal drainages		Death Valley System	Colorado River basin			
		Pecos River[2]	Upper[3]	Lower[4]	Rio Conchos			Baja California[6]	Pacific Coast[7]		Lower[9]	Gila River	Middle[10]	Upper[11]
Priatella phreatophila (Musquiz Blind Catfish)	–	–	–	X[b]	–	–	–	–	–	–	–	–	–	–
Pylodictis olivaris (Flathead Catfish)	X[a]	X	X	X	X	–	–	–	–	–	–	–	–	–
Cyprinodontidae (Pup- and Killfishes)														
Cyprinodon bovinus (Leon Spring Pupfish)	–	X	–	–	–	–	–	–	–	–	–	–	–	–
C. diabolis (Devil's Hole Pupfish)	–	–	–	–	–	–	–	–	–	X	–	–	–	–
C. elegans (Comanche Springs Pupfish)	–	X	–	–	–	–	–	–	–	–	–	–	–	–
C. eximius (Conchos Pupfish)	–	–	X	X	X	–	–	–	–	–	–	–	–	–
C. fontinalis (Carbonaria Pupfish)	–	–	–	–	–	X	–	–	–	–	–	–	–	–
C. macroplepis (Big Scale Pupfish)	–	–	–	–	X	–	–	–	–	–	–	–	–	–
C. macularius (Desert Pupfish)	–	–	–	–	–	–	–	X	–	–	X	X	–	–
C. milleri (Cottonball Marsh Pupfish)	–	–	–	–	–	–	–	–	–	X	–	–	–	–
C. nevadensis (Amargosa Pupfish)	–	–	–	–	–	–	–	–	–	X	–	–	–	–
C. pecosensis (Pecos Pupfish)	–	X	–	–	–	–	–	–	–	–	–	–	–	–
C. radiosus (Owens Pupfish)	–	–	–	–	–	–	–	–	–	X	–	–	–	–
C. rubrofluviatilis (Red River Pupfish)	X	–	–	–	–	–	–	–	–	–	–	–	–	–
C. salinus (Salt Creek Pupfish)	–	–	–	–	–	–	–	–	–	X	–	–	–	–
Cyprinodon sp. (A) (Big-head Pupfish)	–	–	–	X	–	–	–	–	–	–	–	–	–	–
Cyprinodon sp. (B) (Casas Grandes Pupfish)	–	–	–	–	–	X	–	–	–	–	–	–	–	–
Cyprinodon sp. (C) (Whitefin Pupfish)	–	–	–	–	–	X	X	–	–	–	–	–	–	–
Cyprinodon sp. (D) (Monkey Spring Pupfish)	–	–	–	–	–	–	–	–	–	–	–	X	–	–
C. tularosa (Tularosa Pupfish)[15]	–	–	X	–	–	–	–	–	–	–	–	–	–	–
Crenichthys baileyi (White River Springfish)	–	–	–	–	–	–	–	–	–	–	–	–	X	–
C. nevadae (Railroad Valley Springfish)	–	–	–	–	–	–	–	–	–	–	–	–	X	–
Empetrichthys latos (Pahrump Poolfish)	–	–	–	–	–	–	–	–	–	X	–	–	–	–
E. merriami (Ash Meadows Poolfish)	–	–	–	–	–	–	–	–	–	X	–	–	–	–
Fundulus parvipinnis (California Killifish)	–	–	–	–	–	–	–	–	X	–	–	–	–	–
F. kansae (Plains Killifish)	X	–	–	–	–	–	–	–	–	–	–	–	–	–
F. zebrinus (Rio Grande Killifish)	–	X	–	X	–	–	–	–	–	–	–	–	–	–
Poeciliidae (Livebearers)														
Gambusia affinis (Mosquitofish)	X[a]	X	–	X	–	–	–	–	–	–	–	–	–	–
G. alvarezi	–	–	–	–	X	–	–	–	–	–	–	–	–	–
G. amistadensis (Amistad Gambusia)	–	–	–	X[b]	–	–	–	–	–	–	–	–	–	–

Species	Plains[1]	Rio Grande basin				Guzman Basin[5]	Rio Yaqui	Coastal drainages		Death Valley System	Colorado River basin			
		Pecos River[2]	Upper[3]	Lower[4]	Rio Conchos			Baja California[6]	Pacific Coast[7]		Lower[9]	Gila River	Middle[10]	Upper[11]
G. gaigei (Big Bend Gambusia)	-	-	-	X	-	-	-	-	-	-	-	-	-	-
G. hurtadoi	-	-	-	-	X	-	-	-	-	-	-	-	-	-
G. krumholzi (Krumholz Gambusia)	-	-	-	X[b]	-	-	-	-	-	-	-	-	-	-
G. nobilis (Pecos Gambusia)	-	X	-	X	-	-	-	-	-	-	-	-	-	-
G. senilis (Blotched Gambusia)	-	-	-	X	X	-	-	-	-	-	-	-	-	-
Poeciliopsis monacha-occidentalis	-	-	-	-	-	-	X	X	-	-	-	-	-	-
P. occidentalis (Sonora Topminnow)	-	-	-	-	-	-	X	X	-	-	-	X	-	-
P. prolifica	-	-	-	-	-	-	X	-	-	-	-	-	-	-
Atherinidae (Silversides)														
Menidia audens (Mississippi Silverside)	X[a]	-	-	-	-	-	-	-	-	-	-	-	-	-
M. beryllina (Tidewater Silverside)	-	-	-	X[b]	-	-	-	-	-	-	-	-	-	-
Gasterosteidae (Sticklebacks)														
Gasterosteus aculeatus (Threespine Stickleback)	-	-	-	-	-	-	-	-	X	-	-	-	-	-
Percichthyidae (Temperate Basses)														
Morone chrysops (White Bass)	X[a]	X	-	X	-	-	-	-	-	-	-	-	-	-
Centrarchidae (Sunfishes)														
Chaenobryttus gulosus (Warmouth)	X[a]	X	-	X	-	-	-	-	-	-	-	-	-	-
Lepomis cyanellus (Green Sunfish)	X	X	-	-	-	-	-	-	-	-	-	-	-	-
L. megalotis (Longear Sunfish)	X[a]	X	-	X	-	-	-	-	-	-	-	-	-	-
L. macrochirus (Bluegill)	X	X	-	X	-	-	-	-	-	-	-	-	-	-
L. humilis (Redspotted Sunfish)	X[a]	-	-	-	-	-	-	-	-	-	-	-	-	-
Micropterus salmoides (Largemouth Bass)	X	-	-	X	-	-	-	-	-	-	-	-	-	-
Percidae (Perches and Darters)														
Etheostoma australe (Conchos Darter)[16]	-	-	-	-	X	-	-	-	-	-	-	-	-	-
E. grahami (Rio Grande Darter)	-	X	-	X[b]	-	-	-	-	-	-	-	-	-	-
E. lepidum (Greenthroat Darter)	-	X	-	-	-	-	-	-	-	-	-	-	-	-
E. spectabile (Orangethroat Darter)	X[a]	-	-	-	-	-	-	-	-	-	-	-	-	-
Percina macrolepida (Rio Grande Logperch)	-	X	-	X	-	-	-	-	-	-	-	-	-	-
Sciaenidae (Drums and Croakers)														
Aplodinotus grunniens (Freshwater Drum)	X[a]	X	-	X	-	-	-	-	-	-	-	-	-	-
Mugillidae (Mullets)														
Agonostomus monticola (Mountain Mullet)	-	-	-	X[b]	-	-	X	-	-	-	-	-	-	-
Mugil cephalus (Striped Mullet)	-	-	-	X[b]	-	-	X	-	-	-	X	X	-	-
M. curema (White Mullet)	-	-	-	-	-	-	X	-	-	-	-	-	-	-

Species	Plains[1]	Rio Grande basin				Guzman Basin[5]	Rio Yaqui	Coastal drainages		Death Valley System	Colorado River basin			
		Pecos River[2]	Upper[3]	Lower[4]	Rio Conchos			Baja California[6]	Pacific Coast[7]		Lower[9]	Gila River	Middle[10]	Upper[11]
Eleotridae (Sleepers)														
Dormitator latifrons (Fat Sleeper)	-	-	-	-	-	-	X	-	-	-	-	-	-	-
Eleotris picta (Spotted Sleeper)	-	-	-	-	-	-	X	-	-	-	X	-	-	-
Gobiomorus maculatus	-	-	-	-	-	-	X	-	-	-	-	-	-	-
Gobiidae (Gobys)														
Awaous transandeanus	-	-	-	-	-	-	X	-	-	-	-	-	-	-
Eucyclogobius newberryi (Tidewater Goby)	-	-	-	-	-	-	-	-	X	-	-	-	-	-
Cichlidae (Chichlids, Mojarras)														
Cichlasoma beani (Sinaloan Cichlid)	-	-	-	-	-	-	X	-	-	-	-	-	-	-
C. cyanoguttatum (Rio Grande Perch)	-	X	-	X	-	-	-	-	-	-	-	-	-	-
Cichlasoma sp.	-	X	-	X	-	-	-	-	-	-	-	-	-	-
Cottidae (Sculpins)														
Cottus aleuticus (Coastrange Sculpin)	-	-	-	-	-	-	-	-	X	-	-	-	-	-
C. asper (Prickly Sculpin)	-	-	-	-	-	-	-	-	X	-	-	-	-	-
C. bairdi (Mottled Sculpin)	-	-	-	-	-	-	-	-	-	-	-	-	-	X[c]
C. gulosus (Riffle Sculpin)	-	-	-	-	-	-	-	-	X	-	-	-	-	-
Leptocottus armatus (Staghorn Sculpin)	-	-	-	-	-	-	-	-	X	-	-	-	-	-
Pleuronectidae (Righteye Flounders)														
Platichthys stellatus (Starry Flounder)	-	-	-	-	-	-	-	-	X	-	-	-	-	-
Soleidae (Soles)														
Trinectes fonsecensis	-	-	-	-	-	-	X	-	-	-	-	-	-	-

[1]Includes representative species of streams flowing into the Mississippi River and directly to the Gulf of Mexico (principally tributaries of the Arkansas and Red rivers); occurrences marked with an "a" are outside the area of present coverage.

[2]Closed basins (e.g., near Toyahvale, Texas) intimately related to the Pecos River are included.

[3]Rio Grande mainstream and tributaries (excluding Pecos River, see above) and associated closed basins (e.g., Tularosa Basin) upstream from the inflow of the Rio Conchos (Miller, 1978).

[4]Rio Grande mainstream and tributaries downstream from inflow of the Rio Conchos (Trevino-Robinson, 1959; Hubbs et al., 1977); excluding the formerly-isolated Cuatro Cienegas Basin (Minckley, 1969; 1978). Occurrences marked with a "b" are outside the area of present coverage.

[5]Rios Carmen, Santa Maria, and Casas Grandes in Mexico, and Mimbres River, New Mexico (Miller, 1978; Hendrickson et al., 1981).

[6]Minor drainages between the Rio Yaqui and the Colorado River delta, draining directly into the Sea of Cortez (Miller, 1959; Follett, 1960).

[7]Minor drainages of western Baja California and southern California and San Diego and Los Angeles drainages, Santa Maria and Santa Inez rivers of the latter area, draining directly into the Pacific Ocean (Follett, 1960; Moyle, 1976).

[8]Includes drainage basins of Mono Lake (originally fishless; Moyle, 1976), Owens, Amargosa, and Mojave rivers (Miller, 1948; Hubbs and Miller, 1948; Soltz and Naiman, 1978).

[9]Includes Laguna Salada, Salton Sea, and distributaries and sloughs on the Colorado River delta (Rio Hardy, New River, Santa Clara Slough, etc.).

[10]Mainstream Colorado River (mostly within Grand Canyon), Pluvial White River (including Meadow Valley Wash), Moapa River, Virgin River, minor tributaries to Grand Canyon, and Little Colorado River).

[11]Defined as upstream from Glen Canyon Dam; occurrences marked with a "c" are outside the area of present coverage.

[12]Numbers of other members of this speciose genus occur in the Arkansas and Red rivers downstream from the area of present coverage, especially in clearer waters of the Ozarkian Province (Metcalf, 1966; Cross, 1967; Pfleiger, 1975).

[13]Mojave Chub is sometimes considered a full species.

[14]Mexican catfishes are poorly understood (Miller, 1976; 1978) and it is likely that more than one species is included here.

[15]Restricted to the isolated basin of Pleistocene Lake Otero (Tularosa Basin), New Mexico (Miller and Echelle, 1975).

[16]See footnote 12.

Arctic-Boreal Wetlands

Within and adjacent to subalpine forests and grasslands are numerous perennial streams and other aquatic situations bordered by shrub willows (*Salix monticola, S. scouleriana, S. bebbiana, S. lorrata*), and other winter deciduous scrub: e.g., Red Elderberry (*Sambucus racemosa*), Shrubby Cinquefoil (*Potentilla fruticosa*), Goose-berry Currant (*Ribes* spp.), Raspberry (*Rubus* spp.), and at lower boreal and cold temperate elevations, Thin-leaf Alder (*Alnus tenuifolia*) (Fig. 153). While these alpine and subalpine riparian scrublands may be punctuated by Blue Spruce (*Picea pungens*), Aspen (*Populus tremuloides*), and other tree species of the subalpine conifer forest, distinctive riparian tree life forms (and hence riparian forests), are generally absent from this thermal zone.

Except for the highest elevations of the Sangre de Cristo, San Juan, San Pedro, White, and Mogollon mountains, the number and length of subalpine streams in the Southwest is limited by the relatively small watershed areas of sufficient elevation, and also sometimes by geologic situations—e.g., the porosity of respective volcanic and limestone structures of San Francisco Mountains and the Kaibab Plateau in Arizona. Nonetheless, a few informative examples may be found in most mountain ranges approaching or exceeding 3,000 m elevation—including those in southern California (e.g., upper Snow Creek in the San Jacinto Mountains and the upper South Fork of the Santa Ana River in the San Bernardino Mountains). Certain north-flowing streams at lower elevations may also have many characteristics of subalpine systems, such as Workman Creek in the Sierra Ancha of Arizona (1,980 m; Fig. 155).

These boreal scrublands along watercourses are important and distinctive biomes, and in the Southwest provide the southern-most breeding habitats for a characteristic northern avifauna such as Lincoln's Sparrow (*Melospiza lincolni*), White-crowned Sparrow (*Zonotrichia leucophrys*), and MacGillivray's Warbler (*Oporornis tolmiei*). The streams themselves are used as breeding grounds for a few amphibians such as treefrogs (*Hyla* spp.) and also are the past or present home of a number of relicted, native salmonid fishes (e.g., *Salmo clarki* ssp., *S. apache, S. gilae*). These trouts are now largely replaced by introduced species such as Rainbow Trout (*S. gairdneri*) and Brook Trout (*Salvelinus fontinalis*).

Many subalpine grassland meadows possess high water tables, so that small marshy ponds or ciénegas are common features. Beaver dams are locally present, and streamsides are frequently marshy where gradients are not too steep. These streamside ciénega habitats support microtine rodents (*Microtus* spp.), Western Jumping Mouse (*Zapus princeps*), and Water Shrew (*Sorex palustris*), among other small mammals. Both *Rocky Mountain* and *Sierran* subalpine marshlands are most often vegetated by high-elevation or other cold-climate sedges (*Carex* spp., *Cyperus* spp., *Eleocharis* spp. and *Scirpus* spp.) and rushes (*Juncus* spp. and *Luzula* spp.). Taller life-forms such Roundstem Bulrush (*Scirpus pallidus*) and other aquatics such as mannagrass (*Glyceria* spp., including *G. borealis*) occupy deeper (and hence more permanent) marshes and lakesides (Fig. 154), providing nesting habitat for several species of waterfowl including the Mallard (*Anas platyrhynchos*), and Eared and Pied-billed Grebes (*Podiceps caspicus, Podilymbus podiceps*). Denser stands of bulrush also provide nesting habitat for Sora (*Porzana carolina*) and Coot (*Fulica americana*). High altitude amphibians, treefrogs (*Hyla eximia, H. regilla*), Cricket Frog (*Pseudacris triseriata*), and Tiger Salamander (*Ambystoma tigrinum nebulosum*), are characteristic of both temporary and permanent ciénega habitats.

Except in the San Juan Mountains, natural lakes are few and greatly outnumbered by small, artifical reservoirs and tanks designed for recreational fishing and livestock use. Although some of these water bodies may be extremely fertile, long periods of snow cover and resulting oxygen depletion in winter, or elevated pH in summer as a result of special chemical and biological conditions, result in only a few providing a substantial fishery for introduced salmonids. They are, however, extensively used by birds and, if fishes are absent, support large numbers of amphibians.

Figure 153. *Subalpine riparian scrub along North Fork of the White River, Fort Apache Indian Reservation, Apache County, Arizona. Prevalent species at this locality near the lower limits of subalpine scrub are willows* (Salix bebbiana, S. scouleriana), *Thinleaf Alder* (Alnus tenuifolia), *Blueberry Elder* (Sambucus glauca) *and Hawthorn* (Crataegus erythropoda). *Trees along the stream are Blue Spruce* (Picea pungens), *and Engelmann Spruce* (P. engelmanii), *with Ponderosa Pine* (Pinus ponderosa) *upslope above the temperature inversion layer. Elevation ca. 2,400 m.*

Figure 154. *Subalpine marshland and submergents at Carnero Lake, Apache National Forest, Apache County, Arizona. The marshland emergent is Mannagrass* (Glyceria borealis). *Elevation ca. 2,750 m.*

Cold-Temperate Wetlands

Riparian and other wetland communities in the montane, Plains, and Great Basin biotic provinces of the Southwest are characterized by winter-deciduous trees as well as shrubs and aquatic plants. These ecosystems, while often structurally diverse, are of relatively simple species composition when compared to some riparian communities in warm-temperate and subtropical climates. Adapted to spring flooding after snowmelt, the riparian associations typically are in a successional stage to deciduous forest unless arrested or eliminated by impacts of upstream impoundments and/or grazing by livestock. A large proportion of the Southwest's "live" streams are in this climatic zone (see e.g. Brown et al., 1978). Unlike rivers of moister zones that increase in size and volume throughout the year in a downstream direction, southwestern rivers flowing from mountains across deserts often have diminution of discharge both up- and downstream from intermediate elevations. Moderate to smaller streams now hold the most diverse native fauna, principally because of habitat diversity provided by channels intermediate between highly erosive, montane regions and aggrading conditions of lowland rivers. In the arid Southwest, such places also allow the greatest degree of permanence because of alternating bedrock and alluvial fill, the former inducing scour of pools and other deep places, and the latter providing shallow, high nutrient, productive reaches in periods of sustained flow.

Montane Riparian Wetlands

In montane regions of the Rocky Mountains and in the highest parts of the Sierra Madre Occidental, a "canyon bottom forest" (Fig. 155) may occur along perennial and near-perennial streams from ca. 2,100-2,300 m down to ca. 1,700 m, and locally to as low as 1,350 m. Narrowleaf Cottonwood (*Populus angustifolia*), maple (*Acer grandidentatum*), Box Elder (*Acer negundo*), alder (*Alnus oblongifolia*) and willows (*Salix* spp.) form a riparian series, in which the trees, shrubs, and grasses of adjacent montane coniferous forest are lesser (yet common) participants (e.g., *Quercus gambelii, Pinus ponderosa, Abies concolor, Populus tremuloides, Robinia neomexicana, Rhus glabra, Blepharoneuron tricholepsis*, etc.). Any of several other riparian short-statured or scrub trees such as Water Birch (*Betula occidentalis*), Rocky Mountain Maple (*Acer glabrum*), American Plum (*Prunus americana*), and Bitter Cherry (*Prunus emarginata*) may be locally important. The vine, Virginia Creeper (*Parthenocissus vitacea*), can add a dash of scarlet to an already colorful autumn woods. An analogous series of Black Cottonwood (*Populus trichocarpa*), White Alder (*Alnus rhombifolia*), Bigleaf Maple (*Acer macrophyllum*), and willow (*Salix scouleriana*) finds limited representation along montane streams in some higher Southern California mountains (Sierran riparian deciduous forest).

The dominant aspect of many of these montane streamsides, however, is often one of shrubbery (= scrubland). This may be a result of the singular presence of younger age classes of riparian trees and/or the heavy and sometimes exclusive representation of one or several shrub willows (*Salix irrorata, S. lasiolepis*, etc.). Other riparian scrub species such as Blueberry Elder (*Sambucus glauca*), Red-osier Dogwood (*Cornus stolonifera*), Thin-leaf Alder (*Alnus tenuifolia*), or hawthorn (*Crataegus* spp.) may complement or essentially replace the willows.

The Beaver (*Castor canadensis*) is perhaps the best known montane riparian resident, and Raccoon (*Procyon lotor*) commonly finds its upper elevational limits here. In wider parts of canyons with streamside stands of sedge-grass ciénega, microtine rodents (*Microtus* spp.) often are common. Some of the more densely wooded streamsides within Rocky Mountain and Madrean conifer forests are important to the distribution of White-tailed Deer (*Odocoileus virginianus*) and Wild Turkey (*Meleagris gallopavo*). The Water Ouzel or Dipper (*Cinclus mexicanus*) is a most characteristic nesting bird. Bleak in winter, these habitats are important in other seasons to a number of colorful migratory songbirds such as the Lazuli Bunting (*Passerina amoena*), Yellow Warbler (*Dendroica petechia*), Black-headed Grosbeak (*Pheucticus melanocephalus*), and Yellow-breasted Chat (*Icteria virens*). Other characteristic birds are the Broad-tailed Hummingbird (*Selasphorus platycercus*), Belted Kingfisher (*Megaceryle alcyon*), Warbling Vireo (*Vireo gilvus*) and Western Flycatcher (*Empidonax difficilis*).

A list of indicative or well represented montane reptiles would include the garter snakes (*Thamnophis elegans, T. couchi, T. cyrtopsis, T. eques*), the Narrow-headed Garter Snake (*T. rufipunctatus*), and Alligator Lizard (*Gerrhonotus kingi*). Amphibians may be represented in the appropriate mountains by either of several treefrogs (*Hyla arenicolor, H. regilla, H. eximia*) and in California by the mountain Yellow-legged Frog (*Rana mucosa*). Salamanders are the Tiger Salamander in the north, and the Tarahumara Salamander

Figure 155. *Rocky Mountain montane riparian forest. "Canyon bottom" habitat along Workman Creek in the Sierra Ancha Experimental Forest, Sierra Ancha Mountains, Gila County, Arizona. Deciduous trees are mostly Big-tooth Maple* (Acer grandidentatum), *Arizona Alder* (Alnus oblongifolia), *Narrow-leaf Cottonwood* (Populus angustifolia) *and Gambel Oak* (Quercus gambelii). *Ponderosa Pine* (Pinus ponderosa) *and White Fir* (Abies concolor) *are prevalent conifers from the adjacent coniferous forest. The shrub-vine Canyon Grape* (Vitis arizonica) *is an important participant in forest openings. Elevation ca. 1,950 m.*

(*Ambystoma rosaceum*) in the Sierra Madre Occidental of northwestern México.

Waters of this zone are inhabited by relatively few species of "cold water" fishes. Prior to widespread introductions of Rainbow Trout, most montane streams above 1,800 m were populated by locally endemic species of trouts such as *Salmo apache* in Arizona, *S. gilae* in Arizona and New Mexico, and in the cold-temperate segments of the Sierra Madre Occidental in Mexico, one or more yet-to-be-described species of *Salmo* and the Mexican Golden Trout (*S. chrysogaster*). Many of these native populations are now reduced or extirpated through hybridization with *S. gairdneri* and predation/competition interactions by other introduced salmonids

(Rinne et al., 1981). Another widely introduced species, the European Brown Trout (*Salmo trutta*), descends to somewhat lower elevations where it may be accompanied by the local species of mountain-sucker (genus *Pantosteus*) and such suckers of the genus *Catostomus* that are appropriate to the watershed. Minnows such as spinedace (*Lepidomeda* spp.) and Speckled Dace (*Rhinichthys osculus*) are in the Colorado River basin, and Longnose Dace (*R. cataractae*) occurs in the Rio Grande system. Introduced fishes other than trouts are relatively uncommon in our higher elevation waters, with the exception of the Golden Shiner (*Notemigonus crysoleucus*) in Arizona (Minckley, 1973), which has created substantial management problems by overpopulating fishing lakes.

Plains and Great Basin Riparian Wetlands

Monotypic gallery forests and woodlands of Plains Cottonwood (*Populus deltoides* ssp. *sargentii*), Rio Grande Cottonwood (*P. wislizenii*), Peachleaf Willow (*Salix amygdaloides*), or Narrowleaf Cottonwood are the dominant climax vegetation along streamsides east of the Rocky Mountain and in much of the Great Basin (Figs. 156, 157). Most riparian reaches are in successional stages, and scrublands interrupted by an occasional cottonwood grove or tree and composed chiefly of scrub willows (*Salix exigua* and others; Fig. 158), or less commonly, scrubby trees such as Red-osier Dogwood, are typical riparian communities.

In the western, warmer portions of the Great Basin biotic province, disclimax riparian scrublands and strands populated by the introduced Saltcedar (*Tamarix chinensis*) now comprise many miles of river and stream channels, including ephemeral tributaries (Fig. 159). This situation is especially prevalent in areas of manipulated discharge below storage reservoirs, as along the San Juan River near Farmington, New Mexico. Other exotic plants such as Russian Olive (*Elaeagnus angustifolia*) and Camelthorn (*Alhagi camelorum*) have become naturalized and contribute increasingly to the composition of scrublands along these and other Great Basin drainages.

Although these environments are important to a number of riparian animal species of more general distribution (e.g., Yellow-breasted Chat), at least one bird, the Black-billed Magpie (*Pica pica*), is centered here. Also, several species of the Eastern deciduous forest find their southwestern limits in local associations of Plains and Great Basin riparian communities. These include Catbird (*Dumatella carolinensis*), American Redstart (*Setophaga ruticilla*), Veery (*Catharus fuscescens*), Eastern Phoebe (*Sayornis phoebe*), and Red-headed Woodpecker (*Melanerpes erythrocephalus*). Woodhouse's Toad (*Bufo woodhousei*), spadefoot toads (*Scaphiopus intermontanus, S. bombifrons, S. hammondi*), Leopard Frogs (*Rana pipiens* complex), and garter snakes (*Thamnophis radix*) are amphibians and reptiles well represented in these riparian environments.

Of the piedmont and alluvial rivers with cold temperate waters, only those draining to the Gulf of Mexico have relatively diversified fish faunas (Table 32). The largest southwestern ichthyofauna is that of the Plains rivers and of the Rio Grande system. The upper portion of Plains streams, e.g., the uppermost Red River system (Fig. 146), supports numerous minnows characteristic of smaller or moderate-sized habitats: Plains Stoneroller (*Campostoma anomalum*), Creek Chub (*Semotilus atromaculatus*), Plains Minnow (*Hybognathus placitus*), Flathead Chub (*Platygobio gracilis*), Sand Shiner (*Notropis stramineus*), Fathead Minnow (*Pimephales promelas*), and others. White Sucker (*Catostomus commersoni*) also is common, and has been introduced into some western drainages. Plains Killifish (*Fundulus kansae*) lives in shallow, saline, more severe places. An infusion of Plains species occurred into the Rio Grande basin as the Pecos River cut northward through unconsolidated sediments along the southeastern flank of the Rocky Mountains and pirated headwaters of the Brazos, Colorado (of Texas), and Canadian rivers (Belcher, 1975; Leonard and Frye, 1975). Likely examples of this event are Sand Shiner, Flathead, and Creek chubs, and such pairs as Plains and Rio Grande killifishes (*Fundulus kansae* and *F. zebrinus*) and Red River and Pecos pupfishes (*Cyprinodon rubrofluviatilis* and *C. pecosensis*; Echelle and Echelle, 1978). Remnants of an "old" (Tertiary) fauna also

Figure 156. *Cottonwood* (Populus deltoides ssp. sargentii) *forest along the Cimarron River near the Colorado-New Mexico-Oklahoma boundaries. It is December and this deciduous flood plain forest, although appearing bleak and uninviting, is actually a haven for wildlife during the wind swept storms of winter.*

Figure 157. *A "linear" forest of Narrow-leaf Cottonwood (*Populus angustifolia*) along irrigation ditches near Springerville, Apache County, Arizona. Note the scrub understory that is an important cover-type for wildlife during the cold of winter. Elevation ca. 2,150 m.*

Figure 158. *Successional scrubland of scrub willows (*Salix spp.*) and young cottonwood (*Populus sargentii*) along the Vermejo River, Colfax County, New Mexico. Given time these communities will pass into forest and woodland – unless, and until, interrupted by high intensity flooding. Elevation ca. 1,800 m.*

Figure 159. *Great Basin riparian strand along Hamblin Wash, Coconino County, Arizona. An open stand of Saltcedar* (**Tamarix chinensis**) *and smaller shrubs on the floodplain of an ephemeral stream (dry wash) within the Great Basin Desert. Elevation ca. 1,500 m.*

persist in the upper Rio Grande basin, e.g., Pecos Chub (*Gila pandora*) and a mountain-sucker (*Pantosteus plebeius*).

The Colorado River system physically dominates the Southwest and hosts an impressive array of endemic genera and species of fishes. Tributaries to the middle Colorado River now range from series of springs rising from the intermittent channel of Pluvial White River and Meadow Valley Wash in southern Nevada (Hubbs and Miller, 1948; LaRivers, 1962), to larger streams with local perennial flow such as the Virgin and Little Colorado rivers. Extreme isolation has led to differentiation. Each major stream has its own species or subspecies of spinedace: *Lepidomeda vittata* in the Little Colorado, *L. mollispinis mollispinis* in the Virgin River, *L. mollispinis pratensis* in Meadow Valley Wash, *L. altivelis* in lower White River, and *L. albivallis* in the isolated

upper White River (Miller and Hubbs, 1960). Mountain-suckers, although all referred to *Pantosteus clarki* by Smith (1966), show similar differentiation (Minckley, 1973), as do local populations of chubs (*Gila robusta jordani* in the White River system and *G. robusta seminuda* in the Virgin River) and Speckled Dace (Williams, 1978). Thermal endemics of the Moapa River (the springfish, *Crenichythys baileyi*), and Moapa Dace (*Moapa coriacea*) will be discussed later under springs and marshlands, although the last is characteristic of both pools and relatively swift runs.

The upper Colorado River system, mostly north of our area of coverage, supports special big-river fishes to be covered below, plus Roundtail Chub (*G. robusta robusta*), Speckled Dace, and tributary forms of Blue-head Mountain Sucker (*Pantosteus discobolus*).

Montane, Plains and Great Basin Marshlands

Marshes, nowhere extensive in the Southwest, occupy only a small area in this climatic zone so that their wildlife values are particularly high. Some larger, natural examples include Mormon and Stoneman lakes in Arizona, and Buford (Stinking), Boulder, and Horse lakes in New Mexico. Other sizable areas of natural and "managed" marshlands occur near Las Vegas, New Mexico (e.g., on Monte Vista and Alamosa National Wildlife refuges). Smaller examples are clustered in poorly-drained portions of the Mogollon Rim in Arizona, especially where sinkholes have developed through subsurface solution, and occasionally elsewhere within montane forests as well as in the Plains and Great Basin (Fig. 160; Wright, 1964; Wright and Bent, 1968).

Marsh vegetation is characteristically "zoned" along a littoral gradient. Depending upon seasonal water depth, water chemistry, time and "chance," it may be composed largely of emergent plants such as cattail (*Typha latifolia*), bulrush or tule (*Scirpus acutus*), rushes (*Juncus* spp.), sedges (*Carex* spp.), Three-square (*Scirpus americanus*), Salt Grass (*Distichlis stricta*), etc., or be mostly submergent, e.g., series of Water Milfoil (*Myriophyllum spicatum*), pondweeds (*Potamogeton* spp.), introduced water-weed (*Elodea* spp.), manna grasses (*Glyceria* spp.), or charophytes (*Chara* spp., *Nitella* spp.). Often there is some interspersion with trees and shrubs, particularly willows. Spike-rushes (*Eleocharis* spp.) are characteristic emergents in marshes subject to desiccation, and often occur there in monospecific stands (Figs. 161, 162).

These marshlands provide feeding and watering habitat for a number of migratory bats. Muskrats (*Ondatra zibethicus*) may be common all year. The Western Jumping Mouse is best represented in interior marshes, while Mink (*Mustela vison*) is rare, and apparently restricted in the Southwest to a few wetlands in and near the Sangre de Cristo Mountains.

Almost all cold-temperate marshlands host some nesting as well as migrating waterfowl. The principal species are Mallard, Pintail (*Anas acuta*), Cinnamon Teal (*A. cyanoptera*), Redhead (*Aythya americana*), and Ruddy Duck (*Oxyura jamiacensis*). Although no longer a nesting bird in the Southwest, the Sandhill Crane (*Grus canadensis*) still relies heavily on several of those areas for staging sites during migration. The short-statured and more open plant associations are used by cranes, waterfowl, and numerous shorebirds, and the taller structured emergents such as bulrushes (*Scirpus americanus, S. acutus,* etc.) may provide nesting sites for American Bittern (*Botaurus lentiginosus*), Virginia Rail (*Rallus limicola*), Sora, Common Yellow Throat (*Geothlypis trichas*), Yellow-headed Blackbird (*Xanthocephalus xanthocephalus*), Red-winged Blackbird (*Agelaius phoeniceus*), and Long-billed Marsh Wren (*Cistotherus palustris*).

Western Garter Snake is the most commonly encountered reptile, and Leopard Frogs (*Rana pipiens* complex) and Tiger Salamander are amphibians found throughout the region. Cricket Frogs, although indicative, are local in distribution.

Fishes are rarely present in these habitats, other than the young of minnows and suckers that are seasonally present in marshy areas adjacent to streams. Marshlands with open water have been stocked with numerous exotics, however, including salmonids for seasonal fisheries. Others include eastern centrarchids and cyprinids, many of which are now

Figure 160. *Montane (Rocky Mountain) marshland of waterweed (Sagittaria spp.), a widespread, cold temperate taxon, at Mormon Lake, Coconino National Forest, Coconino County, Arizona. Elevation ca. 2,150 m.*

Figure 161. *Montane marshland of Spikerush (Eleocharis parvula) at Sunflower Flat, Kaibab National Forest, Coconino County, Arizona. A seasonally flooded environment that provides waterfowl nesting habitat on an interim basis. Elevation ca. 2,150 m.*

Figure 162. *Interior marsh within Great Basin desertscrub. Except for small patches of Saltcedar (*Tamarix chinensis; *dark group in left center background), Saltgrass (*Distichlis stricta) *almost exclusively dominated this now-drained wetland (Obed Meadows) south of Saint Johns, Apache County, Arizona. Elevation ca. 2,000 m. This cosmopolitan halophyte constitutes the principal vegetation of many alkali wetlands in the Southwest from sea level to more than 2,150 m.*

considered noxious; e.g., Green Sunfish (*Lepomis cyanellus*) and Golden Shiner, and predators such as Northern Pike (*Esox lucius*). The last species may exert pressure on young of nesting waterfowl, and will certainly have an undesirable impact on other native aquatic animals if it becomes established in streams.

Warm-Temperate Wetlands

Included in this climatic zone are some of the Southwest's most extensive, and yet more endangered wetlands. Ranging from open mud flats to complex broadleaf deciduous forests with canopies more than 30 m above the ground, these habitats provide a wide diversity to an enormous variety of aquatic and semi-aquatic inhabitants.

Biotic communities represented include several fasciations of riparian deciduous forest, riparian scrub, and both interior and coastal marshes and strands (Fig. 163). Also present, but outside the scope of this discussion, are an as-yet-undetermined number of inland and marine submergent communities. Moreover, these last biomes with their complex plant and animal reefs, kelp "forests," eelgrass "meadows," and a myriad of other underseascapes, support at least seasonally a more or less pelagic mammalian fauna of Gray Whale (*Eschrictius robustus*), Harbor Porpoise (*Phoecoena vomerina*), dolphins (*Lagenorhynchus obliquidens,*

Delphinus bairdi, Tursiops gilli) and formerly, the Southern Sea Otter (*Enhydra lutris nereis*), as well as providing feeding habitat for numerous oceanic birds.

Marine reptiles are absent or poorly represented in these waters. Reproducing populations of sea turtles and sea snakes generally are restricted to tropic-subtropic waters farther south. The fish fauna in coastal marine habitats is, however, large and diversified. Because of the California current, temperate marine environments extend south along the Pacific Coast to below the 27th Parallel of Baja California (Fitch and Lavenberg, 1975). Waters as far south as Bahía Magdalena (between the 24th and 25th parallels) remain cold enough to form a barrier to tropical coastal fishes. The Golfo de California (Sea of Cortez) is protected from these currents, and has a tropic-subtropic fish fauna clearly derived from the south (Walker, 1960; Thomson et al., 1979).

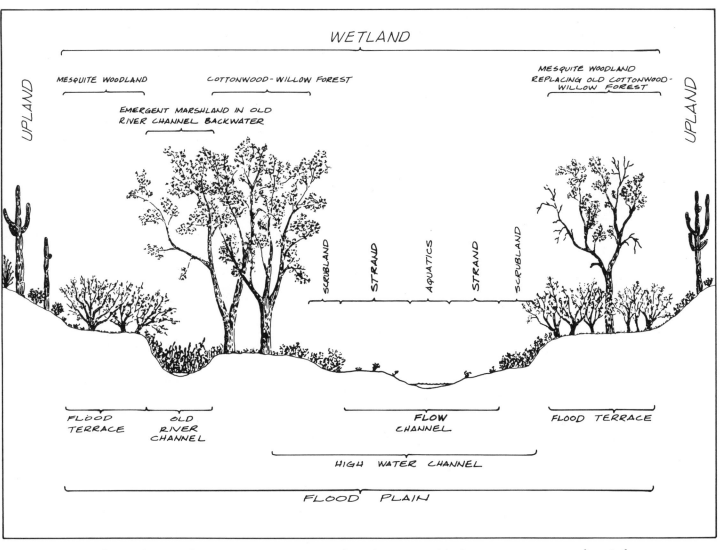

Figure 163. *Semi-diagrammatic representation of riparian communities in warm temperate to subtropical habitats of the American Southwest.*

Interior and Californian Riparian Deciduous Forests and Woodlands

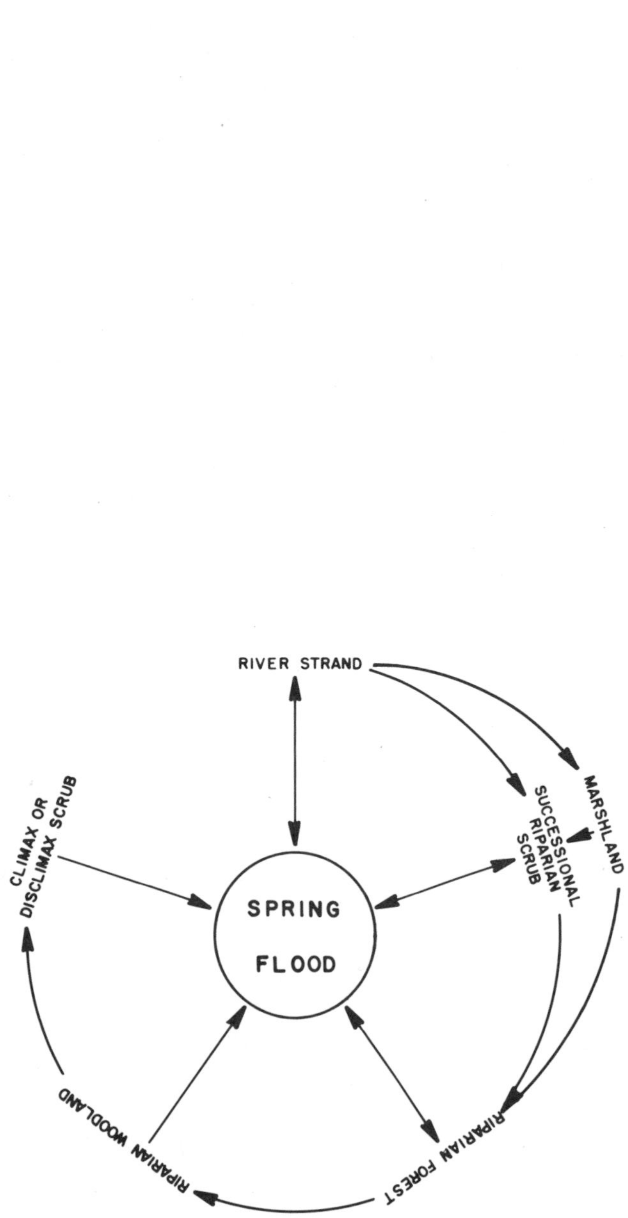

These winter deciduous communities are diverse, because high altitude species penetrate downslope to occur among lowland forms (Fig. 145). Originally, they occupied many of the major as well as secondary drainages in the Californian, Mohavian, sub-Mogollon (=Arizonan), Madrean, and Chihuahuan biotic provinces, where they are now greatly reduced because of reductions in stream flow. These communities are maintained along perennial or seasonally intermittent streams and springs, and may be divided into two major vegetation-types (series) — cottonwood-willow and mixed broadleaf.

These are relictual communities. The present distributions of the two major vegetation types, particularly the mixed broadleaf, reflect a contraction of the formerly widespread, Early Tertiary mixed mesophytic forest. These riparian forests are vernally adapted to Early Tertiary climates and have retreated to pockets where the warm temperate (ancient) climate persists.

Where streamflows are intermittent, well-developed gallery forest can be expected only where surface flow reliably occurs during winter-spring months, because the onset of the spring growing season can be expected prior to April 15 (Zimmerman, 1969; Hibbert et al., 1974). After mid-April, increased evaporation and phytotranspiration often result in only subsurface flow in all but the larger streams, at least during daytime. The Southwest's warm-temperature riparian forests and woodlands, therefore, require abundant water during March and April when most arboreal species leaf, set seed, and germinate (Zimmerman, 1969). Summer precipitation usually does not result in sustained streamflow in seasonal stream channels (Zimmerman, 1969; Hibbert, 1971; Hibbert et al., 1974), and riparian forests in the Southwest have therefore remained vernally adapted. Probably for this reason, these forests are poorly represented and often absent from western pediments of the Sierra Madre Occidental where winter-spring precipitation and runoff is characteristically low.

Communities of cottonwood (*Populus fremontii* and others; in portions of the Chihuahuan biotic province, *P. fremontii* ssp. *mesetae*), and willows (*Salix gooddingii, S. exigua, S. bonplandiana*, and others) are characteristic stream features throughout the Southwest's warm-temperature zones. These short-lived associations are typically encountered and reach their best development in alluvial sands, clays, and gravels on flood plains. The forest canopy may be from 15 to 30 m or more in height and with open and park-like understories in mature groves, or populated by thickets of young cottonwoods or willow depending upon stage and grazing intensity.

Indications are that cottonwood-willow associations are maintained and depend on periodic spring floods (Fig. 164). Evidence for the winter-spring flood adaptation of these communities are "new" forests along undammed portions of the Verde and San Pedro rivers following spring floods in 1962 and 1967, respectively, a result of fortuitous timing of floods with the narrowly defined germination requirements of these riparian salices. Also, the presence of Californian and Mohavian cottonwood-willow fasciations would indicate that they too are vernal-adapted, and that summer precipitation and runoff is of little or no importance to their regeneration and only encourages competitors such as Saltcedar.

Stabilized flows from storage reservoirs and summer flooding in today's wider channels facilitates dissemination of Salt-

Figure 164. *Hypothetical successional cycle of Southwestern riverine wetlands based on and around the natural flooding phenomena of the region.*

cedar, allowing that species to complement or replace cotton-wood and willow along many miles of the Rio Grande, Pecos, upper Gila, and other rivers where such conditions now occur. Stabilized flow below dams seems especially to result in decadent stands of native trees in which reproduction is lacking. Unlike other riparian trees, such as sycamores (*Platanus* spp.) that reproduce almost entirely by sprouting into clones, cottonwood and willow are mainly disseminated by seed. Studies by Horton et al. (1960) and Zimmerman (1969) have shown that these species germinate in spring, and that seeds remain viable for less than 7 weeks. If receding waters fail to provide a suitable seed bed, little or no reproduction of these short-lived species will occur.

Saltcedar produces seed from March through October in the Southwest, and may colonize seed beds similar to those used by cottonwood and willow (Horton, 1977). Summer flooding therefore, may enhance that species, except in canyon-bound rivers where the violence of scour appears to exclude it (Minckley and Clark, 1979; Turner and Karpiscak, 1980). Wider places in stream channels provide ameliorating effects on such flooding, so that only major channel-straightening events (Burkham, 1972) can remove dense Saltcedar cover and allow germination and successful recolonization by native trees. The advent of extensive water storage with attendant regulation of streamflow and reduction of floods, coupled with invasion of Saltcedar, have resulted in a great reduction of interior cottonwood-willow communities. Intensity of livestock predation along many stream channels adds another factor often alone sufficient to preclude survival of the limited numbers of seedings.

Mixed broadleaf series of *Interior* and *Californian riparian deciduous forest* occur along rubble-bottomed perennial and near-perennial streams. In the interior Southwest (sub-Mogollon Arizona and New Mexico, northwestern Chihuahua and northeastern Sonora; Figs. 165, 166), aboreal constituents may be admixtures of stands of regional species or ecotypes of such Holarctic genera, as Arizona Sycamore (*Platanus wrightii*), Velvet Ash (*Fraxinus pennsylvanica* var. *velutina*), Fremont Cottonwood, Arizona Alder, Arizona Walnut (*Juglans major*) and willows (*Salix exigua* and others). At some of the higher elevations (ca. 1,400 to 1,800 m), Boxelder, Bigtooth Maple, Narrowleaf Cottonwood, and cherries (*Prunus* spp.) may make their appearance and even dominate locally. At lower elevations (1,100 to 1,500 m), a number of trees of more southern distribution are often present: e.g., Western Soapberry (*Sapindus saponaria* var. *drummondii*), Texas Mulberry (*Morus microphylla*), Netleaf Hackberry (*Celtis reticulata*), and Mexican Elder (*Sambucus mexicana*). Arizona Cypress (*Cupressus arizonica*) is not uncommon, and the forest or woodland often contains oaks (*Quercus gambelii, Q. emoryi, Q. arizonica*) and conifers (*Pinus ponderosa, Juniperus deppeana*) from upstream and adjacent uplands. Near settlements and locally elsewhere, such exotic trees as the Tree-of-heaven (*Ailanthus altissima*), Catalpa (*Catalpa bignonioides*), Osage-orange (*Maclura pomifera*), and even fruit trees may be present. Some of the more noticeable understory species include Bracken Fern (*Pteridum aquilinum*), Smooth Sumac (*Rhus glabra*), Poison-ivy (*Rhus toxicodendron*) and several deciduous vines, especially Canyon Grape (*Vitis arizonica*).

Californian mixed broadleaf forests and woodlands (Figs. 167, 168) have much the same appearance and share many of the same species and genera as their interior counterparts.

Figure 165. *Summer aspect of Interior riparian deciduous forest (mixed broadleaf series) along Gap Creek, Yavapai County, Arizona. A "gallery" forest of the interior riparian "big six"—sycamore (*Platanus wrightii*), alder (*Alnus oblongifolia*), willows (*Salix gooddingii, S. bonplandiana*), walnut (*Juglans major*), ash (*Fraxinus pennsylvanica* var. velutina*), cottonwood (*Populus fremontii*)—and other winter deciduous trees along a boulder-bottomed, perennial stream. Elevation ca. 1,200 m.*

While the sycamore and alder here are *Plantanus racemosa* and *Alnus rhombifolia*, the widespread *Populus fremontii, Fraxinus pennsylvanica* var. *velutina, Sambucus mexicana, Salix gooddingii*, and *Salix nigra* are represented in both fasciations. Canyon Live Oak (*Quercus chrysolepis*) and Coast Live Oak (*Q. agrifolia*) are often important arboreal constituents and Poison Oak (*Rhus toxicodendron*) is an important understory component in southern California and northern Baja California del Norte riparian communities, as is California Blackberry (*Rubus vitifolius*).

As in cottonwood-willow associations, lowered streamflow has reduced a number of mixed broadleafed forests and woodlands to scattered individual trees, opening the canopy and reducing its desirability for some species of wildlife. Because of previous watershed practices, flash flooding of a destructive nature now too often destroys many miles of these streamside forests. Timbering practices and grazing by livestock has further reduced and affected the forest understory by curtailing or eliminating some forest-associated species.

Figure 166. *Mature broadleaf series of Interior riparian deciduous forest along Cajón Bonito, northern Sonora, Mexico. Species present are the same as for Gap Creek (excluding alder; Fig. 165). Elevation ca. 1,100 m; photograph by Dean A. Hendrickson, June, 1978.*

Numerous wildlife species are totally or largely dependent on these riparian deciduous communities and many others reach their greatest densities there. Two tree squirrels, Arizona Gray Squirrel (*Sciurus arizonensis*) and Apache Fox Squirrel (*S. nayaritensis*), are largely confined to mixed broadleaf forests within their respective Mogollon and Madrean provinces. The Western Gray Squirrel (*S. griseus*) and introduced Fox Squirrel (*S. niger*) use analogous broadleaf forests in California. Now extremely rare, the River Otter (*Lutra canadensis*) was at one time found in the interior Southwest within warm temperate streams. This species, like the beaver, is more a stream obligate than in need of forest per se.

Other tree-requiring species found in riparian deciduous forests are the Raccoon, and, in California, the introduced Opossum (*Didelphis marsupialis*). Cliffs typically associated with warm temperate canyon streams support small carnivores such as the Ringtailed Cat (*Bassariscus astutus*) and skunks (*Mephitus* spp., *Spilogale putorius*). In drier places, burrowing mammals such as pocket gophers (*Thomomys* spp.) may also be largely restricted to this mesic zone. Numerous bats roost in riparian trees (e.g., Red Bat, *Lasiurus borealis*), or in crevices and holes in cliffs (*Myotis* spp., *Pipistrellus hesperus*), preying on the rich aquatic and riparian insect fauna.

Several nesting birds are obligate to either riparian deciduous trees, cliffs, or the streams themselves. Warm-temperate-inhabiting examples in the interior Southwest include Summer Tanager (*Piranga rubra*), Mississippi Kite (*Ictinia misisippiensis*), Zone-tailed Hawk (*Buteo albonotatus*), and Black Hawk (*Buteogallus anthracinus*); the White-tailed Kite (*Elanus leucurus*) nests in Californian riparian deciduous forest, while the Yellow Warbler, Yellow-billed Cuckoo (*Coccyzus americanus*), Bullock's Oriole (*Icterus bullocki*), and numerous other insectivores such as the Cliff Swallow (*Petrochelidon pyrrhonota*) are common to both biomes. Many others are well represented, and the Madrean fasciations often are host to several neotropical raptors, numerous hummingbirds, (i.e., Blue-throated [*Lampornis clemenciae*], Violet-crowned [*Amazilia verticalis*], Lucifer [*Calothorax lucifer*] and Broad-billed [*Cynanthus latirostris*]), and songbirds e.g., Sulphur-bellied Flycatcher (*Myiodynastes luteiventris*), Rose-throated Becard (*Platypsaris aglaiae*), and Coppery-tailed Trogon (*Trogon elegans*). These last and some sub-Mogollon communities also provide important habitats for common game species, as the White-tailed Deer, Black Bear (*Ursus americanus*), and Wild Turkey, also found in Eastern deciduous forests.

Riparian deciduous forests and stream channels are of equal importance to cold-blooded life forms, including the

Figure 167. *Exterior view of Californian riparian deciduous forest* (Platanus racemosa, Salix gooddingii *series*) *along Arroyo San Rafael, Baja California del Norte, Mexico. The change in life-form height between this linear wetland community and the adjacent upland chaparral is readily apparent. Note the extensive but discontinuous areas of coastalscrub on hillsides, which are subclimax here to chaparral and the result of recent fires. Elevation ca. 1,350 m.*

California Newt (*Taricha torosa*), Ensatina (*Ensatina eschscholtzi*), Pacific and California Slender Salamanders (*Batrachoceps pacificus, B. attenuatus*), and California Toad (*Bufo boreas holophilus*). Also present are the arboreal or boulder-inhabiting Pacific Treefrog and California Treefrog (*H. cadaverina*); Canyon Treefrog and Arizona Treefrog (*H. wrightorum*) are in the interior. Species of the *Rana pipiens* complex also are common in the interior, and in Mexico the Tarahumara Frog (*Rana tarahumarae*) and stream-adapted Tarahumara Salamander (Collins, 1979) are present at warm-temperate elevations.

Leaf litter along streams may provide habitat for alligator lizards, *Gerrhonotus multicarinatus* in California and *G. kingi* in the interior. Where suitable loose soils are present along Californian streams one may find the California Legless Lizard (*Anniella pulchra*). Numerous snakes such as kingsnakes (*Lampropeltis* spp.) are well represented in streamside environments, and some, such as Ringnecked Snake (*Diadophis punctatus*), are most often found within riparian forest and woodland. In Madrean fasciations one may encounter the Green Rat Snake (*Elaphe triaspis*), Vine Snake (*Oxybelis aeneus*), and Huachuca Blackhead Snake (*Tantilla wilcoxi*

wilcoxi). Streams provide hunting grounds and escape areas for several species of the more or less aquatic garter snakes, depending on biotic province and microhabitat characteristics.

As noted before, diversity in fishes is high at intermediate elevations in warm temperate habitats. Most characteristic in the Gila River Basin are Roundtail Chub, Longfin Dace (*Agosia chrysogaster*), Sonoran Sucker (*Catostomus insignis*), and Desert Mountain-sucker (*Pantosteus clarki*). Spikedace (*Meda fulgida*) and Loach Minnow (*Tiaroga cobitis*), each a member of a monotypic genus endemic to the Gila River, may also occur, as may Speckled Dace in cooler, well-shaded streams or in large rivers at intermediate elevations. This same assemblage, less the two Gila River endemics, also is present in the Bill Williams River, tributary to the lower Colorado River mainstream.

The north-flowing Río Conchos, originating in the rugged and inaccessible Sierra Madre Occidental of western Mexico, is occupied by a few fishes of Plains origins, e.g., Longnose Dace (*Rhinichthys cataractae*) and Fathead Minnow. However, Mexican Stoneroller (*Campostoma ornatum*) replaces Plains Stoneroller, and Mexican species with little relationships to the north are dominant, e.g., Chihuahuan Shiner (*Notropis chihuahua*), Ornate Minnow (*Codoma ornata*), and others (Table 32). The influence of the Plains fauna in the Río Conchos drainage diminishes rapidly to the West, as does overall faunal diversity. The fathead minnow is native to the basin of Lago de Guzmán. The Beautiful Shiner (*Notropis formosus*) found there is a close relative of the Red Shiner. All other species are of the "old" fauna, e.g., Chihuahua Chub (*Gila nigrescens*) and a mountain-sucker (*Pantosteus plebeius*), or are of Mexican origins.

The Río Yaqui watershed is a composite of sub-basins derived from the Río Casas Grandes, Río Conchos, Gila River, and drainages to the south and west (Hendrickson et al., 1981). Colorado River fishes such as Roundtail Chub appear here, but species characteristic of north-central Mexico, Ornate Minnow, Mexican Stoneroller, Mesa del Norte Chub (*Gila pulchra*), etc., mostly prevail. The coarse-scaled Yaqui Sucker (*Catostomus bernardini*) occurs from mountains through deserts, and is scarcely separable from the Conchos Sucker (*C. conchos*) to the east or the Sonoran Sucker (*C. insignis*) of the Gila River basin to the north (Miller 1976). The Longfin Dace, a minnow adapted to severe conditions of the Sonoran Desert (Minckley and Barber, 1971), but also moving into tropical and temperate habitats, also is shared by the Río Yaqui and Gila River, as is the Sonoran Topminnow (*Poeciliopsis occidentalis*).

To the west, coastal drainages of southern California have few native fishes, and those that persist are under severe pressure from human population growth and use of existing water. The omnipresent Speckled Dace occurs there along with Arroyo Chub (*Gila orcutti*) and Santa Ana Mountain-sucker (*Pantosteus santaanae*) in tributaries to the Los Angeles Basin, implying by their presence a former connection of that drainage with the Colorado River system (Smith, 1966). The only native fishes to occupy all coastal drainages in southern California are those able to disperse through sea water, e.g., Threespine Stickleback (*Gasterosteus aculeatus*) and California Killifish (*Fundulus parvipinnis*) (Moyle, 1976). In the interior, the Death Valley system contains only remnants of fluviatile fishes persisting in springs and marshes connected by short reaches of flowing water; these relicts of wetter times are discussed later.

Figure 168. *Interior view of Californian riparian deciduous forest (mixed broadleaf series) of alder* (Alnus rhombifolia) *along South Fork of San Felipe Creek, San Diego County, California. It is early spring and the alders are just leafing out; immediately downstream was a mixture of such typical Californian riparian species as sycamore* (Platanus racemosa), *cottonwood* (Populus fremontii), *willow* (Salix gooddingii) *and Canyon Live Oak* (Quercus chrysolepis) *Elevation ca. 1,375 m.*

Riparian Scrublands

Although riparian scrublands cover many kilometers of floodplain and stream channels within the Mohavian, Chihuahuan, Madrean, and Californian provinces, there has been a tendency to ignore these biotic communities in favor of the richer biota of more structurally diverse assemblages. These communities may also constitute a riparian understory or prelude for the riparian forests just discussed, grade into marshlands or, in the more xeric places where salts accumulate, form alkali-loving communities of Saltgrass (*Distichlis spicata*) and chenopods (Gary, 1965; Bloss and Brotherson, 1979). There is ample evidence for vast increases in this formation type in past years, largely at the expense of cottonwood forests and woodland (see e.g. Turner, 1974).

Riparian scrub may range from simple disclimax consociations of introduced Saltcedar to complex and diversified associations containing dozens of species (Fig. 169). Some of the most extensive warm-temperate riparian scrublands in the Southwest are along the Río Grande from Belen, Texas, to Big Bend National Park, along portions of the Pecos River, and along the Colorado River where it traverses the Mohave desert.

Common and Giant Reeds (*Phragmites australis* and *Arundo donax*) form tall stands of "cane" along the immediate banks of the Río Grande, often extending into the stream and forming floating mats (Fig. 170). The large Goodding Willow (*S. gooddingii*) and smaller Black (*S. nigra*) and Sandbar willows (*S. interior*) also are present, along with Seepwillow (*Baccharis salicifolia*), Saltcedar, and other scrub. Other species present in riparian scrublands here (and elsewhere) include:

Aster spinosus	Aster
Baccharis sarothroides	Desert Broom
Equisetum spp.	horsetails
Heliotropium curassavicum	Heliotrope
Hymenoclea spp.	burrobrushes
Pluchea camphorata	Camphor-weed
Verbesina encelioides	Cowpen Daisy

Protected aquatic habitats such as cut-off ponds may support cattail, sedges, and other emergent marshland plants in addition to scrub. Pondweeds and submergent aquatic plants occur intermittently, but such communities have scarcely been described. Terraces near the Río Grande support a diverse scrubland community of Honey and Screwbean Mesquite (*Prosopis glandulosa* and *P. pubescens*), Catclaw (*Acacia greggii*), Black-brush (*A. rigidula*), Huisache (*A. farnesiana*), Desert-willow (*Chilopsis linearis*), Tree Tobacco (*Nicotiana glauca*), Common Buttonbush (*Cephalanthus occidentalis*), and Texas Paloverde (*Cercidium texanum*). Bermuda Grass (*Cynodon dactylon*) has become a major stabilizing ground cover at places where rivers scour rocky shorelines, forming extensive, lush sods (Wauer, 1973, 1977).

Canyon segments of large rivers in this climatic zone, now subjected to upstream controls, were too heavily scoured during flooding to support other than the most rudimentary riparian scrublands. This classically occurred in the Grand Canyon of the Colorado River prior to the closure of Glen Canyon Dam, wherein most riparian vegetation was a shrubby border of mesquite above the level of most annual floods (Turner and Karpiscak, 1980; Fig. 148). Similar conditions may still be found in "box" canyons of the Salt and Gila rivers in Arizona (Minckley and Clark, 1979), and of course in the central gorges of the Río Grande within Big Bend National

Figure 169. *Diverse riparian scrubland in Grapevine Springs Canyon, within the Mohave Desert, Mohave County, Arizona. A few of the riparian plants present are* **Tamarix chinensis, Phragmites australis, Prosopis glandulosa** *var.* **torreyana, Acacia greggii, Equisetum** *spp. and* **Baccharis salicifolia.** *Elevation ca. 1,220 m.*

Figure 170. *Riparian scrubland of reeds* (Phragmites australis), *Saltcedar* (Tamarix chinensis), *Screwbean* (Prosopis pubescens), *Buttonbush* (Cephalanthus occidentalis) *and Texas Honey Mesquite* (Prosopis glandulosa) *along the Rio Grande, in Big Bend National Park, Brewster County, Texas. Elevation ca. 750 m.*

Park (Denyes, 1956). Some of these segments have as yet resisted incursions by the aggressive Saltcedar (see Robinson 1965).

No mammals appear particularly restricted to this vegetation, although a number of bats (Easterla, 1973) as well as Cotton Rat (*Sigmodon hispidis*), White-footed Mouse (*Peromyscus leucopus*), Desert Pocket Mouse (*Perognathus penicillatus*), Beaver, and Raccoon are often well represented there (Boerr and Schmidly, 1977). A few of the nesting birds strongly associated with riparian scrub in their appropriate biotic provinces are Crissal Thrasher (*Toxostoma dorsale*), Verdin (*Auriparus flaviceps*), Black-tailed Gnatcatcher (*Polioptila melanura*), Phainopepla (*Phainopepla nitens*), Black Phoebe (*Sayornis nigricans*), and Lucy's Warbler (*Vermivora luciae*).

The Western Spadefoot and Red-spotted Toad (*Bufo punctatus*) are, if not characteristic, two widespread and common amphibians, and Woodhouse's Toad enters the Chihuahuan Desert region only along the rivers (Conant, 1978). In the more open scrub, the Side-blotched Lizard (*Uta stansburiana*) is perhaps the most commonly encountered reptile. Aquatic species include Spiny Softshelled Turtle (*Trionyx spiniferus*

emoryi), Pond Slider (*Chrysemys scripta*), and in the Big Bend Region, the Plain-bellied Water Snake (*Natrix erythrogaster*) (Conant, 1963, 1969, 1978).

From the mouth of the Rio Conchos downstream in the Rio Grande, and in the lowermost Pecos River, the Mississippi River fish fauna dominates. Hubbs et al. (1977) termed this the Rio Conchos-Rio Grande faunal assemblage, characterized by species such as Speckled Chub (*Hybopsis aestivalis*), Blue Sucker (*Cycleptus elongatus*), River Carpsucker (*Carpiodes carpio*), buffalofishes (*Ictiobus* spp.), Channel and Blue Catfishes (*Ictalurus punctatus* and *I. furcatus*), Red Shiner (*Notropis lutrensis*), and others. Interestingly, a tributary faunal assemblage in the same region consists of mostly species of Mexican derivation, e.g., Conchos Pupfish (*Cyprinodon eximius*), Mexican Stoneroller, Chihuahua Shiner, Mexican Tetra (*Astyanax mexicanus*), Mosquitofish (*Gambusia affinis*), Roundnose Minnow (*Dionda episcopa*), and Tamaulipas Shiner (*Notropis braytoni*), along with sunfishes (*Lepomis* spp.) and the more ubiquitous Fathead Minnow and Red Shiner.

Californian Maritime and Interior Marshlands

Marshlands of this climatic zone are represented by tidal marsh as well as a number of interior marsh biomes. In estuaries and in some protected lagoons and sheltered bays, a *Californian maritime marshland* occurs intermittently southward along the Pacific Coast to the Vizcaíno Peninsula (Fig. 171). Even as far south as Scammon's Lagoon, mangroves and other tropic-subtropic species are lacking or poorly represented, and both the marshland flora and fauna have a decidedly temperate aspect.

Occurring just inland from the intertidal zone (=strand) of mud flats, sand, or (rarely) rock, these marshes are maintained and flushed by tidal action; the emergent vegetation is at or just above the high water line. As is the case with tidal marsh ecosystems almost everywhere, most of these wetlands have been greatly reduced and disturbed by human activities. Figures showing reductions in areas of marshlands in Mission Bay, San Diego Bay, and Tijuana Estuary are especially impressive (MacDonald, 1977). Much of these and other coastal marshlands once found in southern California have been destroyed; small marshes, such as the one at Newport Bay, have been preserved only after considerable effort. The best examples of warm temperate maritime marshland remaining in the Southwest are in Baja California del Norte, the most extensive in the vicinity of Scammon's Lagoon.

Emergent vegetation in these marshes is most often consociations of either cordgrass (*Spartina foliosa*), or at slightly higher sites subject to less inundation, Pickleweed (*Allenrolfea occidentalis*), or Glasswort (*Salicornia virginica*). This latter species colors an otherwise sedate marsh, turning reddish orange in autumn and harboring the bright orange parasitic Dodder (*Cuscuata salina*) in spring and summer. Other beach and marsh plants as Batis (*Batis maritima*), Saltgrass (*Distichlis spicata*), Marsh-rosemary or Sea-lavender (*Limonium californicum*), Seep-weed (*Suaeda californica*), *Monanthochloë littoralis*, Arrowgrass (*Triglochin maritima*), *Salicornia subterminalis, Jaumea carnosa*, and *Anemopsis californica*, while widespread, are only rarely dominant. Although communities of Giant Bulrush (*Scirpus californicus*), rushes, and even cattail may be present in the brackish waters of some estuaries, such situations are uncommon in the Southwest. Saltbushes (*Atriplex patula* and others) and Alkali-heath (*Frankenia grandifolia*) may occupy mounds and other relatively xeric locales within the marsh. The only annual of consequence is *Salicornia bigelovii* (Macdonald, 1977).

Few birds nest in Californian maritime marshlands, the endangered Light-footed Clapper Rail (*Rallus longirostris levipes*) and Black Rail (*Laterallus jamaicensis*) being two important exceptions. Nonetheless, a great variety of migrating and wintering shore and marsh birds make these wetlands particularly attractive to bird watchers. Species especially found here or in adjacent strands include:

Calidris melanotos	Pectoral Sandpiper
C. minutilla	Least Sandpiper
Catoptrophorus semipalmatus	Willet
Himantopus mexicanus	Black-necked Stilt
Limnodromus griseus	Short-billed Dowitcher
L. scolopaceus	Long-billed Dowitcher
Limosa fedoa	Marbled Godwit
Numenius americanus	Long-billed Curlew
N. phaeopus	Whimbrel
Nycticorax nycticorax	Black-crowned Night Heron

Recurvirostra americana	American Avocet
Totanus flavipes	Lesser Yellowlegs
T. melanoleucus	Greater Yellowlegs
Tringa solitaria	Solitary Sandpiper

The larger marshes are (or were) also used by numerous wintering waterfowl, as for nighttime resting and feeding by wintering Black Brant (*Branta nigricans*).

Amphibians are understandably absent from saline maritime marshes, and reptiles are scarcely represented in brackish waters by Western Pond Turtle (*Clemmys marmorata*) and two garter snakes (*Thamnophis sirtalis* and *T. elegans*), near inflowing streams. Fishes are variously represented depending on salinity, water depth, and permanency. Marine species that characteristically move into brackish Californian marshes include Shiner Perch (*Cymatogaster aggregata*), Arrow and Tidewater Gobies (*Clevelandia ios* and *Eucyclogobius newberryi*), Longjaw Mudsucker(*Gillichthys mirabilis*), and Starry Flounder (*Platichthys stellatus*). Three-spine Stickleback and California Killifish also are characteristic of marshy areas along southern California coastlines, and the killifish successfully occupies hypersaline waters to 128 g/l (Miller and Lea, 1972).

Inland, other warm-temperate marshlands occur in old river oxbows, on poorly drained lands, at springs and other shallow water sites in the Californian, Mohavian, Madrean, and Chihuahuan biotic provinces (Figs. 172, 176). Many of these represent remnants of once large aquatic systems, the great Pluvial lakes and rivers of the west (Hubbs and Miller, 1948; Reeves, 1969; Strain, 1970; Hubbs et al., 1974), now scarcely perpetuated by far lower precipitation and meager outflows of groundwater. Many of these environments are saline, a result of evaporative concentration of salts in endorheic basins, and plant communities often are reminiscent of seaside marshes.

Remnant springs and marshes of pluvial Lake Manly in Death Valley are generally surrounded by saltgrass (Fig. 173) and scattered clumps of Common Reed. Small rushes (*Juncus cooperi, Nitrophila occidentalis*) are associated with saltgrass in wetter areas. The edges of open water are occupied by bulrushes, principally *Scirpus americanus* and with lesser frequency *S. maritimus* var. *macrostachyus*. Lizardtail (*Anemopsis californica*) is a characteristic species near seeps (Bradley, 1970). In other parts of the Death Valley system, as in Ash Meadows where salts are less concentrated, marshes in spring outflows are backed by extensive scrublands of mesquite and other woody plants (Beatley, 1971). These areas now are suffering invasion by Saltcedar in a pattern similar to that described for riparian wetlands.

Inland marshes are most extensive in the Chihuahuan biotic province, in the remarkable number of bolsones of that region (Figs. 174, 175, 176). These Chihuahuan interior marshlands are subject to great variation in the length and frequency of inundation, and some may be dry for periods of a year or more (Henrickson, 1978), with broad strands surrounding zones of permanent water.

Deeper and better watered locales and springs (e.g., the Cuatro Ciénegas Basin in Coahuila, Mexico), where salinities are not too high, support complex communities of cattail (*Typha angustifolia*), bulrushes, and Common Reed. Lesser sedges and grasses (*Eleocharis cellulosa, E. rostellata, E. caribaea, Carex pringlei, Spartina spartinae*, and *Setaria geniculata*) may develop as an understory or as monospecific stands. Drier

shorelines are covered by grasses and sedge including *Fimbristylis thermalis, Fuirena simplex*, and *Schoenus nigricans* (see e.g., Fig. 175).

Herbaceous plants of this assemblage include Heliotrope, Water Hyssop (*Bacopa monnieri*), Water Primrose (*Ludwigia octovalvis*), Lizard Tail, Water Parsnip (*Berula erecta*), and many others. In open water, Waterlily (*Nymphaea ampla*), Bladderwort (*Utricularia obtusa*), and charophytes often are common, as are species of pondweed (*Potamoegeton nodosus, P. pectinatus*), Holly-leaf Naiad (*Najas marina*), Widgeon-grass (*Ruppia maritima*), and Common Pondmat (*Zannichellia palustris*) (Pinkava, 1978).

The few and, therefore, particularly valuable cattail and other marshlands within the Western warm-temperate provinces provide "oases" to widely separated populations of nesting and migrating marsh birds. Even the smallest of these seem to support a family of Red-winged Blackbirds, Coots, and Long-billed Marsh Wrens (*Cistothorus palustris*). Both the larger natural wetlands, such as those in Chihuahua (lagunas Bavícora, Santa María, Mexicanos, and Patos) and "managed" marshes such as at Bosque del Apache National Wildlife Refuge, Willcox Playa, Lake McMillan, and San Simón Ciénega in New Mexico and Arizona, may provide nesting habitat for the endemic Mexican Duck (*Anas platyrhynchos diazi*) (O'Brien, 1975). Other waterfowl, such as Snow Geese (*Chen caerulescens*), Pintails (*Anas acuta*), Green-winged Teal (*A. crecca*), American Widgeon (*Anas americana*), and Shovelers (*Spatula clypeata*), use these marshes during migration, and there is some nesting by Blue-winged Teal (*Anas discors*), Cinnamon Teal (*Anas cyanoptera*), Ruddy Ducks (*Oxyura jamaicensis*), and Avocets and other marsh birds. During winter months these marshlands and adjacent playa strands may provide roosting sites for large concentrations of Sandhill Cranes.

Marshlands sometimes support dense populations of aquatic amphibians, especially after heavy runoff expands available water and/or dilutes saline conditions. Included there are many desert and grassland species, spadefoot toads (*Scaphiopus* spp.), true toads (*Bufo cognatus, B. debilis*), and Tiger Salamander, which are preyed upon by the ever-present garter snakes.

Few mammals are restricted to these interior marshlands, but springfed pools and seeps should be as important to bats moving over arid lands as they are to migrating birds (O'Farrell and Bradley, 1970). In Death Valley, it has been demonstrated that the moister areas near marshes enhance populations of desert rodents, including Round-tailed Ground Squirrel (*Spermophilus tereticaudus*), Desert Kangaroo Rat (*Dipodomys deserti*), Harvest Mouse (*Reithrodontomys megalotis*), and Desert Woodrat (*Neotoma lepida*) (Bradley and Deacon, 1971; Sulley et al., 1972). Mesic marshlands also may allow northern or high-altitude mammalian species to persist in lowlands, e.g., Vagrant Shrew (*Sorex vagrans*) on isolated mountains in southern Arizona (Cockrum, 1960) and Meadow Vole (*Microtus pennsylvanicus*) in springfed marshes near Galeana, Chihuahua (Cockrum and Bradley, 1968).

Warm temperate wetlands support an inordinate number of fishes in the Southwest. This largely results from extreme isolation and substantial security, afforded by desert springs, and the presence of a remarkable family of fishes, the Cyprinodontidae, which has members adapted for survival in some of the most severe aquatic habitats yet described. Well-

Figure 171. *Californian maritime marshland at Laguna Manuela, Baja California del Norte, Mexico. Emergent vegetation is a consociation of Cordgrass (***Spartina foliosa***).*

Figure 172. *Interior (Sonoran) marshland, woodland, and submergents at Arivaca Slough, Pima County, Arizona (winter aspect). Although there is often much integration, there also is often only one or two representative dominants in each formation-class—in this case cattail (***Typha domingensis***) marshland and willow (***Salix gooddingii***) woodland. Elevation ca. 1,100 m.*

Figure 173. *Interior (Mohavian) marshland along Salt Creek, Death Valley National Monument, San Bernardino County, California. Consociations of Salt Grass (**Distichlis stricta**; drier areas) and Bulrush (**Scirpus olneyi** in the water), with few associates, line this marsh-stream. Elevation ca. −25 m; photograph by Stuart G. Fisher, November 1979.*

Figure 174. *Sub- and emergent growth-forms of **Nymphaea ampla** in clear, 1.5-m-deep water of Laguna Tío Candido, Cuatro Ciénegas Basin, Coahuila, Mexico; Elevation ca. 750 m; photograph by W.L. Minckley, December 1979.*

Figure 175. *Laguna Tío Candido and associated marshes. A complex submergent and marshland community, including* Typha angustifolia, Scirpus olneyi, *and lesser aquatics, with drier shorelines vegetated by* Fimbristylis thermalis, Fuirena simplex, Schoenus nigricans, *and* Sporobolus *sp. Elevation ca. 750 m; photograph by W.L. Minckley, December 1979.*

Figure 176. *Interior (Chihuahuan) marshland near the inflow of Rio Churince to Laguna Grande, Cuatro Ciénegas Basin, Coahuila, Mexico; flooded Salt Grass* (Distichlis stricta) *dominates the saline flats surrounding this barrial lake, with clumps of sedges* (Scirpus olneyi *and others) occurring in less severe areas near the inflowing stream; dunes in backround result from wind transport from the lake margin during drier periods (see also Fig. 179). Elevation ca. 750 m.*

watered Pluvial environments allowed for dispersal of these fishes, along with a few other groups (coarse-scaled species of *Gila*, some *Catostomus*, and Speckled Dace), and when those rivers and lakes dried, the fishes remained to exploit meager remnants. If the riverine Pecos, Conchos, and Red River pupfishes are excluded, the other pupfish species listed in Table 32 from the Rio Grande-Pecos River systems are presently restricted to springs and spring-fed habitats, as are most of the mosquitofishes (excluding *Gambusia affinis* and *G. senilis*). Thus each of these genera is today prepared for natural desiccation with representatives positioned to survive all but human-caused catastrophe, such as pumpage resulting in lowering of water tables.

In the Death Valley system, the Mojave River has a single endemic—the Mohave Chub (*Gila bicolor mohavensis*). The higher-elevation Owens River has four: the Owens Tui Chub (*G. bicolor snyderi*), Owens Sucker (*Catostomus fumeiventris*), Owens Pupfish (*Cyprinodon radiosus*), and a distinctive Speckled Dace (Miller, 1973). The Amargosa River basin, including Death Valley itself and Ash Meadows, had a distinctive Speckled Dace and seven species of pup- and poolfishes: Amargosa Pupfish (*Cyprinodon nevadensis*), Salt Creek Pupfish (*C. salinus*), Cottonball Marsh Pupfish (*C. milleri*), Devil's Hole Pupfish (*C. diabolis*), Pahrump Poolfish (*Empetrichthys latos*), and the recently extirpated Ash Meadows Poolfish (*E. merriami*). These are separated into no fewer than 14 subspecies (Miller, 1948; Soltz and Naiman, 1978).

The remarkable Cuatro Ciénegas Basin in Coahuila, Mexico, is the only other desert bolsón to demonstrate a similar degree of endemism as Death Valley. Comparison of Tables 32 and 33 reveals the distinctiveness of the Cuatro Ciénegas fish fauna, which includes 15 of a total of 20 species that are known only from, or are differentiated within, that small area (Minckley, 1969, 1978). Endemism in invertebrates (Taylor, 1966; Minckley, 1978), other vertebrates as the aquatic Coahuilan Box Turtle (*Terrapene coahuila*) (Brown, 1974), other endemic turtles (*Trionyx ater, Chrysemys scripta taylori*) (Conant, 1978), terrestrial forms such as *Gerrhonotus lugoi* (McCoy, 1970), and numerous vascular plants (Pinkava, 1978), plus a remarkable assemblage of relict species (Milstead, 1960), further attest to the antiquity of aquatic and terrestrial habitats of that area (Meyer, 1973). Most other basins of the Chihuahuan Desert have lesser, but nonetheless significant, levels of endemism: the basin of Lago de Guzmán in northern Chihuahua and southern New Mexico contains at least three pupfishes and the endemic Chihuahua Chub (*Gila nigrescens*), and Laguna de Bavicora supports a species of *Gila*.

Some species of fishes in desert springs are stenotherms, adapted to narrow temperature ranges of their special environments. The most spectacular of these is perhaps the Moapa Dace (*Moapa coriacea*) and some populations of springfish (*Crenichthys baileyi, C. nevadae*) of the Pluvial White River of Nevada. These animals live within a few degrees of what is considered lethal maxima for many fishes, between 30.5° and 34° C for the dace and up to 35° for *C. baileyi*, and at least *M. coriacea* does not venture into cooler waters (Deacon and Bradley, 1972). Indeed, certain populations of these animals are protected from depredations and competition from introduced fishes by high temperatures that only they can withstand; most introduced species simply cannot invade the thermal environments enjoyed by these natives.

Development of water resources adjacent to marshlands in the interior Southwest spells doom for many of the special habitats just discussed. Lowering of water tables obviously destroys surface waters of a permanent nature, and obligate aquatic communities disappear. If waters persist, non-native organisms provide other problems. Severity of this situation was recently pointed out by Pister (1974) as follows (brackets ours):

> "During the past 35 years, man's activities apparently have caused the extinction of 4 species and 6 subspecies in 6 genera [of fishes] within California, Nevada and Arizona. In addition, at least 50 species and subspecies in 26 genera within 8 Great Basin States and northern Mexico are considered threatened, 19 of which are currently listed as endangered by the Secretary of the Interior... This situation has resulted primarily from agricultural pumping and diversion of watercourses and has been aggravated by the introduction of predaceous game fishes and other piscine competitors."

At this writing his statement stands true, and additional taxa may soon be added to Pister's ominous statistics.

Table 33. Fishes of the Cuatro Ciénegas basin, Coahuila, Mexico. Symbols: *=restricted to the basin; **=differentiated to the subspecific level within the basin when compared to materials from the lower Rio Grande into which Cuatro Ciénegas water now flows as a result of canal connection *via* Rio Salado (from Minckley, 1978).

Characidae
  Astyanax mexicanus (Mexican Tetra)

Cyprinidae
  *Notropis xanthicara (Cuatro Ciénegas Shiner)
  **Dionda episcopa (Roundnose Minnow)

Ictaluridae
  **Ictalurus lupus (Headwater Channel Catfish)
  Pylodictis olivaris (Flathead Catfish)

Cyprinodontidae
  *Lucania interioris (Interior Killifish)
  *Cyprinodon bifasciatus (Laguna Pupfish)
  *Cyprinodon atrorus (Ciénega Pupfish)

Poeciliidae
  *Gambusia longispinis (Cuatro Ciénegas Mosquitofish)
  Gambusia marshi (Río Salado Mosquitofish)

Poeciliidae (cont'd)
  *Xiphophorus gordoni (Cuatro Ciénegas Platyfish)

Centrarchidae
  **Lepomis megalotis (Longear Sunfish)
  **Micropterus salmoides (Largemouth Bass)

Percidae
  *Etheostoma sp. (Cuatro Ciénegas Darter)

Cichlidae
  Cichlasoma (Herichthys) cyanoguttatum (Rio Grande Perch)
  Cichlasoma (Herichtys) sp.
  *Cichlasoma (Parapetenia) sp. "A" ("Lugo's Cichlid")
  *Cichlasoma (Parapetenia) sp. "B" ("Caracole Cichlid")
  *Cichlasoma (Parapetenia) sp. "C" ("Longhead Cichlid")
  *Cichlasoma (Parapetenia) sp. "D" ("Unexpected Cichlid")

Californian Maritime Strands

Although strand vegetation occupies numerous, if not extensive, interior habitats, this formation is most often considered in context of those beach communities occurring between low tide and the influence of the highest tide (=intertidal zone). There the barenness of vegetation caused by wave action, saline water tables, shifting sandy substrates, salt spray, and sand blast, corresponds to the desertscrub formation of upland vegetation. As in interior habitats, substrates may be rock, sand, or mud. In the warm temperate Southwest, coastal vegetation is represented by *Californian maritime strand*, which has been well described by Barbour and Johnson (1977).

The Californian coastline is characterized by rocky cliff and tidepool environments, sandy beaches and dunes, and more rarely, salt marsh and mud flats (Fig. 177). Dunes and sandy beaches are confined to bays (e.g., Long Beach). Beaches tend to build seaward in summer and erode during winter months, and the spring tidal range is moderate (Hinton, 1969).

Strand vegetation in California varies greatly from north to south—forbs dominate south of 34°N and the plants are mostly prostrate, herbaceous, evergreen, and succulent. Vegetation increases landward toward the foredunes; dune vegetation commonly grades into coastal scrub and may be composed of open communities of the same species. The species as well as density of vegetation also follows a zonation based on environmental tolerances. In the Southwest, the number of species in the littoral strip is few, some of the most common and characteristic being the Saltbush (*Atriplex leucophylla*), Sand-verbena (*Abronia maritima*), Silver Beachweed (*Ambrosia chamissonis*), Evening-primrose (*Oenothera cheiranthifolia*), Goldenweed (*Haplopappus venetus*), Mock Heather (*Haplopappus ericoides*), *Lupinus chamissonis, Abronia umbellata,* and the Ice Plant (*Mesembryanthemum crystallinum*). Introduced species such as *Mesembryanthemum chilense, Monanthochloë littoralis,* and the Beach Morning Glory (*Calystegia soldanaella*) are now important in the strand makeup, and the annual *Cakile maritima* is commonly the last vascular plant to seaward before submergents are encountered. On dunes north of San Diego one may find the introduced Beachgrass (*Ammophila arenaria*).

Although the number and variety of sandpipers, gulls, stilts, plovers, and other shorebirds during migration and in winter is especially large in this habitat, the numbers of nesting species that use such places in the Southwest is not. Furthermore, many of these are now, because of human disturbance, confined to islands and the few available remote sites. In the Southwest, some of the most common or indicative species are:

Charadrius alexandrinus	Snowy Plover
Endomychura hypoleuca	Xantus' Murrelet
Haematopus bachmani	Black Oystercatcher
Larus occidentalis	Western Gull
Phalacrocorax auritis	Double-crested Cormorant
P. penicillatus	Brant's Cormorant
Sterna albifrons	Least Tern
Thalasseus elegans	Elegant Tern

Some of the rockier headlands and offshore habitats support rookeries of California Sea Lion (*Zalophus californianus*), and sand bars, reefs, and spits are calving sites for the Harbor Seal (*Phoca vitulina*). Two former breeding pinnipeds of the mainland strand of southern California and

Figure 177. *Californian maritime strand at Scammon's Lagoon, Baja California del Norte, Mexico. These "mud flats" at low tide at first appear unvegetated, but closer inspection reveals prostrate seagrasses, diverse algae, and other plant life in addition to a rich, depression- and burrow-inhabiting invertebrate and vertebrate fauna.*

northern Baja California, the Elephant Seal (*Mirounga angustirostris*) and Guadalupe Fur Seal (*Arctocephalus townsendi*), are making a comeback and are now in the process of reoccupying many of their former breeding islands (Hubbs et al., 1965, 1968). These large animals, plus other mammals and fishes, provided carrion in strand areas which served as food for special raptors such as California Condor (*Gymnogyps californianus*).

Rocky bays, shorelines, and tidepools of the California coast support a diversity of fishes (Miller and Lea, 1972; Fitch and Lavenberg, 1975). Bay, Rockpool, and Mussel Blennies (*Hypsoblennius gentilis, H. gilberti, H. jenkinsi*) are common, as are clinid blennies (*Gibbonsia* spp.), and cottids or sculpins (*Clinocottus analis, Oligocottus rubellio*), Zebra and Bluebanded Gobies (*Lythrypnus zebra, L. dalli*), Yellowfin Goby (*Acanthogobius flavemanos*), and the unique Blind Goby (*Typhlogobius californiensis*) that lives in close-spaced habitats beneath rocks. Clingfish (*Gobieosox meandricus*) occur on wave-washed rocks and can withstand several hours out of water if in a moist place. The Rockweed Gunnel (*Xererpes fucorum*) was said by Barnhart (1936) to move freely over stones or sand out of water at low tides.

Shore fishes that move into strand areas at high tide include Needlefish (*Strongylura exilis*), Jack Smelt (*Atherinopsis californiensis*), White Seabass and Orangemouth Corvina (*Cynoscion nobilis, C. xanthulus*) and Yellowfin Croaker (*Umbrina roncador*). Other species are more common in bays,

e.g., Cabezón (*Scorpaenichthys marmoratus*), Silver Mojarra (*Eucinostomus argenteus*), or have young that inhabit rockpools and adults that move to deeper, inshore areas: Gopher Rockfish (*Sebastes auriculatus*), Black-and-Yellow Rockfish (*S. chrysomelas*), and Opaleye (*Girella nigricans*). The unique surfperches (Embiotocidae) of the California coast live in shallow bays and along sandy shorelines. This live-bearing group includes numerous species (Tarp, 1952), among which the Shiner and Silver Surfperches (*Cymatogaster aggregata, Hyperprosopon ellipticum*) are perhaps most familiar; one species of this large group is freshwater (Tule Perch, *Hysterocarpus traskii*) in the Sacramento-San Joaquin system north of our area. Perhaps the most indicative strand fish of southern California is the Grunion (*Leuresthes tenuis*), which reproduces at high tide in spring and summer. The female buries in the sand, tail down to its pectoral fins. Eggs are spawned and fertilized by sperm that moves through the sand. The eggs remain for 2 weeks longer, until dislodged by the next series of high tides, and embryos break free upon contact with water. Other, related species such as Jack Smelt spawn on eelgrass within the zone of the lowest tides.

Soft bottoms in bays are inhabited by similar shoreline fishes at high tide, and by cryptic forms as Worm-hole or Shadow Goby (*Quietula y-cauda*) that inhabits burrows of invertebrates, or Arrow Goby (*Clevelandia ios*) and Longjaw Mudsucker (*Gillichthys mirabilis*) that persist at low tide in shallow depressions or holes in wet mud.

Warm-Temperate Interior Strands

Strand habitats also occur in flood channels of rivers, along banks of streams below hydroelectric dams, along receding reservoirs, and around intermittent fluctuating lakes. Vegetation of these harsh environments is made up of either short-lived successional species or plants adapted to periodic flooding, scouring or soil deposition, and in the case of playa lakes, high salinities or other special chemical conditions associated with evaporation of inland waters. Depending on substrate and frequency and type of inundation, interior strand communities along streams may be composed of sparse, open stands of riparian scrub species (e.g., *Baccharis salicifolia* and *Tamarix chinensis*), seedlings of riparian trees such as willows and cottonwood, or any of a number of characteristic annuals, biennials, and short-lived perennials, or barren except for simple plants such as algae (Fig. 178).

A narrow strand occurs along the mainstream Colorado River, within Grand Canyon, downstream from Glen Canyon Dam, where stability is provided by upstream dams that produce an almost-tidal, daily fluctuation in water levels. Sandbars and other available substrates are vegetated sparsely by Saltcedar, Coyote Willow, Seepwillow, and the introduced Camelthorn. This zone is an example of a remarkable exchange of materials from aquatic to terrestrial systems, with great numbers of aquatic invertebrates, principally the Scud (*Gammarus lacustris*) stranded in drifting algae, feeding ants and other streamside invertebrates as well as a myriad of vertebrates ranging from iguanid lizards (e.g., *Sceloperus magister, Urosaurus ornatus*), to insectivorous birds and raccoons.

Upstream from dams, along the margins (strands) of fluctuating reservoirs, Bermuda Grass may occupy a large percentage of the soil surface, growing rapidly downslope as waters are removed for irrigation or power production. It provides meadows heavily used by cottontail, jackrabbits and, in season, flocks of geese and ducks. Flooding of these grassy areas adds substantial nutrients to aquatic systems when high waters occur after snowmelt or summer monsoons. When southwestern reservoirs are relatively stabilized, with daily and annual fluctuations of less than a few meters, clumps of Common Reed (*Phragmites australis*) establish, floating on the water surface on the lakeward side during high water and settling to root at the water's edge when lake levels are down.

A number of shore and other coastal strand birds are equally adapted to interior strands along rivers and reservoirs. These include Spotted Sandpiper (*Actitus macularis*)—rocky stream channels during migration and in winter (grasses and marshlands used for nesting); Killdeer (*Charadrius vociferus*)—open ground, gravel bars; Rough-winged Swallow (*Stelgidopteryx ruficollis*)—cutbanks of streams; and Black Phoebe (*Sayornis nigricans*)—overhanging walls of canyon-bound creeks and rivers. Isolated sand bars are heavily used by wintering Canada Geese (*Branta canadensis*) and other waterfowl as a nighttime resting place and as a source of gravel. Raccoons are otherwise perhaps the most frequent homeotherms occupying inland strands while foraging at night, and a number of other scavengers, as along seacoasts, depend upon this zone for carrion, e.g., the Bald Eagle (*Haliaeetus leucocephalus*) and lesser, stream-dependent raptors such as the Black Hawk.

Inland river strands are also depended upon by the Spiny Softshell Turtle for basking and oviposition, and are similarly

Figure 178. *Riparian strand along the San Pedro River within the Chihuahuan Desert south of Cascabel, Cochise County, Arizona. A seasonally inundated mud-sand substratum populated by a few plants of Burrobrush* (Hymenoclea monogyra) *and nightshade* (Solanum *spp.). Elevation ca. 1,000 m.*

used by other aquatic turtles in areas where they occur. Garter snakes, toads (*Bufo woodhousei, B. microscaphus*), and other, less aquatic adapted species, forage in strand. Debris and other sparse cover along streams and lakes may be occupied by frogs, principally the Leopard Frog; they are rarely found on open beaches or gravel bars where they would be highly vulnerable to aerial predators. No fishes regularly use the strand zone. However, the substantial nutrients derived from periodically flooded vegetation and debris undoubtedly enhance fishes and other aquatic organisms. Flood-deposited debris is processed by terrestrial reducers in this zone when dry, converting it to smaller particles that can more easily be used by the aquatic system (Bruns and Minckley, 1980).

Within the fluctuation zones of interior playa lakes even greater selection pressures come to bear. Salts accumulated in the vast internal drainage basins of the Southwest, especially in the Chihuahuan Desert where well over 300,000 km² are endorheic, often are sufficiently concentrated to affect water uptake by plants. Because these saline accumulations are periodically inundated, they qualify as strand, and many of the plants occurring there are the same as in maritime

situations. Vegetation, if present, consists of one or a few pioneer species including *Allenrolfea occidentalis*, Salt Grass, Sea-purslane (*Sesuvium verrucosum*), species of *Suaeda* (*S. jacoensis, S. palmeri, S. fructicosa*), *Salicornia rubra*, and the halophytic Lovegrass (*Eragrostis obtusiflora*). A single species at low density often comprises the sole vegetation of such areas. Common species on higher, less frequently flooded and less saline habitats include *Atriplex acanthocarpa, A. obovata*, and other saltbushes, sacatón grasses (*Sporobolus arioides, S. wrightii*) and *Suaeda* spp. Even the most salt tolerant plants often grow at low to non-saline conditions, thus the "pioneers" listed above also live in other communities. *Allenrolfea occidentalis*, although characteristic of playa habitats (Ungar, 1974), is among the desertscrub on bajada slopes where individual plants may reach 2 m in height (Pinkava, 1978). However, severe saline strand habitats appear to have selected for an unique inland flora. Of 40 taxa considered halophytic in the Chihuahuan Desert, 25 (including 3 genera) are endemic to such areas (Henrickson, 1978).

Many species germinate and develop on these saline sites only after rains temporarily decrease soil salts, then die as salinity again increases. Annual grasses in this category

Figure 179. *Dune field west of Laguna Grande (see also Fig. 176) in the Cuatro Ciénegas Basin, Coahuila, Mexico. Such dunes of gypsum or other materials are characteristic landscape features downwind of Pluvial lakes and are populated by both gypsophilous-adapted terrestrial and saline-adapted wetland plants. Elevation ca. 750 m.*

include *Leptochloa fascicularis* and *L. uninervia*, various *Panicum* spp., *Bouteloua karwinskii*, etc. Herbs, either annual or perennial, are *Allionia incarnata*, *Heliotropium curassavicum*, *Salsola iberica*, *Hoffmanseggia glauca*, and *Tidestromia lanuginosa*, among many others. Aquatic macrophytes are rare. Wigeongrass (*Ruppia maritima*) sometimes becomes dense, but disappears as salinities rise. Other aquatic species (e.g., *Zannichellia palustris* and *Potamogeton pectinatus*) are even more ephemeral, appearing with temporary freshening of these brackish and saline environments.

As along the sea, larger playas have associated dune fields, some of which reflect the special nature of their waters in being of almost pure gypsum (e.g., White Sands, New Mexico, from Pluvial Lake Otero, and dunes in the Cuatro Ciénegas Basin, Coahuila, Mexico; Fig. 179). Under these circumstances, gypsophilous plant species have evolved to complement those especially adapted for existence in saline, interior strand.

Inland playa strands are remarkably inhospitable for animals. Available water is often too salty to drink and plants are sparse and unpalatable. Black-tailed Jackrabbits (*Lepus californicus*) range into strand to feed when new growth or germination cycles follow freshening. Similarly, seed-eating birds (e.g., Mourning Dove, Black-throated Sparrow, *Amphispiza bilineata*) invade such areas for food, and avoid them when seed production is low (Raitt and Pimm, 1978). Some marsh and shore birds use these habitats, notably Snowy Plover and Avocet, which are characteristic of alkali playas at inland locales, and large flocks of Sandhill Cranes use Willcox Playa and other shallow water "dry" lakes as wintering ground, along with numerous waterfowl.

Shorelines and grassy flats have a small herpetofauna, principally species of whiptail lizards (*Cnemidophorus* spp.) and their predators such as Whipsnake (*Masticophis flagellum*). Fishes rarely occur, but pupfishes (*Cyprinodon* spp.) occasionally invade and form substantial populations in Laguna Grande and Laguna Salada in Coahuila, Mexico (*C. atrorus*; Minckley, 1969), or in shorepools as along the Salton Sea (*C. macularius*; Barlow, 1958).

Tropical-Subtropical Wetlands

Although a number of Tamaulipan elements extend up the Río Grande to beyond Big Bend National Park, only those southwestern wetlands within the Sinaloan and Sonoran biogeographic provinces are considered to belong within this climatic zone (Fig. 3). Even in these provinces the wetland flora and fauna is a mixture of Nearctic and Neotropical species, the former contributing far more to the Sonoran province than to the Sinaloan. As is the case with subtropical uplands, there is a wide range of habitat types, some of which exhibit exceptional plant and animal diversity. Included are consociations of mesquite bosques and palm oases as well as associations of deciduous and evergreen riparian forests, riparian and maritime scrublands, marshland, and strands, and an enormous variety of freshwater, brackish, alkali, and marine submergent communities.

Human alteration of almost all of the regions' rivers coupled with introductions of non-native species has resulted in major modification and displacement of the original stream biota. The relatively few endemic fishes have been reduced in numbers with impoundments and other stream modifications. These changes in streamflow have been remarkably effective in bringing the large and sometimes bizarre fishes native to the lower Colorado River system to the brink of extinction. Some smaller species persist in good numbers, but other species are becoming increasingly restricted in range. With the passing of the major river also goes a number of delta-dependent fish species from the Gulf of California (e.g., the Giant Totoaba, *Cynoscion macdonaldi*).

It is beyond the scope of this publication to present a discussion of the vast offshore resources of the Gulf of California, although they belong to this thermal zone. A definitive work on the biota of this important marine region is needed. A start in this direction covering reef fishes recently appeared (Thomson et al., 1979). Certainly the abundance and importance of food and game fishes in the Gulf alone warrants such an attempt, and a discussion of its varied migratory and sedentary inhabitants would be most interesting and informative.

Sonoran Riparian Deciduous Forest and Woodlands

Centered in the Sonoran biotic province are streamside associations of tropic-subtropic subspecies of willow (*Salix gooddingii* var. *variabilis*), cottonwood or alamo (*Populus fremontii* var. *macdougalii*), *P. dimorpha,* and/or Velvet Mesquite (*Prosopis veluntina*). Winter deciduous, these biomes are nonetheless subtropical riparian where they are restricted to streams and springs below 1,100 to 1,200 m elevation in and immediately adjacent to, the Sonoran Desert. While now much reduced in extent, these forests are still represented by impressive examples and may contain individual trees of great size (Fig. 180).

Willow and cottonwood forests were, and remain, largely restricted to the immediate flood plains of perennial, or at least spring-flowing streams, where they are maintained by periodic winter-spring flooding (Fig. 181). As such, southwestern tropic-subtropic examples are largely restricted to the lower Colorado River and Arizona Upland subdivisions of the Sonoran Desert, which possess watersheds of sufficient winter precipitation and hence the spring discharges necessary to support them. Mesquite *"bosques"* (Spanish for woodlands) attain their maximum development on alluvium of old dissected flood plains, especially those laid down at the confluence of major watercourses and their larger tributaries (Fig. 182).[1] Consequently, these higher "secondary" flood plains are commonly 1.5 to 6.0 m above the river channel.

Many of the more famous bosques referred to in the literature are today mostly of historical interest—e.g., the mesquite forests at San Xavier, Komatke (New York Thicket), and the mesquite and cottonwood forests along the lower Gila and Colorado rivers. Nonetheless, some excellent examples still occur as scattered remnants along the Santa María, Verde, middle Gila, San Pedro, San Miguel, Magdalena, Sonora, and other "desert" river systems. These remaining bosques are, however, all threatened by a variety of human-related causes,—clearing for agriculture and pasture, water diversion, flood control and water storage projects, cutting of trees for fuel, and most importantly, the lowering of groundwater tables. Because mesquite cannot reach groundwater much below 14 m, this last factor has been responsible for the almost total ruin of mesquite forests at San Xavier, Casa Grande Ruins National Monument, Komatke, and elsewhere in Arizona where live streams no longer persist as a result of groundwater pumping. Conversion to agriculture continues to greatly reduce the once extensive bosques along the lowermost Colorado, San Pedro, and Gila rivers. While several thousands of acres of mesquite woodlands have been withdrawn for purposes of preservation, high demands for fuel wood and groundwater threaten all remaining bosques both north and south of the U.S. and Mexican border.

With some notable exceptions, willow-cottonwood forests have been reduced to isolated groves and are now scattered along the Colorado River where they once were extensive (see

---

[1] *Although mesquite bosques have been described as occurring in riparian situations within the Mohave and Chihuahuan deserts (e.g. Wauer, 1973), the tall (to over 15 m) tree-forming Velvet Mesquite is lacking in these biomes, mesquite being represented here by the shorter-statured and multi-trunked Western Honey Mesquite (P.* **glandulosa** *var.* **torreyana**), *Screwbean (P.* **pubescens**) *or Texas Honey Mesquite (P.* **glandulosa** *var.* **glandulosa**). *Moreover many of the subtropic plants and animal associates of the bosques are lacking in these warm temperate* scrublands. *Both Fremont Cottonwood and Goodding Willow also occur in warm temperate biomes; again, however, different plants and animal associates are to be expected.*

e.g. Ohmart et al., 1977) (Fig. 183). In many places, such areas now are vegetated by *Sonoran riparian scrubland*. Often, the remaining groves are open woodlands of over-mature individuals that are lacking in reproduction and may be expected soon to disappear because of stream regulation. Gallery forest of willow and cottonwood can nonetheless still be found along reaches of undammed and more "natural" portions of the Verde River, middle Gila River, the Hassayampa River below Wickenburg, Arizona, the San Pedro River, and the Río Bavispe and Río Yaqui, all flood-prone ecosystems where Goodding Willow frequently outnumbers cottonwood 100 to 1. Understories may be a tangle of young trees, especially mesquite, or be relatively open.

Historically, annual and perennial grasses (e.g., Vine-mesquite Grass, *Panicum obtusum*), forbs (e.g., Careless Weed, *Amaranthus palmeri*), and in more saline areas, saltbushes (*Atriplex polycarpa, A. lentiformis, Suaeda torreyana*) constituted the understory in mature bosques. The interior of mesquite bosques was typically open and park-like. Old, fire-scarred trees predominated. Today, because of grazing and other disturbances, a number of introduced forbs and grasses such as Filaree (*Erodium cicutarium*), mustards (Cruciferae), including Yellow Rocket (*Sisymbrium irio*), Red Brome (*Bromus rubens*), and in more open places, Schismus (*Schismus barbatus*) and Bermuda Grass, are frequently encountered understory species. Vines as the climbing milkweeds (*Sarcostemma* spp.), gourds (*Cucurbita* spp.), and Canyon Grape are often common and conspicuous constituents where grazing has not been too severe. Hackberry or Cumero (*Celtis reticulata*), Mexican Elder (*Sambucus mexicana*), and Screwbean Mesquite may be important arboreal associates, at least locally. At higher elevations (760 to 1,100 m), an individual Velvet Ash (*Fraxinus pennsylvanica* var. *velutina*) or other temperate species, e.g., sycamore, may occupy a particular site, and Blue Paloverde, Catclaw, and Ironwood (*Olneya tesota*) can be common in more arid locations. Graythorn (*Zizyphus obtusifolia*) or one or more of the allthorns (*Lycium fremontii, L. andersonii, L. berlandieri*) frequently occupy an occasional opening or sunny place along with young mesquites.

The continued clearing of riparian communities along the lower Gila and Colorado Rivers (and in other areas) has resulted in type conversions other than to farmland. It has been noted that where intermittent flooding occurs during the long Southwestern growing season, Saltcedar or Tamarisk (*Tamarix chinensis*) tends to replace mesquite and other native riparian vegetation (Horton, 1977). This tendency is particularly prevalent in saline areas after the native woodlands have been cleared or burned and the water table is at or near the surface. Saltcedar duff is highly flammable and fire initiates a cycle to a disclimax scrub populated only by Saltcedar. Whether this replacement is partially dependent on changes in water and soil chemistry, or entirely a result of the inherent ability of Saltcedar to rapidly repopulate floodplains, is imperfectly known (Everitt, 1980), but the aggressive ability of Saltcedar to displace native riparian species has been well demonstrated by Horton et al. (1960), Turner (1974c), and Warren and Turner (1975). It suffices to say that Saltcedar continues to replace willows, cottonwoods and Mesquite in those bosques where these trees are destroyed, spring floods are controlled, and a saline water table is at or near the surface.

Nesting use of these riparian communities by colonies of White-winged Dove (*Zenaida asiatica*) and Mourning Dove (*Zenaida macroura*) is well documented (Neff, 1940; Arnold, 1943; Wigal, 1973; Brown, 1977). Their importance to other avian species as Lucy's Warbler, Vermilion Flycatcher (*Pyrocephalus rubinus*), Abert's Towhee (*Pipilo aberti*), Cardinal (*Cardinalis cardinalis*), Pyrrhuloxia (*Cardinalus sinuatus*), Phainopepla, Varied Bunting (*Passerina versicolor*) and others has been discussed by Brandt (1951), Phillips et al. (1964) and more recently by Hubbard (1977b), Anderson et al. (1977), and Clark (1979). Cottonwood-willow forests where they still occur, determine the northern nesting distribution of a number of Neotropical raptors, such as the Gray Hawk (*Buteo nitidus*), in addition to a number of other tropic-subtropic species e.g., Rose-throated Becard and Thick-billed Kingbird (*Tyrannus crassirostris*); these trees also provide nesting sites for the southernmost Bald Eagles. Other localized nesting species are the Yellow-billed Cuckoo, Great Blue Heron, and Mississippi Kite.

Although the few mammals here are not distinctive, numerous bats are characteristic of streamsides in this region. The Silver-haired Bat (*Lasionycteris noctivagens*) and Big Brown Bat (*Eptesicus fuscus*) often roost in trees such as cottonwood. The Desert Pocket Mouse often occupies mesquite bosques as a result of its preference for deep, sandy, rock-free soils (Porter, 1962). Beaver crop cottonwood-willow communities heavily when abundant, and the Raccoon is locally common. It is hypothesized that the former limited occurrence and incursions of Jaguar (*Felis onca*) into Arizona were largely through these subtropic riparian woodlands.

Several other species of tree-requiring vertebrates, such as the Sonoran Spiny Lizard (*Sceloporus clarki*) and Tree Lizard are well represented in riparian woodlands, although not centered there. Open places support whiptail lizards, and some all-female "species" of that complex extend far into deserts along permanent streams. Streamside amphibians other than frogs of the *Rana pipiens* complex, which are common, include the unusual Colorado River Toad (*Bufo alvarius*), a large species that is the scourge of smaller animals in mesquite bosques on rainy summer nights.

The original fish community of the lowermost Colorado and Gila River mainstreams was small, consisting of four minnows, two suckers, two or three marine species, a cyprinodontid and a poeciliid. Of these, Woundfin (*Plagopterus argentissimus*), Roundtail Chub, and Flannelmouth Sucker (*Catostomus latipinnis*) were rare, being recorded only a few times in the reach from the present Hoover Dam to the Colorado Delta. The marine Machete (*Elops affinis*) and Striped Mullet (*Mugil cephalus*), and the Desert Pupfish and Sonoran Topminnow (*Poeciliopsis occidentalis*) were only in the lowermost parts of the two rivers (the last only in the Gila). Both of the last species must have inhabited only margins and backwaters. This left a unique assemblage of riverine fishes, consisting only of Colorado Squawfish (*Ptychocheilus lucius*), Razorback Sucker (*Xyrauchen texanus*) and Bonytail Chub (*Gila elegans*), in the main channel of the largest river of the American Southwest.

Backwater sloughs and marshes along the rivers also are known to have been used by Squawfish and Razorback, both of which moved as adults from the channel into such ancillary habitats (Miller, 1961; Minckley, 1965; Seethaler, 1978). Such places also provided refuge areas for protection,

Figure 180. *Sonoran riparian deciduous forest of Fremont Cottonwood* (Populus fremontii *var.* macdougalii) *and Goodding Willow* (Salix gooddingii *var.* variabilis) *with Common Reed* (Phragmites australis) *and Saltcedar* (Tamarix chinensis) *at the immediate water's edge. Once extensive, these forests have been reduced to "island" remnants as this one on the Colorado River 20 km west of Yuma, Yuma County, Arizona. Elevation ca. 40 m.*

Figure 181. *Sonoran riparian deciduous forest of* Populus dimorpha, Salix gooddingii, Celtis reticulata, *and* Prosopis velutina *along the Rio Yaqui near Highway 15, Sonora, Mexico.*

Figure 182. *Sonoran riparian deciduous woodland (interior view) – a Mesquite (*Prosopis velutina*) bosque along the San Pedro River near Redington, Pima County, Arizona. Such examples of these magnificent woodlands are becoming rare with declining water tables and increasing demands for cooking and heating fuel. The recent absence of fire is indicated by the abundant presence of young Mesquite. Elevation 900 m.*

feeding, and growth by young of these fishes, if such were required, and adults of these and Bonytail Chub used such areas to avoid floods (Minckley, 1973). All evidence indicates a great abundance of Colorado Squawfish and Razorback Sucker, and that Bonytail were common and widespread (Minckley, 1973, 1979). These fishes are now almost gone from the area, or in the case of the sucker and Bonytail persist as large adults in reservoirs, with little evidence of successful reproduction.

Minor backwaters, shallows, and shoals over sandbars once available for young native fishes now are occupied by a myriad of introduced predators. Where young squawfish, suckers, or chubs once might have been preyed upon only by water birds such as Great Blue Heron, a host of young Large- and Smallmouth Bass (*Micropterus salmoides, M. dolomieui*), lesser sunfishes (e.g., *Chaenobryttus gulosus, Lepomis cyanellus, L. macrochirus* and *L. microlophus*), cichlids (*Tilapia zilli, T. mossambicus, T. aureum*), and the voracious Mosquitofish (Meyers, 1965) now occur. The channel and deeper back-

waters are occupied by adults of these species, plus large and specialized predators such as Flathead Catfish (*Pylodictis olivaris*) and Striped Bass (*Morone saxatilis*). Additional fishes, although perhaps not capable of predation on native forms, certainly increase competitive interactions. Threadfin Shad (*Dorosoma petenense*) feed on plankton and detrital materials, and a bit on benthic invertebrates. Red Shiner (*Notropis lutrensis*) swarms in currents, feeding on drifting and surface materials. Exotic mollies (*Poecilia latipinna* and *P. mexicana*) eat detrital materials and algae. These pressures, accompanied by dewatering of the lower Gila River, channelization which reduced the shallow, quiet backwaters within the strand zone, dredging, and construction of reservoirs and other stabilizing features which enhanced the predominantly lentic-adapted introduced fishes, all have contributed to extinction of this remarkable component of the Southwest's aquatic communities.

Filling of the Salton Sea (Pluvial Lake Cahuilla or LeConte) in 1905-07 by flood flows of the Colorado River, after it broke through frontworks of irrigation channels and was diverted from its normal distributaries to the Gulf of California, created an artifical condition worthy of our treatment.[1] Initially (Evermann, 1916), native fishes from the Colorado River, Bonytail Chub (*Gila elegans*), Razorback Sucker (*Xyrauchen texanus*), Striped Mullet (*Mugil cephalus*), a few trout (*Salmo clarki*), and the introduced Carp were present. Desert Pupfish (*Cyprinodon macularius*) may have entered from the river, or may have spread from springs in the basin that were inundated by rising waters. Machete (*Elops affinis*) appeared later (Dill and Woodhull, 1942).

Evaporative concentration of the Salton Sea gradually increased salinity to exceed seawater (> 35 g/l) in 1945 (Carpelan, 1961), perhaps accelerated by diversion of saline irrigation return waters to the basin. Freshwater fishes disappeared prior to the 1950s, with the exception of Pupfish; marine fishes also failed to reproduce and began to disappear. From 1929 through the late 1940s, numerous invertebrates and fishes were introduced in an attempt to create an inland marine fishery, but by 1949, only pupfish, Mosquitofish, a few Striped Mullet and Longjaw Mudsucker (*Gillichthys mirabilis*) remained. After 1950, introductions of Sargo (*Anisotremus davidsoni*), Bairdiella (*Bairdiella icistius*), and predatory Orangemouth Corvina (*Cynoscion xanthulus*) spawned successfully (Walker et al., 1961). This fishery persists today, but introductions of numerous salt-tolerant freshwater fishes such as Red Shiner, African cichlids (*Tilapia mossambicus, T. zilli* and others), and a myriad of Central American poeciliids (*Poecilia latipinna, P. mexicana, Poeciliopsis gracilis, Xiphophorus variatus, X. helleri*) is now endangering the native

[1] *This situation obviously happened before, since Wilke (1980) has demonstrated that a Neolithic fishing culture existed on the shores of the Cahuilla Basin ca. 1,500 years before present. Rock wiers were constructed annually along shorelines to intercept aggregations of Razorback Sucker and Bonytail Chub, both of which were recovered from middens in the area. Wiers were progressively constructed downslope, at a rate commensurate with projected annual evaporative decline in water levels, and the fishery collapsed in about 55-60 years, at a time when salinities would likely have resulted in decreased populations of freshwater fishes.*

Figure 183. *Riparian deciduous forest of cottonwood (*Populus fremontii *var.* macdougalii*) along the Colorado River near Yuma, Yuma County, Arizona, near the turn of this century. The river and its magnificent riparian ecosystem are still "untamed" as construction of the Imperial Canal is underway. Elevation ca. 100 m. Photograph courtesy of the U.S. Bureau of Reclamation and Robert D. Ohmart.*

pupfish (Schoenherr, 1979) and the valuable game fishes as well.

The few, smaller streams of lowlands draining into the Gila River once supported a simple community of Longfin Dace, Sonoran Topminnow, and occasionally Desert Pupfish. Pupfish populations now are extinct throughout the vast basin, and are approaching extinction in the Río Sonoyta in northern Sonora as a result of pumping along that stream; a remnant is protected at Organ Pipe Cactus National Monument in Quitobaquito Spring. The Sonoran Topminnow has been destroyed in the Gila River basin except for a few populations in springs, isolated stream segments, or artificial refugia (Minckley et al., 1977); however, as noted above, it remains abundant in the Río Yaqui and lesser drainages of northwest Sonora.

Coastal rivers between the Colorado and Río Yaqui generally enter sands at their lower reaches before entering the sea; their fauna is thus restricted to middle and highland portions. Longfin Dace is the only species that has succeeded in occupying all these desert rivers, from the Río Sonora to the Río Sonoyta; Sonoran Topminnow is found in all but the latter stream. The Yaqui Chub (*Gila purpurea*) occurs in the ríos Matape and Sonora along with Mexican Stoneroller. The Opata Sucker (*Catostomus wigginsi*) is restricted to Río Sonora and Sonora Chub (*Gila ditaenia*) holds forth in the Río Concepción. Desert Pupfish is shared by Río Sonoyta and the lower Colorado-Gila rivers, reflecting in its presence the intimate relationships of the first stream to the Colorado River delta prior to its diversion south by the Pinacate lava flows (Hubbs and Miller, 1948).

Sonoran Oasis Forest and Woodlands

These evergreen, Miocene and Pliocene relicts are restricted to certain isolated, permanent springs, seeps and moist canyons, in and at the western edge of the Sonoran Desert. In Arizona and California, natural oases are represented by groves of the California Fan Palm (*Washingtonia filifera*), and in Baja California del Norte by California Fan Palm and/or Blue Fan Palm (*Erythea armata*). Further south in Baja California, the Sky-duster Palm (*Washingtonia robusta*) makes its appearance as do feral populations of Date Palm (*Phoenix dactylifera*). The last species now dominates the oases at Mulegé and San Ignacio and has so for many years (Shreve, 1951; Moran, 1977; Fig. 184).

Although numerous localities for *Washingtonia filifera* have been reported for California and Baja California del Norte, the only native palms in Arizona are those in the Kofa Mountains, at and near Alkali Springs near Castle Hot Springs, and possibly at Ciénega Spring northeast of Parker. This is the palm of such famous southern California oases as Palm Springs, 29 Palms, and Palm Desert. There it grows both in arroyo habitats and at seeps, particularly those along the San Andreas fault zone(s) (Vogl and McHargue, 1966). In many places, the groves have been destroyed; yet this species has been introduced and cultivated to the extent that it is an ubiquitous ornamental throughout the Southwest's subtropical and warm-temperate regions.

Blue Palms are endemic to northern Baja California (including Isla de la Guarda); one population on the eastern slope of the Sierra Juarez is within 24 km of the International Boundary (Moran, 1977). This species is also found in arroyo habitats where it is frequently accompanied by California Fan Palms.

Neither *Washingtonia* nor *Erythea armata* occur naturally in Sonora. Groves of the Mexican Blue Palm (*Sabal uresana*) and *Erythea aculeata* within and near Sinaloan thornscrub in that state are perhaps best considered as northern consociations of Sinaloan riparian mixed evergreen forest.

Plant associates are the more mesic and riparian species found within the Sonoran desert. Except for Fremont Cottonwood, the associates are of considerably shorter stature than the palms. Some common residents in fan palm oases in California and Arizona according to Vogl and McHargue (1966) and Brown et al. (1976) are:

Acacia greggii	*Cynodon dactylon*
Ambrosia ambrosioides	*Hymenoclea monogyra*
Atriplex spp.	*Phragmites australis*
Baccharis salicifolia	*Prosopis velutina*
Baccharis sarothroides	*Tamarix chinensis*
Carex ultra	*Tessaria sericea*
Cercidium floridum	*Zizyphus obtusifolia*

California Fan Palms are alkali tolerant and, once established, appear to increase or maintain themselves against potential competitors if water supplies are adequate.

No vertebrate species are known to be exclusively associated with these communities, but the presence of dense foliage and moisture in the arid environments undoubtedly enhances and concentrates local animal inhabitants. Ryan (1968) speculated that the rare occurrence of the Western Yellow Bat (*Dasypterus egaxanthinus*) in the southwestern United States was closely associated with palm oases.

Figure 184. *Sonoran oasis of native California Fan Palms (*Washingtonia filifera*) and feral Date Palm (*Phoenix dactylifera*) at San Ignacio, Baja California del Sur, Mexico. Elevation ca. 250 m.*

Sinaloan Riparian Evergreen Forest and Woodland

These diverse tropic-subtropic streamside communities extend northward into our area to as far north as the lowermost Río Yaqui and its tributaries. There, at their northern extremity, they are mostly below 760 m elevation, but may extend to as high as 1,100 or 1,500 m farther south in the region around Alamos (Gentry, 1942; Felger, 1971). Contained within subtropical deciduous forest and thornscrub, the contrast in winter between these evergreen and semievergreen bottomland communities with adjacent, bare hillsides is striking.

Overstory species include a number of stately, tropicsubtropic trees, occurring alone or as mixtures (Figs. 185, 186). In the Southwest, these include several species of wild figs (*Ficus* spp.), a cottonwood (*Populus dimorpha*), Goodding Willow, palms (*Sabal uresana, Erythea aculeata, Clethra lanata*), and the stately Montezuma Cypress or Cedros (*Taxodium mucronatum*); overstories may be closed (forest) or open (woodland). At higher elevations several warm temperate trees, such as sycamore, the Evergreen Magnolia (*Magnolia schiediana*) and oaks may join the forest assemblage (Felger, 1971). Lianas (*Arrabidaea litoralis, Marsdenia edulis, Gouania mexicana*) and the climbing *Pisonia capitata* increase diversity. Tropical epiphytes are represented by the orchid *Oncidium cebolleta* and bromeliads (e.g., *Tillandsia inflata, Hechtia* spp.). A characteristic understory shrub is Garabato (*Celtis iguanaea*). Other understory species according to Gentry (1942) and Felger (1971) are:

Bacopa monnieri	*Prosopis juliflora*
Begonia spp.	*Rotala ramosior*
Eustoma exaltatum	*Samolus ebracteatus*
Fuirena simplex	*Sartwellia mexicana*
Gratiola breviofolia	*Sassafridium macrophyllum*
Guazuma ulmifolia	*Selaginella* spp.
Heteranthera limosa	*Sesbania sesban*
Oreopanax salvinii	*Stanhopea* spp.
Pithecellobium spp.	*Vallesia glabra*
	Vitex mollis

These communities host a large and varied animal community. This fauna has been poorly investigated but includes such large and spectacular species as the Jaguar and the smaller Ocelot (*Felis pardalis*). Heteromyid rodents (*Dipodomys* spp. and *Perognathus* spp.) are common in the region but not particularly characteristic of riparian zones. Cotton rats (*Sigmodon hispidis, S. minimus*) tend to be more abundant near streams and other wetlands, and troops of Coati (*Nasua nasua*) forage extensively within the riparian corridors (Burt, 1938).

Birds include the spectacular Military Macaw (*Ara militaris*), Black-bellied Tree Duck (*Dendrocygna autumnalis*), and Magpie Jay (*Calocitta formosa*). Other colorful species are the Lilac-crowned Parrot (*Amazonia finschi*), Blue-rumped Parrotlet (*Forpus cyanopygius*), Green Parakeet (*Aratinga holochlora*), Coppery-Tailed Trogon, Berylline Hummingbird (*Amazilia beryllina*), Black-chinned Hummingbird (*Archilochus alexandri*), the unique Russet-crowned Motmot (*Momotus mexicanus*), and a host of songbirds and other species more typical of riparian communities to the north (e.g., Gila Woodpecker, *Melanerpes uropygialis*; Van Rossem, 1945).

Reptiles and amphibians of these communities are rich and diversified—73% of the overall fauna of 74 species recorded

Figure 185. *Sinaloan riparian evergreen woodland of Montezuma Cypress* (Taxodium mucronatum) *on the Rio Cuchajaqui, a tributary of the Rio Fuerte, near Alamos, Sonora, Mexico. Elevation ca. 760 m.*

from near Alamos, Sonora, had northern affinities and the remainder was from the tropics (25%) or endemic to Mexico (2%). Unfortunately, Heringhi (1969), who detailed the herpetofauna, did not provide critical information on habitat relations, but amphibians associated with watercourses certainly include Colorado River Toad, Giant Toad (*Bufo marinus*), a tree-frog (*Pachymedusa dacnicolor*), and Leopard Frog. Turtles, too, are far more common in this area than elsewhere in the Southwest. Mud turtles (*Kinosternon hirtipes* and *K. integrum, K. alamosae*), although semi-aquatic, are frequent in streams and arroyos, and larger waters support the pond turtles *Chrysemys picta, Pseudemys scripta mayae,* and *Rhinoclemys pulcherrima*. Snakes directly associated with the water are Boa Constricor (*Constrictor constrictor*), Watersnake (*Natrix valida*), and Indigo Snake (*Drymarchon corias*); others are associated with streamside trees (Vine Snake, *Oxybelis aeneus*), with fine soils on stream terraces (Blackhead Snake, *Tantilla planiceps yaquiae*), or are attracted to mesic habitats.

Creeks and rivers of the southern Río Yaqui drainage at lower elevations, and the Río Mayo to the south, support tropical fishes capable of dipersing through brackish water, e.g., Pacific Shad (*Dorosoma smithi*), topminnow (*Poeciliopsis prolifica*), and the all-female *P. monacha-occidentalis* that depends on sperm of bisexual species of *Poeciliopsis* for its unique forms of reproduction (gynogenesis and "hybridogenesis"; Schultz, 1977), and, where pools are present, Sinaloan Cichlid (*Cichlasoma beani*). Forms such as Mexican Stoneroller, Longfin Dace, and in the Río Yaqui proper, Beautiful Shiner, Yaqui Sucker, Yaqui Catfish (*Ictalurus pricei*), and Roundtail Chub, all enter this tropical zone, which approaches to southern limits of natural range for most of these northern genera on the west coast of Mexico (Meek, 1904; Miller, 1959; Stuart, 1964). Marine fishes penetrate these lowland rivers far above tidal influence. A goby (*Awaous transandeanus*) has been taken near Movas and almost to Presa Novillo on the Río Yaqui, and Mountain

Figure 186. *Rio Chico near the town of Rio Chico, Sonora, Mexico. Cottonwood (Populus monticola), willow (Salix spp.), wild fig (Ficus spp.) and thornscrub (Acacia spp.) as a mixed riparian community along the boulder-bottomed stream. Elevation ca. 200 m; photograph by Dean A. Hendrickson, June 1978.*

Mullet (*Agonostomus monticola*) also has such capability (Hendrickson et al., 1981). Introduced species such as Largemouth Bass, White Crappie (*Pomoxis annularis*), Bluegill (*Lepomis macrochirus*) and Redear Sunfish (*L. microlophus*) are generally restricted to reservoirs, but the stream-adapted Carp (*Cyprinus carpio*), River Carpsucker (*Carpiodes carpio*), and Green Sunfish (*Lepomis cyanellus*) have invaded and become established in local creeks and rivers.

Sonoran Riparian Scrubland

In and along drainages within the Sonoran Desert are scrublands of low to medium height (1.5 to 3.0 m), too dense to be considered desertscrub or strand. Although these scrublands usually contain plant species also found in adjacent desertscrub (e.g., *Lycium brevipes, Acacia greggii, Celtis pallida,* and especially the highly facultative mesquite), the actual stream channel dominants are usually distinctive riparian species. Seepwillow (*Baccharis salicifolia*) is abundant nearest water, with Desert Broom (*B. sarothroides*) in drier places and Mule Fat (*B. viminea*) in desert washes. Arrow-weeds (*Tessaria sericea, Pluchea camphorata,* and *P. purpurascens*) and Burro-brush may dominate on sandy soils (Fig. 187). These and other evergreen shrubs have adapted to successional situations as befits their restricted occurrence to flood-prone areas. The deciduous Desert-willow (*Chilopsis linearis*) is a common arboreal component, as is the increasingly prevalent, deciduous Saltcedar.

Along the saline portions of the lower Colorado and Gila rivers and in the Salton Sea basin, are dense and taller (to 11 m or higher) "thickets" of introduced Saltcedar and the evergreen Athel (*Tamarix aphylla*). In the less disturbed sites, these may be accompanied by native Screwbean Mesquite, Lenscale, or Quailbush (*Atriplex lentiformis*), Arrow-weed, Western Honey Mesquite (*P. glandulosa* var. *torreyana*), and such purely salt-shrub species as *Suaeda torreyana, Atriplex polycarpa* and *Allenrolfea occidentalis* (Fig. 188). These communities are highly flammable because of deciduous and other properties of Saltcedar, and are now typically in a fire-succession stage. Each fire (or clearing) increases the prevalence of the root sprouting Saltcedar at the expense of more valuable native vegetation. Consequently, fire disclimax consociations of Saltcedar now exclusively occupy extensive areas along the lower Colorado River, its delta, tributaries, distributaries (e.g., Rio Hardy, Alamo, and New rivers), agricultural drains and sumps, and other poorly-drained, alkaline places (Fig. 189).

The value of these thickets to game species is well known, and such places often support a high density of Desert Cottontail (*Sylvilagus auduboni*) and Gambel's Quail (*Lophortyx gambeli*), and if of sufficient height (3+ m), nesting Mourning and White-winged Doves. Other birds well represented in Sonoran riparian scrub are the Crissal Thrasher, Abert's Towhee, Brown Towhee (*Pipilo fuscus*), Say's Phoebe (*Sayornis saya*), and Black-tailed Gnatcatcher (Anderson and Ohmart, 1977; Anderson et al., 1977). If standing water is present, such scrublands may also often be inhabited by the Yuma Clapper Rail (*Rallus longirostris yumaensis*).

Figure 187. *Sonoran riparian scrubland, strand, and woodland on the Verde River, Tonto National Forest, Maricopa County, Arizona. Scrubland composed largely of Burrobush* (Hymenoclea monogyra), *Arrowweed* (Tessaria sericea) *and Velvet Mesquite* (Prosopis velutina) *in scrub form. Elevation ca. 450 m.*

Figure 188. *Sonoran "lower Colorado River" scrub of Arrowweed* (Tessaria sericea, Pluchea camphorata), *Saltcedar* (Tamarix chinensis), *Western Honey Mesquite* (Prosopis glandulosa var. torreyana), *Screwbean* (Prosopis pubescens), *Lenscale or Quailbush* (Atriplex lentiformis) *and Seepweed* (Suaeda torreyana), *with an occasional willow* (Salix exigua, S. gooddingii), *near Cibola, Yuma County, Arizona, along the Colorado River. Elevation ca. 120 m.*

Figure 189. *Sonoran "lower Colorado River" riparian scrub in the Colorado River delta below Riito, Sonora, Mexico. A disclimax community dominated by Saltcedar* (Tamarix chinensis). *The remaining few mesquites* (Prosopis glandulosa var. torreyana, Prosopis pubescens) *will eventually be displaced with increasing incidence of fire brought on by the flammable properties of Saltcedar. Elevation ca. 5 m.*

Sinaloan Maritime Scrubland

Mangrove swamps of considerable extent occur in the Boca del Yaqui region of the Sonoran Coast, i.e., from due south of Ciudad Obregón to west of Potám. Other small, discontinuous areas are in protected bays, lagoons, and estuaries on the coast northward to near Punta Sargento (29° 18'N) and on the Gulf Coast of Baja California northward to just south of Bahía de Los Angeles (29° 05'N) (Felger and Lowe, 1976). Mangroves are rare or absent north of 27°00'N on the Pacific side of Baja California (Hastings et al., 1972).

Species present may be all or any of Black Mangrove (*Avicennia germinans*), Red Mangrove (*Rhizophora mangle*), and White Mangrove (*Languncularia racemosa*). The lack of other plant associates is often conspicuous, and delineation of these tideland communities is abrupt, with little integration with adjacent strand, tidemarsh, and/or desertscrub (saltbush) communities (Fig. 190, 191).

As many of their names imply, several bird species are closely associated with Sinaloan maritime scrub—the Mangrove Cuckoo (*Coccyzus minor*), Mangrove Swallow (*Iridoprocne albilinea*), Mangrove Warbler (*Dendroica erithachorides*), and others such as Tiger Bittern (*Tigrisoma mexicanum*), Wood Stork (*Mycteria americana*), Anhinga (*Anhinga anhinga*), and Roseate Spoonbill (*Ajaia ajaia*). Two particularly abundant nesting birds in mangrove swamps are White-winged Dove and Clapper Rail.

The value of these wetlands as nurseries and feeding grounds to a host of marine life is well known. Juveniles and young adults of commercial and sport fishes are common, e.g., snappers (*Lutjanus* spp.). The Giant Jewfish (*Epinephelus itajaro*), which may achieve weights of 450 kg in open waters of the Gulf, occurs as young only in mangrove-vegetated esteros (Thomson, 1973).

Figure 190. *Sinaloan maritime scrubland and "barrier" strand bordering the Sonoran Desert, Sonora, Mexico. A northern outlier of the more extensive mangrove swamps farther south at Punta Sargento, this scrubland is composed mainly of Black Mangrove (**Avicennia germinans**). Photograph by Richard L. Todd.*

Figure 191. *Interior view of Sinaloan maritime strand and scrubland near Punta Sargento, Sonora, Mexico. Note the "chaparral-like" landscape physiognomy and simplicity in overstory species composition (**Avicennia germinans**).*

Sonoran and Sinaloan Interior Marshlands and Submergent Communities

Emergent vegetation varies from pure stands (=consociations) of such short statured and alkali resistant species as Saltgrass, Alkali Bulrush (*Scirpus maritimus* var. *paludosus*), and Three-square (*Scirpus americanus*), to dense, impenetrable communities of reed (*Phragmites australis*) and Giant Bulrush (*Scirpus californicus*), locally called "tules." Often, however, the most prevalent and characteristic species is the cattail, principally represented in these parts of the Southwest by *Typha domingensis*. At the edge of the marsh there is typically much intermingling with adjacent scrublands of Saltcedar, Arrow-weed, Quailbush (*Atriplex lentiformis*), and mesquite (Fig. 192). In more seasonally flooded areas the communities are often mosaics of the shorter marsh species (i.e., *Juncus* spp., *Eleocharis* spp., *Cynodon dactylon*, *Distichlis spicata*) and taller scrub (e.g., Saltcedar) depending on slight variations in hydrology or successional stage. As with all marshlands, hydrosoil mud bottoms are characteristic.

Emergent aquatic vegetation along the channel of the now stabilized Colorado River mainstream includes larger sedges such as Giant Bulrush and Three-square (Minckley, 1979). These plants form thick stands which rise as high as 3 m above the surface, creating a broad, 1-5 m zone of quiet, shaded water to 1.5 m deep. Cattail also forms beds on sloping, stabilized or aggrading banks that extend as far as 15 m from shore in water to a meter deep, especially on the quiet side of bends. When currents contact such beds, dense roots and rhizomes hold as dense mats, and undercuts of more than 2.5 m may occur. Stands of Giant Reed also are present, living as large clumps along less hygric shorelines.

Numerous small, semi-aquatic plant taxa form understories within marsh communities and along the banks. Included here are Pennywort (*Hydrocotyle verticillata*), Water-hyssop, Smartweed (*Polygonum fusiforme*), Spearmint (*Mentha spicata*), and a diversity of sedges and grasses (*Cyperus strigosus*, *C. erythrorhizos*, *Eleocharis parvula*, *E. caribaea*, *Leptochloa uninervia*, and *Paspalum dilatatum*).

Present-day submergent communities of the lower Colorado River channel are obviously new since drastic fluctuations in water levels and scour prior to dams scarcely allowed their development in more than a periodic or rudimentary way. Today there are large, monospecific stands of Sago Pondweed (*Potamogeton pectinatus*) with Water Milfoil (*Myriophyllum spicatum*) and the introduced Parrot-feather (*M. brasiliense*) collectively second in abundance (Minckley, 1979). The pondweed is most common in deeper water (to 4.5 m) and often in current that exceeds a meter per second. Milfoil and Parrot-feather form dense beds in shallower (to 2.5 m) water that moves at less than 0.5 m/second. Charophytes are in eddies or other places with slower currents, but are also interspersed with other taxa in the channel. Shorelines and quiet backwaters support some Hornwort (*Ceratophyllum demersum*), but more commonly these areas are choked with Holly-leafed Naiad (*Najas marina*). Shallow waters are inhabited by stands of Leafy Pondweed (*Potamogeton foliosus*), Common Pondmat (*Zannichellia palustris*), and sparse stands of Water Nymph (*Najas guadalupensis*). Bladderwort (*Utricularia* spp.) and duckweeds (*Lemna* spp.) commonly inhabit quiet backwaters, and often are entangled with other plants in the channel.

Although these marshlands and aquatic communities are justifiably considered important wintering grounds for water-

Freshwater marshes are rare in these biogeographic provinces. This is because of their dependence on old oxbows of large rivers as the Colorado, Yaqui, Mayo, etc. Today, many of the larger marshlands occur where rivers enter large reservoirs (e.g., the "delta" of the Bill Williams River in Lake Havasu, Arizona). A very few are associated with natural springs or intersect groundwater tables, the last as at Laguna Prieta (Fig. 151). More common today are brackish and saltwater marshlands dependent for their existence on wastewater discharges, agricultural drains, and silt-laden reservoirs. These include the managed marshes at the edges of the Salton Sea in Imperial County, California, Santa Clara Slough near the Gulf of California, and Picacho Lake in Pinal County, Arizona.

Figure 192. *Topock Marsh, an interior marshland on the Colorado River at the northernmost edge of the Sonoran Desert, Mohave County, Arizona. The cattail is* **Typha domingensis;** *the shrub in foreground is* Atriplex lentiformis. *Elevation ca. 260 m; photograph by Richard L. Todd.*

fowl, they also possess (or possessed) a distinctive nesting avifauna, some distinctly Neotropical. Examples of the latter are Fulvous Whistling Duck (*Dendrocygna bicolor*), Purple Gallinule (*Porphyrula martinica*), Least Grebe (*Podiceps dominicus*), and Snowy-egret (*Egretta thula*). Other species such as Sora, Coot (*Fulica americana*), Black-crowned Night Heron, Least Bittern (*Ixobrychus exilis*), Red-winged Blackbird, and Yellow-throat are widespread species in both temperate and tropical North America. Nesting populations of the Yuma Clapper Rail, a fresh- or brackish-water race of the species, are restricted to spring-wet Sonoran marshlands along the Colorado River, and in more interior locales whenever several hectares of marsh vegetation approaches or exceeds a meter in height.

The Muskrat (*Ondatra zibethicus*) is the common mammalian inhabitant, foraging on vegetation and on the now abundant, introduced Asiatic Clam (*Corbicula fluminea*). The Sonoran Mud Turtle (*Kinosternon sonoriense*) and other mud turtles are still locally common, as are the Colorado River and Giant Toads (*Bufo alvarius, B. marinus*) within their respective ranges. Nonetheless, the former distribution of the Sonoran Mud Turtle once included the lowermost Colorado and Verde

rivers where it now is rare or absent, possibly replaced by the Spiny Softshell Turtle, introduced into the Colorado River at around the turn of the century (Miller, 1946). Another introduction, the Bullfrog (*Rana catesbeiana*), is now widespread throughout many mudbottomed, freshwater habitats within the Sonoran and Sinaloan provinces, where it contributes to the present day rarity of the native Checkered Garter Snake (*Thamnophis marcianus;* see also Moyle, 1973 and Conant, 1978), and at least locally, native frogs.

Populations of small fish species, such as Desert Pupfish and Sonoran Topminnow, now are extirpated from much of their former ranges in this zone because of interactions with introduced fishes and dewatering of streamside marshes. Salton Sea agricultural drains and a few spring-fed marshes still support small numbers of the pupfish, but African cichlids and a myriad of other, tropical fishes (e.g., Sailfin and Mexican Mollies, *Poecilia latipinna* and *P. mexicana*) now threaten even these refugia. These non-native forms also are present along the lower Colorado River and are exerting inexorable pressure; the pupfish remains common only in hypersaline parts of Santa Clara Slough on the Colorado River Delta.

Sonoran Maritime Strand and Submergent Communities

Both the Baja California and Sonora coasts of the Gulf of California possess tide flat, beach dune, beach rubble, and sea cliff habitats. The mud substrates of the quieter waters may be inhabited by tide-influenced communities of Iodine Bush (*Allenrolfea occidentalis*), Seepweed, Sea-lavender (*Limonium californicum*), saltbushes (*Atriplex barclayana,* and others), Batis (*Batis maritima*), alkali-heath (*Frankenia* spp.), Ice Plant, glassworts (*Salicornia* spp.), Saltgrass, and other halophytes (e.g., *Tricerma phyllanthoides*). The plant cover varies from almost nil to open to relatively dense tidalscrub, and is dependent on frequency of inundation and edaphic conditions (Felger and Lowe, 1976). The upper, inland associations commonly grade into desertscrub and there may be much intermingling of the two, both in aspect and composition. Beach dunes are normally thinly vegetated, if at all, by deep-rooted, mat-like, or otherwise short-statured forbs, grasses being poorly represented on these subtropical beaches (Fig. 193). Examples given by Felger and Lowe (1976) of strand vegetation on dunes along the Sonora coast are: *Abronia maritima, A. villosa, Astragalus magdalenae, Dicoria canescens, Euphorbia leucophylla, Monanthochlöe littoralis, Jouvea pilosa, Helianthus niveus,* and *Croton californicus.* As elsewhere, the lateral dunes closer to the sea are more subject to wind and spray and so, possess less vegetative cover. The beaches themselves are essentially free of vegetation.

Even harsher environments for plants are rock rubble and sea cliff shores. The vegetation on these sites is typically extremely sparse and may be composed of nonvascular species, annuals and/or the hardier cliff-dwelling desertscrub and thornscrub perennials found inland. Examples, again from Felger and Lowe (1976), are: *Amaranthus watsonii, Nicotiana trigonophylla, Ficus petiolaris, F. palmeri, Hofmeisteria crassifolia, H. fasciculata, Eucnide rupestris, Pleurocoronis laphamioides,* and the Sweet Mangle (*Maytenus phyllanthoides*).

Rocky coastal environments are sometimes the hauling sites for large herds of California Sea Lions. While the sand "hauling" beaches used by Green Sea Turtle are now rare, sandy beaches and mud flats provide winter feeding habitats for migrating curlew, sandpipers, dowitchers, phalaropes, and other shore birds. Nesting species using these habitats are often restricted to islands (e.g., Isla Raza) and some of the less disturbed mainland beaches. These include the Royal Tern (*Thalasseus maximus*), Elegant Tern (*T. elegans*), Least Tern, Snowy Plover, and Wilson's Plover (*Charadrius wilsonia*). Rocks and sea cliffs, especially on offshore islands, often host such colonial nesting pelagic and shore foraging birds as the Manx Shearwater (*Puffinus puffinis*), Western Gull (*Larus occidentalis*), Heermann's Gull (*L. heermanni*), Laughing Gull (*L. atricilla*), Brown Booby (*Sula leucogaster*), Blue-footed Booby (*S. nebouxii*), Blue-faced Booby (*S. dactylatra*), Least Storm Petrel (*Halocyptena microsoma*), Black Storm Petrel (*Oceanodroma melania*), Brown Pelican (*Pelecanus occidentalis*), Double-crested Cormorant, Black Skimmer (*Rynchops nigra*), and Red-billed Tropicbird (*Phaethon aethereus*). Other more solitary and habitat specific nesting species include the American Oyster Catcher (*Haematopus pallitus*) (rock rubble), and Osprey (*Pandion haliaetus*) (pinnacle or other elevated structure).

Fish communities of the intertidal (littoral) zone of the Gulf of California have recently been discussed by Thomson et al. (1979). In the upper Gulf, above the "midriff" islands of

Figure 193. *Sonoran maritime strand just north of Cruz Piedra, Sonora, Mexico, looking toward Empalme. An open "beach" strand of Sandverbena (*Abronia maritima*), Beach Sunflower (*Helianthus niveus*) and Pickleweed (*Salicotnia spp.*).*

Ángel de la Guarda and Tiburón, communities consist of relatively few species that show great seasonal population fluctuations. About 60 species of fishes occupy this zone near Puerto Peñasco, Sonora. The most abundant kind is the Panamic Sargeant Major (*Abudefduf troscheli*) followed by the Gulf Opaleye (*Girella simplicidens*), two clinid blennies (*Paraclinus sini, Malacoctenus gigas*), and the Sonoran Goby (*Gobisoma chiquita*). The strand inhabiting clingfishes (e.g., *Tomicodon humeralis, Gobiesox pinninger,* etc.) are also common, rarely occurring below mid-tidal zones, and daily exist for a period of hours above the level of the sea. A major piscivore is the Spotted Sand Bass (*Paralabrax maculatosfasciatus*). Over the long term, warm-temperate species dominate the community in numbers and biomass—Gulf Opaleye, Spotted Sand Bass, Rock Wrasse (*Halichoeres semicinctus*), Sargo (*Anisotremus davidsoni*), and Bay Blenny (*Hypsoblennius gentilis*). These are cold-tolerant species able to survive occasional low sea temperatures that decimate several tropical species in this region, especially Panamic Sargeant Major (Thomson and Lehner, 1976).

The central Gulf has about twice as many species as the upper portion and they are far more colorful than the drab, cryptic fishes characterizing upper-gulf shorelines. Warm-temperate species so common nearer the Colorado Delta are absent or uncommon here. The Panamic Sargeant Major remains abundant, but the Cortez Damselfish (*Eupomacentrus rectifraenum*) becomes one of the more conspicuous forms. A dominant piscivore is the Leopard Cabrilla (*Mycteroperca rosacea*) (Hobson, 1968). Angelfishes (*Pomacanthus zonipectus* and *Holacanthus passer*) and butterfly-fishes (*Chaetodon humeralis* and *Heniochus nigrirostris*) are frequent, as are several species of wrasses (*Halichoeres nicholsi, H. dispilus, Bodianus diplotaenia, Thalassoma lucansanum*). Larger prey items are taken by Baja Grouper (*Mycteroperca jordani*), while

Spotted Sand Bass is replaced by Flag Cabrilla (*Epinephelus labriformes*) as a major predator on smaller species. Moray eels (*Gymnothorax castaneus* and *Muraena lentiginosa*) become common, occupying the predatory niche at night. Bumphead Parrotfish (*Scarus perrico*) and Yellowtail Surgeonfish (*Prionurus punctatus*) are common herbivores in the system, and schools of grunts (*Microlepidotus inornatus* and *Haemulon sexfasciatum*) cruise over reefs and feed on sandy areas nearby. Further south, out of the area, species of Indo-west Pacific origins have become established, and on the peninsula side of the Gulf, the shoreline fauna is equally as rich as any in that region (Thomson et al., 1979).

Numerous other small shore fishes, related to those discussed for the California coast, also occur in profusion in the Sea of Cortez. Soft bottoms are occupied by about half the known fauna of ca. 30 species of gobies. Especially characteristic are the estuarine Guaymas Goby (*Quietula guaymasiae*) and Longtail Goby (*Gobionellus sagittula*). Mullets (*Mugil cephalus, M. curema*) feed on detrital materials associated with muds, and in shallows fall prey to nets of local fishermen. Mojarras (*Gerres cinereus, Diapterus peruvianus, Eucinostomus* spp.) also are abundant. The Longjaw Mudsucker, shared by both Gulf and California coasts, can use atmospheric oxygen when stranded by low tides (Todd and Ebeling, 1966). And, an important counterpart of the Californian Grunion is the strand-spawning Gulf Grunion (*Leuresthes sardina*), the young of which migrate to soft- or sand-bottomed areas for food and shelter. Muds grade into sandy substrates, where Stingrays (*Urolophus halleri*), mojarras, mullets, Grunts (*Pomadasys branicki*), various croakers (*Cynoscion parvipinnis, Bairdiella icistius*), Bonefish (*Albula vulpes*), and some flounders (*Etropus crossotus, Achirus mazatlanus, Symphurus melanorum,* and others), hold forth (Thomson, 1973, Thomson et al., 1979).

Sonoran Interior Strands

Stream channels and other interior strands of tropic Sinaloan and sub-tropic Sonoran zones are typically occupied by open stands of scrub (e.g., *Baccharis salicifolia*), shrubs (e.g., Tree Tobacco, *Nicotiana glauca*), and weeds (Careless Weed (*Amaranthus palmeri*), Thorn Apple (*Datura* spp.), nightshade (*Solanum* spp.), sunflower (*Helianthus* spp.), and dock (*Rumex* spp.)). Wetter sites have a correspondingly greater herbaceous cover and may present a dense stand of annuals, particularly Cocklebur (*Xanthium strumarium*), Rabbit's Foot Grass (*Polypogon monspeliensis*), and diverse composites. Other less-watered basins and channels, or those subject to frequent scours, may be populated only by algae or only very early successional species. As is the case with strands everywhere, the substrate may be of mud, sand, rock, or rubble (Figs. 143, 194, 195).

Plant-animal relationships within these linear and basin communities remain largely unstudied and therefore are poorly known. Smaller desert streams often meander in aggraded, braided channels through sandy beds where change is constant (Figs. 143, 147). Over a period of a year, fluctuations in water levels are pronounced, so that aquatic and semi-aquatic animals may simply survive in periods of drought in greatly reduced, permanent segments, and fulfill their principal biological function of reproduction in winter months of higher flow or after spates produced by summer rains. The concept of strands, therefore, may be applied even to some fishes that have become adapted to such extremes, e.g., Longfin Dace and Sonoran Topminnow. The remarkable Longfin Dace has been recorded to survive partial desiccation beneath mats of algae when evaporation lowered stream levels (Minckley and Barber, 1971), and the livebearing topminnows have a similarly remarkable tenacity of life, persisting in drying pools at high temperatures and in foul conditions. Survival of a single female topminnow may insure population of an area as a single insemination may be used for consecutive broods. Numbers of embryos appear food related so that in an expanded habitat following rainfall, a female may produce many young and employ superfetation to increase her reproductive rate (Schoenherr, 1977). Growth rates of both these "desert-adapted" fishes is rapid, and reproductive individuals can appear in a few weeks.

Other stream animals (e.g., aquatic insects) have recently been demonstrated to have remarkably short life cycles, so that vagaries of the environment are circumvented by aerial life-history stages at all times of year. Death from desiccation in isolated channels or by flash flood is balanced by continuous reproduction and development from egg to adult in as few as 7 days (Gray, 1980). Perhaps, as suggested by Gray, this remarkably high turnover coupled with very high rates of production in desert streams (Busch and Fisher, 1981) aids in explaining high densities of insectivorous birds (and bats) along their courses. Expansion of research on water-land interactions in strands should provide information far out of proportion to their physical size and apparent importance in southwestern arid zones.

Figure 194. *Riparian strand within the Sonoran Desert on the Salt River, Gila County, Arizona. Desiccated algae on rock rubble habitat occupies the more frequently inundated channel of the stream; sand substratum of the periodically flooded plain supports an open population of Saltcedar (**Tamarix chinensis**). Elevation ca. 750 m.*

Figure 195. *The almost unvegetated strand of Laguna Salada, Baja California del Norte, Mexico. Inundated infrequently, these playa habitats were nonetheless wetlands and should be considered as such. This "dry lake" is now being filled with return water from agricultural drains through the Rio Hardy—in effect a managed repetition of the Salton Sea experience in the United States. Elevation ca. −3 m.*

References

Aitchison, W.W. and M.E. Theroux. 1974. *A Biotic Inventory of Chevelon Canyon. Coconino and Navajo Counties. Arizona.* 69 pp. Museum of Northern Arizona. Flagstaff.

Aldon, E.F. and H.W. Springfield. 1973. *The Southwestern Pinyon-Juniper Ecosystem: A bibliography.* USDA Forest Service General Technical Report RM-4. 20 pp. Rocky Mountain Forest and Range Experiment Station. Fort Collins, Colo.

Aldous, A.E. and H.L. Shantz. 1924. Types of vegetation in the semiarid portion of the United States and their economic significance. *Journal of Agricultural Research* 28(2):99-128.

Aldrich, J.W. and A.J. Duvall. 1955. *Distribution of American Gallinaceous Game Birds.* Fish and Wildlife Service Circular 34:1-23. USDI Fish and Wildlife Service. Washington, D.C.

Alexander, R.R. 1974. *Silviculture of Subalpine Forests in the Central and Southern Rocky Mountains. The Status of Our Knowledge.* USDA Forest Service Research Paper RM-121. 88 pp. Rocky Mountain Forest and Range Experiment Station. Fort Collins, Colo.

Allee, W.C., A.E. Emerson, O. Park, T. Park and K.P. Schmidt. 1949. *Principles of Animal Ecology.* W.B. Saunders Co. Philadelphia.

Alley, D.W., D.H. Dettman, H.W. Li, and P.B. Moyle. 1977. Habitats of native fishes in the Sacramento River Basin. pp. 87-94. In: Sands, A. (ed.), *Riparian Forests in California.* Institute of Ecology Publication 15. University of California, Davis.

Allred, Donald M., D. Elden Beck, and Clive D. Jorgensen. 1963. Biotic communities of the Nevada test site. *Brigham Young University Science Bulletin* 2(2):1-52.

Alvarez, Ticul. 1963. The recent mammals of Tamaulipas, Mexico. *University of Kansas Publications. Museum of Natural History* 14(15):363-473.

American Ornithologists' Union. 1973. Thirty-second supplement to the AOU checklist of North American birds. *Auk* 90(2):411-419.

American Ornithologists' Union. [1957] 1975. *Checklist of North American Birds.* Fifth edition. 691 pp. Port City Press, Inc. Baltimore, Md.

American Ornithologists' Union. 1976. Thirty-third supplement to the AOU checklist of North American birds. *Auk* 93(4):875-879.

Anderson, Bertin W., Alton E. Higgins, and Robert D. Ohmart. 1977. Avian use of saltcedar in the lower Colorado River Valley. pp. 128-136. In: Johnson, R.R. and D.A. Jones (technical coordinators), *Importance. Preservation. and Management of Riparian Habitat: A Symposium.* USDA Forest Service General Technical Report RM-43. 217 pp. Rocky Mountain Forest and Range Experiment Station. Fort Collins, Colo.

Anderson, Bertin W. and Robert D. Ohmart. 1977. Vegetation structure and bird use in the lower Colorado River Valley. pp. 23-24. In: Johnson, R.R. and D.A. Jones (technical coordinators), *Importance. Preservation. and Management of Riparian Habitat: A Symposium.* USDA Forest Service General Technical Report RM-43. 217 pp. Rocky Mountain Forest and Range Experiment Station. Fort Collins, Colo.

Anderson, S. 1972. Mammals of Chihuahua: Taxonomy and distribution. *Bulletin of the American Museum of Natural History* 148:1-410.

Arizona Interagency Technical Committee. 1963. *Vegetative units of the Major Land Resource Areas.* Text and map. University of Arizona, Agricultural Extension Service, Tucson.

Arizona Water Commission and U.S. Department of Agriculture. 1972. *Vegetation. Croplands. Urban and Mining Area—Santa Cruz-San Pedro River Basins. Arizona.* Map M7-EN-23185-10. Phoenix, Ariz.

Armstrong, David M. 1972. *Distribution of Mammals in Colorado.* Monograph of the Museum of Natural History 3. 415 pp. University of Kansas.

Arnold, J.F., D.A. Jameson, and E.H. Reid. 1964. *The Pinyon-Juniper Type of Arizona: Effect of Grazing. Fire. and Tree Control.* USDA Forest Service Production Research Report 84. 28 pp.

Arnold, L.W. 1943. *A Study of the Factors Influencing the Management of and a Suggested Management Plan for the Western White-winged Dove in Arizona.* 103 pp. Arizona Game and Fish Commission. Phoenix.

Aschmann, Homer. 1959. *The Central Desert of Baja California: Demography and Ecology.* 315 pp. Ibero-Americana 42. University of California Press. Berkeley.

Austin, M.E. 1965. *Land Resource Regions and Major Land Resource Areas of the United States (Exclusive of Alaska and Hawaii).* USDA Soil Conservation Service. Agricultural Handbook 196. 82 pp. Washington, D.C.

Axelrod, D.I. 1950. Evolution of desert vegetation in western North America. pp. 215-306. In: Axelrod, D.I. (ed.), *Studies in Late Tertiary Paleobotany.* Carnegie Institution of Washington Publication 590.

Axelrod, D.I. 1956. Mio-Pliocene floras from west-central Nevada. *University of California Publications in Geological Science* 33:1-316.

Axelrod, D.I. 1957. Late Tertiary floras and the Sierra Nevadan uplift. *Geological Society of America Bulletin* 68:19-46.

Axelrod, D.I. 1958a. The Pliocene Verdi flora of western Nevada. *University of California Publications in Geological Science* 34:61-160.

Axelrod, D.I. 1958b. Evolution of the Madro-Tertiary geoflora. *Botanical Review* 24:433-509.

Axelrod, D.I. 1966. The early Pleistocene Soboba flora of southern California. *University of California Publications in Geological Science* 60:1-109.

Axelrod, D.I. 1967. Evolution of the California closed-pine forest. pp. 93-150. In: Philbrick, R.N. (ed.), *Proceedings of the Symposium on Biology of the California Islands.* Santa Barbara Botanical Garden. Santa Barbara, Calif.

Axelrod, D.I. 1970. Mesozoic paleogeography and early angiosperm history. *Botanical Review* 36:277-319.

Axelrod, D.I. 1972. Edaphic aridity as a factor in angiosperm evolution. *American Naturalist* 106:311-320.

Axelrod, D.I. 1973. History of the Mediterranean ecosystem in California. pp. 225-277. In: DiCastri, F. and H.A. Mooney (eds.), *Mediterranean Type Ecosystem: Origin and Structure.* Springer-Verlag.

Axelrod, D.I. 1975. Evolution and biogeography of Madrean-Tethyan sclerophyll vegetation. *Missouri Botanical Garden Annals* 62:280-334.

Axelrod, D.I. 1976. History of the coniferous forests, California and Nevada. *University of California Publications in Botany* 70:1-62.

Axelrod, D.I. 1979a. Age and origin of Sonoran Desert vegetation. *California Academy of Sciences Occasional Papers* 132:1-74.

Axelrod, D.I. 1979b. Desert vegetation, its age and origin. In: Goodin, J.R. and D.K. Northington (eds.), *Arid Land Plant Resources. Proceedings of the International Arid Lands Conference on Plant Resources.* Texas Tech University. Lubbock.

Bailey, A.M. and A.J. Niedrach. 1965. *The Birds of Colorado.* 895 pp. in 2 vols. Denver Museum of Natural History.

Bailey, F.M. (with W.W. Cooke). 1928. *The Birds of New Mexico.* 807 pp. New Mexico Department of Game and Fish. Santa Fe.

Bailey, H.P. 1966. *Weather of Southern California.* 87 pp. University of California Press. Berkeley and Los Angeles.

Bailey, Robert G. 1976. *Ecoregions of the United States.* USDA Forest Service. Ogden, Utah. Map.

Bailey, Robert G. 1978. *Description of the Ecoregions of the United States.* 77 pp. USDA Forest Service. Ogden, Utah.

Bailey, V. 1905. *Biological Survey of Texas.* U.S. Department of Agriculture, North American Fauna 25:222 pp. Washington, D.C.

Bailey, V. 1913. *Life Zones and Crop Zones of New Mexico.* USDA North American Fauna 35. 100 pp. U.S. Department of Agriculture. Washington, D.C.

Bailey, V. 1931. *Mammals of New Mexico.* North American Fauna 53. 412 pp. U.S. Department of Agriculture. Washington, D.C.

Bailey, V. 1936. *The Mammals and Life Zones of Oregon.* USDA North American Fauna 55. 416 pp. U.S. Department of Agriculture. Washington, D.C.

Baird, S.F. and C. Girard. 1853. Fishes. In: Captain L. Sitgreaves, *Report of an Expedition Down the Zuni and Colorado Rivers.* 32nd Congress, 2nd Session. Executive Report 59. Washington, D.C.

Baker, R.H. 1956. Mammals of Coahuila, Mexico. *University of Kansas Publications. Museum of Natural History* 9(7):125-335.

Bakker, E.S. 1971. *An Island Called California.* 357 pp. University of California Press. Berkeley.

Balda, R.P. 1969. Foliage use by birds of the oak-juniper woodland and ponderosa pine forest in southern Arizona. *Condor* 71:399-412.

Balda, R.P. 1974. *Population Densities. Habitat Selection and Foliage Use by Birds of Selected Ponderosa Pine Forest Areas in the Beaver Creek Watershed.* Program Report to Coconino National Forest. Flagstaff, Arizona. 8 April 1974. 24 pp.

Balda, R.P., G. Weisenberger, and M. Strauss. 1970. White-crowned sparrow (*Zonotrichia leucophrys*) breeding in Arizona. *Auk* 87:809.

Bancroft, G. 1926. The faunal areas of Baja California del Norte. *Condor* 28:209-215.

Barbour, M.G. 1969. Patterns of genetic similarity between *Larrea divaricata* of North and South America. *American Midland Naturalist* 81:54-67.

Barbour, M.G. and A.F. Johnson. 1977. Beach and dune. pp. 223-261. In: Barbour, M.G. and J. Major (eds.), *Terrestrial Vegetation of California.* John Wiley and Sons. New York, N.Y.

Barbour, M.G. and Jack Major (eds.). 1977. *Terrestrial Vegetation of California.* 1002 pp. John Wiley and Sons. New York.

Barlow, G.W. 1958. High salinity mortality of desert pupfish, *Cyprinodon macularius. Copeia* 1958:231-232.

Barnhart, P.S. 1936. *Marine Fishes of Southern California.* 209 pp. University of California Press. Berkeley.

Bauer, H.L. 1930. Vegetation of the Tehachapi Mountains, California. *Ecology* 11:263-280.

Beard, J.S. 1953. Savannah vegetation of northern tropical America. *Ecological Monographs* 23:149-215.

Beatley, Janice C. 1966. Ecological status of introduced brome grasses (*Bromus* spp.) in desert vegetation of southern Nevada. *Ecology* 47:548-554.

Beatley, Janice C. 1971. *Vascular Plants of Ash Meadows, Nevada.* Laboratory of Nuclear Medicine and Radiation Biology. Final Report. 59 pp. USAEC Contract AT (04.1) Gen. 2. (processed) University of California. Los Angeles.

Beatley, Janice C. 1974a. Effects of rainfall and temperature on the distribution and behavior of *Larrea tridentata* (creosotebush) in the Mohave Desert of Nevada. *Ecology* 55(2):245-261.

Beatley, Janice C. 1974b. Phenologic events and their environmental triggers in Mohave Desert ecosystems. *Ecology* 55(4):856-863.

Beatley, Janice C. 1975. Climates and vegetation patterns across the Mohave/ Great Basin Desert transition of southern Nevada. *American Midland Naturalist* 93(1):53-70.

Beatley, Janice C. 1976. *Vascular Plants of the Nevada Test Site and Central-southern Nevada: Ecologic and Geographic Distributions.* National Technical Information Service, U.S. Department of Commerce. Springfield, Va.

Beetle, A.A. 1960. A study of sagebrush: The section Tridentatae of Artemisia. *University of Wyoming Agricultural Experiment Station Bulletin* 368:1-83.

Behle, William H. 1976. Mohave desert avifauna in the Virgin River Valley of Utah, Nevada, and Arizona. *Condor* 78:40-48.

Behle, William H. and M.L. Perry. 1975. *Utah Birds—Guide, Checklist and Occurrence Charts.* 144 pp. Utah Museum of Natural History. University of Utah, Salt Lake City.

Belcher, R.C. 1975. The geomorphic evolution of the Rio Grande. *Baylor Geological Studies Bulletin* 29:1-64.

Benson, Lyman. 1957. *Plant Classification.* 688 pp. D.C. Heath and Co. Boston, Mass.

Benson, Lyman. 1969. *The Cacti of Arizona.* Third edition (revised). 218 pp. University of Arizona Press. Tucson.

Benson, Lyman and R.A. Darrow. 1944. *A Manual of Southwestern Trees and Shrubs.* University of Arizona Biological Sciences Bulletin 6. 411 pp. Tucson.

Benson, Lyman and Robert A. Darrow. 1954. *Trees and Shrubs of the Southwestern Deserts.* Second edition. 437 pp. University of New Mexico Press. Albuquerque.

Billings, W.D. 1949. The shadscale vegetation zone of Nevada and eastern California in relation to climate and soils. *American Midland Naturalist* 42(1):87-109.

Billings, W.D. 1951. Vegetational zonation in the Great Basin of western North America. pp. 101-122. In: *Comptes Rendues due Colloque sur les Bases Ecologiques de la Regeneration de la Vegetation des Zones Arides. UISB.*

Billings, W.D. 1973a. Arctic and alpine vegetations: Similarities, differences, and susceptibility to disturbance. *Bioscience* 23:697-704.

Billings, W.D. 1973b. Tundra grasslands, herblands and shrublands and the role of herbivores. In: Kesel, R.H. (ed.), *Grassland Ecology,* Louisiana State University Press. Baton Rouge.

Billings, W.D. 1978. Alpine phytogeography across the Great Basin. *Intermountain Biogeography: A symposium. Great Basin Naturalist Memoirs* 2:105-117.

Billings, W.D. and H.A. Mooney. 1968. The ecology of arctic and alpine plants. *Biological Review* 43:481-529.

Birkenstein, Lillian R. and Roy E. Tomlinson. 1981. *Native Names of Mexican Birds.* USDI Fish and Wildlife Service Resource Publication 139. 159 pp. U.S. Department of the Interior. Washington, D.C.

Biswell, H.H. 1956. Ecology of California grasslands. *Journal of Range Management* 9:19-24.

Blair, W.F. 1940. *A Contribution to the Ecology and Faunal Relationships of the Mammals of the Davis Mountain Region, Southwestern Texas.* University of Michigan Museum of Zoology Miscellaneous Publication 46. 39 pp.

Blair, W.F. 1950. The biotic provinces of Texas. *Texas Journal of Science* 2:93-117.

Blair, W.F. 1952. Mammals of the Tamaulipan Biotic Province in Texas. *Texas Journal of Science* 4:230-250.

Blair, W.F. and T.H. Hubbell. 1938. The biotic districts of Oklahoma. *American Midland Naturalist* 20:425-454.

Bliss, L.C. 1973. Tundra grasslands, herblands, and shrublands and the role of herbivores. In: Kesel, R.H. (ed.), *Grassland Ecology.* Louisiana State University Press. Baton Rouge.

Bloss, D.A. and J.D. Brotherson. 1979. Vegetation response to a moisture gradient on an ephemeral stream in central Arizona. *Great Basin Naturalist* 39:161-176.

Boerr, W.J. and D.J. Schmidly. 1977. Terrestrial mammals of the riparian corridor in Big Bend National Park. pp. 212-218. In: Johnson, R.R. and D.A. Jones (technical coordinators), *Importance, Preservation and Management of Riparian Habitat: A Symposium.* USDA Forest Service General Technical Report RM-43. Rocky Mountain Forest and Range Experiment Station. Fort Collins, Colo.

Booth, E.S. 1968. *Mammals of Southern California.* 99 pp. University of California Press. Berkeley.

Bowns, James E. and Neil E. West. 1976. *Blackbrush (Coleogyne ramosissima Torr.) on Southwestern Utah Rangelands.* 30 pp. Utah Agricultural Experiment Station Research Report 27.

Bradley, W.G. 1964. The vegetation of the desert game range with special reference to the desert bighorn. *Transactions of the Desert Bighorn Council* 8:43-67.

Bradley, W.G. and J.E. Deacon. 1965. *The Biotic Communities of Southern Nevada.* 74 pp. University of Nevada, Desert Research Institute. Preprint 9.

Bradley, W.G. 1965. A study of the blackbrush plant community on the Desert Game Range. *Transactions of the Desert Bighorn Council* 1965:56-61.

Bradley, W.G. 1970. The vegetation of Saratoga Springs, Death Valley National Monument, California. *Southwestern Naturalist* 15:111-129.

Bradley, W.G. and J.E. Deacon. 1971. The ecology of small mammals at Saratoga Springs, Death Valley National Monument, California. *Journal of the Arizona Academy of Science* 6:206-215.

Bradley, W.G. and Roger A. Mauer. 1973. Rodents of a creosote bush community in southern Nevada. *Southwestern Naturalist* 17:333-344.

Brand, D.D. 1936. Notes to accompany a vegetation map of northwestern Mexico. *University of New Mexico Bulletin, Biological Series* 4:5-27.

Brand, D.D. 1937. The natural landscape of northwestern Chihuahua. *University of New Mexico Bulletin* 280:1-27.

Brandt, H. 1951. *Arizona and Its Bird Life.* 723 pp. Bird Research Foundation. Cleveland, Ohio.

Branson, B.A., C.J. McCoy, Jr., and M.E. Sisk. 1960. Notes on the freshwater fishes of Sonora with an addition to the known fauna. *Copeia* 3:217-220.

Branson, F.A., R.F. Miller, and I.S. McQueen. 1967. Geographic distribution and factors affecting the distribution of salt desert shrubs in the United States. *Journal of Range Management* 29(5):287-296.

Branson, F.A., Reuben F. Miller, and I.S. McQueen. 1976. Moisture relationships in twelve northern desert communities near Grand Junction, Colo. *Ecology* 57(6):1104-1124.

Braun, E. Lucy. 1950. *The Deciduous Forests of Eastern North America.* 596 pp. Blakiston Co., Inc. Philadelphia, Penna.

Braun, E. Lucy. 1967. *The Deciduous Forests of Eastern North America.* 596 pp. Hafner Publishing Company, New York.

Braun-Blanquet, J. 1932. *Plant Sociology: The Study of Plant Communities.* (Transl. G.D. Fuller and H.S. Conrad). McGraw-Hill Book Co., Inc. New York, N.Y.

Braun-Blanquet, J. 1964. *Pflanzensoziologie.* Third edition. Springer-Verlag. Berlin.

Britton, N.L. 1889. A list of plants collected at Fort Verde and vicinity in the Mogollon and San Francisco mountains, Arizona 1884-1888, by Dr. E.A. Mearns, U.S.A. *Transactions of the New York Academy of Sciences* 8:61-76.

Brown, David E. 1973a. *The Natural Vegetative Communities of Arizona.* State of Arizona, Arizona Resources Information System (ARIS). Phoenix. Map (scale 1:500,000).

Brown, David E. 1973b. Western range extensions of scaled quail, montezuma quail and coppery-tailed trogon in Arizona. *Western Birds* 4:59-60.

Brown, David E. 1977. White-winged doves (*Zenaida asiatica*). pp. 246-272. In: Sanderson, G.C. (ed.), *Management of Migratory Shore and Upland Game Birds in North America.* International Association of Fish and Wildlife Agencies. Washington, D.C.

Brown, David E. 1978. The vegetation and occurence of chaparral and woodland flora on isolated mountains within the Sonoran and Mohave deserts in Arizona. *Journal of the Arizona-Nevada Academy of Science* 13:7-12.

Brown, David E. 1980. A system for classifying cultivated and cultured lands within a systematic classification of natural ecosystems. *Journal of the Arizona-Nevada Academy of Science* 15:48-53.

Brown, David E., Neil B. Carmony, Charles H. Lowe, and Raymond M. Turner. 1976. A second locality for native California fan palms (*Washingtonia filifera*) in Arizona. *Journal of the Arizona Academy of Science* 11(1):37-41.

Brown, David E., Neil B. Carmony, and R.M. Turner. 1977. The inventory of riparian habitats. In: Johnson, R.R. and D.A. Jones (technical coordinators), *Importance, Preservation and Management of Riparian Habitat: A Symposium.* Technical Report RM-43. 217 pp. USDA Forest Service, Rocky Mountain Forest and Range Experiment Station. Fort Collins, Colo.

Brown, David E., Neil B. Carmony, and R.M. Turner. 1979. *Drainage map of Arizona Showing Perennial Streams and Some Important Wetlands.* Second edition. Arizona Game and Fish Department. Phoenix. (map, scale 1:1,000,000).

Brown, David E. and D.H. Ellis. 1977. *Status Summary and Recovery Plan for the Masked Bobwhite.* U.S. Department of Interior, Fish and Wildlife Service, Office of Endangered Species, Region 2. Albuquerque.

Brown, David E. and Charles H. Lowe. 1973. *A Proposed Classification for Natural and Potential Vegetation in the Southwest with Particular Reference to Arizona.* Arizona Game and Fish Department. Federal Aid to Wildlife Project Rep. W-53R-22-WP4-J-1:1-26.

Brown, David E. and Charles H. Lowe. 1974a. A digitized computer-compatible classification for natural and potential vegetation in the Southwest with particular reference to Arizona. *Journal of the Arizona Academy of Science* 9(Suppl. 2):1-11.

Brown, David E. and Charles H. Lowe. 1974b. The Arizona system for natural and potential vegetation—illustrated summary through the fifth digit for the North American Southwest. *Journal of the Arizona Academy of Science* 9(Suppl. 3):1-56.

Brown, David E. and Charles H. Lowe. [1977] 1978. *Biotic Communities of the Southwest.* USDA Forest Service General Technical Report RM-41. 1 p. map. Rocky Mountain Forest and Range Experiment Station. Fort Collins, Colo.

Brown, David E. and Charles H. Lowe. 1980. *Biotic Communities of the Southwest.* USDA Forest Service General Technical Report RM-78. 1 p. map. Rocky Mountain Forest and Range Experiment Station. Fort Collins, Colo.

Brown, David E., Charles H. Lowe, and Charles P. Pase. 1977. A digitized systematic classification for the natural vegetation of North America with a hierarchical summary of world ecosystems. In: *Symposium on Classification, Inventory and Analysis of Fish and Wildlife Habitat* [Phoenix, Ariz., January 24-27.]

Brown, David E., Charles H. Lowe, and Charles P. Pase. 1979. A digitized classification system for the biotic communities of North America, with community (series) and association examples for the Southwest. *Journal of the Arizona-Nevada Academy of Science* 14, (Suppl. 1):1-16.

Brown, David E. Charles H. Lowe, and Charles P. Pase. 1980. *A Digitized Systematic Classification for Ecosystems With an illustrated Summary of the Natural Vegetation of North America.* USDA Forest Service General Technical Report RM-73. 93 pp. Rocky Mountain Forest and Range Experiment Station. Fort Collins, Colo.

Brown, David E. and R.H. Smith. 1976. Predicting hunting success from call-counts of mourning and white-winged doves. *Journal of Wildlife Management* 40(4):743-749.

Brown, David E. and Paul M. Webb. 1979. A preliminary reconnaissance of the habitat of the peninsular pronghorn (*Antilocapra americana peninsularis*). *Journal of the Arizona-Nevada Academy of Science* 14:30-32.

Brown, H. 1884. *Ortyx virginianus* in Arizona. *Forest and Stream* 22(6):104.

Brown, W.S. 1974. Ecology of the aquatic box turtle, *Terrapene coahuila* (Chelonia, Emydidae) in northern Mexico. *Bulletin of the Florida State Museum, Biological Sciences* 19:1-67.

Brune, G. 1975. *Major and Historical Springs of Texas.* 94 pp. Texas Water Development Board Report 189.

Bruner, W.E. 1931. The vegetation of Oklahoma. *Ecological Monographs* 1:99-188.

Bruns, D.A. and W.L. Minckley. 1980. Distribution and abundance of benthic invertebrates in a Sonoran desert stream. *Journal of Arid Environments* 3:117-131.

Bryan K. 1925a. *The Papago Country, Arizona, a Geographic, Geologic, and Hydrologic Reconnaissance with a Guide to Desert Watering Places.* 436 pp. U.S. Geological Survey Water-Supply Paper 449.

Bryan, K. 1925b. Date of channel trenching (arroyo cutting) in the arid Southwest. *Science* (new series) 62:334-344.

Bryan, K. 1928. Change in plant associations by change in ground-water level. *Ecology* 9:474-478.

Buechner, H.K. 1950. Life history, ecology, and range use of the pronghorn antelope in Trans-Pecos Texas. *American Midland Naturalist* 43(2):257-354.

Burcham, L.T. 1957. *California Rangeland. An Historic-ecological Study of the Range Resources of California.* California Department of Natural Resources of Sacramento.

Bureau of Land Management. 1978a. *Draft Environmental Statement–Proposed Livestock Grazing Program Cerbat/Black Mountain Planning Units.* USDI, BLM Arizona State Office. Phoenix.

Bureau of Land Management. 1978b. *Upper Gila-San Simon Grazing Environmental Statement Draft.* USDI, BLM Arizona State Office. Phoenix.

Burgess, R.L. 1977. The ecological society of America: Historical data and some preliminary analyses. In: Egerton, F.N. III (ed.), *History of American Ecology,* Contribution 184, Eastern Deciduous Forest Biome U.S./I.B.P. Environmental Sciences Division, Oak Ridge National Laboratory. Publication 1037. 24 pp. Arno Press. New York.

Burgess, T.L. 1973. Mammals of the Canyon de Chelly region, Apache County, Arizona. *Journal of the Arizona Academy of Science* 8:21-25.

Burgess, T.L. 1980. *Climate of the Guadalupe Mountains.* Unpublished.

Burk, Jack H. 1977. Sonoran desert. In: Barbour, M.G. and Jack Major (eds.), *Terrestrial Vegetation of California.* 1002 pp. John Wiley and Sons. New York.

Burkham, D.E. 1972. *Channel Changes of the Gila River in Safford Valley, Arizona.* 24 pp. U.S. Geological Survey Professional Paper 655-G.

Burt, W.H. 1932. Description of heretofore unknown mammals from islands in the Gulf of California, Mexico. *Transactions of the San Diego Society of Natural History* 7:161-182.

Burt, W.H. 1938. *Faunal Relationships and Geographic Distribution of Mammals in Sonora, Mexico.* 77 pp. University of Michigan Museum of Zoology Miscellaneous Publication 39. University of Michigan, Ann Arbor.

Busch, D.E. and S.E. Fisher. 1981. Metabolism of a desert stream. *Freshwater Biology* 11:301-307.

Butler, B. Robert. 1976. The evolution of the modern sagebrush-grass steppe biome of the eastern Snake River Plain. pp. 4-39. In: Elston, Robert (ed.), *Holocene Environmental Change in the Great Basin.* Nevada Archeological Survey, Research Paper 6.

Cable, D.R. 1975. Influence of precipitation on perennial grass production in the semidesert Southwest. *Ecology* 56(4):981-986.

Cable, D.R. 1975. *Range Management in the Chaparral Type and its Ecological Basis: The Status of our Knowledge.* USDA Forest Service Research Paper RM-155. 30 pp. Rocky Mountain Forest and Range Experiment Station. Fort Collins, Colo.

Cable, D.R. and S.C. Martin. 1975. *Vegetation Responses to Grazing, Rainfall, Site Conditions, and Mesquite Control on Semidesert Range.* USDA Forest Service Research Paper RM-149. 24 pp. Rocky Mountain Forest and Range Experiment Station. Fort Collins, Colo.

Cain, Stanley A. 1944. *Foundations of Plant Geography.* 556 pp. Harper and Brothers. New York.

Campbell, C.J. and W. Green. 1968. Perpetual succession of stream channel vegetation in a semi-arid region. *Journal of the Arizona Academy of Science* 5:86-98.

Carmichael, R.S., O.D. Knipe, C.P. Pase, and W.W. Brady. 1978. *Arizona Chaparral: Plant Associations and Ecology.* USDA Forest Service Research Paper RM-202. 16 pp. Rocky Mountain Forest and Range Experiment Station. Fort Collins, Colo.

Carothers, S.W., J.R. Haldman, and R.P. Balda. 1973. *Breeding Birds of the San Francisco Mountain Area and the White Mountains, Arizona.* 54 pp. Museum of Northern Arizona Technical Series 12. Flagstaff.

Carpelan, L.H. 1961. Physical and chemical characteristics. pp.17-32. In: Walker, B.W. (ed.), *The Ecology of the Salton Sea.* California Department of Fish and Game, Fisheries Bulletin 113.

Carr, J.N. 1977. *Arizona Game and Fish Department Comprehensive Five Year Plan.* Arizona Game and Fish Department. Federal Aid to Wildlife Project FW-11-R-9, 1:1-12.

Carter, Annetta M. 1974. Evidence for the hybrid origin of *Cercidium sonorae* (Leguminosae: Caesalpinoideae) of northwestern Mexico. *Madroño* 22(5):266-272.

Castetter, E.F. 1956. The vegetation of New Mexico. *New Mexico Quarterly.* 26:257-288.

Chabot, B.F. and W.D. Billings. 1972. Origins and ecology of Sierran alpine flora and vegetation. *Ecological Monographs* 42:163-199.

Chambers, John D. 1974. *Vegetation Classifications for Use in Land Use Planning.* USDA Forest Service, Southwestern Region. 8 pp. Mimeo. Albuquerque.

Choate, G.A. 1966. *New Mexico's Forest Resource.* USDA Forest Service Resource Bulletin INT-5.

Christensen, Earl M. 1959. A comparative study of the climates of mountain brush, pinyon-juniper, and sagebrush communities in Utah. *Proceedings of the Utah Academy of Science* 36:174-175.

Clark, T.O. 1979. Avifauna. pp. 121-194. In: Minckley, W.L. (ed.), *Resource Inventory for the Gila River Complex, Eastern Arizona.* Final report to the USDI Bureau of Land Management. Safford, Ariz.

Clements, F.E. 1916. *Plant Succession–an Analysis of the Development of Vegetation.* Carnegie Institution of Washington Publication 242. 512 pp. Washington, D.C.

Clements, F.E. 1920. Plant indicators. *The Relation of Plant Communities to Process and Practice.* Carnegie Institution of Washington Publication 290. 388 pp. Washington, D.C.

Clements, F.E. 1934. The relict method in dynamic ecology. *Journal of Ecology.* 22:39-68.

Clements, F. E. 1936. The origin of the desert climax and climate. pp. 87-140. In: *Essays in Geobotany in Honor of William Albert Setchell.* University of California Press. Berkeley.

Clements, F.E. and V.E.Shelford. 1939. *Bio-ecology.* John Wiley and Sons. New York.

Clokey, I.W. 1951. Flora of the Charleston Mountains, Clark County, Nevada. *University of California Publications in Botany* 24:1-274.

Clover, E.U. 1937. Vegetational survey of the Lower Rio Grande Valley, Texas. *Madroño* 4:41-66, 77-100.

Cockerell, T.D.A. 1897. Life zones in New Mexico, Part I. *New Mexico Agricultural Experiment Station Bulletin* 24:1-44.

Cockerell, T.D.A. 1898. Life zones of New Mexico, Part II. *New Mexico Agricultural Experiment Station Bulletin* 28:137-179.

Cockrum, E.L. 1960. *The Recent Mammals of Arizona: Their Taxonomy and Distribution.* 276 pp. University of Arizona Press. Tucson.

Cockrum, E.L. 1965. *Mammals of Arizona.* University of Arizona Press, Tucson.

Cockrum, E.L. and W.G. Bradley. 1968. A new subspecies of the meadow vole (*Microtus pennsylvanicus*) from northwestern Chihuahua, Mexico. *American Museum Novitates* 2325:1-7

Cole, G.A. 1963. The American Southwest and Middle America. pp. 393-434. In: Frey, D.G. (ed.), *Limnology in North America.* University of Wisconsin Press. Madison.

Cole, G.A. 1968. Desert limnology. pp. 423-485. In: Brown, W. Jr. (ed.), *Desert Biology.* Volume 1. Academic Press, Inc. New York, N.Y.

Coles, F.H. and J.C. Pederson. 1969. *Utah Big Game Range Inventory–1968.* 164 pp. Utah Department of Natural Resources, Division of Game. F.A. Project W-65-R-16, A-6. Publication 69-2.

Collins, J.C. 1979. Sexually mature larvae of the salamanders *Ambystoma rosaceum* and *A. tigrinum* from Chihuahua, Mexico: Taxonomic and ecological notes. *Journal of Herpetology* 13:351-354.

Collins, Joseph T., James E. Huheey, James L. Knight, and Hobart M. Smith. 1978. *Standard Common and Current Scientific Names for North American Amphibians and Reptiles.* Society for study of amphibians and reptiles, committee on common and scientific names. Herpetological Circular 7.

Comisión Técnico Consultiva Para La Determinación Regional De Los Coeficientes De Agostadero. 1974. *Tipos de Vegetación en el Estado de Sonora.* (mapa). In Coeficientes De Agostadero de la República Mexicana, Estado de Sonora. 133 pp. Secretaría de Agricultura Y Ganadería, México, D.F.

Conant, R. 1963. Semiaquatic snakes of the genus *Thamnophis* from the isolated drainage system of the Rio Nazas and adjacent areas in Mexico. *Copeia* 1963:473-499.

Conant, R. 1969. A review of the water snakes of the genus *Natrix* in Mexico. *Bulletin of the American Museum of Natural History* 142:1-140.

Conant, R. 1978. Semiaquatic reptiles and amphibians of the Chihuahuan Desert and their relationships to drainage patterns of the region. pp. 455-491. In: Wauer, R.H. and D.H. Riskind (eds.), *Transactions of the Symposium on the Biological Resources of the Chihuahuan Desert Region, United States and Mexico. Alpine, Texas, 1974.* USDI National Park Service Proceedings and Transactions Series 3. Washington, D.C.

Cooke, R.V. and R.W. Reeves. 1976. *Arroyos and Environmental Change in the American Southwest.* 213 pp. Oxford University Press. London.

Cooper, W.S. 1922. *The Broadleaf-sclerophyll Vegetation of California-an Ecological Study of the Chaparral and its Related Communities.* 122 pp. Carnegie Institution of Washington. Washington, D.C.

Cope, E.D. 1866. On the Reptilia and Batrachia of the Sonoran Provinces of the Nearctic Region. *Proceedings of the Academy of Natural Sciences of Philadelphia* 18:300-314.

Cope, E.D. and H.C. Yarrow. 1875. Report upon the collections of fishes made in portions of Nevada, Utah, California, Colorado, New Mexico, and Arizona, during the years 1871, 1872, 1873, and 1874. pp. 635-703. In: *Report of the Geographic and Geologic Exploratory Survey West of the 100th Meridian (Wheeler Survey) Volume 5.*

Corle, E. 1951. *The Gila.* 402 pp. University of Nebraska Press. Lincoln.

Correll, Donovan S. and Helen B. Correll. 1972, 1975. *Aquatic and Wetland Plants of the Southwestern United States.* Two volumes. 1,777 pp. Stanford University Press. Stanford, Calif.

Correll, Donovan S. and Marshall Conring Johnston. 1970. *Manual of the Vascular Plants of Texas,* 1881 pp. Texas Research Foundation. Renner, Tex.

Costello, D.F. 1954. In: H.D. Harrington. *Manual of the Plants of Colorado.* 666 pp. Sage Books. Denver, Colo.

Cottle, H.J. 1931. Studies on the vegetation of southwestern Texas. *Ecology* 12:105-155.

Cottle, H.J. 1932. Vegetation on north and south slopes of mountains in southwestern Texas. *Ecology* 13:121-134.

Coues, E. 1866. List of the birds of Fort Whipple, Arizona. *Proceedings of the Academy of Science of Philadelphia* 18:39-100.

Coues, E. 1867. The quadrupeds of Arizona. *American Naturalist* 1:281-292, 351-363, 393-400, 531-541.

Cowardin, L.M., V. Carter, F.C. Golet, and E.T. La Roe. 1977. *Classification of Wetlands and Deep-water Habitats of the United States* (an operational draft). 100 pp. U.S. Department of the Interior, Fish and Wildlife Service.

Cox, B.C., I.N. Healy, and P.D. Moore. 1976. *Biogeography, an Ecological and Evolutionary Approach.* 194 pp. Second edition. Blackwell Science Publications. Oxford-London-Edinburgh-Melbourne.

Coyle, J. and N.C. Roberts. 1975. *A Field Guide to the Common and Interesting Plants of Baja California.* Natural History Publishing Co. La Jolla, Calif.

Critchfield, William B. and Elbert L. Little, Jr. 1966. *Geographic Distribution of the Pines of the World.* 97 pp. USDA Forest Service Miscellaneous Publication 991. Washington, D.C.

Cronemiller, F.P. 1942. Chaparral. *Madroño* 6:199.

Cronquist, Arthur, Arthur H. Holmgren, Noel H. Holmgren, and James L. Reveal. 1972. *Intermountain Flora.* Volume I. New York Botanical Garden, New York, N.Y.

Cross, F.B. 1967. *Handbook of Fishes of Kansas.* 357 pp. University of Kansas Museum of Natural History Miscellaneous Publication 45.

Crosswhite, F.S. 1979. "J.G. Lemmon and Wife," plant explorers in Arizona, California, and Nevada. *Desert Plants* 1:12-21.

Dansereau, P. 1957. *Biogeography.* 394 pp. Ronald Press. New York.

Darlington, P.J., Jr. 1957. *Zoogeography.* 675 pp. John Wiley and Sons, Inc. New York.

Darrow, R.A. 1944. Arizona range resources and their utilization-I. Cochise County. *University of Arizona Agricultural Experiment Station Bulletin* 103:311-366. Tucson.

Dasmann, R.F. 1972. Towards a system for classifying natural regions of the world and their representation by National Parks and reserves. *Biol. Conserv.* 4:247-255.

Dasmann, R.F. 1974. See I.U.C.N., 1974.

Dasmann, R.F. 1976. Biogeographical provinces. *Co-Evolution Quarterly.* 1976 (Fall):32-35.

Daubenmire, R.F. 1952. Forest vegetation of northern Idaho and adjacent Washington and its bearing on concepts of vegetation classification. *Ecological Monographs* 22:303-330.

Daubenmire, R.F. 1954. Alpine timberlines in the Americas and their interpretation. *Butler University Botanical Studies* 11:119-136.

Daubenmire, R.F. 1968. *Plant communities.* Harper and Row. New York.

Daubenmire, R.F. 1969. Ecological plant geography of the Pacific Northwest. *Madroño* 20:11-128.

Daubenmire, R.F. 1970. *Steppe Vegetation of Washington.* 131 pp. Washington State University, Agricultural Experiment Station Technical Bulletin 62.

Daubenmire, R.F. and J. Daubenmire. 1968. *Forest Vegetation of Eastern Washington and Northern Idaho.* 104 pp. Washington Agriculture Experiment Station Technical Bulletin 60.

Dawson, E.Y. 1966. *Seashore Plants of Southern California.* 101 pp. University of California Press. Berkeley.

Davis, E.A. and Charles P. Pase. 1977. Root systems of shrub live oak: Implications for water yield in Arizona chaparral. *Journal of Soil and Water Conservation* 32(4):174-180.

Davis, G.P. Jr. 1982. *Man and Wildlife in Arizona: The American Exploration Period-1824-1865.* Arizona Game and Fish Department. Phoenix.

Davis, W.B. 1966. *The Mammals of Texas.* 267 pp. Texas Parks and Wildlife Department Bulletin 41.

Deacon, J.E. and W.G. Bradley. 1972. Ecological distribution of fishes of Moapa River in Clark County, Nevada. *Transactions of the American Fisheries Society* 101:408-419.

Deacon, J.E. and W.L. Minckley. 1974. Desert fishes. pp. 385-488. In: Brown, G.W. Jr. (ed.), *Desert Biology. Volume II.* Academic Press, Inc. New York.

DeLaubenfels, D.J. 1975. *Mapping the World's Vegetation.* 246 pp. Syracuse University Press, Geography Series 4.

Denyes, H.A. 1956. Natural terrestrial communities of Brewster County, Texas, with special references to the distribution of the mammals. *American Midland Naturalist* 55:289-320.

Desert Bighorn Council. 1957-. Transactions of the Desert Bighorn Council. Las Vegas, Nevada.

Dice, L.R. 1922. Biotic areas and ecological habitats as units for the statement of animal and plant distribution. *Science* 55:335-338.

Dice, L.R. 1939. The Sonoran biotic province. *Ecology* 20:118-129.

Dice, L.R. 1943. *The Biotic Provinces of North America.* 77 pp. University of Michigan Press. Ann Arbor.

Dick-Peddie, W.A. and J.P. Hubbard. 1977. Classification of riparian vegetation. pp. 85-90. In: Johnson, R.R. and D.A. Jones (technical coordinators), *Importance, Preservation and Management of Riparian Habitat: A Symposium.* USDA Forest Service General Technical Report RM-43. 217 pp. Rocky Mountain Forest and Range Experiment Station. Fort Collins, Colo.

Dick-Peddie, W.A. and W.H. Moir. 1970. *Vegetation of the Organ Mountains, New Mexico.* 28 pp. Colorado State University, Range Science Department Science Series 4. Fort Collins, Colo.

Dill, W.A. and C. Woodhull. 1942. A game fish for the Salton Sea, the tenpounder, *Elops affinis. California Fish and Game.* 28:171-174.

Dixon, H. 1935. Ecological studies on the high plateaus of Utah. *Botanical Gazette* 97(2):171-320.

Dominguez, P.G., P.M. Huerta, and M.R. Baez. 1972. *Programa de estudio faunistico de Chihuahua,* Informe 1972. 23 pp. Dirección General de la Fauna Silvestre; Subsecretaría Forestal y de la Fauna S.A.G.

Donart, G.B., D. Sylvester, and W. Hickey. 1978a. A vegetation classification system for New Mexico, U.S.A. pp. 488-490. In: *Proceedings of the First International Rangeland Congress.* Denver, Colorado.

Donart, G.B., D. Sylvester, and W. Hickey. 1978b. *Potential Natural Vegetation-New Mexico.* New Mexico Interagency Range Committee Report II, M7-PO-23846. USDA Soil Conservation Service. (map, scale 1:1,000,000).

Driscoll, R.S., et al. 1976. *MODIFIED ECOCLASS-A Method for Classifying Ecosystems for the Rocky Mountain and Southwestern Regions.* USDA Forest Service. 117 pp. Mimeo. Rocky Mountain Forest and Range Experiment Station, Ad hoc Committee. Fort Collins, Colo.

Driscoll, R.S., John W. Russell, and Marvin C. Meier. 1978. *Recommended National Level Classification System for Renewable Resource Assessments.* USDA Forest Service. 44 pp. mimeo. Rocky Mountain Forest and Range Experiment Station. Fort Collins, Colo.

Duran, V. 1932. The flora of a desert range, the White Mountains. *Madroño* 2: 57.

Durrant, S.D. 1952. *Mammals of Utah: Taxonomy and Distribution.* 549 pp. University of Kansas Publication 6.

Durrenberger, R.W. and X. Murrieta S. 1978. *Clima del Estado de Sonora, México.* 34 pp. Laboratory of Climatology, Mexican Climatology Series No. 3. Arizona State University, Tempe.

Dyksterhuis, E.J. 1957. The savannah concept and its use. *Ecology* 38:435-442.

Easterla, D.A. 1973. Ecology of the 18 species of Chiroptera in Big Bend National Park, Texas. *Northwest Missouri State University Studies* 34:1-165.

Echelle, A.A. and A.F. Echelle. 1978. The Pecos pupfish, *Cyprinodon pecosensis* n. sp. (Cyprinodontidae), with comments on its evolutionary origin. *Copeia* 1978: 569-582.

Ellenberg, H. and D. Mueller-Dombois. 1967. Tentative physiognomic-ecological classification of plant formations of the earth. *Berichte des Geobotanischen Institutes der Eidgenoessischen Technischen Hochschule Stiftung, Ruebel.* Zurich 37:21-55.

Ellis, S.L., C. Fallat, N. Reece and C. Riordan. 1977. *Guide to Land Cover and Use Classification Systems Employed by Western Government Agencies.* USDI Fish and Wildlife Service.

Emerson, F.W. 1932. The tension zone between the grama grass and pinyon-juniper associations in northeastern New Mexico. *Ecology* 13:347-358.

Emory, W.H. 1848. *Notes of a Military Reconnaissance from Fort Leavenworth, in Missouri, to San Diego, in California, Including Parts of Arkansas, Del Norte, and Gila Rivers.* pp. 15-126. Thirtieth Congress, First Session. Executive Document 41.

Evanari, Michael, Leslie Shanon, and Naphtali Tadmor. 1971. *The Negev: The Challenge of a Desert.* 345 pp. Harvard University Press. Cambridge.

Everitt, B.L. 1968. Use of the cottonwood in an investigation of the recent history of a floodplain. *American Journal of Science* 266:417-439.

Everitt, B.L. 1970. *A survey of the Desert Vegetation of the Northern Henry Mountains Region, Hanksville, Utah.* Ph.D. dissertation, John Hopkins University.

Everitt, B.L. 1980. Ecology of saltcedar-a plea for research. *Environmental Geology* 3:77-84.

Evermann, B.W. 1916. Fishes of the Salton Sea. *Copeia* 1916:61-63.

Evermann, B.W. and C. Rutter. 1895. Fishes of the Colorado Basin. *United States Fish Commission Bulletin* 14(1894):473-486.

Fautin, Reed W. 1946. Biotic communities of the northern desert shrub biome in western Utah. *Ecological Monographs* 16:251-310.

Felger, R.S. 1971. The distribution of *Magnolia* in northwestern Mexico. *Journal of the Arizona Academy of Science* 6:251-253.

Felger, R.S., K. Cliffton, and P.J. Regal. 1976. Winter dormancy in sea turtles: Independent discovery and exploitation in the Gulf of California by two local cultures. *Science* 191:283-285.

Felger, R.S. and C.H. Lowe. 1976. *The Island and Coastal Vegetation and Flora of the Northern Part of the Gulf of California.* 59 pp. Natural History Museum of Los Angeles County, Contributions in Science 285.

Felger, R.S. and M.B. Moser. 1973. Eelgrass (*Zostera marina* L.) in the Gulf of California: Discovery of its nutritional value by the Seri Indians. *Science* 181:355-356.

Findley, J.S., A.H. Harris, D.E. Wilson, and C. Jones. 1975. *Mammals of New Mexico.* 360 pp. University of New Mexico Press. Albuquerque.

Fitch, J.E. and R.J. Lavenberg. 1975. *Tidepool and Nearshore Fishes of California.* 156 pp. University of California Press. Berkeley.

Flores Mata, G., J. Jiminez Lopez, X. Madrigal Sanchez, F. Moncayo Ruiz, and F. Takaki. 1971. *Tipos de Vegetación de la República Mexicana.* 59 pp. Subsecretaría de Planeación Dirección General de Estudios, Dirección de Agrología, SRH. México, D.F.

Flores Mata, G., J. Jiminez Lopez, X. Madrigal Sanchez, F. Moncayo Ruiz, and F. Takaki. 1971. *Memoria del mapa de Tipos de Vegetación de la República Mexicana.* Secretaría de Recursos Hidraulicos, Subsecretaría de Planeación, Dirección General de Estudios, Dirección de Agrología, SRH, México, D.F. (manual and map, scale 1:2,000,000).

Follett, W.A. 1960. The freshwater fishes-their origins and affinities. In: *The Biogeography of Baja California and Adjacent Seas.* [*Systematic Zoology* 9:212-232.].

Fonteyn, Paul J. and Mahall, Bruce E. 1978. Competition among desert perennials. *Nature.* 175(5680):544-545.

Fosberg, F.R. 1961. A classification of vegetation for general purposes. *Tropical Ecology* 2:1-28.

Franklin, J.F. 1977. The biosphere reserve program in the United States. *Science* 195:262-267.

Franklin, J.F. and C.T. Dyrness. 1973. *Natural Vegetation of Oregon and Washington.* USDA Forest Service General Technical Report PNW-8. 417 pp. Pacific Northwest Forest and Range Experiment Station. Portland, Oreg.

Franzreb, K.E. 1977. *Bird Population Changes After Timber Harvesting of a Mixed Conifer Forest in Arizona.* USDA Forest Service Research Paper RM-184. 24 pp. Rocky Mountain Forest and Range Experiment Station. Fort Collins, Colo.

Fritts, H.C. 1974. Relationships of ring widths in arid site conifers to variation in monthly temperature and precipitation. *Ecological Monographs* 44:411-440.

Fritts, H.C., G.R. Lofgren, and G.A. Gordon. 1979. Variations in climate since 1602 as reconstructed from tree rings. *Quaternary Research* 12:18-46.

Gaines, D.A. 1977. The valley riparian forests of California: Their importance to bird populations. p. 57-85. In: Sands, A. (ed.), *Riparian Forests in California.* University of California, Institute of Ecology Publication 15. Davis.

Galbraith, William A. and E. William Anderson. 1971. Grazing history of the Northwest. *Journal of Range Management* 24(1):6-12.

Garcia, Enriqueta and Pedro A. Mosiño. 1968. Los climas de la Baja California. pp. 29-56. In: R. del Arnes C., *Comite Nacional Mexicano para el Decenio Hidrologico Internacional, Memoria 1966-1967.* Instituto de Geofisica, Universidad Nacional Autonoma de Mexico.

Garrison, G.A., A.J. Bjugstad, D.A. Duncan, M.E. Lewis, and D.R. Smith. 1977. *Vegetation and Environmental Features of Forest and Range Ecosystems.* 68 pp. USDA Forest Service Agricultural Handbook 475. Washington, D.C.

Gary, H.L. 1965. Some site relations in three floodplain communities in central Arizona. *Journal of the Arizona Academy of Science* 3:209-212.

Gaussen, Henri. 1953. A proposed ecological vegetation map. *Surveying and Mapping* 13:168-173.

Gaussen, Henri. 1955. Les divisions ecologiques du mode. In: *Paris Centre Natimal de la Reecherche Scientifique, 59th International Colloquim.* 1954.

Gehlbach, F.R. 1967. Vegetation of the Guadalupe escarpment, New Mexico-Texas. *Ecology* 48:393-419.

Gentry, Howard Scott. 1942. Rio Mayo plants: *A Study of the Flora and Vegetation of the Valley of the Rio Mayo in Sonora.* 328 pp. Carnegie Institution of Washington, Publication 527.

Gentry, Howard Scott. 1949. *Land Plants Collected by the Valero III, Allan Hancock Pacific Expeditions 1937-1941.* 245 pp. University of Southern California Press. Los Angeles.

Gentry, Howard Scott. 1958. The natural history of jojoba (*Simmondsia chinensis*) and its cultural aspects. *Economic Botany* 12:261-295.

Gentry, Howard Scott. 1972. *The Agave Family in Sonora.* 195 pp. USDA Agricultural Research Service Agricultural Handbook 399. Washington, D.C.

Gentry, Howard Scott. 1978. *The Agaves of Baja California.* 119 pp. California Academy of Science Occasional Papers 130. San Francisco.

Gessner, F. 1955. *Hydrobotanik. I. Energiehaushalt.* 517 pp. Veb. Deutsch. Ver. Wissensch., Berlin.

Gessner, F. 1959. *Hydrobotanik. II. Stoffhaushalt.* 701 pp. Veb. Deutsch. Ver. Wissensch., Berlin.

Gessner, F. 1963. *Hydrobotanik. III. Biozonotik.* 832 pp. Veb. Deutsch. Ver. Wissensch., Berlin.

Gleason, H.A. 1939. The individualistic concept of the plant association. *American Midland Naturalist* 21:91-110.

Glendening, G.E. and C.P. Pase. 1964. Effect of litter treatment on germination of species found under manzanita (*Arctostaphylos*). *Journal of Range Management* 17:265-266.

Glinski, R.L. 1977. Regeneration and distribution of sycamore and cottonwood trees along Sonoita Creek, Santa Cruz County, Arizona. pp. 116-123. In: Johnson, R.R. and D.A. Jones (technical coordinators), *Importance, Preservation and Management of Riparian Habitat: A Symposium.* USDA Forest Service General Technical Report RM-43. 217 pp. Rocky Mountain Forest and Range Experiment Station. Fort Collins, Colo.

Gloyd, H.K. 1937. A herpetological consideration of faunal areas in southern Arizona. *Bulletin of the Chicago Academy of Science* 5:79-136.

Goldman, E.A. 1916. Plant records of an expedition to lower California. *Contr. U.S. National Herbarium* 16:309-371.

Goldman, E.A. 1951. *Biological Investigations in Mexico.* 476 pp. Smithsonian Institution Miscellaneous Publication 115. Washington, D.C.

Goldman, E.A. and R.T. Moore. 1945. The biotic provinces of Mexico. *Journal of Mammalogy* 26:347-360.

Golet, F.C. and J.S. Larson. 1974. *Classification of Freshwater Wetlands in the Glaciated Northwest.* 56 pp. U.S. Bureau of Sport Fisheries and Wildlife. Research Publication 116.

Gordon, A.G. 1968. Ecology of *Picea chihuahuana* (Martinez). *Ecology* 49:880-896.

Gould, F.W. 1951. *Grasses of Southwestern United States.* 343 pp. University of Arizona, Biological Sciences Bulletin 7. Tucson.

Gould, F.W. 1962. *Texas Plants-a Checklist and Ecological Summary.* 97 pp. Texas Agricultural Experiment Station MP-585. College Station, Texas.

Gray, L.J. 1980. *Recolonization Pathways and Community Development of Desert Stream Macroinvertebrates.* Unpublished Ph.D. dissertation, 175 p. Arizona State University. Tempe.

Gregg, R.E. 1963. *The Ants of Colorado.* 792 pp. University of Colorado Press. Boulder.

Griffin, J.R. 1977. Oak woodland. pp. 383-415. In: Barbour, M.G. and J. Major (eds.), *Vegetation of California.* John Wiley and Sons. New York, N.Y.

Grinnell, J. 1908. The biota of the San Bernadino Mountains. *University of California Publication in Zoology* 5:1-170.

Grinnell, J. 1914. An account of the mammals and birds of the lower Colorado valley, with especial reference to the distributional problems presented. *University of California Publications in Zoology* 12:1-217.

Grinnell, J. 1923. Observations upon the bird life of Death Valley. *Proceedings of the California Academy of Sciences* 13:43-109.

Grinnell, J. 1937. Mammals of Death Valley. *Proceedings of the California Academy of Sciences* 23:115-169.

Grinnell, J. and A.H. Miller. 1944. *The Distribution of the Birds of California.* Cooper Ornithology Club, University of California, Berkeley.

Grinnell, J. and H.S. Swarth. 1913. An account of the birds and mammals of the San Jacinto area of southern California, with remarks upon the behavior of geographic races on the margins of their habitats. *University of California Publications in Zoology* 10:197-406.

Gross, F.A. III. 1973. *A Century of Vegetative Change for Northwestern New Mexico.* 50 pp. Natural Resources Conservation Commission of New Mexico.

Gutierrez, R.J. 1975. *A Literature Review and Bibliography of the Mountain Quail, Oreortyx pictus (Douglas).* 33 pp. USDA Forest Service, California Region.

Guzman, G., Jr. 1961. Vegetation zones of the territory of Baja California in relation to wildlife. (trans.) *Transactions of the Desert Bighorn Council* 5:68-74.

Hackman, Robert J. 1973. *Vegetation Map of the Salina Quadrangle, Utah.* U.S.

Geological Survey Map I-591-P. (Map scale 1:250,000).

Hall, E. Raymond. 1946. *Mammals of Nevada.* 710 pp. University of California Press. Berkeley.

Hall, E. Raymond and Keith R. Kelson. 1959. *The Mammals of North America.* 2 Volumes. Ronald Press Co. New York.

Halliday, W.E.D. 1937. A forest classification for Canada. *Canada Department of Mines Research. F.S. Bulletin* 89:1-189.

Halligan, J.P. 1974. Relationship between animal activity and bare areas associated with California sagebrush in annual grassland. *Journal of Range Management* 27:358-362.

Hanes, T.L. 1977. California chaparral. pp. 417-469. In: Barbour, Michael G. and Jack Major (eds.), *Terrestrial Vegetation of California.* John Wiley and Sons, New York.

Hanson, H.C. 1924. *A Study of the Vegetation of Northeastern Arizona.* 94 pp. University of Nebraska Studies 24.

Hardy, L.M. and R.W. McDiarmid. 1969. Amphibians and reptiles of Sinaloa, Mexico. *University of Canada, Museum of Natural History Series* 18(3):39-252.

Hardy, R. 1945. Breeding birds of the pygmy conifers in the Book Cliff region of eastern Utah. *Auk* 62:523-542.

Harrington, H.D. 1954. *Manual of the Plants of Colorado.* Sage Books. Chicago, Ill.

Harrington, H.D. 1964. *Manual of the Plants of Colorado.* Second edition. Sage Books. Denver, Colo.

Harshberger, J.W. 1911. *Phytogeographic Survey of North America.* G.E. Stechert. New York.

Hart, C.M., O.S. Lee, and J.B. Low. 1950. *The Sharp-tailed Grouse in Utah–its Life History, Status, and Management.* 79 pp. Utah State Department of Fish and Game Publication 3.

Hastings, James Rodney. 1959. Vegetation change and arroyo cutting in southeastern Arizona. *Journal of the Arizona Academy of Science* 1:60-67.

Hastings, James Rodney. 1964a. *Climatological Data for Baja California.* 132 pp. University of Arizona Institute of Atmospheric Physics Technical Report 14.

Hastings, James Rodney. 1964b. *Climatological Data for Sonora and Northern Sinaloa.* 152 pp. University of Arizona Institute of Atmospheric Physics Technical Report 15.

Hastings, James Rodney and Robert R. Humphrey. 1969a. *Climatological Data and Statistics for Baja California.* 96 pp. University of Arizona Institute of Atmospheric Physics Technical Reports on the Meteorology and Climatology of Arid Regions 18.

Hastings, James Rodney and Robert R. Humphrey. 1969b. *Climatological Data and Statistics for Sonora and Northern Sinaloa.* 96 pp. University of Arizona Institute of Atmospheric Physics. Technical reports on the Meteorology and Climatology of Arid Regions 19.

Hastings, James Rodney and Raymond M. Turner. 1965a. Seasonal precipitation regimes in Baja California, Mexico. *Geografiska Annaler* 47 (Series A):204-223.

Hastings, James Rodney and Raymond M. Turner. 1965b. *The Changing Mile: An Ecological Study of Vegetation Change With Time in the Lower Mile of an Arid and Semiarid Region.* 317 pp. University of Arizona Press. Tucson.

Hastings, James Rodney, Raymond M. Turner, and Douglas K. Warren. 1972. *An Atlas of Some Plant Distributions in the Sonoran Desert.* 255 pp. University of Arizona Institute of Atmospheric Physics Technical Report 21.

Hayden, F.V. 1873. *Sixth Annual Report of the United States Geological Survey of the Territories, Embracing Portions of Montana, Idaho, Wyoming, and Utah.* Govt. Printing Office. Washington. D.C.

Hayward, C.L. 1948. Biotic communities of the Wasatch chaparral, Utah. *Ecological Monographs* 18:473-506.

Heady, H.F. 1956. Evaluation and measurement of the California annual type. *Journal of Range Management* 9:25-27.

Heady, H.F. 1977. Valley grassland. pp. 491-514. In: Barbour, M.G. and J. Major (eds.), *Terrestrial Vegetation of California.* John Wiley and Sons. New York.

Hendrickson, D.A., W.L. Minckley, R.R. Miller, D.J. Siebert, and P.L. Minckley. 1980. Fishes of the Rio Yaqui basin, Mexico and United States. *Journal of the Arizona-Nevada Academy of Science* 15:1-105.

Henrickson, J.R. 1978. Saline habitats and halophytic vegetation of the Chihuahuan Desert region. pp. 289-314. In: Wauer, R.H. and D.H. Riskind (eds.), *Transactions of the Symposium on the Biological Resources of the Chihuahuan Desert Region, United States and Mexico.* USDI National Park Service Proceedings and Transactions Series 3. Washington, D.C.

Henrickson, J.R. and R.M. Straw. 1976. *A Gazeteer of the Chihuahuan Desert Flora.* 273 pp. California State University. Los Angeles.

Heringhi, H.L. 1969. *An Ecological Survey of the Herpetofauna of Alamos, Sonora, Mexico.* Unpublished M.S. thesis. 56 pp. Arizona State University, Tempe.

Hesse, R., W.C. Allee, and K.P. Schmidt. 1937. *Ecological Animal Geography.* 597 pp. John Wiley and Sons. New York.

Hibbert, A.R. 1971. Increase in streamflow after converting chaparral to grass. *Water Resources Research* 7:71-80.

Hibbert, A.R., E.A. Davis, and D.G. Scholl. 1974. *Chaparral conversion Potential in Arizona, Part 1: Water Yield Response and Effects on Other Resources.* USDA Forest Service Research Paper RM-126. 36 pp. Rocky Mountain Forest and Range Experiment Station. Fort Collins, Colo.

Hinton, S. 1969. *Seashore Life of Southern California.* 181 pp. University of California Press. Berkeley.

Hitchcock, A.S. 1935. *Manual of the Grasses of the United States.* 1051 pp. USDA Miscellaneous Publication 200. Washington, D.C.

Hitchcock, A.S. and Agnes Chase. [1950, 1960] 1971. *Manual of the Grasses of the United States.* Two volumes. Dover Publications, Inc. New York.

Hobson, E.S. 1968. *Predatory Behavior of Some Shore Fishes in the Gulf of California.* 92 pp. USDI Bureau of Sport Fisheries and Wildlife. Research Report 73.

Hoffmeister, D.R. 1956. Mammals of the Graham (Pinaleño) Mountains, Arizona. *American Midland Naturalist* 35:237-388.

Hoffmeister, D.F. 1971. *Mammals of Grand Canyon.* 183 pp. University of Illinois Press. Urbana.

Hoffmeister, D.F. and S.W. Carothers. 1969. Mammals of Flagstaff, Arizona. *Plateau* 41:184-188.

Hoffmeister, D.F. and F.E. Durham. 1971. *Mammals of the Arizona Strip Including Grand Canyon National Monument.* 44 pp. Museum of Natural History, University of Illinois, Technical Series 11.

Hoffmeister, D.F. and W.W. Goodpaster. 1954. 152 pp. *The Mammals of the Huachuca Mountains, Southeastern Arizona.* Illinois Biol. Soc. Monogr. 24.

Holland, R.F. and S.K. Jain. 1977. Vernal pools. pp. 515-533. In: Barbour, M.G. and J. Major (eds.), *Terrestrial Vegetation of California.* John Wiley and Sons. New York.

Holm, T. 1927. The vegetation of the alpine region of the Rocky Mountains in Colorado. *National Academy of Science Memoir* 19(3):1-45.

Hornaday, W.T. 1908. *Camp-fires on Desert and Lava.* 366 pp. Charles Scribner's and Sons. New York.

Horton, J.S. 1960. *Vegetation Types of the San Bernardino Mountains.* 29 pp. USDA Forest Service, Pacific Southwest Forest and Range Experiment Station. Technical Paper 44. Berkeley, Calif.

Horton, J.S. 1977. The development and perpetuation of the permanent tamarisk type in the phreatophyte zone of the Southwest. pp. 128-136. In: Johnson, R.R. and D.A. Jones (technical coordinators), *Importance, Preservation and Management of Riparian Habitat: A Symposium.* USDA Forest Service General Technical Report RM-43. pp. 128-136. Rocky Mountain Forest and Range Experiment Station. Fort Collins, Colo.

Horton, J.S. and C.J. Kraebel. 1955. Development of vegetation after fire in the chamise chaparral of southern California. *Ecology* 36:244-262.

Horton, J.S., F.C. Mounts, and J.M. Kraft. 1960. *Seed Germination and Seedling Establishment of Phreatophyte Species.* 26 pp. U.S. Department of Agriculture, Forest Service, Rocky Mountain Forest and Range Experiment Station. Station Paper 48. Fort Collins, Colo.

Hubbard, J.P. 1965. The summer birds of the forests of the Mogollon Mountains, New Mexico. *Condor* 67:404-415.

Hubbard, J.P. 1970. *Check-list of the Birds of New Mexico.* 103 pp. New Mexico Ornithological Society Publication 3.

Hubbard, J.P. 1971. The summer birds of the Gila Valley, New Mexico. *Delaware Museum of Natural History. Occasional Papers* 2:5-11, 28-30.

Hubbard, J.P. 1974a. Avian evolution in the aridlands of North America. *The Living Bird* 12:155-195.

Hubbard, J.P. 1974b. *The Biota of the Pinos Altos Mountains, Southwestern New Mexico, a Preliminary Report.* 11 pp. Delaware Museum of Natural History.

Hubbard, J.P. 1977a. *A Biological Inventory of the Lower Gila River Valley, New Mexico.* 56 pp. A report jointly prepared by Bureau of Land Management, Bureau of Reclamation, New Mexico Department of Game and Fish, Soil Conservation Service, USDA Forest Service.

Hubbard, J.P. 1977b. Importance of riparian ecosystems: Biotic consideration. p. 14-18. In: Johnson, R.R. and D.A. Jones (technical coordinators), *Importance, Preservation and Management of Riparian Habitats: A Symposium.* USDA Forest Service General Technical Report RM-43. 217 pp. Rocky Mountain Forest and Range Experiment Station. Fort Collins, Colo.

Hubbard, J.P. 1977c. The biological and taxonomic status of the Mexican duck. *New Mexico Department of Game and Fish Bulletin* 16:1-56.

Hubbard, S.P. 1979. Snow grouse. *New Mexico Wildlife* 24:2-5.

Hubbell, S.P. 1979. Tree dispersion, abundance, and diversity in a tropical dry forest. *Science* 203:1299-1309.

Hubbs, C., R.R. Miller, R.J. Edwards, K.W. Thompson, E. Marsh, G.P. Garrett, G.L. Powell, D.J. Morris, and R.W. Zerr. 1977. Fishes inhabiting the Rio Grande, Texas and Mexico, between El Paso and the Pecos confluence. p. 91-97. In: Johnson, R.R. and D.A. Jones (technical coordinators), *Importance, Preservation and Management of Riparian Habitats: A Symposium.* USDA Forest Service General Technical Report RM-43. 217 pp. Rocky Mountain Forest and Range Experiment Station. Fort Collins, Colo.

Hubbs, C.L. 1964. History of ichthyology in the United States after 1850. *Copeia* 1964:42-60.

Hubbs, C.L. and R.R. Miller. 1948. The zoological evidence: Correlation between fish distribution and hydrographic history in the desert basins of western United States. *Bulletin of the University of Utah* 30:17-166.

Hubbs, C.L. and R.R. Miller. 1965. Studies of Cyprinodont fishes, XXII. Variation in *Lucania parva,* its establishment in western United States, and description of a new species from an interior basin in Coahuila, Mexico.

Miscellaneous Publications of the Museum of Zoology, University of Michigan 127:1-111.

Hubbs, C.L., R.R. Miller, and L.C. Hubbs. 1974. Hydrographic history and relict fishes of the North-Central Great Basin. *Memoirs of the California Academy of Sciences* 7:1-259.

Hubbs, C.L., R.S. Peterson, R.L. Gentry, and R.L. Delong. 1968. The Guadalupe fur seal: Habitat, behavior, population size, and field identification. *Journal of Mammalogy* 49:665-675.

Hubbs, C.L., K.W. Radford, and R.T. Orr. 1965. Re-establishment of the northern elephant seal (*Mirounga angustirostris*) *Proceedings of the California Academy of Science* 31:601-612.

Humphrey, R.R. 1950. Arizona range resources–II. Yavapai County. *University of Arizona Agriculture Experiment Station Bulletin* 229:1-55.

Humphrey, R.R. 1953. Forage production on Arizona ranges–III. Mohave County. *University of Arizona Agriculture Experiment Station Bulletin* 244:1-79.

Humphrey, R.R. 1955. Forage production on Arizona ranges–IV. Coconino, Navajo, Apache Counties. A study in range condition. *University of Arizona Agriculture Experiment Station Bulletin* 266:1-84.

Humphrey, R.R. 1958. The desert grassland; a history of vegetational change and an analysis of causes. *Botanical Review* 24:193-252.

Humphrey, R.R. 1960. Forage production on Arizona ranges–V. Pima, Pinal, Santa Cruz Counties. *University of Arizona Agriculture Experiment Station Bulletin* 302:1-138. Tucson.

Humphrey, R.R. 1962. *Range Ecology.* Ronald Press Co. New York.

Humphrey, R.R. 1974. *The Boojum and its Home.* 214 pp. University of Arizona Press. Tucson.

Humphrey, R.R. and David B. Marx. 1980. Distribution of the boojum tree (*Idria columnaris*) on the coast of Sonora, Mexico, as influenced by climate. *Desert Plants* 2(3):183-187.

Hunt, Charles B. 1966. *Plant Ecology of Death Valley, California.* 68 pp. U.S. Geological Survey Professional Paper 509.

Hunt, Charles B., Paul Averitt, and Ralph L. Miller. 1953. *Geology and Geography of the Henry Mountains Region, Utah.* 234 pp. U.S. Geological Survey Prof. Paper 228.

Ibrahim, Kamal M., Neil E. West, D.L. Goodwin. 1972. Phytosociological characteristics of perennial *Atriplex*-dominated vegetation of southeastern Utah. *Vegetatio* 24:13-22.

Instituto de Geografia. 1970. *Cartas de Climas.* map. 1:500,000 scale. Universidad Nacional de Mexico.

International Union for Conservation of Nature and Natural Resources. 1973. *A Working System for Classification of World Vegetation.* 21 pp. IUCN Occasional Paper 5.

International Union for Conservation of Nature and Natural Resources. 1974. *Biotic Provinces of the World–Further Development of a System for Defining and Classifying Natural Regions for Purposes of Conservation.* 57 pp. IUCN Occasional Paper 9.

Ives, R.L. 1949. Climate of the Sonoran Desert Region. *Ann. Ass. American Geography* 39:143-187.

Ives, R.L. 1964. The Pinacate Region, Sonora, Mexico. *California Academy of Sciences Occasional Papers* 47:1-43.

Jaeger, E.C. 1957. *A Naturalist's Death Valley.* (revised edition). 70 pp. Publication 5. North Hollywood Printing, Inc.

Jaeger, E.C. and A.C. Smith. 1966. *Introduction to the Natural History of Southern California.* University of California Press. Berkeley.

Jaeger, E.C. and A.C. Smith. 1971. *Introduction to the Natural History of Southern California.* 104 pp. University of California Press. Berkeley.

Jameson, Donald A., John A. Williams, and Eugene W. Wilton. 1962. Vegetation and soils of Fishtail Mesa, Arizona. *Ecology* 43(3):403-410.

Jensen, H.A. 1947. A system for classifying vegetation in California. *California Fish and Game* 33:199-266.

Johnsgard, P.A. 1973. *Grouse and Quails of North America.* 553 pp. University of Nebraska Press. Lincoln.

Johnson, Bruce K. 1979. *Bighorn Sheep Distribution, Habitat Evaluation, and Food Habits, Pecos Wilderness, New Mexico.* Ph.D. dissertation. Department of Fisheries and Wildlife Biology, Colorado State University, Fort Collins.

Johnson, D.H., M.D. Bryant, and A.H. Miller. 1948. Vertebrate animals of the Providence Mountains area of California. *University of California Publications in Zoology* 48(5):221-376.

Johnson, Hyrum B. 1976. Vegetation and plant communities of southern California–a functional view. pp. 125-164. In: Latting, June (ed.), *Plant Communities of Southern California.* California Native Plant Society Special Publication 2.

Johnson, W.C., R.L. Burgess, and W.R. Keammerer. 1976. Forest overstory vegetation on the Missouri River Floodplain in North Dakota. *Ecological Monographs* 46:59-84.

Johnston, Ivan M. 1924. Expedition of the California Academy of Sciences to the Gulf of California in 1921. *California Academy of Sciences Proceedings* 12(30):951-1218.

Johnston, M.C. 1977. Brief resume of botanical, including vegetational, features of the Chihuahuan Desert Region with special emphasis on their unique-ness. pp. 335-362. In: Wauer, R.H. and D.H. Riskind (eds.), *Transactions of the Symposium on the Biological Resources of the Chihuahuan Desert Region, U.S. and Mexico.* Alpine, Texas. 1974. USDI National Park Service Transactions and Proceedings Series 3. Washington, D.C.

Jones, J. Knox, Jr., Dilford C. Carter, and Hugh H. Genoways. 1975. *Revised Checklist of North American Mammals North of Mexico.* 14pp. Occasional Papers of the Museum of Texas Tech. University 28. Lubbock, Tex.

Jones, J.R. 1974. *Silviculture of Southwestern Mixed Conifers and Aspen: The Status of Our Knowledge.* USDA Forest Service Research Paper RM-122. 44 pp. Rocky Mountain Forest and Range Experiment Station. Fort Collins, Colo.

Jordan, D.S. 1929. *A Manual of the Vertebrate Animals of the Northern United States.* 13th edition. 397 pp. A.C. McCluratt, Co. Chicago, Ill.

Jordan, D.S. and B.W. Evermann. 1896-1901. The fishes of North and Middle America. *Bulletin of the U.S. National Museum* 47:1-3136, in 4 parts.

Jordan, D.S. and C.H. Gilbert. 1883. *Synopsis of the Fishes of North America.* 1,018 pp. Bulletin of the U.S. National Museum 16. Washington, D.C.

Karpiscak, Martin, M. 1980. *Secondary Succession of Abandoned Field Vegetation in Southern Arizona.* Ph.D. dissertation. 219 pp. University of Arizona. Tucson.

Karpiscak, Martin M. and Otto M. Grosz. 1979. Secondary succession of Abandoned Fields in Southern Arizona. Arizona-Nevada Academy of Science 14 (Suppl.):23.

Kaufmann, J.H., D.V. Lanning, and S.E. Poole. 1976. Current status and distribution of the coati in the United States. *Journal of Mammalogy* 57(4):621-637.

Kearney, T.H., L.J. Briggs, H.L. Shantz, J.W. McLane, and R.L. Piemeisel. 1914. Indicator significance of vegetation in Tooele Valley, Utah. *Journal Agricultural Research* 1(5):365-417.

Kearney, T.H., R.H. Peebles, and collaborators. [1951]. *Arizona Flora.* Second edition. 1,085 pp. University of California Press. Berkeley.

Keil, D.J. 1973. Vegetation and flora of the White Tank Mountains Regional Park, Maricopa County, Arizona. *Journal of the Arizona Academy of Sciences* 8:35-39.

Keith, J.O. 1965. The Abert squirrel and its dependence on ponderosa pine. *Ecology* 46:150-163.

Kellog, R. 1932. *Mexican Tail-less Amphibians in the United States National Museum.* 224 pp. U.S. National Museum Bulletin 160. Washington, D.C.

Kelsey, H.P. and W.A. Dayton. 1942. *Standardized Plant Names.* Second edition. 674 pp. J. Horace McFarland Co. Harrisburg, Pa. [For American Joint Committees on Horticultural Nomenclature].

Kendeigh, S.C. 1954. History and evolution of various concepts of plant and animal communities in North America. *Ecology* 35:152-171.

Kendeigh, S.C. 1961. *Animal Ecology.* Prentice-Hall. Englewood Cliffs, New Jersey.

Klipple, G.E. and D.F. Costello. 1960. *Vegetation and Cattle Responses to Different Intensities of Grazing on Short-grass Ranges on the Central Great Plains.* 82 pp. USDA Forest Service Bulletin 1216. Rocky Mountain Forest and Range Experiment Station. Fort Collins, Colo.

Koster, W.J. 1957. *Guide to the Fishes of New Mexico.* 116 pp. University of New Mexico Press. Albuquerque.

Krajina, V.J. 1965. *Biogeoclimatic Zones and Biogeocoenoses of British Columbia.* Department of Botany. Vancouver.

Küchler, A.W. 1964. *The Potential Natural Vegetation of the Conterminous United States.* 116 pp. map. American Geographic Society. Special Publication 361.

Küchler, A.W. 1966. *Potential Natural Vegetation.* (map, scale 1:7,500,000.) U.S. Geological Survey. Washington, D.C.

Küchler, A.W. 1967. *Vegetation Mapping.* 472 pp. Ronald Press. New York.

Küchler, A.W. 1977. Natural Vegetation of California. map. In: Barbour, M.G. and J. Major (eds.), *Terrestrial Vegetation of California.* 1,002 pp. John Wiley and Sons. New York.

Laboratory of Climatology. 1975. *Arizona Precipitation.* map. Arizona Resources Information System Coop. Publ. 5. Arizona State University, Tempe.

Lacey, J.R., P.R. Ogden and K.E. Foster. 1975. *Southern Arizona Riparian Habitat: Spatial Distribution and Analysis.* 148 pp. University of Arizona Office of Arid Lands Studies Bulletin 8.

Lactate, D.S. 1969. *Guidelines for Biophysical Land Classification.* 61 pp. Department of Fisheries and Forestry. Canada Forest Service Publication 1264. Ottawa.

La Marche, V.C. 1973. Guide to the geology. In: Lloyd, R.M. and R.S. Mitchell (eds.), *A Flora of the White Mountains, California and Nevada.* University of California Press. Berkeley.

Lane, J.A. 1968. *A Birder's Guide to Southern California.* 95 pp. L and P Photography. Sacramento.

LaRivers, I. 1962. *Fishes and Fisheries of Nevada.* 782 pp. Nevada State Fish and Game Commission. Carson City.

Larsen, J.A. 1930. Forest types of the northern Rocky Mountains and their climatic controls. *Ecology* (11):631-672.

Larson, R.L., H.W. Menard, and S.M. Smith. 1968. Gulf of California: A result of ocean floor spreading and transforms faulting. *Science* 161:781-783.

Layser, E.F. 1974. Vegetative Classification: Its application to forestry in the northern Rocky Mountains. *Journal of Forestry* 72:354-357.

Lehr, J. Harry. 1978. *A Catalogue of the Flora of Arizona.* 203 pp. Desert Botanical Garden. Phoenix, Ariz.

Lehr, J. Harry and Donald J. Pinkava. 1980. A catalogue of the flora of Arizona–Supplement I. *Journal of the Arizona-Nevada Academy of Science* 15(1):17-32.

Leonard, A.B. and J.C. Frye. 1975. Pliocene deposits and molluscan faunas, east-central New Mexico. *New Mexico Bureau of Mines and Mineral Resources Memoirs* 30:17-93.

Leopold, A. 1924. Grass, brush, timber, and fire in southern Arizona. *Journal of Forestry* 22:1-10.

Leopold, A. 1933. *Game Management.* 481 pp. Charles Scribner's Sons. New York, London.

Leopold, A.S. 1950. Vegetation zones of Mexico. *Ecology* 31:507-518.

Leopold, A.S. 1959. *Wildlife of New Mexico.* 568 pp. University of California Press. Berkeley.

Leopold, L.B. 1976. Reversal of erosion cycle and climatic change. *Journal of Quaternary Research* 6:557-562.

LeSueur, Harde. 1945. *The Ecology of the Vegetation of Chihuahua, Mexico. North of Parallel 28.* 92 pp. University of Texas Publication 452.

Ligon, J.S. 1942. *Masked Bobwhite Quail in Southern Arizona and Notes on Efforts at Restoration.* Arizona Game and Fish Department. Phoenix.

Ligon, J.S. 1952. The vanishing masked bobwhite. *Condor* 54:48-50.

Ligon, J.S. 1961. *New Mexico Birds and Where to Find Them.* 360 pp University of New Mexico Press. Albuquerque.

Little, E.L. 1941. Alpine flora of San Francisco Mountain, Arizona. *Madroño* 6:65-81.

Little, E.L. 1950. *Southwestern Trees, a Guide to the Native Species of New Mexico and Arizona.* 109 pp. USDA Forest Service Handbook 9. Washington, D.C.

Little, E.L. 1971. *Atlas of United States Trees. Vol. 1, Conifers and Important Hardwoods.* 9 pp. 313 maps. USDA Forest Service Miscellaneous Publication 1146.

Little, E.L. 1976. *Atlas of United States Trees. Vol. 3, Minor Western Hardwoods.* 13 pp. 210 maps. USDA Forest Service Miscellaneous Publication 1314.

Little, E.L. 1979. *Checklist of United States Trees.* 375 pp. Agricultural Handbook 541. U.S. Department of Agriculture. Washington, D.C.

Lloyd, Robert M. and Richard S. Mitchell. 1973. *A Flora of the White Mountains, California and Nevada.* University of California Press. Berkeley.

Lowe, C.H., Jr. 1961. Biotic communities in the sub-Mogollon region of the inland Southwest. *Journal of the Arizona Academy of Science* 2:40-49.

Lowe, C.H. (ed.) 1964. *The Vertebrates of Arizona.* 270 pp. University of Arizona Press. Tucson. NOTE: The first 136 pages of this book consisted of a treatment of "Arizona landscapes and habitats" by C.H. Lowe. This separate section was simultaneously released by the publisher as a separate 136 page book entitled *Arizona's Natural Environment: Landscapes and Habitats.*

Lowe, C.H. and D.E. Brown. 1973. *The Natural Vegetation of Arizona.* 53 pp. State of Arizona, Arizona Resources Information System. ARIS Cooperative Publication 2. Phoenix.

Mabry, T.J., J.H. Hunziker, D.R. Difeo, Jr. (eds.) 1977. *Creosotebush: Biology and Chemistry of Larrea in New World Deserts.* 304 pp. U.S./I.B.P. Synthesis Series 6. Academic Press. New York.

MacDonald, K.B. 1977. Coastal salt marsh. pp. 263-294. In: Barbour, M.G. and J. Major (eds.), *Terrestrial Vegetation of California.* 1,002 pp. John Wiley and Sons. New York.

MacKay, H.A. 1970. *A Comparative Floristic Study of the Rio Hondo-Lake Fort-Wheeler Peak Locale, New Mexico and the Huerfano River-Blanca Peak Locale, Colorado.* Ph.D. dissertation. University of New Mexico, Albuquerque.

MacMahon, J.A. 1979. North American deserts: Their floral and faunal components. pp. 21-82. In: Goodall, D.W. and R.A. Perry (eds.), *Arid-land Ecosystems: Structure, Functioning and Management.* Volume I. 881 pp. Cambridge University Press. New York.

Major, J. and D.W. Taylor. 1977. Alpine. pp. 601-675. In: Barbour, M.G. and J. Major (eds.), *Terrestrial Vegetation of California.* John Wiley and Sons. New York.

Markham, Charles G. 1972. Baja California's climate. *Weatherwise* 25(2):66-101.

Marks, John Brady. 1950. Vegetation and soil relations in the lower Colorado desert. *Ecology* 31:176-193.

Marshall, J.T., Jr. 1957. Birds of the pine-oak woodland in southern Arizona and adjacent Mexico. Cooper Ornithological Society. *Pacific Coast Avifauna* 32:1-125.

Martin, P.S. 1963. *The Last 10,000 Years, a Fossil Pollen Record of the American Southwest.* 87 pp. University of Arizona Press. Tucson.

Martin, P.S. 1979. *A survey of potential natural landmarks, biotic themes, of the Mohave-Sonoran desert region.* 358 pp. Heritage Conservation and Recreation Service, USDI, Tucson.

Martin, P.S. and P.J. Merhinger, Jr. 1965. Pleistocene pollen analysis and biogeography of the Southwest. pp. 130-200. In: Wormington, H.H. and D.E. Ellis (eds.), *Pleistocene Studies in Southern Nevada.* Nevada State Museum of Anthropology Paper 13. Carson City.

Martin, P.S. and P.J. Mehringer, Jr. 1965. Pleistocene pollen analysis and biogeography of the Southwest. pp. 433-451. In: Wright, H.E. Jr. and D.G. Frey (eds.), *The Quarternary of the United States.* Princeton University Press. Princeton, New Jersey.

Martin, S.C. 1975. *Ecology and Management of Southwestern Semidesert Grass-shrub Ranges: The Status of Our Knowledge.* USDA Forest Service Research Paper RM-156. 39 pp. Rocky Mountain Forest and Range Experiment Station. Fort Collins, Colo.

Martin, William C. and Charles R. Hutchins. 1980. *A Flora of New Mexico.* R. Gantner Verlag Kommanditgesellschaft. Vaduz, Germany.

Martinez, M. 1945. *Las Pinaceas Mexicanas.* Instituto de Biologia. Mexico, D.F.

Martinez, M. 1948. *Los Pinos Mexicanos.* Botas. Mexico, D.F.

Mason, H.L. 1932. A phylogenetic series of the California closed-cone pines suggested by the fossil record. *Madroño.* 2:39-55.

McCoy, C.L., Jr. 1970. A new alligator lizard (genus *Gerrhonotus*) from the Cuatro Cienegas Basin, Coahuila, Mexico. *Southwestern Naturalist* 15:37-44.

McNaughton, S.J. 1968. Structure and function in California grasslands. *Ecology* 49:962-972.

Mearns, E.A. 1907. *Mammals of the Mexican boundary of the United States. A Descriptive Catalogue of the Species of Mammals Occurring in That Region; With a General Summary of the Natural History, and a List of Trees.* 530 pp. Bulletin of the U.S. National Museum 56.

Meek, S.E. 1902. A contribution to the ichthyology of Mexico. *Field Columbian Museum Zoological Series* 3:63-128.

Meek, S.E. 1903. Distributions of the fresh-water fishes of Mexico. *American Naturalist* 37:771-784.

Meek, S.E. 1904. The freshwater fishes of Mexico north of the Isthmus of Tehuantepec. *Fieldiana, Zoology* 5:1-252.

Merhinger, P.J., Jr. 1967. Pollen analysis of the Tule Springs site, Nevada. pp. 130-200. In: Wormington, H.R. and D.E. Ellis (eds.), *Pleistocene Studies in Southern Nevada.* Nevada State Museum of Anthropology Paper 13. Carson City.

Merkle, J. 1952. An analysis of a pinyon-juniper community at Grand Canyon, Arizona. *Ecology* 33:375-384.

Merkle, J. 1954. An analysis of the spruce-fir community on the Kaibab Plateau, Arizona. *Ecology* 35:316-322.

Merkle, J. 1962. Plant communities of the Grand Canyon area, Arizona. *Ecology* 43:698-711.

Merriam, C.H. 1890. *Results of a Biological Survey of the San Francisco Mountains Region and Desert of the Little Colorado in Arizona.* 136 pp. USDA, North American Fauna 3. Washington, D.C.

Merriam, C.H. 1894a. The geographical distribution of life in North America with special reference to the mammalia. *Proceedings of the Biological Society of Washington* 7:1-64.

Merriam, C.H. 1894b. Laws of temperature control of the geographic distribution of terrestrial animals and plants. *National Geographic Magazine* 6:229-238.

Merriam, C.H. 1898. *Life-zones and Crop-zones of the United States.* 79 pp. USDA, Division of Biological Survey, Bulletin 10.

Metcalf, A.L. 1966. Fishes of the Kansas River system in relation to zoogeography of the Great Plains. *Publications of the University of Kansas Museum of Natural History* 17:23-189.

Meyer, E.R. 1973. Late-quaternary paleoecology of the Cuatro Cienegas basin, Coahuila, Mexico. *Ecology* 54:982-995.

Meyer, E.R. 1975. Vegetation and pollen rain in the Cuatro Cienegas Basin, Coahuila, Mexico. *Southwestern Naturalist* 20:215-224.

Millar, J.B. 1976. *Wetland Classification in Western Canada: A Guide to Marshes and Shallow Open Water Wetlands in the Grasslands and Parklands of the Prairie Provinces.* 38 pp. Canadian Wildlife Service Report 37.

Miller, A.H. 1932. The summer distribution of certain birds in central and northern Arizona. *Condor* 34:96-99.

Miller, A.H. 1946. Vertebrate inhabitants of the pinon association in the Death Valley region. *Ecology* 27:54-60.

Miller, A.H. 1951. An analysis of the distribution of the birds of California. *University of California Publications in Zoology* 50:531-644.

Miller, A.H. 1955. The avifauna of the Sierra del Carmen of Coahuila, Mexico. *Condor* 57:154-178.

Miller, A.H. and L. Miller. 1951. Geographic variation of the screech owls of the deserts of western North America. *Condor* 53:1-177.

Miller, D.J. and R.N. Lea. 1972. *Guide to the Coastal Marine Fishes of California.* 236 pp. California Department of Fish and Game, Fisheries Bulletin 157. (Addendum, 1976. Division of Agricultural Sciences, University of California 4065:237-249.)

Miller, R.R. 1946. The probable origin of the soft-shelled turtle in the Colorado River Basin. *Copeia* 1946:46.

Miller, R.R. 1948. *The Cyprinodont Fishes of the Death Valley System of Eastern California and Southwestern Nevada.* 155 pp. Miscellaneous Publications of the University of Michigan Museum of Zoology 68.

Miller, R.R. 1959. Origin and affinities of the freshwater fish fauna of western North America. pp. 187-222. In: Hubbs, C.L. (ed.), *Zoogeography.* American Association for the Advancement of Science. Special Publication 51.

Miller, R.R. 1961. Man and the changing fish fauna of the American Southwest. *Papers of the Michigan Academy of Science. Arts and Letters* 46:365-404.

Miller, R.R. 1973. *Two New Fishes, Gila bicolor snyderi and Catostomus fumeiventris, from the Owens River Basin, California.* 19 pp. Occasional Papers of the Museum of Zoology 667. University of Michigan, Ann Arbor.

Miller, R.R. 1976. An evaluation of Seth E. Meek's contributions to Mexican ichthyology. *Fieldiana, Zoology* 69:1-31.

Miller, R.R. 1978. Composition and derivation of the native fish fauna of the Chihuahuan Desert region. pp. 365-381. In: Wauer, R.H. and D.H. Riskind (eds.), *Transactions of the Symposium on the Biological Resources of the Chihuahuan Desert Region, United States and Mexico. Alpine, Texas. 1974.* USDI National Park Service Transactions and Proceedings Series 3. Washington, D.C.

Miller, R.R. and C.L. Hubbs. 1960. *The Spiny-rayed Cyprinid Fishes (Plagopterini) of the Colorado River System in Western North America.* 39 pp. Miscellaneous Publications of the University of Michigan Museum of Zoology 115.

Miller, R.R. and A.A. Echelle. 1975. *Cyprinodon tularosa,* a new cyprinodontid fish from the Tularosa basin, New Mexico. *Southwestern Naturalist* 19:365-377.

Milstead, W.W. 1960. Relict species of the Chihuahuan Desert. *Southwestern Naturalist* 5:75-88.

Minckley, W.L. 1965. Native fishes as natural resources. pp. 48-60. In: Gardner, J.L. (ed.), *Native Plants and Animals as Resources in Arid Lands of the Southwestern United States.* Commission on Desert Arid Zones Research, American Association for the Advancement of Science. Contribution 8.

Minckley, W.L. 1969. *Environments of the Bolson of Cuatro Cienegas, Coahuila, Mexico, with Special Reference to the Aquatic Biota.* 65 pp. Science Series 2. Texas Western Press. University of Texas, El Paso.

Minckley, W.L. 1973. *Fishes of Arizona.* 293 pp. Arizona Game and Fish Department. Phoenix.

Minckley, W.L. 1978. Endemic fishes of the Cuatro Cienegas basin, northern Coahuila, Mexico. pp. 383-404. In: Wauer, R.H. and D.H. Riskind (eds.), *Transactions of the Symposium on the Biological Resources of the Chihuahuan Desert Region, United States and Mexico. Alpine, Texas, 1974.* USDI National Park Service Transactions and Proceedings Series 3. Washington, D.C.

Minckley, W.L. 1979. *Aquatic Habitats and Fishes of the Lower Colorado River, Southwestern United States.* 478 pp. Final Report. USDI Bureau of Reclamation. Boulder City, Nev.

Minckley, W.L. and W.E. Barber. 1971. Some aspects of the biology of the longfin dace, a cyprinid fish characteristic of streams in the Sonoran Desert. *Southwestern Naturalist* 15:459-464.

Minckley, W.L. and T.O. Clark. 1979. Terrestrial vegetation. pp. 85-121. In: Minckley, W.L. (ed.), *Resource Inventory for the Gila River Complex, Eastern Arizona.* Final Report. USDI Bureau of Land Management. Safford, Arizona.

Minckley, W.L. and J.E. Deacon. 1968. Southwestern fishes and the enigma of "endangered species." *Science* 159:1424-1432.

Minckley, W.L., J.N. Rinne, and J.E. Johnson. 1977. *Status of the Gila Topminnow and its Co-occurence with Mosquitofish.* USDA Forest Service Research Paper RM-198. 8 pp. Rocky Mountain Forest and Range Experiment Station. Fort Collins, Colo.

Mitchell, R.S. 1973. Phytogeography and comparative floristics. In: Lloyd, R.M. and R.S. Mitchell, *A Flora of the White Mountains, California and Nevada.* University of California Press. Berkeley.

Moir, W.H. 1967. The subalpine tall grass, *Festuca thurberi,* community of Sierra Blanca, New Mexico. *Southwestern Naturalist* 12:321-328.

Mooney, H.A. 1973. Plant communities and vegetation. In: Lloyd, R.M. and R.S. Mitchell, *A Flora of the White Mountains, California and Nevada.* University of California Press. Berkeley.

Mooney, H.A. 1977. Southern coastal scrub. pp. 471-489. In: Barbour, M.G. and J. Major (eds.), *Terrestrial Vegetation of California.* John Wiley and Sons. New York.

Mooney, H.A. and E.L. Dunn. 1970a. Convergent evolution of Mediterranean-climate evergreen sclerophyll shrubs. *Evolution* 24:292-303.

Mooney, H.A. and E.L. Dunn. 1970b. Photosynthetic systems of Mediter-ranean-climate shrubs and trees of California and Chile. *American Naturalist* 104:447-453.

Mooney, H.A., G. St. Andre, and R.D. Wright. 1962. Alpine and subalpine vegetation patterns in the White Mountains of California. *American Midland Naturalist* 68:257-273.

Moore, R.T. 1945. The transverse volcanic biotic province of central Mexico and its relationship to adjacent provinces. *Transactions of the San Diego Society of Natural History* 10:217-235.

Moore, T.C. 1965. Origin and disjunction of the alpine tundra flora on San Francisco Mountain, Arizona. *Ecology* 46:860-864.

Morafka, D.J. 1977. Is there a Chihuahuan Desert? A quantitative evaluation through a herpetofaunal perspective. pp. 437-454. In: Wauer, R.H. and D.H. Riskind (eds.), *Transactions of the Symposium on the Biological Resources of the Chihuahuan Desert Region, U.S. and Mexico. Alpine, Texas.* USDI National Park Service Transactions and Proceedings Series 3.

Moran, R.V. 1977. Palms in Baja California. *Environment Southwest* 478:10-14.

Morris, M. 1935. Natural vegetation of Colorado. map. In: Gregg, R.E. (ed.), *The Ants of Colorado.* University of Colorado Press. Boulder.

Moyle, P.B. 1973. Effects of introduced bullfrogs, *Rana catesbeiana,* on the native frogs of the San Joaquin Valley, California. *Copeia* 1973:18-22.

Moyle, P.B. 1976. *Inland Fishes of California.* 405 pp. University of California Press. Berkeley.

Moyle, P.B. and R. Nichols. 1973. Ecology of some native and introduced fishes of the Sierra Nevada Foothills in central California. *Copeia* 1973:478-490.

Mueller-Dombois, D. and H. Ellenberg. 1974. *Aims and Methods of Vegetation Ecology.* 547 pp. John Wiley and Sons. New York.

Muller, C.H. 1937. Vegetation in the Chisos Mountains, Texas. *Transactions of the Texas Academy of Sciences* 20:5-31.

Muller, C.H. 1939. Relations of the vegetation and climatic types in Nuevo Leon, Mexico. *American Midland Naturalist* 21:687-729.

Muller, C.H. 1940. Plant succession in the Larrea-Flourensia climax. *Ecology* 21:206-212.

Muller, C.H. 1947. Vegetation and climate of Coahuila, Mexico. *Madroño* 9:33-57.

Muller, C.H., W. Muller, and B. Haines. 1964. Chaparral succession in a San Gabriel Mountain area of California. *Ecology* 45:353-360.

Mulroy, Thomas W. and Philip W. Rundel. 1977. Annual plants: Adaptations to desert environments. *Bioscience* 27:109-114.

Munz, P.A. and D.D. Keck. 1949. California plant communities. *El Aliso.* 2:87-105.

Munz, P.A. and D.D. Keck. 1950. California plant communities–supplement. *El Aliso* 2:199-202.

Munz, P.A. and D.D. Keck. 1959. *A California Flora.* 1681 pp. University of California Press. Berkeley.

Munz, P.A. 1935. *A Manual of Southern California Botany.* Claremont College. Claremont, California.

Munz, P.A. 1974. *A Flora of Southern California.* University of California Press. Berkeley.

Munz, P.A. and David D. Keck. 1968. *A California Flora and Supplement.* University of California Press. Berkeley.

Musick, H. Brad. 1975. Bareness of desert pavement in Yuma County, Arizona. *Journal of the Arizona Academy of Science* 10:24-28.

Myers, G.S. 1964. A brief sketch of the history of ichthyology in America to the year 1850. *Copeia* 1964:33-41.

Myers, G.S. 1965. *Gambusia,* the fish destroyer. *Tropical Fish Hobbyist* 1965:31-35.

Nagy, J.G., H.W. Steinoff, and G.M. Ward. 1964. Effect of essential oils of sagebrush on deer rumen microbial functions. *Journal of Wildlife Management* 28:785-790.

Natural Vegetation Committee. 1973. *Landscape with Native Arizona Plants.* 194 pp. Arizona Chapter of the Soil Conservation Society of America. University of Arizona Press. Tucson.

Naveh, Z. 1967. Mediterranean ecosystems and vegetation types in California and Israel. *Ecology* 48:445-459.

Needham, P.R. and R. Grad. 1959. Rainbow trout in Mexico and California-with notes on the cutthroat series. *University of California Publications in Zoology.* 67(1):1-12.

Neff, J.A. 1940. Range, population and game status of the western white winged dove in Arizona. *Journal of Wildlife Management* 4(2):117-127.

Nelson, Edward W. 1922. *Lower California and its Natural Resources.* 194 pp. National Academy of Sciences. Washington, D.C.

New Mexico Interagency Range Committee. 1978. *Potential Natural Vegetation of New Mexico.* map. Interagency Range Committee Report 11.

Nichol, A.A. 1937. The natural vegetation of Arizona. *University of Arizona Agriculture Experiment Station, Technical Bulletin* 68:181-222.

Nichol, A.A. 1952. The natural vegetation of Arizona (Revision of Technical Bulletin 68, 1937). *University of Arizona Agriculture Experiment Station Technical Bulletin* 127:189-230.

Nickerson, Mona F., Glen E. Brink, and Charles Feddema. 1976. *Principal Range Plants of the Central and Southern Rocky Mountains: Names and Symbols.* USDA Forest Service General Technical Report RM-20. 121 pp. Rocky Mountain Forest and Range Experiment Station. Fort Collins, Colo.

Niering, W.A., R.H. Whittaker, and C.H. Lowe. 1963. The saguaro: A population in relation to environment. *Science* 142:15-23.

Norris, K.S. 1974. To Carl Leavitt Hubbs, a modern pioneer naturalist on the occasion of his eightieth year. *Copeia* 1974:581-610.

O'Brien, G.P. 1975. *A Study of the Mexican Duck (Anas diazi) in Southeastern Arizona.* 43 pp. Arizona Game and Fish Department Special Report 5. Phoenix.

Odum, E.P. 1945. The concept of the biome as applied to the distribution of North American birds. *Wilson Bulletin* 57:191-201.

O'Farrell, M.J. and W.G. Bradley. 1970. Activity patterns of bats over a desert spring. *Journal of Mammalogy* 51:18-26.

Ohmart, R.D., W.O. Deason, and C. Burke. 1977. A riparian case history: The Colorado River. pp. 35-47. In: Johnson, R.R. and D.A. Jones (technical coordinators), *Importance, Preservation and Management of Riparian Habitat: A Symposium.* USDA Forest Service General Technical Report RM-43. 217 pp. Rocky Mountain Forest and Range Experiment Station. Fort Collins, Colo.

Oosting, H.J. 1950. *The Study of Plant Communities.* Second edition. 440 pp. W.H. Freeman and Co. San Francisco, Calif.

Orians, G.H. and O.T. Solbrig, editors. 1977. *Convergent Evolution in Warm Deserts: An Examination of Strategies and Patterns in Deserts of Argentina and the United States.* 352 pp. U.S./I.B.P. Synthesis Series, 3. Academic Press. New York.

Ornduff, Robert. 1974. *An Introduction to California Plant Life.* 152 pp. University of California Press. Berkeley.

Orr, R.T. 1972. *Marine Mammals of California.* 64 pp. University of California Press. Berkeley.

Parish, S.B. 1930. Vegetation of the Mohave and Colorado deserts of southern California. *Ecology* 3:481-499.

Parker, K.W. 1945. Juniper comes to the grasslands: why it invades southwestern grassland–suggestions on control. *Amer. Cattle Producer* 27:12-24, 30-32.

Parker, K.W. and S.C. Martin. 1952. *The Mesquite Problem on Southern Arizona Ranges.* 70 pp. U.S. Department of Agriculture Circular 908. Washington, D.C.

Pase, C.P. 1965. *Shrub Seedling Regeneration After Controlled Burning and Herbicidal Treatment of Dense Pringle Manzanita Chaparral.* USDA Forest Service Research Note RM-56. 2 pp. Rocky Mountain Forest and Range Experiment Station. Fort Collins, Colo.

Pase, C.P. and R.R. Johnson. 1968. *Flora and Vegetation of the Sierra Ancha Experimental Forest, Arizona.* USDA Forest Service Research Paper RM-41. 19 pp. Rocky Mountain Forest and Range Experiment Station. Fort Collins, Colo.

Pase, C.P. and E.F. Layser. 1977. Classification of riparian habitat in the Southwest. pp. 5-9. In: Johnson, R.R. and D.A. Jones (technical coordinators), *Importance, Preservation and Management of Riparian Habitat: A Symposium.* USDA Forest Service General Technical Report RM-43. 217 pp. Rocky Mountain Forest and Range Experiment Station. Fort Collins, Colo.

Patric, J.H. and R.L. Hanes. 1964. Chaparral succession in a San Gabriel Mountain area of California. *Ecology* 45:353-360.

Pattie, J.O. 1962. *The Personal Narrative of James O. Pattie.* (ed. W.H. Goetzmann). J.B. Lippincott, Co. Philadelphia, Pa.

Patton, David R. and J.R. Jones. 1977. *Managing Aspen for Wildlife in the Southwest.* USDA Forest Service General Technical Report RM-37. Rocky Mountain Forest and Range Experiment Station. Fort Collins, Colo.

Patton, David R. 1978a. *RUNWILD: A Storage and Retrieval System for Wildlife Habitat Information.* USDA Forest Service General Technical Report RM-51. 8 pp. Rocky Mountain Forest and Range Experiment Station. Fort Collins, Colo.

Patton, David R. 1978b. *Vertebrate Checklist of Arizona and New Mexico.* Wildlife Habitat Bulletin 5. U.S. Department of Agriculture, Forest Service. Albuquerque, N.Mex.

Paulson, H.A. 1975. *Range Management in the Central and Southern Rocky Mountains: A Summary of the Status of our Knowledge of Range Ecosystems.* USDA Forest Service Research Paper RM-154. 34 pp. Rocky Mountain Forest and Range Experiment Station. Fort Collins, Colo.

Paulson, L.J. and J.R. Baker. 1980. Nutrient interactions among reservoirs on the Colorado River. pp. 1647-1656. In *Symposium on Surface Water Impoundments.* Amer. Soc. Chemical Engineers. Minneapolis, Minn.

Pearson, G.A. 1920a. Factors controlling the distribution of forest types. Part I. *Ecology* 1:139-159.

Pearson, G.A. 1920b. Factors controlling the distribution of forest types. Part II. *Ecology* 1:289-308.

Pearson, G.A. 1930. Studies of climate and soil in relation to forest management in the southwestern United States. *Ecology* 18:139-144.

Pearson, G.A. 1931. *Forest Types in the Southwest Determined by Climate and Soil.* 144 pp. U.S. Department of Agriculture Technical Bulletin 247. Washington, D.C.

Peterson, P.V. 1966. *Native Trees of Southern California.* 136 pp. University of California Press. Berkeley.

Peterson, Roger Tory and Edward L. Chalif. 1973. *A Field Guide to Mexican Birds and Adjacent Central America.* 298 pp. Houghton Mifflin Co. Boston, Mass.

Pfister, R.D., and B.L. Kovalchik, S.F. Arno, and R.C. Presby. 1977. *Forest Habitat Types of Montana.* USDA Forest Service. Intermountain Forest and Range Experiment Station. Ogden, Utah.

Pfleiger, W.F. 1975. *Fishes of Missouri.* 343 pp. Missouri Conservation Commission. Jefferson City.

Phillips, A.R. 1939. *The Faunal Areas of Arizona, Based on Bird Distribution.* Masters thesis. University of Arizona. Tucson.

Phillips, A., J. Marshall, and G. Monson. 1964. *The Birds of Arizona.* 212 pp. University of Arizona Press. Tucson.

Piemeisel, R.L. 1938. *Changes in Weedy Plant Cover on Cleared Sagebrush Land and Their Probable Causes.* U.S. Department of Agriculture Technical Bulletin 654. Washington, D.C.

Piemeisel, R.L. 1951. Causes affecting change and rate of change in a vegetation of annuals in Idaho. *Ecology* 32:53-72.

Pike, Z.M. 1810. *Southwestern Expedition.* Lakeside Press. Chicago, Ill.

Pinkava, D.J. 1978. Vegetation and flora of the Cuatro Cienegas basin, Coahuila, Mexico. pp. 327-333. In: Wauer, R.H. and D.H. Riskind (eds.), *Transactions of the Symposium on the Biological Resources of the Chihuahuan Desert Region, United States and Mexico. Alpine, Texas. 1974.* USDI National Park Service Transactions and Proceedings Series 3. Washington, D.C.

Pinkava, D.J. 1979. Vegetation and flora of the Bolson of Cuatro Cienegas region, Coahuila, Mexico, I. *Boletin de la Sociedad Botanica de Mexico.* 38:35-73.

Pister, E.P. 1974. Desert fishes and their habitats. *Transactions of the American Fisheries Society* 103:531-540.

Pister, E.P. 1979. Obituary–Carl Leavitt Hubbs. *Fisheries* 4:28.

Pitelka, F.A. 1941. Distribution of birds in relation to major biotic communities. *American Midland Naturalist* 25:113-137.

Pitelka, F.A. 1943. Review of Dice (1943), The Biotic provinces of North America. *Condor* 45:203-204.

Pitelka, F.A. 1951a. Speciation and ecologic distribution in American jays of the genus *Aphelocoma. University of California Publications in Zoology* 50:195-464.

Pitelka, F.A. 1951b. Ecologic overlap and interspecific strife in breeding populations of Anna and Allen hummingbirds. *Ecology* 32:641-661.

Pond, F.W. 1968. *Changes in Grass Production on Ungrazed Converted Chaparral.* USDA Forest Service Research Note RM-98. 4 pp. Rocky Mountain Forest and Range Experiment Station. Fort Collins, Colo.

Porter, R.D. 1962. *Movements, Populations and Habitat Preferences of Three Species of Pocket Mice (Perognathus) in the Big Bend Region of Texas.* Unpublished Ph.D. dissertation. 347 pp. Texas A & M University. College Station, Texas.

Powell, A.M. and B.L. Turner. 1977. Aspects of the plant biology of the gypsum outcrops of the Chihuahuan Desert. pp. 315-325. In: Wauer, R.H. and D.H. Riskind (eds.), *Transactions of the Symposium on the Biological Resources of the Chihuahuan Desert Region, U.S. and Mexico. Alpine, Texas. 1974.* USDI National Park Service Transactions and Proceedings Series 3. Washington, D.C.

Powell, J.W. 1875. *Report on Exploration of the Colorado River of the West and its Tributaries.* U.S. Government Printing Office. Washington, D.C.

Pyke, Charles B. 1972. *Some Meteorological Aspects of the Seasonal Distribution of Precipitation in the Western United States and Baja California.* Water Resources Center Contribution 139. 205 pp. University of California. Berkeley.

Raitt, R. and S. Pimm. 1978. Temporal changes in northern Chihuahuan bird communities. pp. 579-589. In: Wauer, R.H. and D.H. Riskind (eds.), *Transactions of the Symposium on the Biological Resources of the Chihuahuan Desert Region, United States and Mexico. Alpine, Texas. 1974.* USDA National Park Service Transactions and Proceedings Series 3. Washington, D.C.

Rasmussen, D.I. 1941. Biotic communities of the Kaibab Plateau, Arizona. *Ecological Monographs* 11:229-275.

Raven, Peter H., Donald W. Kyhos, D.E. Breedlove, and W.W. Payne. 1968. Polyploidy in *Ambrosia dumosa* (Compositae: Ambrosieae). *Brittonia* 20(3): 205-211.

Ray, G.C. 1975. *A Preliminary Classification of Coastal and Marine Environments.* 26 pp. IUCN Occasional Paper 14.

Reed, E.L. 1930. Vegetation of the playa lakes in the Staked Plains of western Texas. *Ecology* 11:597-600.

Reeves, C.C., Jr. 1969. Pluvial Lake Palomas, northwestern Chihuahua, Mexico. pp. 143-154. In: *Guidebook of the Border Region.* New Mexico Geological Society, 20th Field Conference.

Reichenbacher, F.W. 1980. Unpublished M.S. thesis. University of Arizona, Tucson.

Reynolds, H.G. and R. Roy Johnson. 1964. *Habitat Relations of Vertebrates of the Sierra Ancha Experimental Forest.* USDA Forest Service Research Paper RM-4. 16 pp. Rocky Mountain Forest and Range Experiment Station. Fort Collins, Colo.

Ridgeway, R. and H. Friedman. 1901-1950. *Birds of North and Middle America.* U.S. National Museum Bulletin 50. 11 parts. Washington, D.C.

Rinne, J.N., W.L. Minckley and J.R. Hanson. 1981. Chemical treatment of Ord Creek, Apache County, Arizona, to re-establish Arizona trout. *Journal of the Arizona-Nevada Academy of Science* 16:74-78.

Robbins, C.S., B. Bruun, and H.S. Zim. 1966. *Birds of North America–a Guide to Field Identification.* 340 pp. Golden Press. New York.

Robbins, W.W. 1917. *Native Vegetation and Climate of Colorado and their Relation to Agriculture.* 56 pp. Colorado Agricultural Experiment Station Bulletin 224.

Roberts, W.G., and J.G. Howe, and J. Major. 1977. A survey of riparian forest flora and fauna in California. pp. 3-19. In: Sands, A. (ed.), *Riparian Forests in California.* Institute of Ecology Publication 15. University of California. Davis.

Robichaux, R. 1977. Geologic history of the riparian forest of California. pp. 21-23. In: Sands, A. (ed.), *Riparian Forest in California.* Institute of Ecology Publication 15. University of California. Davis.

Robinson, M.D. 1968. *Summer Aspects of a High Coniferous Forest in the Chiricahua Mountains, Arizona.* M.S. thesis. 55 pp. Department of Biological Sciences. University of Arizona, Tucson.

Robinson, T.W. 1965. *Introduction. Spread. and Areal Extent of Saltcedar (Tamarix) in the Western States.* 12 pp. U.S. Geological Survey Professional Paper 491-A.

Rowlands, Peter G. 1978. *The Vegetation Dynamics of the Joshua Tree (Yucca brevifolia Engelm.) in the Southwestern United States of America.* Ph.D. dissertation. 192 pp. University of California , Riverside.

Ruffner, J.A. (ed.) 1978. *Climates of the States.* 1183 pp. Gale Research Company. Detroit, Mich.

Rusby, H.H. 1889. General floral features of the San Francisco and Mogollon mountains of Arizona and New Mexico, and their adjacent regions. *Transactions of the New York Academy of Science* 8:76-81.

Rzedowski, J. 1957. Vegetacion de las partes aridas de los Estados de San Luis Potosi y Zacatecas. *Rev. Soc. Mex. Hist. Nat.* 18(1-4):49-101.

Rzedowski, J. 1963. El extrema boreal del bosque tropical siempre verde en Norteamerica continental. Vegetatio. *Acta Geobotanica* 11(4):173-198.

Rzedowski, J. 1965. Vegetacion del Estado de San Luis Potosi. *Act. Cient. Potos. Mex.* 5(1-2):1-291.

Rzedowski, J. 1973. Geographical relationships of the flora of Mexican dry regions. pp. 61-72. In: Graham, Alan (ed.), *Vegetation and Vegetational History of Northern Latin America.* 393 pp. Elsevier. New York.

Rzedowski, J. 1978. *Vegetación de México.* 432 pp. Editorial Limosa S.A., México D.F.

Sands, A. (ed.) 1977. *Riparian Forests in California–Ecology and Conservation.* 122 pp. Institute of Ecology Publication 15. University of California. Davis.

Sauer, C.O. 1950. Grassland climax, fire, and man. *Journal of Range Management* 3:16-21.

Sawyer, J.O., D.A. Thornburgh, and D.R. Griffin. 1977. Mixed evergreen forest. pp. 359-381. In: Barbour, M.G. and J. Major (eds.), *Terrestrial vegetation of California.* John Wiley and Sons. New York.

Sayers, S.S. 1974. Natural vegetation of state, private and Bureau of Land Management lands in Arizona. Arizona State Land Department. *Arizona Land Marks* 4(4):1-45 (includes map). Phoenix.

Schaack, C.G. 1970. *A Flora of the Arctic-alpine Vascular Plants of the San Francisco Mountain, Arizona.* M.S. thesis. Northern Arizona University, Flagstaff.

Schmidly, D.J. 1977. *The Mammals of Trans-Pecos Texas Including Big Bend National Park and Guadalupe Mountains National Park.* 225 pp. Texas A & M University Press. College Station, Texas.

Schmidt, R.H., Jr. 1975. *The Climate of Chihuahua, Mexico.* 50 pp. Institute of Atmospheric Physics, Technical Reports on the Meteorology and Climatology of Arid Regions 23. University of Arizona, Tucson.

Schmidt, R.H., Jr. 1979. A climatic delineation of the 'real' Chihuahuan Desert. *Journal of Arid Environments* 2:243-250.

Schmutz, Ervin M., Charles C. Michaels, and B. Ira Judd. 1967. Boysag Point, a relict area on the north rim of Grand Canyon in Arizona. *Journal of Range Management* 20(6):363-369.

Schoenherr, A.A. 1977. Density dependent and density independent regulation of reproduction in the Gila topminnow, *Poeciliopsis occidentalis* (Baird and Girard). *Ecology* 58:438-444.

Schoenherr, A.A. 1979. Niche separation within a population of freshwater fishes in an irrigation drain near the Salton Sea, California. *Bulletin of the Southern California Academy of Science* 78:46-55.

Schubert, G.H. 1974. *Silviculture of Southwestern Ponderosa Pine: The Status of Our Knowledge.* 71 pp. USDA Forest Service Research Paper RM-123. Rocky Mountain Forest and Range Experiment Station. Fort Collins. Colo.

Schulman, E. 1954. Longevity under adversity in conifers. *Science* 119:395-399.

Schulman, E. 1958. Bristlecone pine, oldest known living thing. *National Geographic Magazine* 113:355-372.

Schultz, R.J. 1977. Evolution and ecology of unisexual fishes. pp. 277-331. In: Hecht, M.K., W.A. Steare and B. Wallace (eds.), *Evolutionary Biology.* Volume 10. Plenum Press. New York.

Scott, B.R. (ed.) 1971. *Vegetal Cover Map.* State of Nevada, Division of Water Resources. map.

Secretaria De Agricultura ya Fomento. 1939. *Atlas Climatologico de México.* Dirección de Geografica Meteorologia Hidrologia. México, D.F.

Seethaler, K. 1978. *Life History and Ecology of the Colorado Squawfish (Ptychocheilus lucius) in the Upper Colorado River Basin.* Unpublished M.S. Thesis. 156 pp. Utah State University, Logan.

Sellers, William D. and Richard H. Hill. 1974. *Arizona Climate 1931-1972.* Second edition. 616 pp. University of Arizona Press. Tucson.

Shantz, H.L. and R.L. Piemeisel. 1924. Indicator significance of the natural vegetation of the southwestern desert regions. *Journal of Agriculture Research* 28(8):721-801.

Shantz, H.L. and R.L. Piemeisel. 1940. *Types of Vegetation in Escalante Valley. Utah. as Indicators of Soil Conditions.* Technical Bulletin 713. U.S. Department of Agriculture. Washington, D.C.

Shantz, H.L. and R. Zon. 1924. *Natural Vegetation.* Atlas of American Agriculture. Part I, Section E (map). U.S. Department of Agriculture. Washington, D.C.

Shelford, V.E. 1932. Life-zones, modern ecology, and the failure of temperature summing. *Wilson Bulletin* 44:144-157.

Shelford, V.E. 1945. The relative merits of the life zone and biome concepts. *Wilson Bulletin* 47:248-252.

Shelford, V.E. 1963. *The Ecology of North America.* University of Illinois Press. Urbana.

Shelford, V.E. and F. Shreve. (eds.) 1926. *Naturalist's Guide to the Americas.* 761 pp. Williams and Wilkins, Co. Baltimore, Md.

Shields, L.M. and L.J. Gardner. 1961. *Bioecology of the Arid and Semiarid Lands of the Southwest.* 69 pp. New Mexico Highlands University Bulletin 212.

Short, H.L. and C.Y. McCulloch. 1977. *Managing Pinyon-juniper Ranges for Wildlife.* USDA Forest Service General Technical Report RM-47. 10 pp. Rocky Mountain Forest and Range Experiment Station. Fort Collins, Colo.

Shreve, F. 1915. *The Vegetation of a Desert Mountain Range as Conditioned by Climatic Factors.* 112 pp. Carnegie Institution of Washington Publication 217.

Shreve, F. 1917. A map of the vegetation of the United States. *Geog. Review* 3:119-125.

Shreve, F. 1917. The physical control of vegetation in rain forest and desert mountain. *Plant World* 20:135-141.

Shreve, F. 1919. A comparison of the vegetational features of two desert mountain ranges. *Plant World* 22:291-307.

Shreve, F. 1922. Conditions indirectly affecting vertical distribution on desert mountains. *Ecology* 3:269-274.

Shreve, F. 1924. Soil temperature as influenced by altitude and slope exposure. *Ecology* 5:128-136.

Shreve, F. 1925. Ecological aspects of the deserts of California. *Ecology* 6:93-103.

Shreve, F. 1927. The vegetation of a coastal mountain range. *Ecology* 8:28-44.

Shreve, F. 1929. Changes in desert vegetation. *Ecology* 10:364-373.

Shreve, F. 1934. Vegetation of the northwestern coast of Mexico. *Bulletin of the Torrey Botanical Club* 61:373-380.

Shreve, F. 1936a. The plant life of the Sonoran Desert. *Scientific Monthly* 42:195-213.

Shreve, F. 1936b. The transition from desert to chaparral in Baja California. *Madroño* 3:257-264.

Shreve, F. 1937. The vegetation of the Cape Region of Baja California. *Madroño* 4:105-113.

Shreve, F. 1937. The vegetation of Sinaloa. *Bulletin of the Torrey Botanical Club.* 64:605-613.

Shreve, F. 1939. Observations on the vegetation of Chihuahua. *Madroño* 5:1-13.

Shreve, F. 1942a. The desert vegetation of North America. *Botanical Review* 8:195-246.

Shreve, F. 1942b. Grassland and related vegetation in northern Mexico. *Madroño* 6:190-198.

Shreve, F. 1942c. The vegetation of Arizona. pp. 10-23. In: Kearney, T.H. and R.H. Peebles. *Flowering Plants and Ferns of Arizona.* U.S. Department of Agriculture Miscellaneous Publication 423. Washington, D.C.

Shreve, F. 1944. Rainfall of norther Mexico. *Ecology* 25:105-111.

Shreve, F. 1951. *Vegetation and Flora of the Sonoran Desert. Volume I. Vegetation.* 192 pp. Carnegie Institution of Washington Publication 591.

Shreve, F. 1964. Vegetation of the Sonoran Desert. Volume 1, Part I. In Shreve, Forrest and Ira L. Wiggins, *Vegetation and flora of the Sonoran Desert.* 2 volumes. Stanford University Press. Stanford, Calif.

Shreve, Forrest and Ira L. Wiggins. 1964. *Vegetation and Flora of the Sonoran Desert.* 2 Volumes. Stanford University Press, Stanford, Calif.

Simmons, N.M. 1966. Flora of the Cabeza Prieta Game Range. *Journal of the Arizona Academy of Sciences* 4:93-104.

Simpson, B.B. (ed.) 1977. *Mesquite: Its Biology in Two Desert Scrub Ecosystems.* 272 pp. U.S./I.B.P. Synthesis Series 4. Academic Press. New York.

Small, A. 1974. *The Birds of California.* Winchester Press. New York.

Smith, A.P. 1908. Some data and records from the Whetstone Mountains, Arizona. *Condor* 10:75-78.

Smith, E.L. 1974. *Established Natural Areas in Arizona—A Guidebook for Scientists and Educators.* 300 pp. Arizona Academy of Sciences, for Office of Economic Planning and Development, State of Arizona. Phoenix.

Smith, G.R. 1966. *Distribution and Evolution of the North American Catostomid Fishes of the subgenus Pantosteus, genus Catostomus.* 132 pp. Miscellaneous Publications of the University of Michigan Museum of Zoology 29.

Smith, H.M. 1940. An analysis of the biotic provinces of Mexico, as indicated by the distribution of the lizards of the genus *Sceloporus.* Escuela Nacional de Ciencias Biologicas 2:95-110.

Smith, H.M. and Rozella B. Smith. 1976. *Synopsis of the Herpetofauna of Mexico.* J. Johnson Co. North Bennington, Vt.

Smith, H.V. 1956. *The Climate of Arizona.* 99 pp. Agricultural Experiment Station Bulletin 279. University of Arizona, Tucson.

Snyder, J.O. 1915. Notes on a collection of fishes made by Dr. E.A. Mearns from rivers tributary to the Gulf of California. *Proceedings of the U.S. National Museum* 49:573-485.

Society of American Foresters. [1954] 1962. *Forest Cover Types of North America.* (exclusive of Mexico). 66 pp. Society of American Foresters. Washington, D.C.

Solbrig, Otto T. 1972. New approaches to the study of disjunctions with special emphasis on the American amphitropical desert disjunctions. pp. 85-100. In: Valentine, D.H. (ed.), *Taxonomy, Phytogeography and Evolution*. Academic Press. London.

Soltz, D.L. and R. Naiman. 1978. *The Natural History of Native Fishes in the Death Valley System*. 76 pp. Natural History Museum of Los Angeles County, Science Series 30.

Southwood, T.R.E. 1966. *Ecological Methods*. 391 pp. Methuen and Co. London.

Spangle, P. and M. Sutton. 1949. The botany of Montezuma Well. *Plateau* 22:11-19.

Spencer, J.S. 1966. *Arizona's Forests*. USDA Forest Service Resource Bulletin INT-6. 55 pp. Intermountain Forest and Range Experiment Station. Ogden, Utah.

Spicer, E.H. 1962. *Cycles of Conquest*. 609 pp. University of Arizona Press. Tucson.

Springfield, H.W. 1976. *Characteristics and Management of Southwestern Pinyon-juniper Rangers: The Status of Our Knowledge*. USDA Forest Service Research Paper RM-160, 31 pp. Rocky Mountain Forest and Range Experiment Station, Fort Collins, Colo.

Stager, K.E. 1954. Birds of the Barranca de Cobre region of southwestern Chihuahua, Mexico. *Condor* 56:21-32.

Standley, P.C. 1915. Vegetation of the Brazos Canyon, New Mexico. *Plant World* 18:179-191.

Standley, P.C. 1920-26. *Trees and Shrubs of Mexico*. 1721 pp. Contributions from the United States National Herbarium 23. Smithsonian Press. Washington, D.C.

Stebbins, R.C. 1954. *Amphibians and Reptiles of Western North America*. 536 pp. McGraw-Hill Book Co. New York.

Stebbins, R.C. 1966. *A Field Guide to Western Reptiles and Amphibians*. 279 pp. Peterson field guide series 16. Houghton Mifflin Co. Boston.

Steenbergh, W.F. and C.H. Lowe. 1976. Ecology of the Saguaro: I. the role of freezing weather in a warm-desert plant population. pp. 49-92. In: *Research in the Parks*. National Park Centennial Symposium. U.S. Department of the Interior, National Park Service Symposium Series 1. Washington, D.C.

Steenbergh, W.F. and C.H. Lowe. 1977. *Ecology of the Saguaro: II. Reproduction. Germination. Establishment. Growth. and Survival of the Young Plant*. 242 pp. U.S. Department of the Interior, National Park Service Scientific Monograph Series 8.

Steenbergh, W.F. and P.L. Warren. 1977. *Preliminary Ecological Investigation of Natural Community Status at Organ Pipe Cactus National Monument*. 152 pp. USDI Cooperative National Park Resources Studies Unit, University of Arizona Technical Report 3.

Stephens, F. 1885. Notes of an ornithological trip in Arizona and Sonora. *Auk* 2:225-231.

Stevens, J.S. 1905. Life areas of California. *Transactions of the San Diego Society of Natural History* 1:1-25.

Stewart, R.E. and H.A. Kontrud. 1971. *Classification of Natural Ponds and Lakes in the Glaciated Prairie Region*. 57 pp. Bureau of Sport Fisheries and Wildlife Resource Publication 92. USDI, Washington, D.C.

Storer, T.I. and R.L. Usinger. 1974. *Sierra Nevada Natural History*. 374 pp. University of California Press. Berkeley.

Strain, W.S. 1970. Bolson integration in the Texas-Chihuahuan border region. pp. 88-90. In: *Geology of the Southern Quitman Mountains Area, Trans-Pecos. Texas. Symposium and Guidebook*. Permian Basin Section, Society of Economic Paleontologists and Murerologists. Midland, Texas.

Strong, Charles W. 1966. An improved method of obtaining density from line-transect data. *Ecology* 47(2):311-313.

Stuart, L.C. 1964. Fauna of Middle America. pp. 316-361. In West, R.C. (ed.), *Handbook of Middle American Indians. Volume 1*. University of Texas Press. Austin.

Sturdevant, G.E. 1927. Flora of the Tonto Platform. *Grand Canyon Nat. Notes* 1:1-2.

Stutz, Howard C. 1978. Explosive evolution of perennial *Atriplex* in western America. pp. 161-168. In: Harper, K.T. and J.L. Reveal, *Intermountain Biogeography: A Symposium*. Great Basin Naturalist Memoirs No. 2.

Sulley, J.M., M.A. Romera, and R.D. Smith. 1972. *Amaragosa Canyon—Dumont Dunes. Proposed Natural Area. California*. Pupfish Habitat Preservation Committee. 118 pp.

Sutton, G.M. and A.R. Phillips. 1942. June bird life of the Papago Indian Reservation, Arizona. *Condor*. 44:57-65.

Sutton, M. 1952. A botanical reconnaissance in Oak Creek Canyon. *Plateau* 25:30-42.

Swarth, H.S. 1914. A distributional list of the birds of Arizona. *Pacific Coast Avifauna* 10:1-133.

Swarth, H.S. 1920. *Birds of the Papago Saguaro National Monument and the Neighboring Region. Arizona*. 63 pp. U.S. Department of the Interior, National Park Service.

Swarth, H.S. 1929. Faunal areas of southern Arizona; a study in animal distribution. *Proceedings of the California Academy of Sciences* 18:267-383.

Sweeney, J.R. 1956. Responses of vegetation to fire. *University of California Publications in Botany* 28:143-250.

Sykes, G. 1937. *The Colorado Delta*. 193 pp. Carnegie Institution of Washington Publication 460.

Tarp, F.H. 1952. A revision of the family Embiotocidae (the surfperches). *Calif. Dept. of Fish and Game Fisheries Bulletin* 88:1-99.

Taylor, D.W. 1966. A remarkable snail fauna from Coahuila, Mexico. *Veliger* 9:151-228, 11 plates.

Taylor, Ronald J. and Thomas F. Patterson. 1980. Biosystematics of Mexican spruce species and populations. *Taxon* 29(4):421-429.

Taylor, W.P. 1934. Significance of extreme or intermittent conditions in distribution of species and management of natural resources, with a restatement of Liebig's Law of the Minimum. *Ecology* 15:374-379.

Taylor, W.P. 1935. *Ecology and Life History of the Porcupine (Erethizon epixanthum) as Related to the Forests of Arizona and the Southwestern United States*. 177 pp. Biological Sciences Bulletin 3. University of Arizona, Tucson.

Taylor, W.P. 1936. The pronghorned antelope in the Southwest. pp. 653-655. In: *North American Wildlife and Natural Resources Conference Transactions Volume 1*.

Taylor, W.P., editor. 1956. *The Deer of North America*. 668 pp. The Stackpole Co. Harrisburg, Penn., and Wildlife Management Institute. Washington, D.C.

Taylor, W.P., W.B. McDougall, C.C. Presnall, and K.P. Schmidt. 1946. The Sierra del Carmen in northern Coahuila (Mexico). *Texas Geographic Magazine* (Spring 1946).

Taylor, W.W. 1962. Tethered nomadism and water territoriality: An hypothesis. *Actas y Memorias. 35th Congress International Americanistas, 1962*:197-203.

Texas Game, Fish and Oyster Commission. 1945. *Principal Game Birds and Mammals of Texas*. Texas Game, Fish and Oyster Commission. Austin, Texas.

Tharp, B.C. 1939. *The Vegetation of Texas*. 74 pp. Texas Academy of Science. Texas Academy Publications in Natural History 1.

Thilenius, J.F. 1975. *Alpine Range Management in the Western United States—Principles. Practices, and Problems: The Status of our Knowledge*. 32 pp. USDA Forest Service Research Paper RM-157. Rocky Mountain Forest and Range Experiment Station. Fort Collins, Colo.

Thomson, D.A. 1973. *Preliminary Ecological Survey of Estero Soldado. Sonora, Mexico*. Final Report. 29 pp. Cella, Barr, Evans and Associates. Tucson, Ariz.

Thomson, D.A. and C.E. Lehner. 1976. Resilience of a rocky intertidal fish community in a physically unstable environment. *Journal of Experimental Marine Biology and Ecology* 22:1-29.

Thomson, D.A., L.T. Findley, and A.N. Kerstitch. 1979. *Reef Fishes of the Sea of Cortez*. 302 pp. John Wiley and Sons, Inc. New York.

Thornber, J.J. 1910. The grazing ranges of Arizona. *University of Arizona Agricultural Experiment Station Bulletin* 65:245-360.

Thornber, J.J. 1911. Plant acclimatization in southern Arizona. *Plant World* 14:15-23.

Thorne, R.J. 1977. Montane and subalpine forests of the Transverse and Peninsular ranges. In: Barbour, M.G. and J. Major (eds.), *Terrestrial Vegetation of California*. John Wiley and Sons. New York.

Tidestrom, I. 1925. *Flora of Utah and Nevada*. 665 pp. Contributions from the United States National Herbarium 25.

Tidestrom, I. and T. Kittell. 1941. *A Flora of Arizona and New Mexico*. Catholic University of America Press. Washington, D.C.

Todd, E.S. and A.W. Ebeling. 1966. Aerial respiration in the longjaw mudsucker, *Gillichthys mirabilis* (Teleostei:Gobiidae). *Biological Bulletin* 130:265-288.

Tomalchev, A.I. 1948. Basic paths of formation of the vegetation of the high mountain landscapes of the northern hemisphere. *Botanicheskii Zhurnal* 33(2):161-180.

Tomlinson, R.E. 1972. *Review of Literature on the endangered Masked Bobwhite*. 28 pp. Research Publication 108. Bureau of Sport Fisheries and Wildlife, U.S. Department of Interior, Washington, D.C.

Townsend, C.H.T. 1893. On the life zones of the Organ Mountains and adjacent region in southern New Mexico, with notes on the fauna of the range. *Science* 22:313-315.

Townsend, C.H.T. 1895-97. On the bio-geography of Mexico, Texas, New Mexico, and Arizona, with special reference to the limits of the life areas, and a provisional synopsis of the bio-geographic division of America. *Transactions of the Texas Academy of Sciences* 1(1895):71-96; 2(1897):33-86.

Trevino-Robinson, D. 1959. The ichthyofauna of the lower Rio Grande, Texas and Mexico. *Copeia* 1959:253-256.

Tucker, J.M. 1960. A range extension for the Chihuahua pine in New Mexico. *Southwestern Naturalist* 5:226.

Turnage, W.V. and T.D. Mallory. 1941. *An Analysis of Rainfall in the Sonoran Desert and Adjacent Territory*. 110 pp. Carnegie Institution of Washington Publication 529.

Turner, D.M. and C.L. Cochran, Jr. 1975. *Wildlife Management Unit 37B — Pilot Planning Study*. Arizona Game and Fish Department, Federal Aid to Wildlife Project FW-11-R-8, J-1:1-128.

Turner, F.B. and R.H. Wauer. 1963. A survey of the herpetofauna of the Death Valley area. *Great Basin Naturalist* 23:119-128.

Turner, G.T. and H.A. Paulsen. 1976. *Management of Mountain Grasslands in the Central Rockies: The Status of our Knowledge*. 24 pp, USDA Forest Service Research Paper RM-161. Rocky Mountain Forest and Range Experiment Station. Fort Collins, Colo.

Turner, R.M. 1974a. *Map Showing Vegetation in the Phoenix Area, Arizona.* U.S. Geological Survey Map I-845-I.

Turner, R.M. 1974b. *Map Showing Vegetation in the Tucson Area, Arizona.* U.S. Geological Survey. Map I-844-H.

Turner, R.M. 1974c. *Quantitative and Historical Evidence of Vegetation Changes along the Upper Gila River, Arizona.* 20 pp. U.S. Geological Survey Professional Paper 655-H.

Turner, R.M., L.H. Applegate, P.M. Bergthold, S. Gallizioli and S.C. Martin. 1980. *Arizona Range Reference Areas.* 34 pp. USDA Forest Service General Technical Report. RM-79. Rocky Mountain Forest and Range Experiment Station, Fort Collins, Colo.

Turner, R.M. and M.M. Karpiscak. 1980. *Recent Vegetation Changes Along the Colorado River Between Glen Canyon Dam and Lake Mead, Arizona.* 125 pp. U.S. Geological Survey Professional Paper 1132. 3 plates.

Twisselman, E.C. 1956. Flora of the Temblor Range and the neighboring part of the San Joaquin Valley. *Wasmann Journal of Biology* 14:161-300.

Udvardy, M.D.F. 1975. *World Biogeographical Provinces.* Co-Evolution Quarterly. Map, scale 1:39,629,000.

Ungar, I.A. 1974. Inland halophytes of the United States. pp. 235-305. In: Reimold, R.J. and W.H. Queen (eds.), *Ecology of Halophytes.* Academic Press, Inc. New York.

Urban, E.K. 1959. Birds from Coahuila, Mexico. *University of Kansas Museum of Natural History Publications* 11: 443-516.

USDA Soil Conservation Service. 1963. *Vegetative Units of the Major Land Resource Areas, Arizona.* Interagency Tech. Committee Range. University of Arizona, Agr. Ext. Serv., Agr. Research Serv., Soil Conservation Service.

U.S. Department of the Interior. 1963. *Natural Resources of Arizona.* 67 pp. U.S. Government Printing Office. Washington, D.C.

U.S. Department of the Interior. 1964. *Natural Resources of New Mexico.* 67 pp. U.S. Government Printing Office. Washington, D.C.

U.S. Environmental Data Service. 1959-1979. *Climatological Data, California.* Volumes 63-83. U.S. Department of Commerce. National Oceanic and Atmospheric Administration. Asheville, N.C.

U.S. Forest Service. 1923. *Plant Distribution of Arizona and New Mexico—Forage Types in Their Relation to the Local Phases of the Range Stock Industry.* USDA Forest Service. Map. Washington, D.C.

U.S. Forest Service. 1949. *Areas Characterized by Major Forest Types in the United States.* (map). USDA Forest Service. Washington, D.C.

Vale, Thomas R. 1975. Presettlement vegetation in the sagebrush-grass area of the Intermountain West. *Journal of Range Management* 28(1): 32-36.

Vallentine, J.F. 1961. *Important Utah Range Grasses.* 48 pp. Utah State University Extension Circular 281.

Van Devender, T.R. 1977. Holocene woodlands in the southwestern deserts. *Science* 198:189-192.

Van Devender, T.R., P.S. Martin, A.M. Phillips, III, and W.G. Spaulding. 1978. Late Pleistocene biotic communities from the Guadalupe Mountains, Culberson County, Texas. pp. 107-113. In: Wauer, R.H. and D.H. Riskind (eds.), *Transactions of the Symposium on the Biological Resources of the Chihuahuan Desert Region. United States and Mexico. Alpine, Texas. 1974.* USDI National Park Service Transactions and Proceedings Series 3. Washington, D.C.

Van Devender, T.R. and W.G. Spaulding. 1979. Development of vegetation and climate in the southwestern United States. *Science* 204:701-710.

Van Devender, T.R., W.G. Spaulding, and A.M. Phillips. 1977. Late Pleistocene plant communities in the Guadalupe Mountains, Culberson County, Texas. pp. 107-118. In: Wauer, R.H. and D.H. Riskind (eds.), *Transactions of the Symposium on the Biological Resources of the Chihuahuan Desert Region. United States and Mexico. Alpine, Texas. 1974.* USDI National Park Service, Transactions and Proceedings Series 3, Washington, D.C.

Van Devender, T.R. and R.D. Worthington. 1978. The herpetofauna of Howell's Ridge Cave and the paleoecology of the northwestern Chihuahuan Desert. pp. 85-106. In: Wauer, R.H. and D.H. Riskind (eds.), *Transactions of the Symposium on the Biological Resources of the Chihuahuan Desert Region. United States and Mexico. Alpine, Texas. 1974.* USDI National Park Service Transactions and Proceedings Series 3. Washington, D.C.

Van Rossem, A.J. 1931. Report on a collection of land birds from Sonora, Mexico. *Transactions. San Diego Society of Natural History* 6:237-304.

Van Rossem, A.J. 1932. The avifauna of Tiburón Island, Sonora, Mexico, with descriptions of four new races. *Transactions. San Diego Society of Natural History* 7:119-150.

Van Rossem, A.J. 1936a. Notes on birds in relation to the faunal areas of south-central Arizona. *Transactions. San Diego Society of Natural History* 8:121-148.

Van Rossem, A.J. 1936b. Birds of the Charleston Mountains, Nevada. *Pacific Coast Avifauna* 24:1-65.

Van Rossem, A.J. 1945. A distributional survey of the birds of Sonora, Mexico. *Occasional Papers of the Louisiana State University Museum of Zoology* 21:3-379.

Vasek, Frank C. and Michael G. Barbour. 1977. Mohave desert scrub vegetation. pp. 835-867. In: Barbour, M.G. and J. Major (eds.), *Terrestrial Vegetation of California.* John Wiley and Sons, Inc. New York.

Vaughan, T.A. 1954. Mammals of the San Gabriel Mountains of California. *University of Kansas Museum of Natural History Publications* 7(9):513-582.

Villa S., A.B. 1968. *La Vegetacion Forestal en el Extreme Meridional de Baja California.* 20 pp. Inv. Nat. Forest. Publ. Num. 10.

Visher, S.S. 1924. *Climatic Laws. Ninety Generalizations With Numerous Corollaries as to the Geographic Distribution of Temperature, Wind, Moisture, etc. A Summary of Climate.* John Wiley and Sons, Inc. New York.

Vogl, R.J. 1966. Salt-marsh vegetation of upper Newport Bay, California. *Ecology* 47:80-87.

Vogl, R.J. 1974. Effects of fire on grasslands. pp. 139-194. In: Kozlowski, T.T. and C.E. Ahlgren (eds.), *Fire and Ecosystems.* Academic Press, Inc. New York.

Vogl, R.J., W.P. Armstrong, K.L. White, and K.L. Cole. 1977. The closed-cone pine and cypresses. pp. 295-358. In: Barbour, M.G. and J. Major (eds.), *Terrestrial Vegetation of California.* John Wiley and Sons. New York.

Vogl, R.J. and L.T. McHargue. 1966. Vegetation of California: Fan palm oases on the San Andreas Fault. *Ecology* 47:532-540.

Vogl, R.J. and B.C. Miller. 1968. The vegetational composition of the south slope of Mt. Pinos, California. *Madroño* 19:225-234.

Walker, B.W. 1960. The distribution and affinities of the marine fish fauna of the Gulf of California. In: *The Biogeography of Baja California and Adjacent Seas. Systematic Zoology* 9:123-133.

Walker, B.W., R.R. Whitney, and G.W. Barlow. 1961. The fishes of the Salton Sea. In: B.W. Walker, *Ecology of the Salton Sea, California, in Relation to the Sports Fishery.* California Fish and Game Bulletin 113:77-91.

Wallace, A.R. 1876. *The Geographical Distribution of Animals, With a Study of the Relations of Living and Extinct Faunas and as Elucidating the Past Changes of the Earth's Surface.* 2 volumes. MacMillan and Co. London.

Wallace, L. 1883. A buffalo hunt in northern Mexico. In: Mayer, A.M. (ed.), *Sport With Rod and Gun.* The Century Co. New York.

Wallmo, O.C. 1955. Vegetation of the Huachuca Mountains, Arizona. *American Midland Naturalist* 54:466-480.

Wallmo, O.C. 1975. Important game animals and related recreation in arid shrubland of the United States. pp. 98-107. In: *Arid Shrublands. Proceedings of the Third Workshop of the U.S./Australia Rangelands Panel. Tucson.* Volume 3.

Walter, H. 1973. *Vegetation of the Earth in Relation to Climate and the Eco-physiological Conditions.* Heidelberg Science Library 15:1-237. English Universities Press, London; Springer-Verlag, New York; Heidelberg, Berlin.

Waring, R.H. and J.R. Franklin. 1979. Evergreen coniferous forests of the Pacific Northwest. *Science* 204:1380-1386.

Warnock, B.H. 1970. *Wildflowers of the Big Bend Country, Texas.* 35 pp. Sul Ross State University. Alpine, Tex.

Warnock, B.H. and W.M. Kittams. 1970. *Plant Communities of Big Bend National Park, Texas.* map. U.S. Department of the Interior, National Park Service.

Warren, D.K. and R.M. Turner. 1975. Saltcedar (*Tamarix chinensis*) seed production, seedling establishment, and response to inundation. *Journal of the Arizona Academy of Science* 10:135-144.

Watson, J.R. 1912. Plant geography of north central New Mexico. *Botanical Gazette* 54:194-217.

Wauer, R.H. 1962. A survey of the birds of Death Valley. *Condor* 64:220-233.

Wauer, R.H. 1964. Ecological distribution of the birds of the Panamint Mountains, California. *Condor* 66:287-301.

Wauer, R.H. 1973. *Birds of Big Bend National Park and Vicinity.* 223 pp. University of Texas Press. Austin.

Wauer, R.H. 1977. Significance of Rio Grande riparian systems upon the avifauna. pp. 165-174. In: Johnson, R.R. and D.A. Jones (technical coordinators), *Importance, Preservation and Management of Riparian Habitat: A Symposium.* USDA Forest Service General Technical Report RM-43. 217 pp. Rocky Mountain Forest and Range Experiment Station. Fort Collins, Colo.

Wauer, R.H. and D.L. Carter. 1965. *Birds of Zion National Park and Vicinity.* 91 pp. Zion Natural History Association.

Wauer, R.H. and J.D. Ligon. 1977. Distributional relations of breeding avifauna of four southwestern mountain ranges. pp. 567-578. In: Wauer, R.H. and D.H. Riskind (eds.), *Transactions of the Symposium on the Biological Resources of the Chihuahuan Desert Region. United States and Mexico. Alpine, Texas. 1974.* USDI National Park Service. Transactions and Proceedings Series 3. Washington, D.C.

Wauer, R.H. and D.H. Riskind. 1977. *Transactions of the Symposium on the Biological Resources of the Chihuahuan Desert Region. United States and Mexico. Alpine, Texas. 1974.* 658. pp. USDI National Park Service. Transactions and Proceedings Series 3. Washington, D.C.

Weaver, J.E. and F.W. Albertson. 1956. *Grasslands of the Great Plains.* 395 pp. Johnsen Publishing Co. Lincoln. Neb.

Weaver, J.W. and F.E. Clements. 1938. *Plant Ecology.* 601 pp. McGraw-Hill Book Co. New York.

Webb, W.L. 1950. Biogeographic regions of Texas and Oklahoma. *Ecology* 31:426-433.

Wells, P.V. 1965. Vegetation of the Dead Horse Mountains, Brewster County

Texas. *Southwestern Naturalist* 10:256-260.

Wells, P.V. 1976. Macrofossil analysis of Wood Rat (Neotoma) middens as a key to the Quarternary vegetational history of arid America. *Quarternary Research* 6:223-248.

Wells, P.V. 1979. An equable glaciopluvial in the West: Pleniglacial evidence of increased precipitation on a gradient from the Great Basin to the Sonoran and Chihuahuan deserts. *Quaternary Research* 12:311-325.

Wells, P.V. 1978. Post-glacial origin of the present Chihuahuan Desert less than 11,500 years ago. pp. 67-83. In: Wauer, R.H. and D.H. Riskind (eds.), *Transactions of the Symposium on the Biological Resources of the Chihuahuan Desert Region, United States and Mexico. Alpine, Texas. 1974.* USDI National Park Service. Transactions and Proceedings Series 3. Washington, D.C.

Wells, P.V. and Juan H. Hunziker. 1976. Origin of the creosotebush (*Larrea*) deserts of southwestern North America. *Ann. Missouri Botanical Gard.* 63:843-861.

Wells, P.V. and Clive D. Jorgenson. 1964. Pleistocene wood rat middens and climatic change in Mohave Desert; a record of juniper woodlands. *Science* 143:1,171-1,174.

Went, Frits W. 1957. *The Experimental Control of Plant Growth.* 343 pp. Chronica Botanica. Waltham, Mass.

West, Neil E. and Kamal I. Ibrahim. 1968. Soil-vegetation relationships in Utah. *Ecology* 49(3):445-456.

White, K.L. 1967. Native bunch grass (*Stipa pulchra*) on Hastings Reservation, California. *Ecology* 48:949-955.

White, Stephen S. 1949. The vegetation and flora of the Rio de Bavispe in northeastern Sonora, Mexico. *Lloydia* 11(4):229-302.

Whitfield, C.J. and H.L. Anderson. 1938. Secondary succession in the desert plains grassland. *Ecology* 19:171-180.

Whitfield, C.J., and E.L. Beutner. 1938. Natural vegetation in the desert plains grassland. *Ecology* 19:26-37.

Whitson, P.D. 1974. *The Impact of Human Use Upon the Chisos Basin and Adjacent Land.* 92 pp. U.S. Department of the Interior, National Park Service Monograph Series 4. Washington, D.C.

Whittaker, R.H. 1951. A criticism of the plant association and climatic climax concepts. *Northwest Science* 25:17-31.

Whittaker, R.H. 1953. A consideration of climax theory: The climax as a population and pattern: *Ecological Monographs* 23:41-78.

Whittaker, R.H. (ed.) 1973. *Ordination and Classification of Communities.* 737 pp. W. Junk. The Hague.

Whittaker, R.H. and W.A. Niering. 1964. Vegetation of the Santa Catalina Mountains, Arizona. I. Ecological classification and distribution of species. *Journal of the Arizona Academy of Science* 3:9-34.

Whittaker, R.H. and W.A. Niering. 1965. Vegetation of the Santa Catalina Mountains, Arizona: a gradient analysis of the south slope. *Ecology* 46:429-452.

Wieslander, A.E. 1932. *Vegetative Types of California (Exclusive of Deserts and Cultivated Lands); San Antonio Quadrangle.* USDA Forest Service, California Forest and Range Experiment Station. map.

Wieslander, A.E. 1934a. *Vegetative Types of California (Exclusive of Deserts and Cultivated Lands); Cucamonga Quadrangle.* USDA Forest Service California Forest and Range Experiment Station. map.

Wieslander, A.E. 1934b. *Vegetative Types of California (Exclusive of Deserts and Cultivated Lands); Pomona Quadrangle.* USDA Forest Service, California Forest and Range Experiment Station. map.

Wieslander, A.E. 1934c. *Vegetative Types of California (Exclusive of Deserts and Cultivated Lands); Pasadena Quadrangle.* USDA Forest Service, California Forest and Range Experiment Station. map.

Wieslander, A.E. 1934d. *Vegetative Types of California (Exclusive of Deserts and Cultivated Lands); Ramona Quadrangle.* USDA Forest Service, California Forest and Range Experiment Station. map.

Wieslander, A.E. 1934e. *Vegetative Types of California (Exclusive of Deserts and Cultivated Lands); San Bernardino Quadrangle.* USDA Forest Service, California Forest and Range Experiment Station. map.

Wieslander, A.E. 1934f. *Vegetative Types of California (Exclusive of Deserts and Cultivated Lands); San Jacinto Quadrangle.* USDA Forest Service, California Forest and Range Experiment Station. map.

Wieslander, A.E. 1935. A vegetation type map of California. *Madroño* 3:140-144.

Wieslander, A.E. 1937a. *Vegetation Types of California (Exclusive of Deserts and Cultivated Lands); Elizabeth Lake Quadrangle.* USDA Forest Service, California Forest and Range Experiment Station. map.

Wieslander, A.E. 1937b. *Vegetative Types of California (Exclusive of Deserts and Cultivated lands); Rock Creek Quadrangle.* USDA Forest Service, California Forest and Range Experiment Station. map.

Wieslander, A.E. 1937c. *Vegetative Types of California (Exclusive of Deserts and Cultivated Lands); Tujunga Quadrangle.* USDA Forest Service, California Forest and Range Experiment Station. map.

Wieslander, A.E. 1938. *Vegetative Types of California (Exclusive of Deserts and Cultivated Lands); Elsinore Quadrangle.* USDA Forest Service, California Forest and Range Experiment Station. map.

Wieslander, A.E. 1940. *Vegetative Types of California (Exclusive of Deserts and Cultivated Lands); Corona Quadrangle.* USDA Forest Service, California Forest and Range Experiment Station. map.

Wigal, D.D. 1973. *A Survey of the Nesting habitats of the White-Winged Dove in Arizona.* 37 pp. Arizona Game and Fish Department Special Report 2. Phoenix.

Wiggins, Ira L. 1969. Observations on the Vizcaíno Desert and its biota. *Proceedings of the California Academy of Sciences* 36(11).

Wiggins, Ira L. 1980. *Flora of Baja California.* 1025 pp. Stanford University Press. Stanford, Calif.

Wilbur, S.R. and R.E. Tomlinson. 1976. *The Literature of the Western Clapper Rail.* 31 pp. U.S. Department of the Interior, Fish and Wildlife Service Special Science Report — Wildlife No. 194. Washington, D.C.

Wilke, P.J. 1980. Prehistoric wier fishing on recessional shorelines of Lake Cahuilla, Salton Basin, Southeastern California. In: *Abstracts. Eleventh Annual Conference. Desert Fishes Council.* Death Valley, Calif.

Williams, J.E. 1978. Taxonomic status of *Rhinichthys osculus* (Cyprinidae) in the Moapa River, Nevada. *Southwestern Naturalist* 23:511-518.

Wilson, R.C. and R.J. Vogl. 1965. Manzanita chaparral in the Santa Ana Mountains, California. *Madroño* 18:47-62.

Woodbury, A.M. 1947. Distribution of pygmy conifers in Utah and northeastern Arizona. *Ecology* 28:113-126.

Woodbury, A.M. and H.N. Russell, Jr. 1945. *Birds of the Navajo Country.* 160 pp. University of Utah Bulletin Biology Series 9.

Woodell, S.R.J., H.A. Mooney, and A.J. Hill. 1969. The behaviour of *Larrea divaricata* (creosote bush) in response to rainfall in California. *Journal of Ecology* 57:37-44.

Woodin, H.E. and A.A. Lindsey. 1954. Juniper-pinyon east of the Continental Divide, as analyzed by the line-strip method. *Ecology* 35:473-489.

Wooton, E.O. and P.C. Standley. 1915. *Flora of New Mexico.* 794 pp. Contributions from the United States National Herbarium No. 19. Washington, D.C.

Workman, John P. and Neil E. West. 1967. Germination of *Eurotia lanata* in relation to temperature and salinity. *Ecology* 48(4):659-661.

Wright, H.E., Jr. 1964. Origin of the lakes in the Chuska Mountains, northwestern New Mexico. *Geological Society of American Bulletin* 75:589-598.

Wright, H.E. Jr. and A.M. Bent. 1968. Vegetation bands around Dead Man Lake, Chuska Mountains, New Mexico. *American Midland Naturalist* 79:8-30.

Wright, J.W. and C.H. Lowe. 1968. Weeds, polyploids, parthenogensis, and the geographical and ecological distribution of all-female species of *Cnemidophorus*. *Copeia* 1968:128-138.

Wright, Robert A. 1970. The distribution of *Larrea tridentata* (D.C.) Coville in the Avra Valley, Arizona. *Journal of the Arizona Academy of Science* 6(1):58-63.

Yang, Tien Wei. 1961. *The Recent Exansion of Creosotebush (Larrea divaricata) in the North American Desert.* 11 pp. Western Reserve Academy Natural History Museum Publication 1.

Yang, Tien Wei. 1970. Major chromosomes races of *Larrea divaricata* in North America. *Journal of the Arizona Academy of Science* 6:41-45.

Yang, Tien Wei and C.H. Lowe. 1956. Correlation of major vegetation climaxes with soil characteristics in the Sonoran Desert. *Science* 123:542.

Yang, Tien Wei and Charles H. Lowe. 1968. Chromosome variation in ecotypes of *Larrea divaricata* in the North American desert. *Madroño* 19:161-164.

York, C.L. 1949. The physical and vegetational basis for animal distribution in the Sierra Vieja range of southwestern Texas. *Texas Journal of Science* 1(3):46-62.

York, J.C. and W.A. Dick-Peddie. 1969. Vegetation changes in southern New Mexico during the past hundred years. pp. 157-166. In: McGinnies, W.G. and B.J. Goldman (eds.), *Arid Lands in Perspective.* University of Arizona Press. Tucson.

Young, James A., Raymond A. Evans, and Jack Major. 1977. Sagebrush steppe. pp. 763-796. In: Barbour, M.G. and J. Major (eds.), *Terrestrial Vegetation of California.* John Wiley and Sons. New York.

Young, James A., Raymond A. Evans, and Paul T. Tueller. 1976. Great Basin plant communities-pristine and grazed. pp. 186-215. In: Elston, Robert (ed.), *Holocene Environmental Change in the Great Basin.* Nevada Archeological Survey Research Paper 6.

Zimmerman, R.C. 1969. *Plant Ecology of an Arid Basin. Tres Alamos—Reddington Area. Southern Arizona.* 51 pp. U.S. Geological Survey Professional Paper 485-D.

Zoltai, S.C., F.C. Pollett, J.K. Jeglum, and G.D. Adams. 1975. Developing a wetland classification for Canada. pp. 497-511. In: *Proceedings of the North American Forest Soils Conference.* Volume 4.

Zwinger, A.H. and B.E. Willard. 1972. *Land Above the Trees: A Guide to American Alpine Tundra.* Harper and Row. New York.

Appendix I. Reprinted from the Journal of the Arizona-Nevada Academy of Science 14 (Suppl. 1):1-16. 1979.

A DIGITIZED CLASSIFICATION SYSTEM FOR THE BIOTIC COMMUNITIES OF NORTH AMERICA, WITH COMMUNITY (SERIES) AND ASSOCIATION EXAMPLES FOR THE SOUTHWEST[1]/

DAVID E. BROWN, Arizona Game & Fish Department, Phoenix

CHARLES H. LOWE, University of Arizona, Tucson

CHARLES P. PASE, USDA Forest Service

INTRODUCTION. — In previous publications on the North American Southwest System we have addressed primarily the North American Southwest region as outlined in Fig. 1 (Brown and Lowe 1973, 1974a,b). Responses to both the classification system and the classification have been favorable in both general interest and use: e.g., Lacey, Ogden, and Foster 1975; Turner and Cochran 1975; Carr 1977; Dick-Peddie and Hubbard 1977; Ellis et al. 1977; Glinski 1977; Hubbard 1977; Pase and Layser 1977; Steenbergh and Warren 1977; Patton 1978; BLM 1978a,b; Turner et al. 1979. In this report we expand the classification nomenclature at digit levels 1-4 to represent the North American continent.

The Southwest System is evolutionary in basis and hierarchical in structure. It is a natural biological system rather than primarily a geography-based one in the sense of Dice 1943; Bailey 1978; and others. The resulting *classifications* are, therefore, *natural hierarchies*.

Because of the open-ended characteristic of a natural hierarchical system, resulting classification provides for orderly change. The inherent accordion-type flexibility provides for expansion and contraction at all levels. It permits accommodation of new information into the classification — addition, transference, and deletion of both (a) ecological taxa, and (b) quantitative data on ecological parameters concerning taxa, as our knowledge accumulates on either or both. Digit levels 7 to n accommodate the latter and digit levels 1-6 accommodate the former (ecological taxa) on a world-wide basis.

The system's potential is the provision of a truly representative picture for biotic environment. It permits but does not require inclusion of any and all biotic criteria in a given classification — animals as well as plants. Thereby included in the system's uses are the mapping of wildlife habitats and the determination and delineation of natural areas on a local to world-wide basis (Brown, Lowe, and Pase 1977). On a local basis, overlapping soil mapping units can provide "habitat-types" with their implied biotic potential for land use planning purposes.

The digitation of hierarchy makes the system computer-compatible; e.g., a system or subsystem for storing and retrieving biotic resource data within or parallel to an overall management system. The Southwest System is currently in use in the RUNWILD program developed for field unit use on remote terminals by Region 3 of the Rocky Mountain Forest and Range Experiment Station, U.S. Forest Service (Patton 1978). The system and classification is similarly incorporated in the State of Arizona Resources Inventory System (ARIS). It is currently used by both industry and agencies for biological studies, resource inventories, and procedures for environmental analysis, for example as required by the National Environmental Policy Act.

The system is responsive to scale. The hierarchical sequence permits mapping at any scale, and various levels of the system have been mapped at 1:1,000,000 (1 inch represents ca. 16 miles), 1:500,000, 1:250,000, 1:62,500 (1 inch represents ca. 1 mile), and others. Moreover, the use of hierarchical sequence permits the needed flexibility for mapping those complex communities where more intensive levels are impractical or needlessly time consuming in a given investigation.

The classification has been expanded to include the major biotic communities of North America (Brown, Lowe, and Pase 1977, 1979). To facilitate communication with potential users, we provide, in addition to some structural modification of the original classification, a number of additional definitions and explanations. Our fourth level (biome) examples for North America are representative; they are not intended as either a definitive or final classification. Examples of the use of the system to the fifth (series = community) and sixth (association) levels are given here for those biomes located wholly or partially within the North American Southwest.

Incorporated in the present classification are contributions from approximately one hundred investigators, primarily biogeographers, wildlife biologists, and ecologists, all of which pertain to or are in general use in the Southwest today. Additional references are given in Brown and Lowe 1974a,b, 1977.

A DIGITIZED HIERARCHY OF THE WORLD'S NATURAL ECOSYSTEMS

Where:

1,000 = Biogeographic (Continental) Realm
  1,100 = Vegetation
    1,110 = Formation-type
      1,111 = Climatic (Thermal) Zone
        1,111.1 = Regional Formation (Biome)
          1,111.11 = Series (Community of generic dominants)
            1,111.111 = Association (Community of specific dominants)
              1,111.1111 = Composition-structure-phase

A number preceeding the comma (e.g., 1,000) refers to the world's *biogeographic realms* (see Table 1). Origin and evolutionary history are recognized as primary in importance in the determination and classification of natural ecosystems. The mapable reality of the world's biogeographic realms is interpretive in part and dependent on criteria used. In those regions where the components of one realm merge gradually with those of another and the assignment of biogeographic origin is difficult, we include such transitional areas (wide ecotones) in both realms. The following seven realms are adapted from Wallace 1876; see also Hesse et al. 1937; Dansereau 1957; Darlington 1957; Walter 1973; I.U.C.N. 1974; DeLaubenfels 1975; Cox et al. 1976:

[1]/A contribution of the Arizona Game and Fish Department (with publication funded by Federal Aid Project W-53R), The University of Arizona Department of Ecology and Evolutionary Biology, and the United States Forest Service, Rocky Mountain Forest and Range Experiment Station.

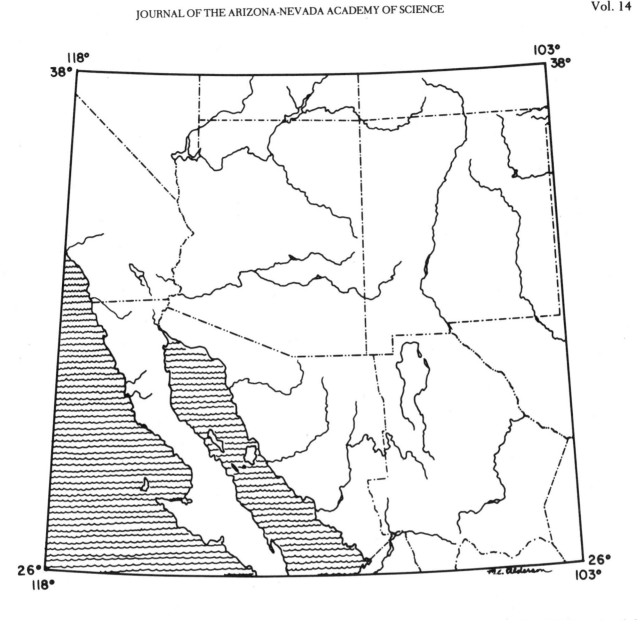

Figure 1. **The Southwest. In delineating a *natural* Southwest region, approximately one half of the area falls in the Republic of Mexico and one half in the United States; the U.S. states of "Arizona and New Mexico" constitute less than half of the "American Southwest." Parts or all of the following states are included: Arizona, Baja California, California, Chihuahua, Coahuila, Colorado, Nevada, New Mexico, Sonora, Texas, Utah. All of Baja California and its associated islands (not completely shown) are included in our concept of a natural North American Southwest region; extreme northern Durango and Sinaloa are also included at Lat. 26° N.**

1000 Nearctic	Continental North America exclusive of the tropics and certain highland areas south of the Tropic of Cancer. We include those tropic-subtropic regions in and adjacent to the North American Southwest and the Caribbean.	
2000 Palaearctic	Eurasia exclusive of the tropics. Africa north of the Sahel.	
3000 Neotropical and Antarctican	Continental South America, Central America, and most of Mexico south of the Tropic of Cancer. Antarctica.	

4000 Oriental	Southeast Asia, the Indian subcontinent; the Phillipines, Indonesia, etc.
5000 Ethiopian	Africa south of the Sahara, Malagasy, and parts of the Arabian peninsula.
6000 Australian	Australia and Tasmania.
7000 Oceanic	Oceanic islands processing a high degree of endemism.

304

First Level. — The first digit after the comma (e.g., 1,100) refers to vegetation, the structural and readily measurable reality of ecosystems. Included are all potential and/or existing plant communities that are presumed to be established naturally under existing climate and the cessation of artificially disruptive (man-caused) influences[2] (Table 1).

Table 1. Summary for the Natural Vegetation of the World to the First Digit level.

Biogeographic Realm	1. Upland Vegetation	2. Wetland Vegetation
1,000. Nearctic	1,100.	1,200.
2,000. Palaearctic	2,100.	2,200.
3,000. Neotropical-Antarctican	3,100.	3,200.
4,000. Oriental	4,100.	4,200.
5,000. Ethiopian	5,100.	5,200.
6,000. Australian	6,100.	6,200.
7,000. Oceanic	7,100.	7,200.

All existing and potential natural vegetation (PNV) is classified as belonging to uplands (1,100) or wetlands (1,200) as in Table 1. Cultivated lands are designated 1,300 (cultivated uplands) and 1,400 (cultivated wetlands). The evolutionary distinctions between plants and animals of terrestrial (upland) ecosystems and those of aquatic or hydric (wetland) ecosystems is recognized by this dichotomy (see Ray 1975).

As discussed here, wetlands include those periodically, seasonally or continually submerged ecosystems populated by species and/or life forms different from the immediately adjacent (upland) climax vegetation, and which are dependent on conditions more mesic than provided by the immediate precipitation. Certain ecosystems having both upland and wetland characteristics and components (e.g., riparian forests) could be properly considered as belonging to both divisions. They are treated in this report as wetlands (1,200).

Second Level. — The second digit after the comma (e.g. 1,110) refers to one of the following recognized ecological formations, which on a worldwide basis are the *formation-types* (biome-types); see Tables 2 and 3. On continents these are referred to as formations, which are vegetative responses (functions) to integrated environmental factors, most importantly plant-available moisture.

UPLAND FORMATIONS

Tundra[3]

Communities existing in an environment so cold that moisture is unavailable during most of the year, precluding the establishment of trees, and in which maximum vegetation development is of herbaceous root perennials, shrubs, lichens and mosses, with grasses poorly represented or at least not dominant.

Forest and Woodland

Forest

Communities comprised principally of trees potentially over 15 meters (50 ft) in height, and frequently characterized by closed and/or multilayered canopies.

Woodland

Communities dominated by trees with a mean potential height usually under 15 meters in height, the canopy of which is usually open — sometimes very open[4] — or interrupted and singularly layered.

Scrubland

Communities dominated by sclerophyll or microphyll shrubs and/or multistemmed trees, generally not exceeding 10 meters (31 ft) in height and usually presenting a closed physiognomy, or, if open, interspersed with other perennial vegetation.

Grassland

Communities dominated actually or potentially by grasses and/or other herbaceous plants.

Desertland

Communities in an arid environment — usually less than 300 mm (12 in) precipitation per annum — in which plants are separated by significant areas devoid of perennial vegetation.

Table 2. Summary for the Natural UPLAND Vegetation of the World to the Second Level (Formation-Type).

Biogeographic Realm	Formation Type					
	1. Tundra	2. Forest	3. Scrubland	4. Grassland	5. Desertland	6. Nonvegetated
1,000 Nearctic	1,110	1,120	1,130	1,140	1,150	1,160
2,000 Palaearctic	2,110	2,120	2,130	2,140	2,150	2,160
3,000 Neotropical-Antarctican	3,110	3,120	3,130	3,140	3,150	3,160
4,000 Oriental	4,110	4,120	4,130	4,140	4,150	4,160
5,000 Ethiopian	5,110	5,120	5,130	5,140	5,150	5,160
6,000 Australian	6,110	6,120	6,130	6,140	6,150	6,160
7,000 Oceanic	7,110	7,120	7,130	7,140	7,150	7,160

---

[2]Our thinking on the complex question of determining climax, successional, and potential vegetation is to consider (and map) ecosystems on the basis of the existing or presumed vegetation of the foreseeable future.

[3]The holistic integrity of a "Tundra" formation is not without question. Treated here, tundra may also be composed of grasslands, scrublands, marshlands (wet tundra), and desertlands in an Arctic-Boreal climatic zone (Billings and Mooney 1968; Billings 1973; and others).

[4]The "savanna" formation (Dansereau 1957; Dyksterhuis 1957; and others) is here recognized (in North America) as an ecotone between *woodland and grassland*. Those homogeneous areas in which the crowns of trees normally cover less than approximately 15 percent of the ground space are classified as grasslands where grasses are actually or potentially dominant (= savanna grassland). Mosaics of grassland and smaller or larger stands of trees and shrubs are "parklands" and are composed of two or more ecologically distinct plant formations (Walter 1973).

Table 3. Summary for the Natural WETLAND Vegetation of the World to the Second Level (Formation-Type).

Biogeographic Realm	Formation Type					
	1. Wet Tundra	2. Forest[1]/	3. Swamp-scrub, Riparian Scrub	4. Marshland	5. Strandland	6. Submergent Aquatic
1,000 Nearctic	1,210	1,220	1,230	1,240	1,250	1,260
2,000 Palaearctic	2,210	2,220	2,230	2,240	2,250	2,260
3,000 Neotropical-Antarctican	3,210	3,220	3,230	3,240	3,250	3,260
4,000 Oriental	4,210	4,220	4,230	4,240	4,250	4,260
5,000 Ethiopian	5,210	5,220	5,230	5,240	5,250	5,260
6,000 Australian	6,210	6,220	6,230	6,240	6,250	6,260
7,000 Oceanic	7,210	7,220	7,230	7,240	7,250	7,260

[1]/Swampforests, bog-forests and riparian forests.

WETLAND FORMATIONS

Wet Tundra[2]/	Wetland communities existing in an environment so cold that available plant moisture is unavailable during most of the year, precluding the establishment of trees and all but a low herbaceous plant structure in a hydric matrix.
Swampforest; Riparian Forest	Wetland communities possessing an overstory of trees potentially over 10 meters (31 ft) in height, and frequently characterized by closed and/or multilayered canopies.
Swampscrub; Riparian Scrub	Wetland communities dominated by short trees and/or woody shrubs, generally under 10 meters (31 ft) in height and often presenting a closed physiognomy.
Marshland	Wetland communities in which the principal plant components are herbaceous emergents which normally have their basal portions annually, periodically, or continually submerged.
Strandland	Beach and river channel communities subject to infrequent but periodic submersion, wind driven waves and/or spray. Plants are separated by significant areas devoid of perennial vegetation.[2]/
Submergent Aquatic	Aquatic communities comprised entirely or essentially of plants mostly submerged or lacking emergent structures.

Some localized upland and wetland areas are essentially without vegetation or are sparingly populated by simple organisms, e.g., on some dunes, lava flows, playas, sinks, etc. For purposes of classification certain of such areas could be considered as belonging to a *nonvegetated formation-type* (Tables 2 and 3).

Third Level. — *The third digit beyond the comma* (e.g., 1,1 1̲1̲) refers to one of four world *climatic zones* (c.f. Walter 1973; Ray 1975; Cox et al. 1976), in which minimum temperature remains a major evolutionary control of and within the zonation and the formation-types (Tables 4 and 5). All four of these broad climatic zones are found in North America and in the "Southwest."

Arctic-Boreal (Antarctic-Austreal)	Characterized by lengthy periods of freezing temperatures, with growing season of short duration (generally 60-150 days), occasionally interrupted by nights of below freezing temperatures.
Cold Temperate	Freezing temperatures of short duration although of frequent occurrence during winter months. Potential growing season generally 100-200 days and confined to spring and summer when freezing temperatures are infrequent or absent.
Warm Temperate	Freezing temperatures of short duration but generally occurring every year during winter months. Potential growing season over 200 days with an average of less than 125-150 days being subject to temperatures lower than 0 °C or to chilling fogs.
Tropical-Subtropical	Infrequent or no 24-hour periods of freezing temperatures, chilling fogs or wind.

Fourth Level. — *The fourth digit beyond the comma* (e.g., 1,111.1̲) refers to a subcontinental unit that is a *major biotic community* (=biome). Biomes are natural communities characterized by a distinctive vegetation physiognamy within a formation; accordingly, the natural geography of biomes is commonly *disjunctive*. A single biome is not to be confused with a single biotic (biogeographic) province; in distribution, a province is always a *continuous* (non-disjunctive) biogeographic area that may include several (e.g., five or more) biomes.[1]/

Our nomenclature at the biome (fourth) level incorporates useful geographic terms in the same sense of Weaver and Clements (1938). While such terms are also associated with biotic provinces (as in Fig. 2) we are classifying biomes, not biotic provinces. Biomes are characterized by a distinctive evolutionary history within a formation; thus they tend to be centered in, but are not restricted to particular biogeographic regions or provinces (e.g., see Weaver and Clements 1938; Clements and Shelford 1939; Pitelka 1941, 1943; Dice 1939, 1943; Odum 1945; Allee et al. 1949; Kendeigh 1954, 1961; Dansereau 1957; Shelford 1963; Daubenmire and Daubenmire 1968; Udvardy 1975; Dasmann 1976).

This fourth level and the fifth level (below) have provided the most successful and useful mapping of states, regions, and continents (e.g., in North America, Harshberger 1911; Shreve 1917, 1951; Shantz and Zon 1924; Bruner 1931; Morris 1935; Wieslander 1935; Brand 1936;

---

[2]/Treated here, tundra may also be composed of grasslands, scrublands, marshlands (wet tundra), and desertlands in an Arctic-Boreal climatic zone; see footnote 3.

[2]/Strand communities are situated in harsh physical environments that produce their characteristic physiognomy. Accordingly, strandland is treated as the wetland equivalent of desertland. While occurring in the usual sense on beaches and other seacoast habitats, freshwater (or interior) strands also occur in river channels, along lake margins, and below reservoir high water lines.

[1]/Originally termed *biotic provinces* by Lee Dice (1943) who developed this biogeographic concept in North America between 1922 (biotic areas) and 1943 (biotic provinces); they have been referred to variously in recent literature as "biotic provinces" (Dasmann 1972, 1974; IUCN 1973), "biogeograph[ic] provinces" (Udvardy 1975; Dasmann 1976), "ecoregions" (Bailey 1976, 1978), and "b[io]gions" (Franklin 1977).

Table 4. Summary for the Natural UPLAND Vegetation of Nearctic and Adjacent Neotropical North America to the Third Level.

Formation	Climatic (Thermal) Zone			
	1. Arctic-Boreal	2. Cold Temperate	3. Warm Temperate	4. Tropical-Subtropical
1,110 Tundra	1,111			
1,120 Forest & Woodland	1,121	1,122	1,123	1,124
1,130 Scrubland	1,131	1,132	1,133	1,134
1,140 Grassland	1,141	1,142	1,143	1,144
1,150 Desertland	1,151	1,152	1,153	1,154
1,160 Nonvegetated	1,161	1,162	1,163	1,164

Table 5. Summary for the Natural WETLAND Vegetation of Nearctic and Adjacent Neotropical North America to the Third Level.

Formation	Climatic (Thermal) Zone			
	1. Arctic-Boreal	2. Cold Temperate	3. Warm Temperate	4. Tropical-Subtropical
1,210 Wet Tundra	1,211			
1,220 Forest[1]/	1,221	1,222	1,223	1,224
1,230 Swampscrub	1,231	1,232	1,233	1,234
1,240 Marshland	1,241	1,242	1,243	1,244
1,250 Strandland	1,251	1,252	1,253	1,254
1,260 Submergent Aquatic	1,261	1,262	1,263	1,264

[1]/Swampforests, bog-forests and riparian forests.

Nichol 1937; LeSueur 1945; Jensen 1947; Leopold 1950; Castetter 1956; Küchler 1964, 1977; Brown 1973; Franklin and Dyrness 1973; Brown and Lowe 1977). Biomes and biogeographic provinces are also the bases for the biosphere reserve program (MAB) in the United States and elsewhere (IUCN 1974; Franklin 1977).

A partial summary of the biotic communities (biomes) for Nearctic and adjacent Neotropical America is given in Tables 6 and 7.

Fifth Level. — *The fifth digit beyond the comma* (e.g., 1,111.1<u>1</u>) refers to the principal plant-animal communities within the biomes, distinguished primarily on taxa that are distinctive climax plant dominants. Daubenmire and Daubenmire (1968) organized their data according to major dominants in climax communities referred to as *climax series.* "Series," or "cover-types" (sensu Society of American Foresters 1954), or "vegetation-types" (sensu Flores et al. 1971), are each composed of one or more biotic associations characterized by shared climax dominants within the same formation, zone, and biome (Oosting 1950; Lowe 1964; Franklin and Dyrness 1973; Pfister et al. 1977). For example, within Rocky Mountain montane conifer forest (122.3), the Pine Series (122.32) includes all of the Rocky Mountain forest associations in which *Pinus ponderosa* is a dominant.

Community diversity of tropical and subtropical upland climax dominants is inherently more complex than in boreal and temperate communities. Moreover, some taxa may exhibit polymorphism to the extent that the same species may be dominant — and ecotypically differentiated — in more than a single formation. As an extreme case in southwestern North America, mesquite *(Prosopis juliflora)* may be a dominant life-form in certain desertland, disclimax grassland, scrubland, woodland, and riparian forest communities, and exhibit phenotypic and presumably genotypic population differentiation across the complex gradient. Facultative growth-form is exhibited by dominant plant taxa in both cold and warm climatic zones.

The distribution of some plant dominants also may span more than a single climatic zone, as in *Larrea*, *Prosopis*, and the introduced

Tamarix. However, important plant and animal associates of these dominant species are usually encountered when passing from one formation or climatic zone to another. When specific and generic dominants are shared by more than one biome, closer investigation may reveal genetic geographic variation within the shared species, as in the chromosome races of creosotebush *(Larrea divaricata,* Yang and Lowe 1968; Yang 1970).

It is clear that the determination of fifth and sixth (below) level communities in particular will require modification and revision in the classification as field data accumulate. Some of the more widely distributed and commonly recognized series in the Southwest are given in Tables 6 and 7 under the appropriate biome.

Sixth Level. — *The sixth digit beyond the comma* (e.g., 1,111.11<u>1</u>) refers to distinctive plant associations, and associes (successional associations), based on the occurrence of particular dominant species more or less local or regional in distribution and generally equivalent to habitat-types as outlined by the Daubenmires (1968), Layser (1974), Pfister et al. (1977), and others. While we give examples for certain communities within southwestern biomes, the enormous numbers of sets precludes presentation here for the treatments given in Tables 6 and 7. Associations may be added at length for regional studies by using a, b, c, sets as is also indicated in the tables in Brown and Lowe (1974a,b).

Seventh Level. — *The seventh digit beyond the comma* (e.g., 1,111.111<u>1</u>) accommodates detailed measurement and assessment of quantitative structure, composition, density and other attributes for dominants, understories, and other associated species. This level and additional ones in the system provide the flexibility required for encompassing data for ecological parameters measured in intensive studies on limited areas (see e.g., Dick-Peddie and Moir 1970).

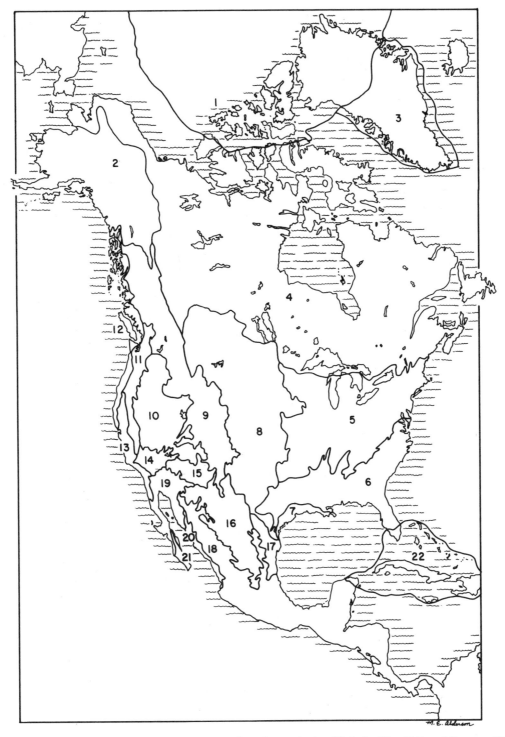

Figure 2. Biogeographic provinces of Nearctic and adjacent Neotropical North America (modified after Dice 1943, and Dasmann 1974), discussed in text under fourth (Biome) digit level.

1. Polar	8. Plains	15. Mogollon (Interior)
2. Alaskan	9. Rocky Mountain	16. Chihuahuan
3. Greenlandian	10. Great Basin	17. Tamaulipan
4. Canadian	11. Sierran-Cascade	18. Madrean
5. Northeastern	12. Sitkan-Oregonian	19. Sonoran
6. Southeastern	13. Californian	20. Sinaloan
7. Gulf Coastal	14. Mohavian	21. San Lucan
		22. Carribean

Table 6. Nomenclature of UPLAND Biotic Communities of Nearctic and Adjacent Neotropical North America with Community (Series) and Association Examples for the North American Southwest.

1,100 Nearctic Upland Vegetation
  1,110 Tundra Formation
    1,111 Arctic Tundras
      1,111.1 Polar (High Arctic) Tundra
      1,111.2 Alaskan (Low Arctic) Coastal Tundra
      1,111.3 Canadian (Barren Ground = Low Arctic) Tundra
      1,111.4 Arctic Alpine Tundra
      1,111.5 Rocky Mountain Alpine Tundra
        1,111.51[1]/ Lichen-Moss Series*
          111.511 *Rhizocarpon geographicum* Association*
        111.52 Mixed Herb Series*
        111.53 Avens-Sedge Series*
          111.531 *Geum turbinatum* Association*
          111.532 *Geum turbinatum-Carex bella* Association*
        111.54 Woodrush Series*
          111.541 *Kobresia bellardi*-grass-forb Association*
      111.6 Sierran-Cascade Alpine Tundra
        111.61 Lichen-Moss Series*
        111.62 Mixed Herb Series*
          111.621 *Selaginella watsoni-Eriogonum umbellatum et al.* Association*
      111.7 Adirondack-Appalachian Alpine Tundra

120 Forest and Woodland Formation
  121 Boreal Forests and Woodlands
    121.1 Canadian Subarctic Conifer Forest and Woodland (North American Taiga)
    121.2 Appalachian Subalpine Conifer Forest
    121.3 Rocky Mountain Subalpine Conifer Forest and Woodland[2]/
      121.31 Engelmann Spruce-Alpine Fir Series*
        121.311 *Picea engelmanni-Abies lasiocarpa* Association*
        121.312 *Picea engelmanni* Association*
        121.313 *Abies lasiocarpa* Association*
        121.314 *Abies lasiocarpa arizonica* Association*
        121.315 *Picea pungens* Association*
        121.316 *Populus tremuloides* subclimax Association*
      121.32 Bristlecone Pine-Limber Pine Series*
        121.321 *Pinus aristata-Pinus flexilis* Association*
        121.322 *Pinus aristata* Association*
        121.323 *Pinus flexilis* Association*
    121.4 Sierran-Cascade Subalpine Conifer Forest
      121.41 Limber Pine-Lodgepole Pine Series*
        121.411 *Pinus flexilis-Pinus contorta murrayana* Association*
        121.412 *Pinus flexilis* Association*
    121.5 Madrean Subalpine Conifer Forest
  122 Cold Temperate Forests and Woodlands
    122.1 Northeastern Deciduous Forest
    122.2 Pacific Coastal (Oregonian) Conifer Forest
    122.3 Rocky Mountain (= Petran) Montane Conifer Forest
      122.31 Douglas-fir-White Fir (= Mixed Conifer) Series*

        122.311 *Pseudotsuga menziesi* Association*
        122.312 *Pseudotsuga menziesi-Abies concolor* Association*
        122.313 *Pseudotsuga menziesi*-mixed conifer (*Abies concolor, Pinus flexilis, Acer glabrum, Populus tremuloides, Pinus ponderosa*) Association*
        122.314 *Populus tremuloides* subclimax Association*
      122.32 Pine Series*
        122.321 *Pinus ponderosa* Association*
        122.322 *Pinus ponderosa*-mixed conifer Association*
        122.323 *Pinus ponderosa-Quercus gambeli* Association*
        122.324 *Pinus ponderosa-Quercus arizonica* Association*
        122.325 *Pinus ponderosa-Juniperus deppeana* Association*
        122.326 *Populus tremuloides* subclimax Association*
        122.327 *Pinus flexilis reflexa* Association*
        122.328 *Pinus ponderosa-Pinus leiophylla* Association*
      122.33 Gambel Oak Series*
        122.331 *Quercus gambeli* Association*
    122.4 Great Basin Conifer Woodland
      122.41 Pinyon-Juniper Series*
        122.411 *Pinus edulis-Juniperus scopulorum* Association*
        122.412 *Pinus edulis* Association*
        122.413 *Juniperus scopulorum* Association*
        122.414 *Pinus edulis-Juniperus monosperma* Association*
        122.415 *Juniperus monosperma* Association*
        122.416 *Pinus monophylla-Juniperus osteosperma* Association*
        122.417 *Pinus monophylla* Association*
        122.418 *Juniperus osteosperma* Association*
        122.419 *Pinus monophylla-Juniperus californica* Association*
        122.411a *Juniperus californica* Association*
        122.412a *Pinus quadrifolia-Juniperus californica* Association*
        122.413a *Pinus quadrifolia* Association*
        122.414a *Pinus monophylla-Juniperus californica*-chaparral Association*
    122.5 Sierran-Cascade Montane Conifer Forest
      122.51 Mixed Conifer Series*
        122.511 *Abies concolor*-mixed conifer (*Pinus contorta murrayana, Pinus jeffreyi* et al.) Association*
      122.52 Pine Series*
        122.521 *Pinus ponderosa* Association*
        122.522 *Pinus ponderosa-P. jeffreyi* Association*
        122.523 *Pinus ponderosa-Quercus kelloggi* Association*
        122.524 *Pinus jeffreyi* Association*
      122.53 Black Oak Series
        122.531 *Quercus kelloggi* Association*
    122.6 Madrean Montane Conifer Forest
      122.61 Douglas-fir-Mixed Conifer Series*
        122.611 *Pseudotsuga menziesi* Association*
        122.612 *Pseudotsuga menziesi-Pinus flexilis, Acer glabrum, Populus tremuloides, Pinus ponderosa* et al. Association*
      122.62 Pine Series*
        122.621 *Pinus flexilis ayacahuite* Association*
        122.622 *Pinus ponderosa* Association*
        122.623 *Pinus ponderosa*-mixed conifer Association*
        122.624 *Pinus ponderosa-Quercus* spp. Association*
        122.625 *Pinus ponderosa-Juniperus deppeana* Association*
  123 Warm Temperate Forests and Woodlands
    123.1 Southeastern Mixed Deciduous and Evergreen Forest
    123.2 Californian Mixed Evergreen Forest
      123.21 Mixed Mesophytic Series*

---

*Examples only.

[1]/The first "1" (in front of comma and representing the Nearctic Realm) is understood, and cropped for tabular convenience only, from this point onward.

[2]/Separation of this biotic community into Rocky Mountain and Great Basin units may be warranted.

123.211 Mixed hardwood Association*
123.212 *Quercus chrysolepis*-mixed hardwood Association*
123.22 Big-cone Spruce Series*
123.221 *Pseudotsuga macrocarpa* Association*
123.3 Madrean Evergreen Forest and Woodland
123.31 Encinal (Oak) Series*
123.311 Mixed *Quercus* (= *Quercus* spp.) Association*
123.312 *Quercus grisea* Association*
123.313 *Quercus emoryi* Association*
123.314 *Quercus chihuahuaensis* Association*
123.315 *Quercus arizonica* Association*
123.316 *Quercus* spp.-*Pinus cembroides*-*Juniperus* spp. Association*
123.317 *Pinus cembroides* Association*
123.318 *Juniperus deppeana* Association*
123.32 Oak-Pine Series*
123.321 *Quercus hypoleucoides*-*Quercus rugosa* Association*
123.322 *Quercus* spp.-*Pinus leiophylla* Association*
123.323 *Quercus* spp.-*Pinus engelmanni* Association*
123.324 *Quercus* spp.-*Pinus* spp. Association*
123.325 *Quercus* spp.-*Arbutus xalapensis*-*Pinus* spp. Association*
123.4 Californian Evergreen Woodland
123.41 Encinal (Oak) Series*
123.411 Mixed *Quercus* Association*
123.412 *Quercus agrifolia* Association*
123.413 *Quercus agrifolia*-*Juglans californica* Association*
123.414 *Quercus engelmanni* Association*
123.42 Walnut Series*
123.421 *Junglans californica* Association*
123.5 Relict Conifer Forest and Woodland
123.51 Closed-cone Pine Series*
123.511 *Pinus attenuata* Association*
123.512 *Pinus muricata* Association*
123.513 *Pinus torreyana* Association*
123.52 Cypress Series*
123.521 *Cupressus arizonica arizonica* Association*
123.522 *Cupressus arizonica glabra* Association*
123.523 *Cupressus arizonica stephensoni* Association*
123.524 *Cupressus arizonica montana* Association*
123.525 *Cupressus forbesi* Association*
123.526 *Cupressus forbesi*-*Pinus muricata* Association*
124 Tropical-Subtropical Forests and Woodlands
124.1 Caribbean Montane Rain Forest
124.2 Caribbean Cloud Forest
124.3 Caribbean Evergreen Forest
124.4 Caribbean Deciduous Forest
124.5 Tamaulipan Deciduous Forest
124.6 Sinaloan Deciduous Forest
124.61 Mixed Short Tree Series*
124.611 Mixed Deciduous Association*
124.612 *Lysiloma watsoni*-mixed deciduous Association*
124.613 *Conzattia sericea* Association*
124.614 *Ceiba acuminata*-mixed deciduous Association*
124.615 *Bursera inopinnata*-mixed deciduous Association*
130 Scrubland Formation
131 Arctic-Boreal Scrublands
131.1 Alaskan (Low Arctic) Coastal Scrub
131.2 Canadian (Low Arctic, Barren Ground) Subpolar Scrub
131.3 Alaskan Alpine and Subalpine Scrub
131.4 Adirondack-Appalachian Alpine and Subalpine Scrub
131.5 Rocky Mountain Alpine and Subalpine Scrub
131.51 Willow Series*
131.511 *Salix brachycarpa* Association*

131.512 *Salix planifolia* Association*
131.513 *Salix nivalis* Association*
131.52 Spruce Elfinwood Series*
131.521 *Picea engelmanni* Association*
131.53 Bristlecone Pine Elfinwood Series*
131.531 *Pinus aristata* Association
131.6 Sierran-Cascade Alpine and Subalpine Scrub
131.61 Limber Pine-Lodgepole Pine Elfinwood Series*
131.611 *Pinus flexilis* Association*
132 Cold Temperate Scrublands
132.1 Great Basin Montane Scrub
132.11 Oak-scrub Series*
132:111 *Quercus gambeli* Association*
132.12 Mountain mahogany Series*
132.121 *Cercocarpus montanus* Association*
132.13 Maple-scrub Series*
132.131 *Acer grandidentatum* Association*
132.14 Serviceberry Series*
132.141 *Amelanchier alnifolia* Association*
132.15 Bitterbush Series*
132.151 *Purshia tridentata* Association*
132.16 Mixed Deciduous Series*
132.161 Mixed Scrub Association*
132.2 Sierran-Cascade Montane Scrub
132.21 Manzanita Series*
132.211 *Arctostaphylos glauca* Association*
132.212 *Arctostaphylos glandulosa* Association*
132.22 Mixed Scrub Series*
132.221 Mixed scrub Association*
132.3 Plains Deciduous Scrub
132.31 Oak-Scrub Series*
132.311 *Quercus harvardi* Association*
132.32 Sumac Series*
132.321 *Rhus lanceolata* Association*
132.33 Mixed Deciduous Series*
132.331 *Prunus americana* et al. Association*
133 Warm Temperate Scrublands
133.1 Californian Chaparral
133.11 Chamise Series*
133.111 *Adenostoma fasciculatum* Association*
133.112 *Adenostoma fasciculatum*-mixed sclerophyll Association*
133.113 *Adenostoma sparsifolium* Association*
133.12 Scrub Oak Series*
133.121 *Quercus dumosa* Association*
133.122 *Quercus dumosa*-mixed sclerophyll Association*
133.123 *Quercus dumosa*-*Quercus wislizeni* Association*
133.13 Manzanita Series*
133.131 *Arctostaphylos glauca* Association*
133.132 *Arctostaphylos glandulosa* Association*
133.133 *Arctostaphylos glandulosa*-*Pinus coulteri* Association*
133.134 *Arctostaphylos glandulosa*-mixed sclerophyll Association*
133.14 Ceanothus Series*
133.141 *Ceanothus cordulatus* Association*
133.142 *Ceanothus* spp.-mixed sclerophyll Association*
133.2 Californian Coastalscrub
133.21 Sage Series*
133.211 *Artemisia californica* Association*
133.212 *Artemisia californica*-*Salvia* spp. Association*
133.22 Mixed Shrub Series*
133.221 *Eriogonum fasciculatum*-*Simmondsia chinensis* et al. Association*
133.222 *Encelia californica*-Mixed shrub Association*

*Examples only.

133.3 Interior Chaparral
    133.31 Scrub Oak Series*
        133.311 *Quercus turbinella* Association*
        133.312 *Quercus turbinella-Cerocarpus breviflorus*
            Association*
        133.313 *Quercus turbinella-Cercocarpus betuloides*
            Association*
        133.314 *Quercus turbinella*-mixed sclerophyll
            Association*
        133.315 *Quercus intricata* Association*
        133.316 *Quercus intricata-Cercocarpus* spp. Association*
        133.317 *Quercus intricata-Quercus* spp. Association*
        133.318 *Quercus intricata*-mixed sclerophyll Association*
        133.319 *Quercus pungens* Association*
        133.311a *Quercus pungens*-mixed sclerophyll
            Association*
    133.32 Manzanita Series*
        133.321 *Arctostaphylos pringlei* Association*
        133.322 *Arctostaphylos pungens* Association*
    133.33 Ceanothus Series*
        133.331 *Ceanothus greggi* Association*
        133.332 *Ceanothus greggi*-mixed sclerophyll Association*
    133.34 Mountain mahogany Series*
        133.341 *Cercocarpus breviflorus* Association*
        133.342 *Cercocarpus montanus* Association*
    133.35 Silktassel Series*
        133.351 *Garrya wrighti* Association*
        133.352 *Garrya ovata* Association*
    133.36 Mixed Evergreen Sclerophyll Series*
        133.361 Mixed sclerophyll Association*
133.4 Southeastern Maritime Scrub
134 Tropical-Subtropical Scrublands
  134.1 Caribbean Thornscrub
  134.2 Tamaulipan Thornscrub
  134.3 Sinaloan Thornscrub
    134.31 Mixed Deciduous Series*
        134.311 Mixed scrub-*Fouquieria macdougali* Association*
        134.312 Mixed scrub-*Ipomoea arborescens* Association*
        134.313 Mixed scrub-*Lysiloma divaricata* Association*
        134.314 Mixed scrub-*Acacia cymbispina* Association*
        134.315 Mixed scrub-*Ceiba acuminata* Association*
        134.316 Mixed scrub-Mixed tree Association*
    134.32 Mesquite Disclimax Series*
        134.321 *Prosopis juliflora velutina*-mixed scrub Association*
140 Grassland Formation
  141 Arctic-Boreal Grasslands
    141.1 Alaskan (Low Arctic) Coastal Grassland
    141.2 Canadian (Low Arctic) Grassland
    141.3 Appalachian Subalpine (Balds) Grassland
    141.4 Rocky Mountain Alpine and Subalpine Grassland
      141.41 Bunchgrass Series*
        141.411 *Festuca thurberi* Association*
        141.412 *Festuca arizonica* Association*
        141.413 Mixed grass-forb Association*
      141.42 Sedge-Forb-Grass Series*
        141.421 *Carex* spp.-mixed forb-grass Association*
    141.5 Sierran Cascade Alpine and Subalpine Grassland
      141.51 Bunchgrass Series*
        141.511 *Sitanion hystrix*-mixed forb Association*
        141.512 Mixed grass-forb Association*
      141.52 Sedge-Forb-Grass Series*
        141.521 *Carex* spp.-mixed forb-grass Association*
    141.6 Madrean Alpine and Subalpine Grassland
  142 Cold Temperate Grasslands

142.1 Plains Grassland
  142.11 Bluestem "tall-grass" Series*
    142.111 *Andropogon scoparius* Association*
    142.112 *Andropogon* spp.-mixed tall-grass Association*
    142.113 *Andropogon* spp.-*Quercus harvardi* Association*
    142.114 Mixed tall-grass Association*
    142.115 *Artemisia filifolia*-mixed scrub disclimax
        Association*
  142.12 Grama "short-grass" Series*
    142.121 *Bouteloua gracilis* Association*
    142.122 *Bouteloua* spp. Association*
    142.123 *Bouteloua* spp.-mixed grass Association*
    142.124 *Bouteloua* spp.-mixed grass-mixed scrub
        Association*
  142.13 Buffalo-grass Series*
    142.131 *Buchloe dactyloides*-mixed grass Association*
  142.14 Mixed "Short-grass" Series*
    142.141 *Aristida* spp.-*Bouteloua gracilis-Buchloe
        dactyloides* Association*
  142.15 Shrub-Grass Disclimax Series*
    142.151 *Gutierrezia sarothrae* Association*
142.2 Great Basin Shrub-Grassland
  142.21 Wheatgrass Series*
    142.211 *Agropyron smithi* Association*
    142.212 *Agropyron smithi*-mixed scrub Association*
    142.213 *Agropyron smithi-Artemisia tridentata*
        Association*
  142.22 Mixed Bunchgrass Series*
    142.221 Mixed grass Association*
    142.222 Mixed grass-*Artemisia tridentata* Association*
  142.23 Ricegrass Series*
    142.231 *Oryzopsis hymenoides* Association*
  142.24 Sacaton Series*
    142.241 *Sporobolus airoides* Association*
    142.242 *Sporobolus airoides-Atriplex canescens*
        Association*
142.3 Pacific Coastal (Oregonian) Grassland
142.4 Rocky Mountain Montane Grassland
  142.41 Mixed Meadow Series*
    142.411 Mixed forb-grass Association*
  142.42 Rush Series*
    142.421 *Juncus* spp. Association*
  142.43 Fern Series*
    142.431 *Pteridium aquilinum* Association*
  142.44 Iris Disclimax Series*
    142.441 *Iris missouriensis* Association*
142.5 Sierran-Cascade Montane Grassland
  142.51 Mixed Meadow Series*
    142.511 Mixed forb-grass Association*
  142.52 Rush Series*
    142.521 *Juncus* spp. Association*
143 Warm Temperate Grasslands
  143.1 Scrub-Grassland (Semidesert Grassland)
    143.11 Grama Grass-Scrub Series*
      143.111 *Bouteloua eriopoda-Yucca elata* Association*
      143.112 *Bouteloua eriopoda-Prosopis juliflora*
        Association*
      143.113 *Bouteloua eriopoda*-mixed grass-mixed scrub
        Association*
      143.114 *Bouteloua* spp.-mixed grass-mixed scrub
        Association*
    143.12 Tobosa Grass-Scrub Series*
      143.121 *Hilaria mutica* Association*
      143.122 *Hilaria mutica-Prosopis juliflora* Association*

*Examples only.

143.123 *Hilaria mutica*-mixed scrub Association*
143.13 Curleymesquite grass-scrub Series*
    143.131 *Hilaria belangeri*-mixed scrub Association*
143.14 Sacaton-Scrub Series*
    143.141 *Sporobolus wrighti* Association*
    143.142 *Sporobolus wrighti-Prosopis juliflora* Association*
143.15 Mixed Grass-Scrub Series*
    143.151 Mixed grass-*Yucca elata* Association*
    143.152 Mixed grass-*Prosopis juliflora* Association
    143.153 Mixed grass-*Acacia greggi* Association*
    143.154 Mixed grass-*Fouquieria splendens* Association*
    143.155 Mixed grass-mixed scrub Association*
143.16 Shrub-Scrub Disclimax Series*
    143.161 *Aplopappus tenuisectus* Association*
    143.162 *Aplopappus tenuisectus-Yucca elata* Association*
    143.163 *Aplopappus tenuisectus-Prosopis juliflora* Association*
    143.164 *Aplopappus tenuisectus*-mixed scrub Association*
    143.165 *Gutierrezia sarothrae-Prosopis juliflora* Association*
143.2 Californian Valley Grassland
  143.21 Annual Disclimax Series*
    143.211 Mixed annual grass Association*
    143.212 *Avena fatua* Association*
    143.213 *Bromus rubens* Association*
    143.214 Mixed forb Association*
144 Tropical-Subtropical Grasslands
  144.1 Caribbean Savanna Grassland
  144.2 Gulf Coastal (Tamaulipan) Grassland
  144.3 Sonoran Savanna Grassland
    144.31 Mixed Root-perennial Grass Series*
      144.311 *Heteropogon contortus-Bouteloua* spp.-*Aristida* spp.-mixed scrub Association*
    144.32 Grama Series*
      144.321 *Bouteloua rothrocki-Prosopis juliflora* Association*
      144.322 *Bouteloua* spp.-mixed scrub Association*
    144.33 Three-awn Series*
      144.331 *Aristida* spp.-*Prosopis juliflora* Association*
      144.332 *Aristida* spp.-mixed scrub Association*
150 Desertland Formation
  151 Arctic-Boreal Desertlands
    151.1 Polar Desertscrub
  152 Cold Temperate Desertlands
    152.1 Great Basin Desertscrub
      152.11 Sagebrush Series*
        152.111 *Artemisia tridentata* Association*
        152.112 *Artemisia tridentata*-mixed scrub-grass Association*
        152.113 *Artemisia nova* Association*
      152.12 Shadscale Series*
        152.121 *Atriplex confertifolia* Association*
        152.122 *Atriplex confertifolia*-mixed scrub Association*
      152.13 Blackbrush Series*
        152.131 *Coleogyne ramosissima* Association*
      152.14 Rabbitbrush Series*
        152.141 *Chrysothamnus nauseosus* Association*
      152.15 Winterfat Series*
        152.151 *Eurotia lanata* Association*
        152.152 *Eurotia lanata*-mixed scrub Association*
      152.16 Mixed scrub Series*
        152.161 *Ephedra viridis-Eriogonum* spp.-mixed scrub Association*
      152.17 Saltbush Series*
        152.171 *Sarcobatus vermiculatus* Association*
        152.172 *Atriplex canescens* Association*
  153 Warm Temperate Desertlands
    153.1 Mohave Desertscrub

153.11 Creosotebush Series*
    153.111 *Larrea divaricata* Association*
    153.112 *Larrea divaricata-Ambrosia dumosa* Association*
    153.113 *Larrea divaricata-Yucca* spp. Association*
153.12 Blackbrush Series*
    153.121 *Coleogyne ramosissima* Association*
    153.122 *Coleogyne ramosissima-Yucca* spp. Association*
153.13 Mesquite Series*
    153.131 *Prosopis juliflora torreyana* Association*
153.14 Bladdersage Series*
    153.141 *Salazaria mexicana* Association*
153.15 Joshuatree Series*
    153.151 *Yucca brevifolio-Acamptopappus sphaerocephalus-Larrea divaricata*-mixed scrub Association*
    153.152 *Yucca brevifolia-Coleogyne ramosissima* Association*
    153.153 *Yucca brevifolia-Larrea divaricata* Association*
153.16 Catclaw Series*
    153.161 *Acacia greggi*-mixed scrub Association*
153.17 Saltbush Series*
    153.171 *Suaeda torreyana* Association*
    153.172 *Atriplex* spp. Association*
153.2 Chihuahuan Desertscrub
  153.21 Creosotebush-Tarbush Series*
    153.212 *Larrea divaricata-Parthenium incanum*-mixed scrub Association*
    153.213 *Larrea divaricata-Flourensia cernua* Association*
    153.214 *Flourensia cernua* Association*
  153.22 Whitethorn Series*
    153.221 *Acacia neovernicosa* Association*
    153.222 *Acacia neovernicosa-Larrea divaricata* Association*
  153.23 Sandpaperbush Series*
    153.231 *Mortonia scabrella* Association*
    153.232 *Mortonia scabrella-Rhus microphylla* Association*
  153.24 Mesquite Series*
    153.241 *Prosopis juliflora glandulosa* (shrub hummock) Association*
    153.242 *Prosopis juliflora glandulosa-Artemesia filifolia* Association*
  153.25 Succulent Series*
    153.251 *Agave lecheguilla* Association*
    153.252 *Agave lecheguilla-Yucca* spp. Association*
    153.253 *Opuntia* spp.-*Agave* spp.-*Larrea divaricata* Association*
  153.26 Mixed Scrub Series*
    153.261 *Fouquieria splendens*-mixed scrub Association*
  153.27 Saltbush Series*
    153.271 *Suaeda torreyana* Association*
    153.272 *Atriplex canescens* Association*
    153.273 *Atriplex* spp.-*Artemisia filifolia* Association*
154 Tropical-Subtropical Desertlands
  154.1 Sonoran Desertscrub
    154.11 Creosotebush-Bursage ("Lower Colorado Valley") et al Series*
      154.111 *Larrea divaricata* Association*
      154.112 *Larrea divaricata-Ambrosia dumosa* Association*
      154.113 *Ambrosia dumosa* Association*
      154.114 *Prosopis juliflora torreyana* (shrub hummock) Association*
      154.115 *Cercidium floridum-Olneya tesota-Dalea spinosa* riparian Association*
      154.116 *Fouquieria splendens-Agave deserti* Association*

*Examples only.

154.117 *Opuntia bigelovi* Association*
154.12 Paloverde-Mixed Cacti ("Arizona Upland") Series*
    154.121 *Ambrosia deltoidea-Cercidium microphyllum*-
        mixed scrub Association*
    154.122 *Ambrosia deltoidea-Carnegiea gigantea*-mixed
        scrub Association*
    154.123 *Simmondsia chinensis*-mixed scrub Association*
    154.124 *Larrea divaricata-Canotia holacantha*
        Association*
    154.125 *Larrea divaricata*-mixed scrub Association*
    154.126 *Encelia farinosa*-mixed scrub Association*
    154.127 Mixed shrub-*Cercidium microphyllum-Olneya*
        tesota-mixed scrub Association
154.13 Brittlebush-Ironwood ("Plains of Sonora") Series*
    154.131 *Encelia farinosa-Olneya tesota* Association*
    154.132 *Encelia farinosa*-mixed scrub Association*
    154.133 Mixed shrub-mixed scrub Association*
    154.134 Mixed shrub-*Prosopis juliflora velutina*
        Association*
    154.135 Mixed shrub-*Forchammeria watsoni* Association*
154.14 Copal-Torote ("Central Gulf Coast") Series*
    154.141 *Jatropha cinerea-Bursera microphylla*
        Association*
    154.142 *Jatropha* spp.-*Bursera microphylla-Pachycereus*
        pringlei Association*
    154.143 *Jatropha* spp.-*Idria columnaris*-mixed scrub
        Association*
154.15 Agave-Bursage ("Vizcaino") Series*
    154.151 *Ambrosia chenopodifolia-Agave shawi*
        Association*
    154.152 *Ambrosia* spp.-*Agave shawi-Pachycormus*
        discolor-Idria columnaris-mixed scrub Association*
    154.153 *Ambrosia* spp.-*Agave shawi-Pachycereus*
        pringlei-mixed scrub Association*
    154.154 Mixed shrub-*Agave shawi* Association*
    154.155 *Eriogonum fasiculatum*-mixed scrub Association*
154.16 Paloblanco-Agria ("Magdalena") Series*
    154.161 *Machaerocereus gummosus*-mixed scrub
        Association*
154.17 Saltbush Series
    154.171 *Suaeda torreyana* Association*
    154.172 *Allenrolfea occidentalis* Association*
    154.173 *Atriplex* spp.-*Prosopis juliflora torreyana*
        Association*
    154.174 *Atriplex polycarpa-Lycium* spp.-*Prosopis juliflora*
        velutina Association*
    154.175 *Frankenia palmeri-Atriplex julacea* Association*

Table 7. Nomenclature of WETLAND Biotic Communities (Fourth Level) of Nearctic and Adjacent Neotropical North America with Some Community (Series) and Association Level Examples for the North American Southwest.

1,200 Nearctic Wetland Vegetation
  1,210 Wet Tundra Formation
    1,211 Arctic Wet Tundra
      211.1 Polar (High Arctic) Wet Tundra[1]/
      211.2 Greenlandian Wet Tundra
      211.3 Alaskan (Coastal) Wet Tundra
      211.4 Canadian (Low Arctic) Wet Tundra
220 Forest Formation
  221 Boreal Swamp and Riparian Forests

*One or more examples only are given for these levels.
[1]/The first "1" (in front of comma and representing the Nearctic Realm) is understood, and cropped for tabular convenience only, from this point onward.

    221.1 Canadian Swampforest
  222 Cold Temperate Swamp and Riparian Forests
    222.1 Northeastern Bog, Swamp and Riparian Forests
    222.2 Plains and Great Basin Riparian Deciduous Forest
      222.21 Cottonwood-Willow Series*
        222.211 *Populus sargenti* Association*
        222.212 *Populus sargenti-Salix amygdaloides* Association*
        222.213 *Populus wislizeni* Association*
        222.214 *Populus* spp.-*Salix* spp. Association*
        222.215 *Salix exigua* Association*
    222.3 Rocky Mountain Riparian Deciduous Forest
      222.31 Cottonwood-Willow Series*
        222.311 *Populus angustifolia-Salix* spp. Association*
      222.32 Mixed Broadleaf Series*
        222.321 *Acer negundo-Populus angustifolia*-mixed
          deciduous Association*
        222.322 *Acer grandidentatum* Association*
    222.4 Sierran-Cascade Riparian Deciduous Forest
      222.41 Cottonwood-Willow Series*
        222.411 *Populus trichocarpa-Salix* spp. Association*
      222.42 Mixed Broadleaf Series*,
        222.412 *Acer macrophyllum-Populus trichocarpa-Alnus*
          rhombifolia-mixed deciduous Association*
  223 Warm Temperate Swamp and Riparian Forests
    223.1 Southeastern Swamp and Riparian Forest
    223.2 Interior Southwestern Riparian Deciduous Forest and
      Woodland
      223.21 Cottonwood-Willow Series*
        223.221 *Populus fremonti-Salix* spp. Association*
        223.212 *Populus fremonti* Association*
        223.213 *Populus wislizeni* Association*
        223.214 *Populus acuminata* Association*
      223.22 Mixed Broadleaf Series*
        223.221 *Platanus wrighti-Fraxinus velutina-Populus*
          fremonti-mixed deciduous Association*
        223.222 *Platanus wrighti* Association*
        223.223 *Fraxinus velutina* Association*
        223.224 *Alnus oblongifolia* Association*
        223.225 *Juglans major* Association*
    223.3 Californian Riparian Deciduous Forest and Woodland
      223.31 Cottonwood-Willow Series
        223.311 *Populus fremonti-Salix* spp. Association*
      223.32 Mixed Broadleaf Series
        223.321 *Platanus racemosa*-mixed deciduous Association*
        223.322 *Alnus rhombifolia* Association*
  224 Tropical-Subtropical Swamp, Riparian and Oasis Forests.
    224.1 Caribbean Interior Swamp and Riparian Forests
    224.2 Caribbean Maritime Swampforest
    224.3 Tamaulipan Interior Swamp and Riparian Forests
    224.4 Sinaloan Interior Swamp and Riparian Forests
      224.41 Mixed Evergreen Series*
        224.411 *Ficus* spp.-mixed evergreen and deciduous
          Association*
        224.412 *Taxodium mucronatum* Association*
        224.413 *Populus* sp.-mixed evergreen and deciduous
          Association*
      224.42 Palm Series*
        224.421 *Sabal uresana* Association*
    224.5 Sonoran Riparian and Oasis Forests
      224.51 Palm Series*
        224.511 *Washingtonia filifera* Association*
        224.512 *Washingtonia filifera-Populus fremonti*
          Association*
        224.513 *Washingtonia filifera-Brahea armata*
          Association*

224.514 *Brahea armata* Association*
224.515 *Phoenix dactylifera-Washingtonia filifera* Association*
224.52 Mesquite Series*
224.521 *Prosopis juliflora velutina* Association*
224.522 *Prosopis juliflora velutina*-mixed short tree
Association*
224.53 Cottonwood-Willow Series*
224.531 *Populus fremonti-Salix gooddingi* Association*
224.532 *Populus fremonti* Association*
224.533 *Salix gooddingi* Association*
230 Swampscrub Formation
231 Arctic-Boreal Swampscrubs
231.1 Polar (High Arctic) Swampscrub
231.2 Greenlandian Swampscrub
231.3 Alaskan Swampscrub
231.4 Canadian Swampscrub
231.5 Adirondack-Appalachian Alpine and Subalpine Swamp
and Riparian Scrub
231.6 Rocky Mountain Alpine and Subalpine Swamp and
Riparian Scrub
231.61 Willow Series*
231.611 *Salix bebbiana* Association*
231.7 Sierran-Cascade Alpine and Subalpine Swamp and
Riparian Scrub
231.71 Willow Series*
231.711 *Salix* spp. Association*
232 Cold Temperate Swamp and Riparian Scrubs
232.1 Northeastern Deciduous Swampscrub
232.2 Plains and Great Basin Swamp and Riparian Scrub
232.21 Willow Series*
232.211 *Salix* spp.-mixed scrub Association*
232.22 Saltcedar Disclimax Series*
232.221 *Tamarix chinensis* Association*
232.3 Rocky Mountain Riparian Scrub
232.31 Willow-Dogwood Series*
232.311 *Salix* spp.-mixed deciduous Association*
232.4 Sierran-Cascade Riparian Scrub
232.41 Willow Series*
232.411 *Salix* spp. Association
232.5 Pacific Coastal (Oregonian) Swamp and Riparian Scrub
233 Warm Temperate Swamp and Riparian Scrubs
233.1 Southeastern Mixed Deciduous and Evergreen
Swampscrub
233.2 Interior Southwestern Swamp and Riparian Scrub
233.21 Mixed Narrowleaf Series*
233.211 *Cephalanthus occidentalis-Baccharis glutinosa*-
mixed scrub Association*
233.22 Saltcedar Disclimax Series*
233.221 *Tamarix chinensis*-mixed deciduous Association*
233.3 Californian Deciduous Swamp and Riparian Scrub
233.31 Mixed Narrowleaf Series*
233.311 *Salix lasiolepis* Association*
234 Tropical-Subtropical Swamp and Riparian Scrub
234.1 Caribbean Interior Swampscrub
234.2 Caribbean Maritime Swampscrub
234.3 Tamaulipan Interior Swampscrub
234.4 Tamaulipan Maritime Swampscrub
234.5 Sinaloan Interior Swamp and Riparian Scrub
234.51 Mixed Evergreen Series*
234.511 *Vallesia glabra-Baccharis glutinosa-Salix
bonplandiana* Association*
234.6 Sinaloan Maritime Swampscrub
234.61 Mangrove Series*

234.611 *Avicennia germinans* Association*
234.612 *Rhizophora mangle* Association*
234.7 Sonoran Deciduous Swamp and Riparian Scrub
234.71 Mixed Scrub Series*
234.711 *Prosopis pubescens-Prosopis juliflora torreyana-
Pluchea sericea* Association*
234.72 Saltcedar Disclimax Series*
234.721 *Tamarix chinensis* Association*
234.722 *Tamarix chinensis*-mixed scrub Association*
240 Marshland Formation
241 Arctic-Boreal Marshlands
241.1 Polar (High Arctic) Marshland
241.2 Greenlandian Marshland
241.3 Alaskan Maritime (Coastal) Marshland
241.4 Canadian Interior Marshland
241.5 Canadian Maritime (Coastal) Marshland
241.6 Adirondack-Appalachian Alpine and Subalpine
Marshland
241.7 Rocky Mountain Alpine and Subalpine Marshland
241.71 Rush Series*
241.711 *Juncus balticus* Association*
241.72 Manna Grass Series*
241.721 *Glyceria borealis* Association*
241.8 Sierran-Cascade Alpine and Subalpine Marshland
241.81 Rush Series*
241.811 *Juncus* spp. Association*
242 Cold Temperate Marshlands
242.1 Northeastern Interior Marshland
242.2 Northeastern Maritime (Coastal) Marshland
242.3 Plains Interior Marshland
242.31 Rush Series*
252.311 *Juncus tenuis* Association*
242.32 Bur-reed Series*
242.321 *Sparganium angustifolium* Association*
242.33 Cattail Series*
242.331 *Typha latifolia* Association*
242.34 Bulrush Series*
242.341 *Scirpus validus* Association*
242.4 Rocky Mountain Montane Marshland
242.41 Rush Series*
242.411 *Juncus saximontanus* Association*
242.5 Great Basin Interior Marshland
242.51 Rush Series*
242.511 *Juncus* spp. Association*
242.52 Saltgrass Series*
242.521 *Distichlis stricta* Association*
242.6 Sierran-Cascade Montane Marshland
242.61 Rush Series*
242.611 *Juncus* spp. Association*
242.7 Pacific Coastal (Oregonian) Interior Marshland
242.8 Pacific Coastal (Oregonian) Maritime Marshland
243 Warm Temperate Marshlands
243.1 Southeastern Interior Marshland
243.2 Southeastern Maritime Marshland
243.3 Chihuahuan Interior Marshland
243.31 Saltgrass Series*
243.311 *Distichlis stricta* Association*
243.4 Mohavian Interior Marshland
243.41 Rush Series*
243.411 *Juncus cooperi* Association*
243.42 Saltgrass Series*
243.421 *Distichlis stricta* Association*
243.5 Madrean Marshland
243.51 Rush Series*
243.511 *Juncus mexicanus* Association*

*Examples only.

243.6 Californian Interior Marshland
  243.61 Cattail Series*
    243.611 *Typha latifolia* Association*
    243.612 *Typha domingensis* Association*
243.7 Californian Maritime Marshland
  243.71 Cordgrass Series*
    243.711 *Spartina foliosa* Association*
  243.72 Glasswort Series*
    243.721 *Salicornia virginica* Association*
244 Tropical-Subtropical Marshland
  244.1 Caribbean Interior Marshland
  244.2 Caribbean Maritime Marshland
  244.3 Tamaulipan Interior Marshland
  244.4 Gulf Coast Maritime Marshland
  244.5 Sinaloan Interior Marshland
    244.51 Cattail Series*
      244.511 *Typha domingensis* Association*
  244.6 Sinaloan Maritime Marshland
    244.61 Glasswort Series*
      244.611 *Salicornia* spp. Association*
  244.7 Sonoran Interior Marshland
    244.71 Cattail Series*
      244.711 *Typha domingensis* Association*
    244.72 Giant Reed Series*
      244.721 *Phragmites communis* Association*
    244.73 Bulrush Series*
      244.731 *Scirpus americanus* Association*
    244.74 Threesquare Series*
      244.741 *Scirpus olneyi* Association*
  244.8 Sonoran Maritime Marshland
    244.81 Saltgrass Series*
      244.811 *Distichlis stricta* Association*
    244.82 Glasswort Series*
      244.821 *Salicornia* spp. Association*
250 Strand Formation
  251 Arctic-Boreal Strands
    251.1 Polar Maritime Strand
    251.2 Greenlandian Strand
    251.3 Alaskan Maritime Strand
    251.4 Canadian Interior (Stream and Lake) Strand
    251.5 Canadian Maritime Strand
    251.6 Adirondack-Appalachian Alpine and Subalpine Stream
      and Lake Strand
    251.7 Rocky Mountain Alpine and Subalpine Stream and Lake
      Strand**
    251.8 Sierran-Cascade Alpine and Subalpine Stream and Lake
      Strand**
  252 Cold Temperate Strands
    252.1 Northeastern Interior (Stream and Lake) Strand
    252.2 Northeastern Maritime Strand
    252.3 Plains Interior (Stream and Lake) Strand
      252.31 Annual Series**
        252.311 *Xanthium saccharatum*-mixed annual Association*
    252.4 Rocky Mountain Montane Stream and Lake Strand*
      252.41 Annual Series**
    252.5 Great Basin Interior Strand*
      252.51 Annual Series**
    252.6 Sierran-Cascade Montane Stream and Lake Strand*
      252.61 Annual Series**
    252.7 Pacific Coastal (Oregonian) Interior Strand
    252.8 Pacific Coastal (Oregonian) Maritime Strand

253 Warm Temperate Strands
  253.1 Southeastern Interior Strand
  253.2 Southeastern Maritime Strand
  253.3 Chihuahuan Interior Strand
    253.31 Annual Series*
      253.311 *Xanthium saccharatum* Association*
  253.4 Mohavian Interior Strand
    253.41 Annual Series*
      253.411 *Xanthium saccharatum* Association*
    253.42 Mixed Scrub Series*
      253.421 *Tamarix chinensis*-mixed shrub Association*
  253.5 Madrean Stream and Lake Strand
    253.51 Annual Series*
      253.511 Mixed annual Association*
  253.6 Californian Stream and Lake Strand
    253.61 Annual Series*
      253.611 *Nicotiana attenuata* Association*
  253.7 Californian Maritime Strand
    253.71 Mixed Scrub Series*
      253.711 *Abronia maritima-Atriplex leucophylla-Cakile
        maritima* et al. Association*
    253.72 Sea-grass Series*
      253.721 *Phyllospadix scouleri* Association*
    253.73 Green Algae Series*
      253.731 *Ulva californica* Association*
    253.74 Brown Algae Series*
      253.741 *Pelvetia fastigiata* Association*
    253.75 Red Algae Series*
      253.751 *Gigartina canaliculata* Association*
254 Tropical-Subtropical Strands
  254.1 Caribbean Interior Strand
  254.2 Caribbean Maritime Strand
  254.3 Tamaulipan Interior Strand
  254.4 Gulf Coast (Tamaulipan) Maritime Strand
  254.5 Sinaloan Interior Strand*
    254.51 Annual Series**
  254.6 Sinaloan Maritime Strand**
    254.61 Mixed Scrub Series*
  254.7 Sonoran Interior Strand
    254.71 Mixed Scrub Series*
      254.711 *Baccharis glutinosa-Solanum nodiflorum-
        Nicotiana* spp.-*Rumex hymenosepalus* et al.
        Association*
    254.72 Annual Series*
      254.721 *Amaranthus palmeri* Association*
  254.8 Sonoran Maritime Strand
    254.81 Mixed Scrub Series
      254.811 *Abronia maritima-Helianthus niveus-Jouvea
        pilosa* et al. Association*
260 Submergent Aquatic Vegetation
  261 Arctic-Boreal Submergent Aquatics
    261.1 Polar Marine Submergents
    261.2 Greenlandian Inland Submergents
    261.3 Alaskan Marine Submergents
    261.4 Canadian Inland Submergents
    261.5 Canadian Marine Submergents
    261.6 Adirondack-Appalachian Alpine and Subalpine
      Submergents
    261.7 Rocky Mountain Alpine and Subalpine Submergents
      261.71 Pondweed Series*
        261.711 *Potamogeton natans* Association*
    261.8 Sierran-Cascade Alpine and Subalpine Submergents
      261.81 Pondweed Series*
        261.811 *Potamogeton alpinus* Association*
  262 Cold Temperate Submergent Aquatics

*Examples only.
**Our incomplete knowledge of these biotic communities precludes presentation of
representative fifth (series) and sixth level (association) examples.

262.1 Northeastern Inland Submergents
262.2 Northeastern Marine Submergents
262.3 Plains Inland Submergents
    262.31 Pondweed Series*
      262.311 *Potamogeton foliosus* Association*
262.4 Rocky Mountain Montane Submergents
    262.41 Pondweed Series*
      262.411 *Potamogeton foliosus* Association*
262.5 Great Basin Inland Submergents
    262.51 Pondweed Series*
      262.511 *Potamogeton foliosus* Association*
262.6 Sierran-Cascade Montane Submergents
    262.61 Pondweed Series*
      262.611 *Potamogeton pusillus* Association*
262.7 Pacific Coastal (Oregonian) Inland Submergents
262.8 Pacific Coastal (Oregonian) Marine Submergents
263 Warm Temperate Submergent Aquatics
  263.1 Southeastern Inland Submergents
  263.2 Southeastern Marine Submergents
  263.3 Chihuahuan Inland Submergents
    263.31 Pondweed Series*
      263.311 *Potamogeton pectinatus* Association*
  263.4 Mohavian Inland Submergents
    263.41 Pondweed Series*
      263.411 *Potamogeton pectinatus* Association*
  263.5 Madrean Inland Submergents
    263.51 Pondweed Series*
      263.511 *Potamogeton pectinatus* Association*
  263.6 Californian Inland Submergents
    263.61 Pondweed Series*
      263.611 *Potamogeton pectinatus* Association*
    263.62 Milfoil Series*

      263.621 *Myriophyllum exalbescens* Association*
  263.7 Californian Marine Submergents
    263.71 Ruppia Series*
      263.711 *Ruppia maritima* Association*
    263.72 Eelgrass Series*
      263.721 *Zostera marina* Association*
    263.73 Giant Kelp Series*
      263.731 *Macrocystis pyrifera* Association*
    263.74 Feather-boa kelp Series*
      263.741 *Egregia laevigata* Association*
    263.75 Southern Sea Palm Series*
      263.751 *Eisenia arborea* Association*
264 Tropical-Subtropical Submergent Aquatics
  264.1 Caribbean Inland Submergents
  264.2 Caribbean Marine Submergents
  264.3 Tamaulipan Inland Submergents
  264.4 Gulf Coastal Marine Submergents
  264.5 Sinaloan Inland Submergents
    264.51 Pondweed Series**
  264.6 Sinaloan Marine Submergents*
    264.61 Phytoplankton Series**
  264.7 Sonoran Inland Submergents**
    264.71 Pondweed Series*
      264.711 *Potamogeton pectinatus* Association*
    264.72 Milfoil Series*
      264.721 *Myriophyllum brasiliense* Association*
  264.8 Sonoran Marine Submergents
    264.81 Ruppia Series*
      264.811 *Ruppia maritima* Association*
    264.82 Eelgrass Series*
      264.821 *Zostera marina* Association*

References

To save space the *Literature Cited* section for Appendix I is not reprinted here since all citations can be found in the *References* section (pgs. 288 to 301) of the present volume.

---

*Examples only.
**Our incomplete knowledge of these biotic communities precludes presentation of sixth level (association examples.

Appendix II. Scientific and equivalent common names of plants and animals arranged by biomes.

Arctic and Alpine Tundras
111.5 Rocky Mountain Alpine Tundra

Scientific Name	Common Name
PLANTS	
Abies lasiocarpa var. *arizonica*	Corkbark Fir
Achillea lanulosa	Western Yarrow
Androsace carinata	Rock Jasmine
A. septentrionalis	Rock Jasmine
Anemone globosa	Pacific Windflower
Aquilegia formosa	Sitka Columbine
A. scopulorum	Columbine
Arabis pendulina	Rockcress
Arenaria filiforum	sandwort
A. rubella	sandwort
A. saxatile	sandwort
Astragulus platytropis	Milk Vetch, Locoweed
Botrychium lunaria	Moonwort
Calyptrydium monosperum	—
Campanula rotundifolia	Bluebell, Harebell
Carex spp.	sedges
C. aquatilis	Water Sedge
C. ebenea	Ebony Sedge
C. foena	Silvertop Sedge
C. mariposana	sedge
C. nova	Black-spiked Sedge
C. petasata	sedge
C. scopulorum	Cliff Sedge
Castilleja haydenii	Hayden Painted Cup
C. miniata	Scarlet Painted Cup
Cerastium beeringianum	Alpine Mouse-ear
Claytonia megarrhiza	Bigroot Spring Beauty
Crepis nana	Tiny Hawksbeard
Cryptogramma acrostichoides	Rockbrake, Parsley Fern
Cystopteris fragilis	Fragile Bladderfern
Deschampsia caespitosa	Tufted Hairgrass
Dodecatheon alpinum	Alpine Shooting Star
Draba crassifolia	Hairy Whitlowgrass
D. jaegeri	draba
Erigeron compositus	Fernleaf Fleabane
E. formosissimus	fleabane
E. simplex	fleabane
Eriogonum kennedyi spp. *alpigeum*	Kennedy Eriogonum
E. saxatile	Hoary Buckwheat
E. umbellatum	Sulfur Buckwheat
Erysimum nivale	Snowy Erysimum
Festuca ovina var. *brachyphylla*	Alpine Fescue
Fragaria ovalis	Wild Strawberry
Gentiana amarella	Felwort
G. barbellata	Bearded Gentian
G. heterosepala	gentian
G. tenella	Alpine Fringed-gentian
Geum turbinatum	Golden Avens
Helenium hoopesii	Orange Sneezeweed
Heuchera versicolor forma *pumila*	Painted Alumroot
Holodiscus microphyllous	Little-leaf Rockspires
Hulsea vestita var. *pygmaea*	—
Hymenoxys brandegei	Brandegee Bitterweed
Ivesia cryptocaulis	—
Juncus albescens	rush
J. castaneus	rush
J. drummondii	Drummond Rush
J. parryi	Parry Rush
Kobresia myosuroides	kobresia
Leptodactylon pungens ssp. *pulchriflorum*	Granite Gilia
Lonicera involucrata	Bearberry Honeysuckle
Luzula parviflora	Millet Woodrush
Mimulus primuloides	Primrose Monkeyflower
M. tilingii	monkeyflower
Monardella odoratissima	Pacific Monardella
Oreonana vestita	—
Oxyria digyna	Alpine Mountain Sorrel
Pedicularis hastata	—
Phleum alpinum	Alpine Timothy
Phlox covillei	Coville Phlox
P. diffusa	Spreading Phlox
Picea engelmannii	Engelmann Spruce
Pinus artistata	Bristlecone Pine
P. flexilis	Limber Pine
Poa arctica	Arctic Bluegrass
P. glauca	Greenland Bluegrass
P. reflexa	Nodding Bluegrass
P. rupicola	Timberline Bluegrass
Polygonum bistortoides	Bistort
Potentilla nivea	Alpine Cinquefoil
Primula angustifolia	Alpine Primrose
Pseudocymopterus montanus	Mountain Parsley
Raillardella argentea	—
Ranunculus eschscholtzii var. *oxynotus*	buttercup
R. inamoenus	buttercup
Ribes cereum	Wax Currant
R. montigenum	Gooseberry Currant
Salix petrophila	Skyland Willow, Alpine Willow
S. planifolia	Planeleaf Willow
Saxifraga spp.	saxifrages
S. rhomboidea var. *francisciana*	Diamondleaf Saxifrage
Sedum rhodanthum	Stonecrop
Selaginella watsonii	Alpine Selaginella
Sibbaldia procumbens	—
Silene acaulis	Silene
Sitanion hystrix	Bottlebrush Squirreltail
Stellaria calycantha	Northern Starwort
S. longipes var. *laeta*	starwort
S. umbellata	starwort
Stipa occidentalis	Western Needlegrass
Taraxacum ceratophorum	Horned Dandelion
T. laevigatum	Red-seeded Dandelion
T. lyratum	dandelion
Thlaspi montanum var. *fendleri*	Wild Candytuft
Trifolium spp.	clovers
T. nanum	Dwarf Clover
Trisetum spicatum	Spike Trisetum
Veronica wormskjoldii	Alpine Speedwell
MAMMALS	
Eutamias minimus	Least Chipmunk
E. quadrivittatus	Colorado Chipmunk
Marmota flaviventris	Yellow-bellied Marmot
Ochotona princeps	Pika
Ovis canadensis	Mountain Sheep
Peromyscus maniculatus	Deer Mouse
Sorex cinereus	Masked Shrew
S. vagrans	Vagrant Shrew
BIRDS	
Anthus spinoletta	Water Pipit
Eremophila alpestris	Horned Lark
Lagopus leucurus	White-tailed Ptarmigan
Leucosticte australis	Brown-capped Rosy Finch
Selasphorus platycercus	Broadtailed Hummingbird
Sialia currucoides	Mountain Bluebird
Zonotrichia leucophrys	White-crowned Sparrow

AMPHIBIANS AND REPTILES
none

Boreal Forests and Woodlands
121.3 Rocky Mountain (Petran) Subalpine Conifer Forest

Scientific Name	Common Name
PLANTS	
Abies concolor	White Fir
A. lasiocarpa	Subalpine Fir
A. lasiocarpa var. *arizonica*	Corkbark Fir
Acer glabrum	Rocky Mountain Maple
Agoseris spp.	mountain-dandelions
Alnus tenuifolia	Thinleaf Alder
Arctostaphylos uva-ursi	Kinnickinick
Artemisia tridentata	Big Sagebrush
Balsamorhiza spp.	balsamroots
Berberis repens	Creeping Mahonia
Betula occidentalis	Water Birch
Bromus anomalus	Nodding Brome
Bromus ciliatus	Fringed Brome
Ceanothus cordulatus	Snow Bush, Mountain Whitethorn
Cercocarpus ledifolius	Curlleaf Mountain-mahogany
Chaemaebataria millifolium	Fernbush, Desert Sweet
Crepis spp.	hawksbeards
Danthonia spp.	oatgrasses
Epilobium spp.	fireweeds
Erigeron formosissimus	Fleabane
Festuca spp.	fescues
Fragaria ovalis	Wild Strawberry
Gentiana spp.	gentians
Gilia aggregata	Scarlet Gilia
Helenium spp.	sneezeweeds
Heuchera spp.	alumroots
Juniperus communis	Dwarf Juniper
Lathyrus arizonicus	Arizona Peavine
Lonicera involucrata	Bearberry Honeysuckle
Oxalis spp.	woodsorrels
Phleum alpinum	Alpine Timothy
Physocarpus alternans	Ninebark
Picea engelmannii	Engelmann Spruce
P. pungens	Blue Spruce
Pinus aristata	Bristlecone Pine
P. flexilis	Limber Pine
Poa spp.	bluegrasses
Populus tremuloides	Quaking Aspen
Potentilla fruticosa	Shrubby Cinquefoil
Primula spp.	primroses
Prunus emarginata	Bitter Cherry
Pseudocymopterus montanus	Mountain Parsley
Pseudotsuga menziesii	Douglas-fir
Ribes spp.	currants
Rubus spp.	raspberries
Salix bebbiana	Bebb Willow
S. scouleriana	Scouler Willow
Sambucus glauca	Blueberry Elder
S. microbotrys	Red Elderberry
Senecio spp.	groundsels
Symphoricarpos spp.	snowberries
Tanacetum canum	Tansy
Taraxacum officinale	Dandelion
Trifolium spp.	clovers
Trisetum spicatum	—
Vaccinium oreophilum	Blueberry
Veratrum californicum	Skunk Cabbage, False Hellebore
Vicia spp.	vetches
Viola spp.	violets
MAMMALS	
Cervus elaphus	Elk, Wapiti
Citellus lateralis	Golden-mantled Ground Squirrel
Clethrionomys gapperi	Gapper's Red-backed Mouse
Eutamias minimus	Least Chipmunk
Lepus americanus	Snowshoe Hare
Martes americana	Marten
Neotoma cinerea	Bushy-tailed Woodrat
Odocoileus hemionus	Mule Deer
Peromyscus maniculatus	Deer Mouse
Sorex nanus	Dwarf Shrew
S. vagrans	Vagrant Shrew
Sylvilagus nuttallii	Nuttall's Cottontail
Tamiasciurus hudsonicus	Red Squirrel
BIRDS	
Cardeulis pinus	Pine Siskin
Carpodacus cassinii	Cassin's Finch
Catharus guttatus	Hermit Thrush
Certhia familiaris	Brown Creeper
Colaptes auratus	Common Flicker
Dendroica coronata	Yellow-rumped Warbler
Dendragapus obscurus	Blue Grouse
Empidonax difficilis	Western Flycatcher
E. hammondii	Hammond's Flycatcher
Junco spp.	juncos
Loxia curvirostra	Red Crossbill
Myadestes townsendi	Townsend's Solitaire
Nucifraga columbiana	Clark's Nutcracker
Nuttallornis borealis	Olive-sided Flycatcher
Parus gambeli	Mountain Chickadee
Perisoreus canadensis	Gray Jay
Picoides tridactylus	Northern Three-toed Woodpecker
Piranga ludoviciana	Western Tanager
Pinicola enucleator	Pine Grosbeak
Regulus calendula	Ruby-crowned Kinglet
R. satrapa	Golden-crowned Kinglet
Selasphorus platycercus	Broad-tailed Hummingbird
Sialia currucoides	Mountain Bluebird
Sitta canadensis	Red-breasted Nuthatch
S. pygmaea	Pygmy Nuthatch
Sphyrapicus thyroideus	Williamson's Sapsucker
Tachycineta thalassina	Violet-green Swallow
Turdus migratorius	American Robin
Vireo gilvus	Warbling Vireo
AMPHIBIANS AND REPTILES	
Plethodon neomexicanus	Jemez Mountain Salamander
Sceloporus graciosus	Sagebrush Lizard

121.4 Sierran Subalpine Conifer Forest

Scientific Name	Common Name
PLANTS	
Abies concolor	White Fir
Allium monticola	Onion
Arabis platysperma	Rockcress
Arceuthobium cyanocarpum	Limber Pine Dwarf Mistletoe
Arctostaphylos patula	Greenleaf Manzanita
Arenaria nuttallii	Sandwort
Calochortus invenustus	Mariposa
Carex spp.	sedges
Castanopsis sempervirens	Bush Chinquapin
Cercocarpus ledifolius	Curlleaf Mountain-mahogany
Corrallorhiza maculata	Spotted Coral Root
Draba corrugata	Draba
Erigeron breweri	Fleabane
Eriogonum kennedyi	Kennedy Buckwheat
E. saxatile	buckwheat
E. umbellatum	buckwheat
Galium parishii	Bedstraw
Holodiscus microphyllous	Rockspirea
Juniperus occidentalis	Western Juniper
Monardella cinerea	—
Oreonana vestita	—
Penstemon spp.	penstemons, beardtongues
Pinus albicaulis	Whitebark Pine
P. balfouriana	Foxtail Pine
P. contorta var. *murrayana*	Lodgepole Pine
P. flexilis	Limber Pine
P. jeffreyi	Jeffrey Pine
P. monticola	Western White Pine
Pterospora andromeda	Pine Drops
Pyrola spp.	wintergreens
Ribes spp.	currants
Sarcodes sanguinea	Snowplant
Silene verecunda	Silene
Sitanion hystrix	Bottlebrush Squirreltail
Stipa parishii	Parish Needlegrass

Scientific Name	Common Name
Tsuga mertensiana	Mountain Hemlock
Viola purpurea	Goosefoot Violet

MAMMALS

Citellus lateralis	Golden-mantled Ground Squirrel
Eutamias speciosus	Lodgepole Pine Chipmunk
Glaucomys sabrinus	Northern Flying Squirrel
Microtus longicaudus	Long-tailed Vole
Tamiasciurus hudsonicus mearnsii	Chickaree

BIRDS

Cardeulis pinus	Pine Siskin
Carpodacus cassinii	Cassin's Finch
Empidonax oberholseri	Dusky Flycatcher
Loxia curvirostra	Red Crossbill
Nucifraga columbiana	Clark's Nutcracker
Parus gambeli	Mountain Chickadee
Regulus satrapa	Golden-crowned Kinglet
Sitta canadensis	Red-breasted Nuthatch

Cold-Temperate Forests and Woodlands
122.3 Rocky Mountain (Petran) and Madrean Montane Conifer Forest

Scientific Name	Common Name

PLANTS

Abies concolor	White Fir
Achillea lanulosa	Western Yarrow
Aconitum columbianum	Columbia Monkshood
Agropyron spp.	Wheatgrasses
Agrostis scabra	Rough Bentgrass
Alnus firmifolia	Alder
Arbutus arizonica	Arizona Madrone
Aster spp.	asters
Berberis repens	Creeping Mahonia
Blepharoneuron tricholepsis	Pine Dropseed
Bromus anomalus	Nodding Brome
B. ciliatus	Fringed Brome
B. marginatus	Mountain Brome
Carex geophila	Dryland Sedge
Ceanothus fendleri	Fendler Ceanothus
C. huichugore	Buckbush
Chimaphila umbellata	Pipsissewa
Cyperus fendlerianus	Fendler Flatsedge
Delphinium tenuisectum	Carrotleaf Larkspur
Deschampsia caespitosa	Tufted Hairgrass
Erigeron concinnus	Tidy Fleabane
E. divergens	Spreading Fleabane
E. flagellaris	Trailing Fleabane
E. formosissimus	—
E. macranthus	—
Festuca arizonica	Arizona Fescue
Fragaria ovalis	Wild Strawberry
Geranium spp.	geraniums, crane's bills
G. caespitosa	Purple Geranium
Goodyera oblongifolia	Western Rattlesnake Plantain
Helenium hoopesii	Orange Sneezeweed
Huechera rubescens	alumroot
H. versicolor	alumroot
Holodiscus dumosus	Bush Rockspirea
Iris missouriensis	Rocky Mountain Iris
Juniperus deppeana	Alligatorbark Juniper
Koeleria cristata	Prairie Junegrass
Lathyrus arizonicus	Arizona Peavine
L. graminifolia	Grassleaf Peavine
Lonicera arizonica	Arizona Honeysuckle
L. involucrata	Bearberry Honeysuckle
Lupinus spp.	lupines
Mertensia spp.	false mountain bluebells
M. franciscana	Mountain Bluebell
Monarda menthaefolia	Mintleaf Beebalm
Muhlenbergia minutissima	Littleseed Muhly
M. montana	Mountain Muhly
M. virescens	Screwleaf Muhly
Pachystima myrsinites	Mountain-clover, Myrtle Boxleaf
Panicum bulbosum	Bulb Panicum
Physocarpus monogynus	Ninebark
Picea chihuahuana	Chihuahua Spruce
P. engelmannii	Engelmann Spruce
P. pungens	Blue Spruce
Pinus ayacahuite	Mexican White Pine
P. engelmannii	Apache Pine
P. flexilis	Limber Pine
P. leiophylla var. chihuahuana	Chihuahua Pine
P. lumholtzii	Pino Triste
P. ponderosa	Ponderosa Pine
P. ponderosa var. scopulorum	Rocky Mountain Ponderosa
P. ponderosa var. arizonica	Arizona Pine
P. strobiformis	Southwestern White Pine
Poa fendleriana	Fendler Bluegrass
P. pratensis	Kentucky Bluegrass
Polygonum sawatchense	Sawatch Knotweed
Populus tremuloides	Quaking Aspen
Potentilla concinna	Elegant Cinquefoil
Pseudocymopterus montanus	Mountain Parsley
Pseudotsuga menziesii	Douglas-fir
Pteridium aquilinum	Bracken Fern
Pterospora andromeda	Pine Drops
Pyrola virens	Wintergreen
Quercus arizonica	Arizona White Oak
Q. fulva	—
Q. gambelii	Gambel Oak
Q. grisea	Gray Oak
Q. hypoleucoides	Silver-leaf Oak
Q. pennivenia	—
Q. rugosa	Netleaf Oak
Q. viminea	—
Rhus glabra	Smooth Sumac
Ribes aureum	Golden Currant
R. pinetorum	Orange Gooseberry
R. viscosissimum	Sticky Currant
Robinia neomexicana	New Mexican Locust
Rosa arizonica	Arizona Rose
Rudbeckia laciniata	Cutleaf Coneflower
Sambucus cerulea	Blue Elderberry
S. velutina	Velvet Elder
Scrophularia parviflora	Figwort
Senecio neomexicanus	New Mexican Groundsel
Sitanion hystrix	Bottlebrush Squirreltail
Smilacina racemosa	Feather Solomonseal
Solidago missouriensis	Missouri Goldenrod
S. sparsiflora	Few-leaved Goldenrod
Stipa pringlei	Pringle Needlegrass
Symphoricarpos longiflorus	Longflowered Snowberry
S. oreophilus	Mountain Snowberry
S. rotundifolius	Roundleaf Snowberry
S. utahensis	Utah Snowberry
Taraxacum officinale	Dandelion
Thalictrum fendleri	Meadowrue
Trifolium rusbyi	Rusby Clover
Valeriana arizonica	Valerian, Tobacco Root
Vicia americana	American Vetch
Viola adunca	Hook Violet
V. canadensis	Canadian Violet

MAMMALS

Canis lupus	Gray Wolf, Lobo
Cervus elaphus	Elk, Wapiti
Citellus lateralis	Golden-mantled Ground Squirrel
Eptesicus fuscus	Big Brown Bat
Erethizon dorsatum	Porcupine
Eutamias canipes	Gray-footed Chipmunk
E. cinereicollis	Gray-collared Chipmunk
E. quadrivittatus	Colorado Chipmunk
E. umbrinus	Uinta Chipmunk
Microtus longicaudus	Long-tailed Vole
M. mexicanus	Mexican Vole
M. montanus	Montane Vole
Mustela frenata	Long-tailed Weasel
Myotis auriculus	Southwestern Myotis
M. evotus	Long-eared Myotis
M. volans	Long-legged Myotis
Neotoma mexicana	Mexican Woodrat
Odocoileus hemionus	Mule Deer
O. virginianus	White-tailed Deer
Peromyscus maniculatus	Deer Mouse
Sciurus aberti	Abert's Squirrel

Sorex merriami	Merriam's Shrew
S. nanus	Dwarf Shrew
S. vagrans	Vagrant Shrew
Sylvilagus floridanus	Eastern Cottontail
S. nuttallii	Nuttall's Cottontail
Tamiasciurus hudsonicus	Red Squirrel

BIRDS

Accipiter gentilis	Goshawk
Aegolius acadicus	Saw-whet Owl
Campephilus imperialis	Imperial Woodpecker
Cardellina rubrifrons	Red-faced Warbler
Carduelis pinus	Pine Siskin
Catharus occidentalis	Russet Nightingale Thrush
Certhia familiaris	Brown Creeper
Columba fasciata	Band-tailed Pigeon
Contopus pertinax	Coues' Flycatcher
Cyanocitta stelleri	Steller's Jay
Dendroica coranata	Yellow-rumped Warbler
D. graciae	Grace's Warbler
Empidonax affinis	Pine Flycatcher
E. difficilis	Western Flycatcher
Ergaticus ruber	Red Warbler
Eugenes fulgens	Rivoli's Hummingbird
Euptilotis neoxenus	Eared Trogon
Glaucidium gnoma	Pygmy Owl
Hesperiphona vespertina	Evening Grosbeak
Junco spp.	juncos
Junco phaeonotus	Yellow-eyed Junco
Loxia curvirostra	Red Crossbill
Meleagris gallopavo merriami	Merriam's Turkey
M. gallopavo mexicana	Gould's Turkey
Myadestes obscurus	Brown-backed Solitaire
M. townsendi	Townsend's Solitaire
Otus flammeolus	Flammulated Owl
Parus sclateri	Mexican Chickadee
Peucedramus taeniatus	Olive Warbler
Piranga flava	Hepatic Tanager
P. ludoviciana	Western Tanager
Rhynchopsitta pachyrhyncha	Thick-billed Parrot
Ridgwayia pinicola	Aztec Thrush
Selasphorus platycercus	Broad-tailed Hummingbird
Sialia mexicana	Western Bluebird
Sitta pygmaea	Pygmy Nuthatch
Spizella passerina	Chipping Sparrow
Strix occidentalis	Spotted Owl
Vireo gilvus	Warbling Vireo
V. solitarius	Solitary Vireo

AMPHIBIANS AND REPTILES

Ambystoma tigrinum	Tiger Salamander
Aneides hardyi	Sacramento Mountain Salamander
Crotalus pricei	Twin-spotted Rattlesnake
C. viridis	Western Rattlesnake
Diadophis punctatus	Ringneck Snake
Eumeces callicephalus	Mountain Skink
E. multivirgatus epipleurotus	Many-lined Skink
E. skiltonianus	Western Skink
Gerrhonotus kingi	Arizona Alligator Lizard
Lampropeltis pyromelana	Sonora Mountain Kingsnake
Phrynosoma douglassi	Short-horned Lizard
Pituophis melanoleucus	Gopher Snake
Plethodon neomexicanus	Jemez Mountain Salamander
Salvadora grahamiae	Mountain Patch-nosed Snake
Sceloporus jarrovi	Yarrow's Spiny Lizard
S. scalaris	Bunchgrass Lizard
S. virgatus	Striped Plateau Lizard
Thamnophis elegans	Western Terrestrial Garter Snake

122.5 Sierran Montane Conifer Forest

Scientific Name	Common Name

PLANTS

Abies concolor	White Fir
Arabis repanda	Rockcress
Arctostaphylos glandulosa	Eastwood Manzanita
A. patula	Greenleaf Manzanita
A. pringlei var. drupacea	Pringle Manzanita
Artemisia tridentata	Big Sagebrush
Asclepias eriocarpa	Woolypod Milkweed
Bromus breviaristatus	Slimleaf Brome
B. orcuttianus var. halli	Orcutt Brome
Castanopsis sempervirens	Bush Chinquapin
Castilleja martinii	Paintbrush
Ceanothus cordulatus	Snow Bush, Mountain Whitethorn
C. integerrimus	Deerbrush
Cercocarpus betuloides	Birchleaf Mountain-mahogany
C. ledifolius	Curlleaf Mountain-mahogany
Chaenactis santolinoides	—
Chimaphila menziesii	Menzies Pipsissewa
Chrysothamnus nauseosus	Rubber Rabbitbrush
Clarkia rhomboidea	—
Collinsia childii	—
Convolvulus occidentalis	California Glorybind
Cordylanthus nevinii	—
Eriodictyon trichocalyx	Hairy Yerbasanta
Eriogonum parishii	Parish Buckwheat
E. wrightii var. subscaposum	Wright Buckwheat
Fritillaria pinetorum	Fritillary
Garrya flavescens	Yellowleaf Silktassel
Gilia splendens var. grantii	—
Koeleria cristata	Prairie Junegrass
Libocedrus decurrens	Incense Cedar
Linanthus ciliatus	Mustang Clover
Lotus davidsonii	Deer Vetch, Bird's Foot Trefoil
Melica imperfecta	Coastrange Melic
M. stricta	Rock Melic
Mimulus johnstonii	Monkeyflower
Penstemon spp.	penstemons, beardtongues
P. bridgesii	Bridges Penstemon
P. caesius	San Bernardino Penstemon
P. grinnellii	Grinnell Penstemon
P. labrosus	Longlips Penstemon
Pinus coulteri	Coulter Pine
P. jeffreyi	Jeffrey Pine
P. lambertiana	Sugar Pine
P. monophylla	Singleleaf Pinyon, One-needle Pinyon
P. ponderosa	Ponderosa Pine
Poa scabrella	Pine Bluegrass
Pseudotsuga macrocarpa	Big-cone Douglas-fir
Quercus chrysolepis	Canyon Live Oak
Q. kelloggii	California Black Oak
Rhamnus californica var. cuspidata	California Buckthorn, Coffeeberry
Ribes nevadense	Sierra Currant
R. roezli	Sierra Gooseberry
Rubus parviflorus	Western Thimbleberry
Salix coulteri	Coulter Willow
Sambucus cerulea	Blue Elderberry
Sarcodes sanguinea	Snowplant
Sitanion hystrix	Bottlebrush Squirreltail
Solanum xanti	Purple Nightshade
Stipa parishii	Parish Needlegrass
Streptanthus bernardinus	—
Symphoricarpos parishii	Parish Snowberry
Tetradymia canescens	Gray Horsebrush
Viola purpurea	Goosefoot Violet

MAMMALS

Eutamias merriami	Merriam's Chipmunk
Mustela frenata	Long-tailed Weasel
Odocoileus hemionus	Mule Deer
Peromyscus maniculatus	Deer Mouse
Scapanus latimanus	Broad-footed Mole
Sciurus griseus	Western Gray Squirrel
Thomomys umbrinus	Southern Pocket Gopher

BIRDS

Accipiter striatus	Sharp-shinned Hawk
Carduelis pinus	Pine Siskin
Carpodacus purpureus	Purple Finch
Certhia familiaris	Brown Creeper
Columba fasciata	Band-tailed Pigeon
Contopus sordidulus	Western Wood Peewee
Cyanocitta stelleri	Steller's Jay
Dendroica coronata	Yellow-rumped Warbler
Glaucidium gnoma	Pygmy Owl
Junco hyemalis	Dark-eyed Junco, Oregon Junco
Loxia curvirostra	Red Crossbill
Myadestes townsendi	Townsend's Solitaire

Otus flammeolus	Flammulated Owl
Parus gambeli	Mountain Chickadee
Picoides albolarvatus	White-headed Woodpecker
P. villosus	Hairy Woodpecker
Piranga ludoviciana	Western Tanager
Sialia mexicana	Western Bluebird
Sitta pygmaea	Pygmy Nuthatch
Sphyrapicus thyroideus	Williamson's Sapsucker
S. varius	Yellow-bellied Sapsucker
Stellula calliope	Calliope Hummingbird
Strix occidentalis	Spotted Owl
Turdus migratorius	American Robin
Vireo huttoni	Hutton's Vireo

AMPHIBIANS AND REPTILES

Aneides lugubris	Arboreal Salamander
Charina bottae umbricata	Southern Rubber Boa
Crotalus viridis helleri	Southern Pacific Rattlesnake
Diadophis punctatus	Ringneck Snake
Ensatina eschscholtzi	Ensatina
Eumeces gilberti	Gilbert's Skink
Lampropeltis getulus	Common Kingsnake
L. zonata	California Mountain Kingsnake
Pituophis melanoleucus	Gopher Snake
Sceloporus graciosus	Sagebrush Lizard
S. occidentalis	Western Fence Lizard
Thamnophis elegans	Western Terrestrial Garter Snake

122.4 Great Basin Conifer Woodland

Scientific Name	Common Name

PLANTS

Agropyron smithii	Western Wheatgrass
Amelanchier alnifolia	Saskatoon Serviceberry
Artemisia arbuscula spp. nova	Black Sagebrush
A. ludoviciana	Louisiana Sagebrush
Atriplex canescens	Fourwing Saltbush
A. confertifolia	Shadscale
Berberis fremontii	Fremont Mahonia, Barberry
B. haematocarpa	Red Mahonia
Bouteloua gracilis	Blue Grama
Bromus spp.	brome grasses
Calochortus nuttallii	Sego-lily
Canotia holocantha	Crucifixion Thorn, Canotia
Ceratoides lanata	Winterfat
Cercocarpus intricatus	Littleleaf Mountain-mahogany
C. ledifolius	Curlleaf Mountain-mahogany
C. montanus	Alderleaf Mountain-mahogany
Chaemaebataria millifolium	Fernbush, Desert Sweet
Chrysothamnus spp.	rabbitbrushes
Coleogyne ramosissima	Blackbrush
Coryphantha vivipara var. arizonica	—
C. missouriensis	—
Cowania mexicana	Cliffrose
Echinocereus engelmannii var. variegatus	Hedgehog Cactus
E. fendleri	
E. triglochidiatus var. melanacanthus	Red Hedgehog Cactus
Ephedra viridis	Mountain Joint-fir
Eriogonum spp.	buckwheats
Fallugia paradoxa	Apache Plume
Garrya wrightii	Wright Silktassel
Gilia spp.	gilias
Gutierrezia sarothrae	Broom Snakeweed
Hilaria jamesii	Galleta
Juniperus californica	California Juniper
J. monosperma	One-seed Juniper
J. osteosperma	Utah Juniper
J. scopulorum	Rocky Mountain Juniper
Koeleria cristata	Prairie Junegrass
Lupinus spp.	lupines
Mammillaria wrightii	Wright Pincushion
Muhlenbergia spp.	muhlies
Opuntia basilaris var. aurea	Yellow Beavertail
O. erinacea	Mohave Prickly Pear
O. fragilis	Little Prickly Pear
O. imbricata	Tree Cholla
O. macrorhiza	Plains Prickly Pear
O. phaeacantha	Engelmann Prickly Pear
O. polycantha	Plains Prickly Pear

O. whipplei	Whipple Cholla
Oryzopsis hymenoides	Indian Ricegrass
Pediocactus papyracanthus	Grama Grass Cactus
P. simpsonii	—
Penstemon spp.	penstemons, beardtongues
Pinus cembroides	Mexican Pinyon
P. edulis	Rocky Mountain Pinyon
P. monophylla	Singleleaf Pinyon, One-needle Pinyon
P. quadrifolia	Parry Pinyon
Purshia tridentata	Antelope Bitterbrush
Quercus arizonica	Arizona White Oak
Q. emoryi	Emory Oak
Q. gambelii	Gambel Oak
Q. grisea	Gray Oak
Q. turbinella	Shrub Live Oak
Rhamnus crocea	Hollyleaf Buckthorn
Rhus trilobata	Squawbush, Skunkbush Sumac
Ribes spp.	currants
Sclerocactus whipplei var. intermedius	—
Senecio longilobus	Threadleaf Groundsel
Shepherdia spp.	buffaloberries
Sphaeralcea coccinea	Scarlet Globe Mallow
S. digitata	Juniper Globe Mallow
S. marginata	
Sporobolus spp.	dropseeds
Symphoricarpos spp.	snowberries
Yucca baccata	Banana Yucca, Datil
Y. glauca	Small Soapweed

MAMMALS

Cervus elaphus	Elk, Wapiti
Neotoma cinerea arizonae	Arizona Bushy-tailed Woodrat
Odocoileus hemionus	Mule Deer
Peromyscus truei	Pinyon Deer Mouse

BIRDS

Dendroica nigrescens	Black-throated Gray Warbler
Empidonax wrightii	Gray Flycatcher
Gymnorhinus cyanocephalus	Pinyon Jay
Icterus parisorum	Scott's Oriole
Vireo vicinior	Gray Vireo

AMPHIBIANS AND REPTILES

Cnemidophorus velox	Plateau Striped Whiptail
Sceloporus graciosus	Sagebrush Lizard

Warm-Temperate Forests and Woodlands
123.3 Madrean Evergreen Woodland

Scientific Name	Common Name

PLANTS

Agave palmeri	Palmer Agave
A. parryi	Parry Agave
Arbutus arizonica	Arizona Madrone
A. texana	Texas Madrone
Arctostaphylos pungens	Mexican Manzanita
Artemisia ludoviciana	Louisiana Sagebrush
Bothriochloa barbinodis	Cane Bluestem
Bouteloua curtipendula	Sideoats Grama
B. gracilis	Blue Grama
B. hirsuta	Hairy Grama
Brickellia spp.	bricklebushes
Cassia leptocarpa	Slimpod Senna
Ceanothus huichugore	Buckbush
Cercocarpus montanus	Alderleaf Mountain-mahogany
Coryphantha recurvata	Hens-and-chicks Cactus
Cowania mexicana	Cliffrose
Cupressus arizonica	Arizona Cypress
Cyperus spp.	flatsedges
Dalea spp. (Psorothamnus spp.)	indigo-bushes
Dodonaea viscosa	Hopbush
Echinocereus ledingii	Hedgehog Cactus
E. pectinatus var. rigidissimus	Rainbow Cactus
E. triglochidiatus	Red Hedgehog Cactus

Elyonurus barbiculmis	Wolfspike
Eragrostis intermedia	Plains Lovegrass
Ericameria laricifolia	Larchleaf Goldenweed
Eriogonum spp.	buckwheats
Erythrina flabelliformis	Southwestern Coralbean
Eysenhardtia orthocarpa	Kidneywood
Ferocactus wislizenii	Barrel Cactus
Garrya wrightii	Wright Silktassel
Heteropogon contortus	Tanglehead
Hibiscus spp.	rose-mallows
Juniperus deppeana	Alligatorbark Juniper
J. monosperma	One-seed Juniper
Leptochloa dubia	Green Sprangletop
Lupinus spp.	lupines
Lycurus phleoides	Wolftail
Mammillaria gummifera	Cream Cactus
M. orestera	Pincushion
Mimosa biuncifera	Wait-a-minute, Cat-claw
M. dysocarpa	Velvet-pod Mimosa
Muhlenbergia emersleyi	Bullgrass
M. porteri	Bush Muhly
M. torreyi	Ring-grass
Nolina microcarpa	Beargrass, Sacahuista
Opuntia phaeacantha	Engelmann Prickly Pear
O. spinosior	Cane Cholla
O. violacea var. santarita	Purple Prickly Pear
Oxalis spp.	woodsorrels
Penstemon spp.	penstemons, beardtongues
Phaseolus spp.	beans
Pinus cembroides	Mexican Pinyon
Pinus cooperi	Cooper Pine
P. durangensis	Durango Pine
P. engelmannii	Apache Pine
P. leiophylla var. chihuahuana	Chihuahua Pine
P. lumholtzii	Pino Triste
P. ponderosa var. arizonica	Arizona Pine
Quercus albocincta	—
Q. arizonica	Arizona White Oak
Q. chihuahuensis	Chihuahua Oak
Q. chuchuichupensis	—
Q. durifolia	—
Q. emoryi	Emory Oak, Bellota
Q. epileuca	—
Q. fulva	—
Q. grisea	Gray Oak
Q. hypoleucoides	Silver-leaf Oak
Q. oblongifolia	Mexican Blue Oak
Q. pennivenia	Hand Basin Oak
Q. rugosa	Netleaf Oak
Q. santaclarensis	Santa Clara Oak
Q. toumeyi	Toumey Oak
Q. viminea	—
Rhamnus betulaefolia	Birchleaf Buckthorn
Rhus choriophylla	Mearns Sumac
R. trilobata	Squawbush, Skunkbush Sumac
Salvia spp.	sages
Schizachyrium scoparium	Little Bluestem
Tillandsia recurvata	Ballmoss, Gallitos
Vauquelinia sp.	rosewood
Yucca schottii	Schott Yucca, Hairy Yucca
Y. baccata var. thornberi	Thornber Yucca

MAMMALS

Nasua nasua	Coati
Odocoileus virginianus	White-tailed Deer
Perognathus baileyi	Bailey's Pocket Mouse
Sciurus nayaritensis	Apache Squirrel
Sigmodon ochrognathus	Yellow-nosed Cotton Rat
Sylvilagus floridanus	Eastern Cottontail
Thomomys umbrinus	Southern Pocket Gopher
Ursus arctos	Mexican Grizzly Bear

BIRDS

Amazilia violiceps	Violet-crowned Hummingbird
Aphelocoma ultramarina	Mexican Jay
Columba fasciata	Band-tailed Pigeon
Crytonyx montezumae	Montezuma Quail
Dendroica nigrescens	Black-throated Gray Warbler
Empidonax fulvifrons	Buff-breasted Flycatcher
Hylocharis leucotis	White-eared Hummingbird
Melanerpes formicivorus	Acorn Woodpecker

Meleagris gallopavo mexicana	Gould's Turkey
Otus trichopsis	Whiskered Owl
Parus sclateri	Mexican Chickadee
P. wollweberi	Bridled Titmouse
Picoides arizonae	Arizona Woodpecker
Piranga flava	Hepatic Tanager
Psaltriparus minimus	Bushtit
Sialia mexicana	Western Bluebird
Trogon elegans	Coppery-tailed Trogon
Vermivora crissalis	Colima Warbler
Vireo huttoni	Hutton's Vireo

AMPHIBIANS AND REPTILES

Crotalus lepidus	Rock Rattlesnake
C. pricei	Twin-spotted Rattlesnake
C. willardi	Ridgenose Rattlesnake
Elaphe triaspis	Green Rat Snake
Eumeces callicephalus	Mountain Skink
Hylactophryne augusti	Barking Frog
Lampropeltis pyromelana	Sonora Mountain Kingsnake
Phrynosoma ditmarsi	Ditmar's Horned Lizard
Rana tarahumarae	Tarahumara Frog
Salvadora grahamiae	Mountain Patch-nosed Snake
Sceloporus clarki	Clark's Spiny Lizard
S. jarrovi	Yarrow's Spiny Lizard
S. scalaris	Bunchgrass Lizard
S. virgatus	Striped Plateau Lizard
Tantilla wilcoxi wilcoxi	Huachuca Blackhead Snake
Thamnophis eques	Mexican Garter Snake

123.4 Californian Evergreen Forest and Woodland

Scientific Name	Common Name

PLANTS

Arbutus menziesii	Menzies Madrone, Madroño
Arctostaphylos spp.	manzanitas
Heteromeles arbutifolia	Toyon, Christmas-berry
Juglans californica	California Black Walnut
Libocedrus decurrens	Incense Cedar
Pinus coulteri	Coulter Pine
Pseudotsuga macrocarpa	Big-cone Douglas-fir
Quercus agrifolia	Coast Live Oak
Q. chrysolepis	Canyon Live Oak
Q. engelmannii	Engelmann Oak
Q. kelloggii	California Black Oak
Rhus integrifolia	Squawbush Sumac
R. ovata	Sugar Sumac
Umbellularia californica	California Bay

MAMMALS

Citellus beecheyi	California Ground Squirrel
Odocoileus hemionus californicus	California Mule Deer
Procyon lotor	Raccoon
Sciurus griseus	Western Gray Squirrel
Thomomys bottae	Western Pocket Gopher

BIRDS

Columba fasciata	Band-tailed Pigeon
Melanerpes formicivorus	Acorn Woodpecker
Oreortyx picta	Mountain Quail
Parus inornatus	Plain Titmouse
Picoides nuttallii	Nuttall's Woodpecker
P. villosus	Hairy Woodpecker
Sialia mexicana	Western Bluebird
Sitta carolinensis	White-breasted Nuthatch

AMPHIBIANS AND REPTILES

Aneides lugubris	Arboreal Salamander
Bufo boreas	Western Toad
Ensatina eschscholtzi	Ensatina
E. eschscholtzi klauberi	Large Blotched Ensatina
Lampropeltis getulus	Common Kingsnake
L. zonata	California Mountain Kingsnake
Phrynosoma coronatum	Coast Horned Lizard
Pituophis melanoleucus	Gopher Snake
Sceloporus occidentalis	Western Fence Lizard

123.5 Relict Conifer Forests and Woodlands

Scientific Name	Common Name
PLANTS	
Cupressus arizonica	Arizona Cypress
C.a. glabra	Smooth-barked Arizona Cypress
C. montana	San Pedro Mártir Cypress
C. stephensonii	Guyamaca Cypress
C. forbesii	Tecate Cypress
Pinus attenuata	Knobcone Pine
P. muricata	Bishop Pine
P. torreyana	Torrey Pine

124.6 Sinaloan Deciduous Forest

Scientific Name	Common Name
PLANTS	
Acacia coulteri	Coulter Acacia, Guayavilla
Ambrosia ambrosioides	Bursage, Canyon Ragweed, Chicura
Amoreuxia palmatifida	Saiya
Arrabidaea littoralis	Bejuco Vaquero
Arundinaria longifolia	Longleaf Cane, Otate
Baccharis salicifolia	Seep Willow, Batamote, Jara
Bursera confusa	Torote Papelio
B. epinnata	—
B. fragilis	Tacamahaca, Torote Prieto
B. grandifolia	Chutama, Palo Mulato
B. inopinnata	Torote Copal
B. stenophylla	Torote Blanco
Caesalpinia platyloba	Palo Colorado
C. pulcherrima	Red Bird of Paradise, Tabachin
C. standleyi	—
Calliandra rupestris	—
Cassia biflora	Twin-flower Cassia, Ejotillo-del-Monte
C. emarginata	Palo de Zorillo
C. occidentalis	Coffee Senna, Ejotillo Grande
Ceiba acuminata	Pochote
Celtis iguanaea	Garabato
Cephalocereus alensis	Billygoat Cactus, Pitahaya Barbón
Cestrum lanatum	Jessamine
Cochlospermum vitifolium	Palo Barril
Conzattia sericea	Navío, Palo Joso
Coutarea latiflora	Palo Amargo, Copalqiun
C. pterosperma	Caparche, Copalquin
Croton fragilis	Vara Blanca
Dioscorea convolvulacea var. *grandifolia*	Wild Yam, Chichiwo
Drypetes laterifolia	—
Exogonium bracteatum	Morning glory, Bejuco Blanco
Ficus spp.	figs
F. cotinifolia	Amate Prieto, Nacopuli
F. padifolia	Nacapuli, Camachín, Chuna
Gouania rosei	—
Guazuma ulmifolia	Guacimilla, Guácima
Haematoxylon brasiletto	Brasil
Hechtia sp.	Guapilla
Hybanthus mexicanus	Green Violet
Hymenocallis sonorensis	Sonoran Spiderlily
Hymenoclea monogyra	Burrobush, Jécota
Ipomoea arborescens	Tree Morningglory, Palo Santo
Jarilla chocola	Chócola
Jatropha cordata	Copalillo, Miquelito
J. platanifolia	Ensangregrado
Lemaireocereus montanus	—
Leucaena lanceolata	Bolillo, Guaje
Lysiloma divaricata	Mauto
L. watsoni	Tepeguaje
Manihot angustiloba	Cyanide Plant, Pata de Gallo
M. isoloba	Pata de Gallo
Marsdenia edulis	Talayote, Tonchi
Montanoa rosei	Batayáqui
Oncidium cebolleta	—
Opuntia spp.	prickly pears, chollas, tunas
Pachycereus pecten-aboriginum	Hecho, Cardón Hecho
Phaseolus caracala	

Scientific Name	Common Name
Piscidia mollis	Palo Blanco
Pisonia capitata	Garabato Prieto
Pithecellobium dulce	Guamuchil
P. mexicanum	Palo Chino
P. undulatum	Palo Fierro
Platanus racemosa	Sycamore, Aliso
Randia echinocarpa	Papache
Salpianthus macrodontus	Guayavilla
Sassafridium macrophyllum	Lauretón, Bebelama
Solanum madrense	Sierra Madre Nightshade
S. tequilense	—
S. umbellatum	Berenjena
S. verbascifolium	Cornetón del Monte
Stemmadenia palmeri	Baraco, Tapaco
Stenocereus thurberi	Organ Pipe Cactus, Pitahaya
Tabebuia spp.	Amapa
T. chrysantha	Amapa Amarilla
T. palmeri	Amapa, Amapa Colorado
Taxodium mucronatum	Ahuehuete, Cedro, Sabino
Tigridia pringlei	Chaqual
Tillandsia inflata	Mescalito
Tithonia fruticosa	Mexican Sunflower, Mirasol
Trichilia hirta	Garbancillo, Jumay
Urera caracasana	Ortiguilla
Vincetoxicum caudatum	Talayote
Vitex mollis	Obalamo, Urulama
Willardia mexicana	Nesco, Palo Piojo, Taliste
Wimmeria mexicana	Cedilla, Papelio

MAMMALS	
Felis pardalis	Ocelot, Tigrillo
Nasua nasua	Coati
Odocoileus virginianus	White-tailed Deer
Sciurus truei	Sonoran Squirrel, Aborilla
Sylvilagus cunicularius	Mexican Cottontail

BIRDS
See listing at end of Appendix, page 341.

AMPHIBIANS AND REPTILES
See listing at end of Appendix, page 341.

Arctic-Boreal Scrublands
131.5 Subalpine Scrub

Scientific Name	Common Name
PLANTS	
Abies concolor	White Fir
Juniperus communis	Dwarf Juniper
Picea engelmannii	Engelmann Spruce
Pinus aristata	Bristlecone Pine
P. flexilis	Limber Pine
Salix spp.	willows

MAMMALS	
Lepus americanus	Snowshoe Hare

BIRDS	
Lagopus leucurus	White-tailed Ptarmigan
Zonotrichia leucophrys	White-crowned Sparrow

Cold-Temperate Scrublands
132.1 Great Basin Montane Scrubland

Scientific Name	Common Name
PLANTS	
Acer grandidentatum	Bigtooth Maple
Achillea millefolium	Western Yarrow
Amelanchier alnifolia	Saskatoon Serviceberry
A. utahensis	Utah Serviceberry
Anemone spp.	anemones, windflowers
Antennaria spp.	pussytoes
Arctostaphylos patula	Greenleaf Manzanita
Artemisia arbuscula	Black Sagebrush
A. tridentata	Big Sagebrush

Aster spp.	asters
Balsamorhiza sagittata	Arrowleaf Balsamroot
Berberis spp.	mahonias, barberries
Bromus spp.	bromes
Ceanothus fendleri	Fendler Ceanothus
C. velutinus	Snowbrush, Buckbrush
Celtis reticulata	Netleaf Hackberry
Cercocarpus ledifolius	Curlleaf Mountain-mahogany
C. montanus	Alderleaf Mountain-mahogany
Chrysothamnus viscidiflorus	Sticky-leaved Rabbitbrush
Cowania mexicana	Cliffrose
Erigeron spp.	fleabanes
Eriogonum spp.	buckwheats
Fallugia paradoxa	Apache Plume
Geranium spp.	geraniums, crane's bills
Lupines spp.	lupines
Penstemon spp.	beardtongues, penstemons
Pinus edulis	Pinyon Pine
P. ponderosa	Ponderosa Pine, Arizona Pine
Poa pratensis	Kentucky Bluegrass
Populus tremuloides	Quaking Aspen
Prunus virginianus var. melanocarpa	Black Western Chokecherry
Pseudotsuga menziesii	Douglas-fir
Ptelea spp.	hoptrees, water-ashes
Purshia tridentata	Antelope Bitterbrush
Quercus gambelii	Gambel Oak
Rhus trilobata	Squawbush, Skunkbush Sumac
Ribes spp.	currants
Robinia neomexicana	New Mexican Locust
Rosa woodsii	Wild Rose
Sambucus cerulea	Blue Elderberry
Senecio serra	Butterweed Groundsel
Symphoricarpos spp.	snowberries
Taraxacum officinale	Dandelion
Thalictrum fendleri	Meadowrue
Thermopsis montana	Golden Pea

MAMMALS

Felis rufus	Bobcat
Odocoileus hemionus	Mule Deer
Spermophilus variegatus	Rock Squirrel

BIRDS

Aphelocoma coerulescens	Scrub Jay
Passerina amoena	Lazuli Bunting
Pipilo chlorurus	Green-tailed Towhee
P. erythrophthalmus	Rufous-sided Towhee
Spizella passerina	Chipping Sparrow
Vermivora celata	Orange-crowned Warbler
Vireo gilvus	Warbling Vireo

AMPHIBIANS AND REPTILES

Charina bottae	Rubber Boa
Hypsiglena torquata	Night Snake
Sceloporus graciosus	Sagebrush Lizard

Warm-Temperate Scrublands
133.2 California Coastalscrub

Scientific Name	Common Name

PLANTS

Aesculus parryi	Parry Buckeye
Agave shawii	Maguey
Ambrosia chenopodifolia	San Diego Bursage
Artemisia californica	California Sagebrush
Bergerocactus emoryi	Velvet Cactus
Cercocarpus betuloides	Birchleaf Mountain-mahogany
Dudleya spp.	siemprevivos
Encelia californica	California Encelia
Eriogonum fasciculatum	California Buckwheat
Eriophyllum confertiflorum	Golden Yarrow
Fraxinus trifoliata	Chaparral Ash
Happlopappus squarrosus	Sawtooth Goldenweed
H. venetus	Damiana Goldenweed
Mimulus longiflorus	Southern Monkeyflower
Opuntia basilaris	Beavertail Prickly Pear
O. echinocarpa	Silver Cholla

Prunus fremontii	Desert Apricot, Fremont Peachbush
P. ilicifolia	Hollyleaf Cherry
Quercus dumosa	California Scrub Oak
Rhus integrifolia	Lemonade Sumac
R. ovata	Sugar Sumac
Rosa minutifolia	Wild Rose
Salvia apiana	White Sage
S. leucophylla	White-leaf Sage
S. mellifera	Black Sage
Simmondsia chinensis	Jojoba, Coffeeberry, Goat Nut
Stenocereus gummosus	Pitaya Agria
Yucca whipplei	Our Lord's Candle

MAMMALS

Dipodomys agilis	Nimble Kangaroo Rat
Perognathus fallax	San Diego Pocket Mouse

BIRDS

Amphispiza belli	Sage Sparrow
Lophortyx californicus	California Quail

AMPHIBIANS AND REPTILES

Arizona elegans occidentalis	California Glossy Snake
Cnemidophorus hyperythrus	Orangethroat Whiptail
C. tigris mundus	California Whiptail
Crotalus ruber	Red Diamond Rattlesnake
C. viridis helleri	Southern Pacific Rattlesnake
Hypsiglena torquata klauberi	San Diego Night Snake
Lichanura trivirgata roseofusca	Coastal Rosy Boa
Masticophis lateralis lateralis	California Striped Racer
Phrynosoma coronatum blainvillei	San Diego Horned Lizard
Pituophis melanoleucus annectans	San Diego Gopher Snake
Salvadora hexalepis virgultea	Coast Patchnose Snake
Uta stansburiana elegans	California Side-blotched Lizard

133.1 Californian (Coastal) Chaparral

Scientific Name	Common Name

PLANTS

Adenostoma fasciculatum	Chamise
A. sparsifolium	Red Shanks
Arctostaphylos glandulosa	Eastwood Manzanita
A. glauca	Bigberry Manzanita
Castanopsis sempervirens	Chinquapin
Ceanothus spp.	ceanothuses
C. crassifolius	Hoary Leaf Ceanothus
C. leucodermis	Chaparral Whitethorn
Cercocarpus betuloides	Birchleaf Mountain-mahogany
Eriogonum fasciculatum	California Buckwheat
Fraxinus dipetala	Two-petal Chaparral Ash
Fremontia californica	California Fremontia, Flannelbush
Garrya spp.	silktassels
G. fremontii	Fremont Silktassel
Heteromeles arbutifolia	Toyon
Juniperus spp.	junipers
Lonicera spp.	honeysuckles
Pinus coulteri	Coulter Pine
Prunus ilicifolia	Hollyleaf Cherry
Quercus chrysolepis	Canyon Live Oak
Q. dumosa	California Scrub Oak
Rhamnus californica	California Buckthorn
R. crocea	Hollyleaf Buckthorn
Rhus diversiloba	Poison Oak
R. laurina	Laurel Sumac
R. ovata	Sugar Sumac
Salvia apiana	White Sage
S. mellifera	Black Sage

MAMMALS

Dipodomys agilis	Nimble Kangaroo Rat
Eutamias merriami	Merriam's Chipmunk
Neotoma fuscipes	Dusky-footed Woodrat
Odocoileus hemionus	Mule Deer
Perognathus californicus	California Pocket Mouse
Peromyscus boylei	Brush Mouse
P. californicus	California Mouse
Sylvilagus bachmani	Brush Rabbit
Urocyon cinereoargenteus	Gray Fox

BIRDS

Scientific Name	Common Name
Aimophila ruficeps	Rufous-crowned Sparrow
Aphelocoma coerulescens	Scrub Jay
Calypte anna	Anna's Hummingbird
Chamaea fasciata	Wrentit
Empidonax oberholseri	Dusky Flycatcher
Oreotyx pictus	Mountain Quail
Passerina amoena	Lazuli Bunting
Pipilo erythrophthalmus	Rufous-sided Towhee
P. fuscus	Brown Towhee
Psaltriparus minimus	Bushtit
Spizella atrogularis	Black-chinned Sparrow
Thyromanes bewickii	Bewick's Wren
Toxostoma redivivum	California Thrasher
Vermivora celata	Orange-crowned Warbler

AMPHIBIANS AND REPTILES

Scientific Name	Common Name
Arizona elegans occidentalis	California Glossy Snake
Crotalus ruber	Red Diamond Rattlesnake
C. viridis helleri	Southern Pacific Rattlesnake
Gerrhonotus multicarinatus webbi	San Diego Alligator Lizard
Lichanura trivirgata roseofusca	Coastal Rosy Boa
Masticophis lateralis	Striped Racer
Phrynosoma coronatum	Coast Horned Lizard
Salvadora hexalepis	Western Patchnose Snake
Sceloporus graciosus	Sagebrush Lizard
S. occidentalis	Western Fence Lizard
Tantilla planiceps	Western Blackhead Snake
Trimorphodon biscutatus vandenberghi	California Lyre Snake
Xantusia henshawi	Granite Night Lizard

133.3 Interior Chaparral

Scientific Name	Common Name

PLANTS

Scientific Name	Common Name
Acacia greggii	Cat-claw
Agave spp.	agaves
Arbutus arizonica	Arizona Madrone
A. texana	Texas Madrone
Arctostaphylos pringlei	Pringle Manzanita
A. pungens	Pointleaf Manzanita
Aristida fendleriana	Fendler Three-awn
A. orcuttiana	Beggar-tick Three-awn, Single Three-awn
A. ternipes	Spider Grass
Artemisia ludoviciana	Louisiana Sagebrush
Berberis fremontii	Fremont Mahonia
B. trifoliata	—
Boerhaavia coccinea	Scarlet Spiderling
B. erecta	Erect Spiderling
Bothriochloa barbinodis	Cane Bluestem
Bouteloua curtipendula	Sideoats Grama
B. hirsuta	Hairy Grama
Brickellia californica	Bricklebush
Bromus rubens	Red Brome
Calliandra eriophylla	False-mesquite
Canotia holacantha	Crucifixion Thorn, Canotia
Ceanothus spp.	ceanothuses
C. fendleri	Fendler Ceanothus
C. greggii	Desert Ceanothus
C. integerrimus	Deerbrush
Cercocarpus betuloides	Birchleaf Mountain-mahogany
C. brevifolius	Hairy Mountain-mahogany
Cowania mexicana	Cliffrose
C. plicata	—
Cupressus arizonica	Arizona Cypress
C.a. glabra	Arizona Smooth-bark Cypress
Dalea albiflora	White Dalea
Dasylirion wheeleri	Sotol
Eragrostis intermedia	Plains Lovegrass
Eriodictyon angustifolium	Narrowleaf Yerbasanta
Eriogonum wrightii	Wright Buckwheat
Erodium cicutarium	Filaree, Heron's Bill
Fallugia paradoxa	Apache Plume
Fendlera linearis	Fendlerbush
Forestiera neomexicana	Desert Olive
Fraxinus anomala var. *lowellii*	Lowell Ash
F. greggii	Chaparral Ash, Gregg Ash
Fremontia californica	California Fremontia, Flannelbush
Garrya flavescens	Yellowleaf Silktassel
G. obovata	Bigleaf Silktassel
G. wrightii	Wright Silktassel
Gutierrezia sarothrae	Broom Snakeweed
Heterotheca subaxillaris	Telegraph Plant
Ipomoea coccinea	Scarlet Starglory
Juniperus deppeana	Alligatorbark Juniper
J. monosperma	One-seed Juniper
Leptochloa dubia	Green Sprangletop
L. filiformis	Red Sprangletop
Lycurus phleoides	Wolftail
Marrubium vulgare	Hoarhound
Mimosa biuncifera	Wait-a-minute
Nolina erumpens	Beargrass, Sacahuista
N. microcarpa	Beargrass, Sacahuista
Penstemon spp.	beardtongues, penstemons
P. barbatus	Beardlip Penstemon
P. eatoni	Eaton Firecracker
P. linarioides	Toadflax Penstemon
P. palmeri	Palmer Penstemon, Scented Penstemon
P. pseudospectabilis	Canyon Penstemon
Pinus edulis	Pinyon Pine
P. ponderosa	Ponderosa Pine, Arizona Pine
Poa longiligula	Longtongue Mutton Bluegrass
Prunus virginiana var. *melanocarpa*	Black Western Chokecherry
Ptelea trifoliata	Hoptree
Quercus arizonica	Arizona White Oak
Q. chrysolepis	Canyon Live Oak
Q. emoryi	Emory Oak
Q. grisea	Gray Oak
Q. intricata	Coahuila Scrub Oak
Q. invaginata	Wavy Leaf Oak
Q. laceyi	Lacey Oak
Q. organensis	—
Q. pringlei	Pringle Oak
Q. pungens	Vasey Oak, Sandpaper Oak
Q. turbinella	Shrub Live Oak
Q. tinkhamii	—
Q. toumeyi	Toumey Oak
Rhamnus californica	California Buckthorn
R. crocea	Hollyleaf Buckthorn
Rhus choriophylla	Mearns Sumac
R. ovata	Sugar Sumac
R. trilobata	Squawbush, Skunkbush Sumac
Salvia ramosissima	—
S. regla	—
S. roemeriana	Roemer Sage
Simmondsia chinensis	Jojoba, Coffeeberry, Goat Nut
Solanum xanti	Purple Nightshade
Solidago sparsiflorus	Few-flowered Goldenrod
Sophora spp.	sophoras
Trichachne californica	Arizona Cottontop
Vauquelinia angustifolia	Narrowleaf Rosewood
V. californica	Arizona Rosewood
Verbena wrightii	Wright Verbena, Desert Verbena
Yucca baccata	Banana Yucca, Datil

MAMMALS

Scientific Name	Common Name
Eutamias dorsalis	Cliff Chipmunk
Neotoma albigula	White-throated Woodrat
Odocoileus hemionus	Mule Deer
Peromyscus boylei	Brush Mouse
P. difficilis	Rock Mouse
P. leucopus	White-footed Mouse
Sylvilagus floridanus holzeri	Eastern Cottontail

BIRDS

Scientific Name	Common Name
Aimophila ruficeps	Rufous-crowned Sparrow
Aphelocoma coerulescens	Scrub Jay
Catherpes mexicanus	Canyon Wren
Pipilo erythrophthalmus	Rufous-sided Towhee
P. fuscus	Brown Towhee
Psaltriparus minimus	Bushtit
Spizella atrogularis	Black-chinned Sparrow
Toxostoma dorsale	Crissal Thrasher

AMPHIBIANS AND REPTILES

Scientific Name	Common Name
Arizona elegans	Glossy Snake
Crotalus viridis	Western Rattlesnake
Gerrhonotus kingi	Arizona Alligator Lizard

Scientific Name	Common Name
Hypsiglena torquata	Night Snake
Lampropeltis pyromelana	Sonora Mountain Kingsnake
Leptotyphlops humilis	Southwestern Blind Snake
Masticophis bilineatus	Sonora Whipsnake
M. taeniatus	Desert Striped Whipsnake
Sceloporus occidentalis	Western Fence Lizard
S. undulatus	Eastern Fence Lizard
Tantilla planiceps	Western Blackhead Snake
Trimorphodon biscutatus lambda	Sonoran Lyre Snake
T. b. vilkinsoni	Texas Lyre Snake
Uta stansburiana	Side-blotched Lizard
Xantusia arizonae	Arizona Night Lizard

Tropical-Subtropical Scrublands
134.3 Sinaloan Thornscrub

Scientific Name	Common Name

PLANTS

Scientific Name	Common Name
Abutilon spp.	Indian mallows
Acacia angustissima	Whiteball Acacia
A. constricta	Whitethorn, Mescat Acacia
A. cymbispina	Espino, Chirowi
A. farnesiana	Sweet Acacia, Vinorama
A. millefolia	Many-leaflet Acacia
A. occidentalis	Desota, Teso
A. pennatula	Feather Acacia, Algarroba
A. willardiana	Palo Blanco
Agave schottii	Schott Agave
A. ocahui	Ocahui
Albizzia sinaloensis	Bolillo, Arellano, Palo Joso
Aloysia palmeri	Lippia, Orégano
Ambrosia ambrosioides	Bursage, Canyon Ragweed, Chicura
A. cordifolia	Chicurilla
Antigonon leptopus	Queen's Wreath, Rosa de Montana, San Miquelito
Atamisquea emarginata	Sonoran Caper
Ayenia spp.	ayenias
Boerhaavia spp.	spiderlings
Brickellia coulteri	Bricklebush
Brongniartia alamosana	Vara Prieta
B. palmeri	Vara Prieta
Bumelia occidentalis	Gum Bumelia, Bebelama
Bursera spp.	—
B. confusa	Torote Papelio
B. laxiflora	Copal, Torote Prieto
B. odorata	Torote, Chutama
Caesalpinia platyloba	Palo Colorado
C. pumila	—
Carlowrightia spp.	—
Cassia biflora	Twin-flower Cassia, Ejotillo-del-Monte
C. emarginata	Mora Hedionda, Viche, Palo de Zorillo
Ceiba acuminata	Pochote, Kapok
Celtis iguanaea	Garabato
C. pallida	Desert Hackberry, Granjeno
Cercidium floridum spp. peninsulare	Blue Paloverde
C. sonorae	Sonoran Paloverde, Palo Estribo
Commicarpus scandens	Artclub
Condalia spathulata	Knifeleaf Condalia, Guichutilla
Cordia sonorae	Palo de Asta
Coursetia glandulosa	Chino, Samo Prieto
Croton fragilis	Vara Blanca
Desmodium spp.	tick clovers, tick trefoils
Dodonaea viscosa	Hopbush, Aria
Elytraria spp.	scaly stems
Erythea roezlii	—
Erythrina flabelliformis	Southwestern Coralbean, Chilicote
Eysenhardtia orthocarpa	Kidneywood, Palo Dulce
Forchammeria watsoni	Jito
Fouquieria macdougalii	Tree Ocotillo, Torote Verde
F. splendens	Ocotillo, Coachwhip
Guaiacum coulteri	Guayacán
Guazuma ulmifolia	Guacimilla, Guácima
Haematoxylon brasiletto	Brasil
Hymenoclea monogyra	Burrobush, Jecota
Ipomoea arborescens	Tree Morningglory, Palo Santo
Jacobinia ovata	Espuela de Caballero
Jacquinia pungens	San Juan, San Juanito, San Juanico
Janusia spp.	janusias
Jatropha cardiophylla	Limber Bush, Sangre-de-Cristo
J. cinerea	Zapo, Lomboy
J. cordata	Jiotillo, Copalillo, Miguelito
Karwinskia parviflora	Coyotillo
Lantana velutina	Lantana
Lophocereus schottii	Senita, Old Man, Sina
Lycium berlandieri	Cilindrillo, Barchata
Lysiloma candida	Lysiloma, Palo Blanco
L. divaricata	Mauto
L. watsonii	Tepeguaje
Mimosa laxiflora	Gatuña
Nolina matapensis	Palmita
Olneya tesota	Ironwood, Palofierro
Opuntia comonduensis	Prickly Pear, Tuna
O. fulgida	Jumping Cholla, Cholla
Pachycereus pecten-aboriginum	Hecho, Cardón, Hecho
P. pringlei	Pringle Cardón, Cardón Pelón
Parthenium stramonium	Ocotillo
Passiflora spp.	passion flowers
Phaseolus spp.	beans
Piscidia mollis	Palo Blanco
Pisonia capitata	Garabato Prieto
Pithecellobium dulce	Guamuchil
P. mexicanum	Palo Chino
P. sonorae	Uña de Gato, Palo Gato
Plumeria acutifolia	Frangipani, Cacalosúchil
Prosopis juliflora	Mesquite
P. palmeri	Palo Hierro
Randia echinocarpa	Papache
R. laevigata	Crucecilla de la Sierra
R. obcordata	Papachillo
Ruellia spp.	—
Salvia spp.	sages
Sapindus saponaria	Soap Berry, Jaboncillo
Sapium biloculare	Mexican Jumping Bean, Yerba de la Flecha
Solanum amazonium	Mala Mujer, Berenjena Silvestre
Stegnosperma sp.	Ojo de Zanate
Stenocereus alamosensis	Octopus Cactus, Sina
Stenocereus thurberi	Organ Pipe Cactus, Pitahaya
Talinum spp.	flame flowers
Tecoma stans	Yellow Trumpet Bush, Gloria
Turnera spp.	—
Vallesia glabra	Cacarahue, Otatave, Sitavaro
Vitex mollis	Obalamo, Uvulama
Zizyphus obtusifolia	Gray-leaved Abrojo
Z. sonorensis	Jujube

MAMMALS

Scientific Name	Common Name
Dicotyles tajacu	Javelina
F. rufus baileyi	Bobcat
Lepus alleni	Antelope Jackrabbit
Liomys pictus	Painted Spiny Pocket Mouse
Odocoileus virginianus couesi	Coues' White-tailed Deer
Oryzomys couesi	Coues' Rice Rat
Perognathus pernix	Sinaloan Pocket Mouse

BIRDS

Scientific Name	Common Name
Camptostoma imberbe	Beardless Flycatcher
Glaucidium brasilianum	Ferruginous Owl
Lophortyx douglassii	Elegant Quail
Melanerpes uropygialis	Gila Woodpecker
Micrathene whitneyi	Elf Owl
Myiarchus tyrannulus	Wied's Crested Flycatcher
Parabuteo unicinctus	Harris' Hawk
Polioptila nigriceps	Black-capped Gnatcatcher
Thryothorus sinaloa	Sinaloa Wren
Zenaida asiatica	White-winged Dove

AMPHIBIANS AND REPTILES

Scientific Name	Common Name
Cnemidophorus costatus	—
Crotalus basiliscus	—
Leptodeira punctata	Cat-eyed Snake
Masticophis striolatus	—
M. valida	—
Urosaurus ornatus	Tree Lizard

Arctic-Boreal Grasslands
141.4 Alpine and Subalpine Grasslands

Scientific Name	Common Name
PLANTS	
Achillea spp.	yarrows
Agoseris spp.	mountain-dandelions
Agropyron ssp.	wheatgrasses
Artemisia tridentata	Big Sagebrush
Aster spp.	asters
Carex spp.	sedges
Cyperus spp.	flat sedges
Delphinium spp.	larkspurs
Deschampsia caespitosa	Tufted Hairgrass
Erigeron spp.	fleabanes
Festuca spp.	fescues
Juncus spp.	rushes
Muhlenbergia spp.	muhlies
Phleum alpinum	Mountain Timothy
Populus tremuloides	Quaking Aspen
Potentilla spp.	cinquefoils
Stipa spp.	needlegrasses
Taraxacum spp.	dandelions
Trifolium spp.	clovers
Vicia spp.	vetches
MAMMALS	
Canis latrans	Coyote
C. lupus	Gray Wolf
Cervus elaphus	Elk, Wapiti
Eutamias cinereicollis	Gray-Collared Chipmunk
E. minimus	Least Chipmunk
Lasionycteris noctivagans	Silver-haired Bat
Lasiurus cinereus	Hoary Bat
Lepus townsendii	White-tailed Jackrabbit
Marmota flaviventris	Yellow-bellied Marmot
Microtus longicaudus	Long-tailed Vole
M. mexicanus	Mexican Vole
M. montanus	Montane Vole
M. pennsylvanicus	Meadow Vole
Mustela frenata	Long-tailed Weasel
M. erminea	Ermine
Myotis volans	Long-legged Myotis
Odocoileus hemionus	Mule Deer
Ovis canadensis	Mountain Sheep, Bighorn Sheep
Peromyscus maniculatus	Deer Mouse
Phenocomys intermedius	Heather Vole
Sorex cinereus	Masked Shrew
S. nanus	Dwarf Shrew
S. vagrans	Vagrant Shrew
Spermophilus lateralis	Golden-mantled Ground Squirrel
Taxidea taxus	Badger
Thomomys bottae	Botta's Pocket Gopher
T. talpoides	Northern Pocket Gopher
Ursus arctos	Grizzly Bear
Vulpes vulpes	Red Fox
Zapus princeps	Western Jumping Mouse
BIRDS	
Buteo jamaicensis	Red-tailed Hawk
Chordeles minor	Common Nighthawk
Corvus corax	Raven
Dendrogapus obscurus	Blue Grouse
Dendroica coronata	Yellow-rumped Warbler
Eremophila alpestris	Horned Lark
Falco sparverius	American Kestrel
Passerculus sandwichensis	Savannah Sparrow
Sialia currucoides	Mountain Bluebird
Turdus migratorius	American Robin
AMPHIBIANS AND REPTILES	
Ambystoma tigrinum	Tiger Salamander
Bufo boreas	Western Toad
Phrynosoma douglassii	Short-horned Lizard
Pituophis melanoleucus	Gopher Snake
Pseudacris triseriata	Striped Chorus Frog
Rana pipiens	Northern Leopard Frog
Thamnophis elegans	Wandering Garter Snake

Cold-Temperate Grasslands
142.4 Montane Meadow Grassland

Scientific Name	Common Name
PLANTS	
Astragalus spp.	milk vetches, locoweeds
Blepharoneuron tricholepsis	Hairy Dropseed
Bromus marginatus	Mountain Brome
Carex spp.	sedges
Cyperus spp.	flat sedges
Helianthus spp.	sunflowers
Iris missouriensis	Rocky Mountain Iris
Juncus spp.	rushes
Lathyrus spp.	peavines
Lotus spp.	deer vetches, bird's foot trefoils
Lupinus spp.	lupines
Mimulus nasutus	Monkeyflower
Muhlenbergia montana	Mountain Muhly
Penstemon spp.	penstemons, beardtongues
Poa pratensis	Kentucky Bluegrass
Pteridium aquilinum	Bracken Fern
Rudbeckia hirta	Black-eyed Susan
Senecio spp.	groundsels
Solidago spp.	goldenrods
Sphaeralcea spp.	mallows
Veratrum californicum	Corn-lily, Falsehellebore
Verbascum thapsus	Mullein
Vicia spp.	vetches
Viola spp.	violets
MAMMALS	
Antilocapra americana	Pronghorn
Cervus elaphus	Elk, Wapiti
Microtus californicus	California Vole
M. mexicanus	Mexican Vole
Odocoileus hemionus	Mule Deer
O. virginianus	White-tailed Deer
Thomomys bottae	Botta's Pocket Gopher
BIRDS	
Empidonax difficilis	Western Flycatcher
Falco sparverius	American Kestrel
Meleagris gallopavo	Wild Turkey
Sialia mexicana	Western Bluebird
Sturnella neglecta	Western Meadowlark
S. magna	Eastern Meadowlark
AMPHIBIANS AND REPTILES	
Ambystoma tigrinum	Tiger Salamander
Hyla eximia	Arizona Treefrog, Mountain Treefrog
Pseudacris triseriata	Striped Chorus Frog
Rana pipiens	Northern Leopard Frog
Thamnophis elegans	Wandering Garter Snake

142.1 Plains and Great Basin Grassland

Scientific Name	Common Name
PLANTS	
Agropyron smithii	Western Wheatgrass
Andropogon gerardi	Big Bluestem
A. gerardi var. *paucipilus*	Sand Bluestem
Argemone spp.	prickly poppies
Aristida longiseta	Red Three-awn
Artemisia spp.	sagebrushes
A. filifolia	Sand Sagebrush
A. tridentata	Big Sagebrush
Aster spp.	asters
Atriplex canescens	Fourwing Saltbush
Bahia spp.	bahias
Bouteloua chondrosioides	Sprucetop Grama
B. curtipendula	Sideoats Grama
B. eriopoda	Black Grama
B. gracilis	Blue Grama
B. hirsuta	Hairy Grama
Brickellia spp.	bricklebushes

Scientific Name	Common Name
Buchloë dactyloides	Buffalo Grass
Ceratoides lanata	Winterfat
Chrysothamnus spp.	rabbitbrushes
Cirsium spp.	thistles
Cleome spp.	spiderflowers
Echinocereus engelmannii var. variegatus	Hedgehog Cactus
E. fendleri	Fendler Hedgehog
Eragrostis intermedia	Plains Lovegrass
Festuca arizonica	Arizona Fescue
Gaura spp.	gauras
Gutierrezia spp.	snakeweeds
G. sarothrae	Broom Snakeweed
Helianthus spp.	sunflowers
Hilaria jamesii	Galleta
Juniperus monosperma	One-seed Juniper
J. osteosperma	Utah Juniper
J. scopulorum	Rocky Mountain Juniper
Koeleria cristata	Prairie Junegrass
Lycurus phleoides	Wolftail, Texas Timothy
Mammillaria wrightii	Wright Pincushion
Mirabilis spp.	four o'clocks
Oenothera spp.	primroses
Opuntia spp.	prickly pears, chollas
O. arbuscula	Pencil Cholla
O. clavata	Club Cholla
O. imbricata	Tree Cholla
O. macrorhiza	Plains Prickly Pear
O. phaeacantha	Engelmann Prickly Pear
O. polycantha	Plains Prickly Pear
O. whipplei	Whipple Cholla
Oryzopsis hymenoides	Indian Ricegrass
Panicum obtusum	Vine Mesquite Grass
P. virgatum	Switchgrass
Pediocactus papyracanthus	Grama Grass Cactus
Prosopis glandulosa	Honey Mesquite
Psoralea spp.	scurf-peas
Quercus havardii	Shinnery Oak, Midget Oak
Ratibida spp.	coneflowers, Mexican hats
Rhus copallina var. lanceolata	Prairie Sumac
Rosa spp.	wild roses
Schizachyrium scoparium	Little Bluestem
Senecio spp.	groundsels
Setaria macrostachya	Plains Bristlegrass
Sorghastrum nutans	Indian Grass
Sphaeralcea spp.	mallows
Sporobolus airoides	Alkali Sacaton
S. cryptandrus	Sand Dropseed
Stipa comata	Needle and Thread Grass
Viguiera spp.	Goldeneye
Yucca glauca	Soapweed

MAMMALS

Scientific Name	Common Name
Antilocapra americana	Pronghorn
Bison bison	Bison, Buffalo
Cynomys gunnisoni	Gunnison's Prairie Dog
C. ludoviciana	Plains Prairie Dog
Geomys bursarius	Plains Pocket Gopher
Reithrodontomys montanus	Plains Harvest Mouse
Spermophilus tridecemlineatus	Thirteen-lined Ground Squirrel
Vulpes velox	Swift Fox

BIRDS

Scientific Name	Common Name
Ammodramus savannarum	Grasshopper Sparrow
Athene cunicularia	Burrowing Owl
Bartramia longicauda	Upland Sandpiper
Calamospiza melanocorys	Lark Bunting
Charadrius montana	Mountain Plover
Colinus virginianus	Bobwhite, Bobwhite Quail
Falco mexicanus	Prairie Falcon
Numenius americanus	Long-billed Curlew
Pedioceces phasianellus	Sharp-tailed Grouse
Tympanuchus pallidicinctus	Lesser Prairie Chicken

AMPHIBIANS AND REPTILES

Scientific Name	Common Name
Bufo cognatus	Great Plains Toad
Cnemidophorus sexlineatus viridis	Prairie Lined Racerunner
Crotalus viridis viridis	Prairie Rattlesnake
Diadophis punctatus arnyi	Prairie Ringneck Snake
Elaphe guttata	Corn Snake
Eumeces obsoletus	Great Plains Skink
Heterodon nasicus nasicus	Plains Hognose Snake
Holbrookia maculata	Lesser Earless Lizard
Lampropeltis triangulum celaenops	Mexico Milk Snake
Masticophis flagellum testaceus	Western Coachwhip
Pituophis melanoleucus	Gopher Snake
Scaphiopus bombifrons	Plains Spadefoot
Sceloporus undulatus consobrinus	Southern Prairie Lizard
Sonora episcopa episcopa	Great Plains Ground Snake
Tantilla nigriceps	Plains Blackhead Snake
Terrapene ornata	Western Box Turtle

Warm-Temperate Grasslands
143.1 Semidesert Grassland

Scientific Name	Common Name
PLANTS	
Acacia greggii	Cat-claw Acacia
A. neovernicosa	Viscid Acacia
Agave lechuguilla	Lechugilla
A. parryi	Parry Agave, Mescal
A. parviflora	Little flowered Agave
A. scabra	—
A. schottii	Schott Agave
Aloysia wrightii	Wright Lippia, Oreganillo
Amaranthus spp.	amaranths, pigweeds
Anisicanthus thurberi	Chuparosa, Desert Honeysuckle
Aristida divaricata	Poverty Three-awn
A. longiseta	Red Three-awn
A. purpurea	Purple Three-awn
A. wrightii	Wright Three-awn
Artemisia ludoviciana	Louisiana Sagebrush
Avena spp.	oats
Berberis trifoliata	Agarito, Algerita
Boerhaavia spp.	spiderlings
Bouteloua breviseta	Chino Grama
B. chondrosioides	Sprucetop Grama
B. curtipendula	Sideoats Grama
B. eriopoda	Black Grama
B. filiformis	Slender Grama
B. gracilis	Blue Grama
B. hirsuta	Hairy Grama
Bromus rubens	Red Brome
Buchloë dactyloides	Buffalo Grass
Bumelia lanuginosa	Gum Bumelia
Calliandra eriophylla	False Mesquite
Canotia holacantha	Crucifixion Thorn, Canotia
Celtis pallida	Desert Hackberry
C. reticulata	Netleaf Hackberry
Chilopsis linearis	Desert Willow
Condalia spathulata	Crucillo
C. ericoides	Javelina-bush
Coursetia glandulosa	—
Dasylirion leiophyllum	Sotol, Desert Spoon
D. wheeleri	Sotol, Desert Spoon
Dodonaea viscosa	Hopbush
Echinocactus horizonthalonius	Turk's Head
Echinocereus spp.	hedgehog cacti
E. pectinatus var. rigidissimus	Rainbow Cactus
Ephedra antisyphilitica	Mormon Tea
E. trifurca	Long-leaved Joint-fir, Mormon Tea
Eragrostis intermedia	Plains Lovegrass
E. lehmanniana	Lehmann Lovegrass
Ericameria laricifolia	Turpentine Bush
Eriogonum spp.	buckwheats
Erodium spp.	filarees
Eysenhardtia orthocarpa	Kidneywood
Ferocactus wislizenii	Barrel Cactus
Flourensia cernua	Tarbush
Fouquieria splendens	Ocotillo, Coachwhip
Gossypium thurberi	Desert Cotton, Algodoncillo
Gutierrezia sarothrae	Broom Snakeweed
Heteropogon contortus	Tanglehead
Hilaria belangeri	Curly Mesquite Grass
H. mutica	Tobosa
Isocoma heterophylla	Jimmyweed
I. tenuisecta	Burroweed
Juniperus monosperma	One-seed Juniper
Koeberlinia spinosa	Allthorn, Junco
Larrea tridentata	Creosote Bush

Scientific Name	Common Name
Lupinus spp.	lupines
Lycurus phleoides	Wolftail, Texas Timothy
Lysiloma microphylla	Littleleaf Lysiloma
Mammillaria grahami	—
M. gummifera	Cream Cactus
M. mainiae	—
M. wrightii	—
Martynia spp.	devil's claw
Mimosa biuncifera	Wait-a-while
M. dysocarpa	Velvet Pod
Muhlenbergia porteri	Bush Muhly, Hoe Grass
Neolloydia erectocentra	—
N. intertexta	—
Nolina erumpens	Beargrass, Sacahuista
N. microcarpa	Beargrass, Sacahuista
N. texana	Bunchgrass
Opuntia chlorotica	Pancake Prickly Pear
O. imbricata	Tree Cholla
O. kleiniae	Klein Cholla
O. leptocaulis	Desert Christmas Cactus
O. phaeacantha	Engelmann Prickly Pear
O. spinosior	Cane Cholla
O. violacea	Purple Prickly Pear
O. violacea var. macrocentra	Santa Rita Prickly Pear
Panicum obtusum	Vine Mesquite Grass
Pappophorum vaginatum	Pappus Grass
Parthenium incanum	Mariola
Prosopis glandulosa	Honey Mesquite
P. juliflora	Mesquite
P. velutina	Velvet Mesquite
Quercus chihuahuensis	Chihuahua Oak
Q. emoryi	Emory Oak, Bellota
Q. grisea	Gray Oak
Q. oblongifolia	Mexican Blue Oak
Q. toumeyi	Toumey Oak
Q. turbinella	Shrub Live Oak
Rhus microphylla	Littleleaf Sumac
Sageretia wrightii	Spiny Sageretia
Sapindus saponaria	Western Soapberry
Scleropogon brevifolius	Burrograss
Schizachyrium scoparium	Little Bluestem
Senecio spp.	groundsels
Setaria spp.	bristlegrasses
Sphaeralcea spp.	mallows
Sporobolus wrightii	Wright Sacaton
Tidestromia spp.	white-mats
Trichachne californica	Arizona Cottontop
Tridens muticus	Slim Tridens
T. pilosus	Hairy Tridens
T. pulchellus	Fluffgrass
Yucca baccata	Banana Yucca, Datil
Y. carnerosana	Spanish Bayonet
Y. elata	Soaptree Yucca, Palmillo
Y. macrocarpa	Bigleaf Yucca
Y. rostrata	Beaked Yucca
Y. torreyi	Torrey Yucca
Zinnia acerosa	Desert Zinnia
Zizyphus obtusifolia	Graythorn

MAMMALS

Scientific Name	Common Name
Antilocapra americana	Pronghorn
Canis latrans	Coyote
Dicotyles tajacu	Javelina
Dipodomys merriami	Merriam's Kangaroo Rat
D. ordii	Ord's Kangaroo Rat
D. spectabilis	Banner-tailed Kangaroo Rat
Lepus californicus	Black-tailed Jackrabbit
Neotoma albigula	White-throated Woodrat
N. micropus	Southern Plains Woodrat
Odocoileus hemionus crooki	Mule Deer
O. virginianus	White-tailed Deer
Onychomys torridus	Southern Grasshopper Mouse
Perognathus hispidus	Hispid Pocket Mouse
Peromyscus leucopus	White-footed Mouse
Sigmodon fulviventer	Tawny-bellied Cotton Rat
S. hispidus	Hispid Cotton Rat
Spermophilus spilosoma	Spotted Ground Squirrel
Taxidea taxus	Badger

BIRDS

Scientific Name	Common Name
Aimophila cassinii	Cassin's Sparrow
Amphispiza bilineata	Black-throated Sparrow
Athene cunicularia	Burrowing Owl
Auriparus flaviceps	Verdin
Buteo swainsoni	Swainson's Hawk
Callipepla squamata	Scaled Quail
Campylorhynchus brunneicapillus	Cactus Wren
Carpodacus mexicanus	House Finch
Chondestes grammacus	Lark Sparrow
Corvus cryptoleucus	White-necked Raven
Eremophila alpestris	Horned Lark
Falco mexicanus	Prairie Falcon
F. sparverius	American Kestrel
Geococcyx californianus	Roadrunner
Hirundo rustica	Barn Swallow
Icterus parisorum	Scott's Oriole
Lanius ludovicianus	Loggerhead Shrike
Lophortyx gambelii	Gambel's Quail
Mimus polyglottos	Mockingbird
Molothrus ater	Brown-headed Cowbird
Myiarchus cinerascens	Ash-throated Flycatcher
Phalaenoptilus nuttallii	Poor-will
Picoides scalaris	Ladder-backed Woodpecker
Polioptila melanura	Black-tailed Gnatcatcher
Sayornis saya	Say's Phoebe
Sturnella neglecta	Western Meadowlark
S. magna	Eastern Meadowlark
Toxostoma curvirostra	Curve-billed Thrasher
Tyrannus verticalis	Western Kingbird
Zenaida macroura	Mourning Dove

AMPHIBIANS AND REPTILES

Scientific Name	Common Name
Bufo debilis insidior	Western Green Toad
Cnemidophorus uniparens	Desert Grassland Whiptail
Ficimia cana	Western Hooknose Snake
Heterodon nasicus kennerlyi	Mexican Hognose Snake
Holbrookia texana scitula	Southwestern Earless Lizard
Terrapene ornata luteola	Desert Box Turtle

143.2 Californian Valley Grassland

Scientific Name	Common Name
PLANTS	
Avena spp.	oats
Bromus spp.	brome grasses
Dichelostemma pulchellum	Blue Dick
Erodium spp.	filagrees
Festuca spp.	fescues
Medicago spp.	—
Poa spp.	bluegrasses
Quercus agrifolia	Coast Live Oak
Q. engelmannii	Interior Live Oak
Stipa spp.	needlegrasses
S. pulchra	Purple Needle Grass
MAMMALS	
Antilocapra americana	Pronghorn
Cervus elaphus nannodes	Tule Elk
Ursus arctos	Grizzly Bear
BIRDS	
Gymnogyps californianus	California Condor
Pica nuttallii	Yellow-billed Magpie

Tropical-Subtropical Grasslands

144.3 Sonoran Savanna Grassland

Scientific Name	Common Name
PLANTS	
Acacia angustissima	Whiteball Acacia
A. farnesiana	Sweet Acacia
Amaranthus palmeri	Palmer Amaranth, Bledo, Quelito

Ambrosia spp.	ragweeds
Aristida californica	—
A. hamulosa	Poverty Three-awn
A. ternipes	Spider Grass
A. wrightii	Wright Three-awn
Atamisquea emarginata	Sonoran Caper
Boerhaavia spp.	spiderlings
Bouteloua aristidoides	Needle Grama
B. barbata	Six-weeks Grama
B. curtipendula	Sideoats Grama
B. filiformis	Slender Grama
B. parryi	Parry Grama
B. radicosa	—
B. rothrockii	Rothrock Grama
Caesalpinia pumila	—
Carnegiea gigantea	Saguaro
Cassia spp.	sennas
Cathestecum erectum	False Grama
Celtis pallida	Desert Hackberry
Cercidium floridum	Blue Paloverde
C. microphyllum	Littleleaf Paloverde
C. praecox	Palo Brea
Chloris spp.	windmill grasses, finger grasses
Coursetia glandulosa	
Croton sonorae	Croton
Euphorbia spp.	spurges
Eysenhardtia orthocarpa	Kidneywood
Forchammeria watsoni	Jito
Fouquieria macdougalii	Tree Ocotillo
Guaiacum coulteri	Guayacan
Heteropogon contortus	Tanglehead
Hilaria belangeri	Curly Mesquite Grass
Isomeris spp.	bladderpods
Janusia gracilis	Janusia
Jatropha cardiophylla	Limberbush
Lophocereus schottii	Senita, Old Man
Lycium brevipes	Tomatillo
Olneya tesota	Ironwood, Palofierro
Opuntia arbuscula	Pencil Cholla
O. fulgida	Jumping Cholla, Chain Fruit Cholla
O. leptocaulis	Desert Christmas Cactus
O. thurberi	—
Panicum obtusum	Vine Mesquite Grass
Parkinsonia aculeata	Retama
Portulaca spp.	purslanes
Stenocereus alamosensis	Sina, Octopus Cactus

MAMMALS

Dicotyles tajacu	Javelina
Lepus alleni	Antelope Jackrabbit

BIRDS

Aimophila carpalis	Rufous-winged Sparrow
Buteo albicaudatus	White-tailed Hawk
Caracara cheriway	Caracara
Colinus virginianus ridgwayi	Masked Bobwhite

AMPHIBIANS AND REPTILES

Bufo retiformis	Sonoran Green Toad

Cold-Temperate Desertlands
152.1 Great Basin Desertscrub

Scientific Name	Common Name

PLANTS

Artemisia arbuscula ssp. nova	Black Sagebrush
A. bigelovii	Bigelow Sagebrush
A. filifolia	Sand Sagebrush
A. parryi	Parry Sagebrush
A. spinescens	Bud Sagebrush
A. tridentata	Big Sagebrush
Atriplex canescens	Fourwing Saltbush
A. confertifolia	Shadscale
A. gardneri	Gardner Saltbush
A. nuttallii	Nuttall Saltbush
Bassia hyssopifolia	Fivehook Bassia
Bouteloua spp.	grama grasses

Bromus rubens	Red Brome, Foxtail Chess
B. tectorum	Downy Chess, Cheatgrass Brome
Ceratoides lanata	Winterfat
Chrysothamnus spp.	rabbitbrushes
C. greenei	Greene Rabbitbrush
C. nauseosus	Rubber Rabbitbrush
Coleogyne ramosissima	Blackbrush
Descurainia pinnata	Tansy Mustard
Distichlis spicata var. stricta	Desert Saltgrass
Echinocactus polycephalus var. xeranthemoides	Manyheaded Barrel Cactus
Echinocereus fendleri var. fendleri	Fendler Hedgehog
E. triglochidiatus var. melanacanthus	Red-flowered Hedgehog
Elaeagnus angustifolia	Russian Olive
Ephedra nevadensis	Rough Joint-fir
E. torreyana	Torrey Joint-fir
Eriogonum fasciculatum	Bush Buckwheat
Erodium cicutarium	Filaree, Heron's Bill
Forestiera neomexicana	Tanglebrush, Adelia
Grayia spinosa	Spiny Hopsage
Gutierrezia microcephala	Three-leaved Snakeweed
G. sarothrae	Broom Snakeweed
Hilaria jamesii	Galleta
Opuntia erinacea	Mohave Prickly Pear
O. gracilis	Narrow Cholla
O. polyacantha	Plains Prickly Pear
O. pulchella	Sand Cholla
O. ramosissima	Diamond Cholla
O. whipplei	Whipple Cholla
Oryzopsis hymenoides	Indian Rice Grass
Pediocactus spp.	
Salsola iberica	Russian Thistle
S. kali	Russian Thistle, Tumbleweed
S. paulsenii	—
Sarcobatus vermiculatus	Black Greasewood
Sclerocactus spp.	—
Sisymbrium altissimum	Tumble Mustard
Sitanion hystrix	Bottlebrush Squirreltail
Sporobolus airoides	Alkali Sacaton
Stenopsis linearifolius	—
Stipa speciosa	Desert Needlegrass
Suaeda spp.	seep weeds, sea blights
S. fruticosa	Seep Weed
Tamarix chinensis	Salt Cedar
Tetradymia canescens	Gray Horsebrush

MAMMALS

Antilocapra americana	Pronghorn
Canis latrans	Coyote
Dipodomys microps	Chisel-toothed Kangaroo Rat
D. ordii	Ord's Kangaroo Rat
Lagurus curtatus	Sagebrush Vole
Lepus californicus	Black-tailed Jackrabbit
Microdipodops megacephalus	Dark Kangaroo Mouse
M. pallidus	Pallid Kangaroo Mouse
Microtis montanus	Montane Vole
Perognathus parvus	Great Basin Pocket Mouse
Sorex merriami	Merriam's Shrew
Spermophilus townsendi	Townsend's Ground Squirrel

BIRDS

Alectoris chukar	Chukar
Amphispiza belli	Sage Sparrow
Centrocercus urophasianus	Sage Grouse
Oreoscoptes montanus	Sage Thrasher

AMPHIBIANS AND REPTILES

Cnemidophorus tigris septentrionalis	Northern Whiptail
C. tigris tigris	Western Whiptail
Crotalus viridis luteosus	Great Basin Rattlesnake
C. viridis nuntius	Hopi Rattlesnake
Crotaphytus collaris	Collared Lizard
C. wislizenii	Leopard Lizard
Phrynosoma platyrhinos platyrhinos	Northern Desert Horned Lizard
Pituophis melanoleucus deserticola	Great Basin Gopher Snake
Scaphiopus intermontanus	Great Basin Spadefoot Toad
Sceloporus graciosus	Sagebrush Lizard
S. occidentalis biseriatus	Great Basin Fence Lizard
S. undulatus elongatus	Northern Plateau Lizard
Thamnophis elegans vagrans	Wandering Garter Snake
Uta stansburiana	Side-blotched Lizard

Warm-Temperate Desertlands
153.1 Mohave Desertscrub

Scientific Name	Common Name

PLANTS

Scientific Name	Common Name
Acacia greggii	Cat-claw Acacia, Uña de Gato
Acamptopappus shockleyi	Goldenhead
A. sphaerocephalus	Rayless-goldenhead
Allenrolfea occidentalis	Pickleweed
Ambrosia dumosa	White Bursage
Aristida adscensionis	Six-weeks Three-awn
Artemisia tridentata	Big Sagebrush
Atriplex canescens	Fourwing Saltbush
A confertifolia	Shadscale
A. hymenelytra	Desert Holly
A. polycarpa	All-scale, Cattle Spinach
Bassia hyssopifolia	Fivehook Bassia
Cassia armata	Desert Senna
Ceratoides lanata	Winterfat
Coleogyne ramosissima	Blackbrush
Coryphantha vivipara **var.** *deserti*	Desert Coryphantha
Distichlis spicata **var.** *stricta*	Desert Saltgrass
Dudleya **spp.**	siemprevivos
Echinocactus polycephalus	Many-headed Barrel Cactus
Echinocereus engelmanii **var.** *chrysocentrus*	—
Encelia farinosa	White Brittlebush, Incienso
Ephedra funerea	—
E. nevadensis	Rough Joint-fir, Popotillo
E. torreyana	Torrey Joint-fir
Eriogonum fasciculatum	Bush Buckwheat
Erodium cicutarium	Filaree, Heron's Bill
E. taxanum	Large-flowered Stork's Bill
Euphorbia **spp.**	spurges
Grayia spinosa	Spiny Hopsage
Haplopappus cooperi	Cooper Burroweed
Hilaria rigida	Big Galleta
Hymenoclea salsola	Cheesebrush, White Burrobrush
Juniperus californica	California Juniper
Krameria parvifolia	Little-leaved Ratany
Larrea tridentata	Creosotebush
Lepidospartum latisquamum	Scalebroom
Lycium andersonii	Anderson Thornbush
L. pallidum	Rabbit Thorn
L. shockleyi	Schockley Thornbush
Mendora spinescens	Spiny Menodora
Monoptilon bellioides	Mohave Desert Star
Muhlenbergia porteri	Bush Muhly
Neolloydia johnsonii	—
Nitrophila **spp.**	—
N. occidentalis	Alkali Weed
Opuntia acanthocarpa **var.** *acanthocarpa*	Buckhorn Cholla
O. basilaris	Beavertail Cactus
O. echinocarpa	Silver Cholla, Golden Cholla
O. erinacea	Mohave Prickly Pear
O. ramosissima	Diamond Cholla
O. stanlyi **var.** *parishii*	Parish Cholla
O. whipplei **var.** *multigeniculata*	—
Psorothamnus arborescens	—
P. fremontii	Fremont Dalea
Salazaria mexicana	Paper Bag Bush, Bladder Sage
Salicornia **spp.**	glassworts
S. subterminalis	—
Salvia columbariae	Chia
S. funerea	—
S. mohavensis	Mohave Sage
Sarcobatus vermiculatus	Black Greasewood
Sphaeralcea ambigua	Desert Globe Mallow
Suaeda **spp.**	seep weeds, sea blights
Tetradymia canescens	Gray Horsebrush
Thamnosma montana	Mohave Desert Rue, Turpentine Broom
Yucca baccata	Banana Yucca, Datil
Y. brevifolia	Joshua Tree
Y. schidigera	Mohave Yucca, Spanish Dagger

MAMMALS

Scientific Name	Common Name
Ammospermophilus leucurus	White-tailed Antelope Ground Squirrel
Canis latrans	Coyote
Dipodomys merriami	Merriam's Kangaroo Rat
Equus asinus	Feral Burro
Neotoma lepida	Desert Woodrat
Odocoileus hemionus	Mule Deer
Onychomys torridus	Southern Grasshopper Mouse
Ovis canadensis nelsoni	Desert Bighorn Sheep
Perognathus formosus	Long-tailed Pocket Mouse
P. longimembris	Little Pocket Mouse
Peromyscus crinitus	Canyon Mouse
P. eremicus	Cactus Mouse

BIRDS

Scientific Name	Common Name
Toxostoma curvirostra	Curve-billed Thrasher
T. lecontei	LeConte's Thrasher

AMPHIBIANS AND REPTILES

Scientific Name	Common Name
Arizona elegans eburnata	Desert Glossy Snake
Callisaurus draconoides	Zebratail Lizard
Chionactis occipitalis occipitalis	Mohave Shovelnose Snake
C. occipitalis talpina	Nevada Shovelnose Snake
Cnemidophorus tigris tigris	Western Whiptail
Coleonyx variegatus	Banded Gecko
Crotalus cerastes	Sidewinder
C. mitchelli	Speckled Rattlesnake
C. scutulatus	Mohave Rattlesnake
Crotaphytus collaris	Collared Lizard
C. wislizenii	Leopard Lizard
Dipsosaurus dorsalis	Desert Iguana
Gopherus agassizi	Desert Tortoise
Heloderma suspectum cinctum	Banded Gila Monster
Hypsiglena torquata deserticola	Desert Night Snake
Lampropeltis getulus californiae	California Kingsnake
Leptotyphlops humilis	Western Blind Snake
Lichanura trivirgata gracia	Desert Rosy Boa
Masticophis flagellum	Coachwhip
M. taeniatus	Striped Whipsnake
Phrynosoma platyrhinos calidiarum	Southern Desert Horned Lizard
P. solare	Regal Horned Lizard
Phyllorhynchus decurtatus perkinsi	Western Leafnose Snake
Pituophis melanoleucus deserticola	Great Basin Gopher Snake
Rhinocheilus lecontei lecontei	Western Longnose Snake
Salvadora hexalepis mojavensis	Mohave Patchnose Snake
Sauromalus obesus	Chuckwalla
Sceloporus magister uniformis	Yellowback Spiny Lizard
Sonora semiannulata	Western Ground Snake
Trimorphodon biscutatus lambda	Sonoran Lyre Snake
Uma scoparia	Mohave Fringe-toed Lizard
Urosaurus graciosus graciosus	Western Brush Lizard
Uta stansburiana stejegeri	Desert Side-blotched Lizard
Xantusia vigilis	Desert Night Lizard

153.2 Chihuahuan Desertscrub

Scientific Name	Common Name

PLANTS

Scientific Name	Common Name
Acacia greggii	Cat-claw Acacia, Uña de Gato
A. neovernicosa	Viscid Acacia
Agave falcata	—
A. lechuguilla	Lechuguilla
A. neomexicana	New Mexico Agave
A. scabra	—
A. striata	—
Allenrolfea **spp.**	iodine bushes
Aloysia wrightii	Wright Lippia, Honey Bush
Ancistrocactus scheerii	Texas Fish-hook Cactus
A. uncinatus	Cat-claw Cactus
Ariocarpus fissuratus	Living Rock, Star Cactus
Artemisia filifolia	Sand Sagebrush
Atriplex **spp.**	saltbushes
A. canescens	Fourwing Saltbush
Bouteloua **spp.**	grama grasses

Coldenia canescens	Spreading Coldenia
C. greggii	Gregg Coldenia
Coryphantha echinus	Spiny Pincushion
C. macromeris	Long-tubercle Coryphantha
C. pottsii	Red-flowered Pincushion
C. ramillosa	Dead Grass Cactus
C. sheerii var. valida	Stately Coryphantha
C. strobiliformis	Cone Cactus
C. vivipara	Spiny Star Cactus
Croton spp.	crotons
Dasylirion leiophyllum	Lesser Sotol
D. wheeleri	Sotol, Desert Spoon
Distichlis spicata var. stricta	Desert Saltgrass
Dyssodia spp.	dogweeds, fetid marigolds
D. acerosa	—
D. pentachaeta	Thurber Dyssodia
Echinocactus horizonthalonius	Blue Barrel
E. platyacanthus	—
Echinocereus chloranthus	Green-flowered Hedgehog
E. enneacanthus var. stramineus	—
E. pectinatus var. neomexicanus	—
E. pectinatus var. rigidissimus	Rainbow Cactus
E. texensis	—
E. triglochidiatus var. melanacanthus	Red-flowering Hedgehog
Ephedra trifurca	Longleaf Ephedra, Cañutillo
Epithelantha micromeris	Button Cactus
Euphorbia spp.	spurges
Euphorbia antisyphilitica	Candelilla
Ferocactus hamatacanthus	Texas Longhorn Cactus, Biznaga de Limilla
F. pringlei	Pringle Barrel Cactus
Flourensia cernua	Tarbush, Hojase, Blackbrush
Fouquieria splendens	Ocotillo, Coachwhip
Hechtia scariosa	Guapilla
Helianthus spp.	sunflowers
Hilaria mutica	Tobosa
Hymenoclea monogyra	Burrobrush
Jatropha dioica	Limberbush
Juniperus monosperma	One-seed Juniper
J. pinchottii	Sabino
Koeberlinia spinosa	Allthorn, Junco
Krameria parvifolia var. glandulosa	Littleleaf Ratany
Larrea tridentata	Creosotebush, Gobernadora
Leucophyllum frutescens	Texas Silverleaf, Purple Sage
L. minus	—
Lophophora williamsii	Peyote
Mammillaria gummifera var. meiacantha	—
M. gummifera var. applanata	—
M. pottsii	Potts Mammillaria
Mortonia scabrella	Sandpaper Bush
Neolloydia intertexta	Texas Cactus, Early Visnagita
Nolina erumpens	Beargrass, Sacahuista
N. microcarpa	Beargrass, Sacahuista
N. texana	Bunchgrass, Sacahuista
Opuntia bradtiana	Grusonia
O. imbricata	Cane Cholla
O. kleiniae	Klein Cholla
O. leptocaulis	Desert Christmas Cactus
O. phaeacantha var. discata	Engelmann Prickly Pear
O. phaeacantha var. major	—
O. schottii	Prostrate Club Cholla
O. tunicata	Abrojo, Clavellina
O. violacea var. macrocentra	Purple Prickly Pear
Oryzopsis hymenoides	Indian Rice Grass
Parthenium argentatum	Guayule
P. incanum	Mariola
Peniocereus greggii	Night Blooming Cereus
Prosopis glandulosa var. torreyana	Western Honey Mesquite
Rhus microphylla	Littleleaf Sumac
Selaginella lepidophylla	Resurrection Plant
Sporobolus airoides	Alkali Sacaton
S. contractus	Spike Dropseed
S. flexuosus	Mesa Dropseed
S. giganteus	Giant Dropseed
Suaeda spp.	seep weeds, sea blights
Thelocactus bicolor	Texas Pride
Tidestromia spp.	—
Varilla mexicana	Jarilla, Varilla
Viguiera stenoloba	—
Yucca carnerosana	Spanish Bayonet
Y. elata	Soaptree Yucca, Palmillo
X. filifera	—
Y. rostrata	—
Y. thompsoniana	Thompson Yucca
Y. torreyi	Torrey Yucca
Zinnia acerosa	Desert Zinnia
Z. grandiflora	Prairie Zinnia
Zizyphus obtusifolia	Graythorn

MAMMALS

Ammospermophilus interpres	Texas Antelope Ground Squirrel
Dipodomys merriami	Merriam's Kangaroo Rat
D. nelsoni	Nelson's Kangaroo Rat
Geomys arenarius	Desert Pocket Gopher
Neotoma goldmani	Goldman's Woodrat
Notiosorex crawfordi	Desert Shrew
Odocoileus hemionus crooki	Desert Mule Deer
Onychomys torridus	Southern Grasshopper Mouse
Ovis canadensis mexicana	Desert Bighorn Sheep
Pappogeomys castanops	Yellow-Faced Pocket Gopher
Perognathus nelsoni	Nelson's Pocket Mouse
P. pencillatus	Desert Pocket Mouse
Sylvilagus auduboni	Desert Cottontail

BIRDS

Amphispiza bilineata	Black-chinned Sparrow
Callipepla squamata	Scaled Quail
Campylorhynchus brunneicapillus	Cactus Wren
Chordeiles acutipennis	Lesser Nighthawk
Corvus cryptoleucus	White-necked Raven
Geococcyx californianus	Roadrunner
Icterus parisorum	Scott's Oriole
Toxostoma curvirostra	Curve-billed Thrasher
Zenaida macroura	Mourning Dove

AMPHIBIANS AND REPTILES

Cnemidophorus inornatus	Little Striped Whiptail
C. neomexicanus	New Mexico Whiptail
C. tesselatus	Colorado Checkered Whiptail
C. tigris marmoratus	Marbled Whiptail
Coleonyx brevis	Texas Banded Gecko
C. reticulatus	Reticulated Gecko
Cophosaurus texanus	Greater Earless Lizard
Crotalus atrox	Western Diamondback Rattlesnake
C. scutulatus	Mohave Rattlesnake
Elaphe subocularis	Trans-Pecos Rat Snake
Ficimia cana	Western Hooknose Snake
Gopherus flavomarginatus	Bolson Tortoise
Masticophis flagellum lineatus	Lined Coachwhip
M. taeniatus	Striped Whipsnake
Phrynosoma modestum	Roundtail Horned Lizard
Sceloporus cautus	—
S. maculosus	—
S. magister bimaculosus	Twin-spotted Spiny Lizard
S. merriami	Merriam's Canyon Lizard
S. ornatus	—
S. poinsetti	Crevice Spiny Lizard
Tantilla atriceps	Mexican Blackhead Snake
Uma exsul	—

Tropical-Subtropical Desertlands

154.1 Sonoran Desertscrub

Scientific Name	Common Name

PLANTS

Abronia villosa	Sand Verbena
Acacia constricta	Whitethorn, Mescat Acacia

Scientific Name	Common Name
A. greggii	Cat-claw Acacia, Uña de Gato
A. neovernicosa	Viscid Acacia
Agave avellanidens	—
A. cerulata	Blue Agave
A. deserti	Desert Agave
A. shawii	Maguey
Ambrosia ambrosioides	Canyon Ragweed, Chicura
A. bryantii	Bryant Bursage
A. camphorata	Estafiate
A. chenopodifolia	San Diego Bursage
A. deltoidea	Triangle-leaf Bursage
A. divaricata	—
A. dumosa	White Bursage
A. magdalenae	Magdalena Bursage
Anisacanthus thurberi	Chuparosa, Desert Honeysuckle
Aristida spp.	three-awns
Atamisquea emarginata	Sonoran Caper
Atriplex canescens	Fourwing Saltbush, Chamiso
A. canescens ssp. linearis	Narrow-leaved Wingscale
A. hymenelytra	Desert Holly
A. julacea	—
A. lentiformis	Quail Brush, Lens Scale
A. polycarpa	All-scale, Cattle Spinach
Baccharis sarothroides	Desert Broom, Romerillo
Berberis trifoliata	Algarito, Algerita
Bromus rubens	Red Brome, Foxtail Chess
Bursera hindsiana	Torchwood, Copal
B. laxiflora	Copal, Torote Prieto
B. microphylla	Elephant Tree, Torote
B. odorata	Torote, Chutama
Caesalpinia pumila	
Calliandra eriophylla	False Mesquite, Fairy Duster
Canotia holacantha	Crucifixion Thorn, Canotia
Carnegiea gigantea	Saguaro
Castela emoryi	Corona-de-Cristo
Celtis pallida	Desert Hackberry, Granjeno
Ceiba acuminata	Pochote
Ceratoides lanata	Winterfat
Cercidium floridum	Blue Paloverde
C. microphyllum	Littleleaf Paloverde, Foothill Paloverde, Dipua
C. praecox	Palo Brea
C. sonorae	Sonoran Paloverde, Palo Estribo
Chilopsis linearis	Desert Willow
Chorizanthe rigida	Rigid Spiny Herb
Citharexylum flabellifolium	—
Cordia parvifolia	Palo de Asta
Coreocarpus parthenioides	—
Coursetia glandulosa	Chino, Samo Prieto
Croton sonorae	Sonoran Croton
Descurainia pinnata	Tansy Mustard
D. sophia	
Dyssodia anthemidifolia	—
Echinocactus horizonthalonius var. nichollii	Eagle Claw Cactus
Echinocereus engelmannii	Engelmann Hedgehog Cactus
E. fasciculatus	
E. fendleri var. fendleri	Fendler Hedgehog
E. maritimus	Seaside Hedgehog
Encelia californica	California Encelia
E. farinosa	White Brittlebush, Incienso
E. farinosa var. phenicodonta	—
Ephedra trifurca	Longleaf Ephedra, Cañutillo
Eriogonum deserticola	Desert Buckwheat
E. fasciculatum	Bush Buckwheat
Erodium cicutarium	Filaree, Heron's Bill
Euphorbia magdalenae	Magdalena Spurge
E. misera	Shrubby Spurge
E. tomentulosa	—
E. xantii	
Eysenhardtia orthocarpa	Kidneywood
Ferocactus acanthodes	Compass Barrel Cactus
F. wislizenii	Fish-hook Barrel Cactus
Forchammeria watsoni	Jito
Fouquieria columnaris	Boojum, Cirio
F. diguetii	
F. macdougalii	Tree Ocotillo, Palo Adán
F. splendens	Ocotillo, Coachwhip
Frankenia palmeri	Palmer Alkali Heath, Yerba Reuma
Guaiacum coulteri	Guayacan
Hilaria rigida	Big Galleta
Hymenoclea salsola var. pentalepis	Cheesebrush, White Burrobrush
Hyptis emoryi	Desert Lavender
Ipomoea arborescens	Tree Morningglory, Palo Santo
Isocoma acradenia	Alkali Goldenbush
I. heterophylla	Jimmy Weed, Rayless Goldenrod
I. tenuisecta	Burroweed
Isomeris arborea	Bladderpod
Jacobinia ovata	Espuela de Caballero
Jatropha cardiophylla	Limberbush, Sangre-de-Cristo
J. cinerea	Lomboy
J. cordata	Jiotillo, Copalillo, Miguelito
J. cuneata	Sangre-de-drago
Justicia californica	Chuparosa, Desert Honeysuckle
Krameria grayi	White Ratany
K. parvifolia	Little-leaved Ratany
Lactuca serriola	Prickly Lettuce, Wild Lettuce
Lantana horrida	Lantana, Héba-de-Cristo
Larrea tridentata	Creosotebush, Gobernadora
Lophocereus schottii	Senita, Old Man, Sina
Lycium andersonii	Anderson Thornbush
L. berlandieri	Cilindrillo, Barchata
L. brevipes	Tomatillo
L. californicum	
L. fremontii	—
Lysiloma candida	Fremont Thornbush
Mammillaria grahamiae var. oliviae	Palo Blanco
M. microcarpa	Fish-hook Pincushion
M. tetrancista	
M. thornberi	Thornber Pincushion
Mesembryanthemum crystallinum	Ice Plant, Flor-de-Sol
Muhlenbergia porteri	Bush Muhly
Myrtillocactus cochal	Cochal
Neoevansia striata	Sacamatraca
Neolloydia erectocentra var. acunensis	
Olneya tesota	Ironwood, Palofierro
Opuntia acanthocarpa var. major	—
O. acanthocarpa var. thornberi	Thornber Buckhorn Cholla
O. arbuscula	Pencil Cholla
O. basilaris	Beavertail Cactus
O. bigelovii	Teddy Bear Cholla
O. cholla	Cholla
O. ciribe	—
O. echinocarpa	Silver Cholla, Golden Cholla
O. fulgida	Jumping Cholla, Chain Fruit Cholla
O. fulgida var. mammillata	Boxing Glove
O. kleiniae	Klein Cholla
O. leptocaulis	Desert Christmas Cactus
O. molesta	Wicked Cholla
O. phaeacantha	Brown-spined Prickly Pear
O. phaeacantha var. discata	Engelmann Prickly Pear
O. phaeacantha var. major	—
O. prolifera	
O. ramosissima	Diamond Cholla
O. spinosior	Cane Cholla
O. stanlyi	Devil's Club Ground Cholla
O. stanlyi var. kunzei	Kunze Ground Cholla
O. stanlyi var. peeblesiana	Peebles Ground Cholla
O. thurberi	Thurber Cholla
O. versicolor	Staghorn Colla
O. wigginsii	
Pachycereus pecten-aboriginum	Hecho, Cardón Hecho
P. pringlei	Cardón, Cardó Peló
Pachycormus discolor	Copalquín
Parkinsonia aculeata	Retama
Parthenium incanum	Mariola
Pedilanthus macrocarpus	Candelilla
Peniocereus greggii	Night Blooming Cereus
P. greggii var. transmontanus	—
Pithecellobium confine	
Plantago insularis	Woolly Plantain
Prosopis glandulosa var. torreyana	Western Honey Mesquite
P. pubescens	Screwbean Mesquite
P. velutina	Velvet Mesquite
Psorothamnus spinosa	Smoketree
P. schottii	Indigo Bush
Randia thurberi	Papache
Rhus laurina	Laurel Sumac
Ribes tortuosum	
Ruellia peninsularis	—
Salsola kali	Russian Thistle, Tumbleweed
Sapium biloculare	Mexican Jumping Bean,

Schismus arabicus	Yerba de la Flecha, Arabian Grass
S. barbatus	Mediterranean Grass
Simmondsia chinensis	Jojoba, Goat Nut, Coffee Berry
Sisymbrium irio	Yellow Rocket
Solanum hindsianum	Hinds Nightshade
Sphaeralcea coulteri	Coulter Globe Mallow
S. fendleri	Fendler Globe Mallow
Stenocereus alamosensis	Octopus Cactus, Sina
S. gummosus	Pitahaya Agria
S. thurberi	Organ Pipe Cactus, Pitahaya
Suaeda torreyana **var.** ramosissima	Desert Seep Weed
Tecoma stans	Yellow Trumpet Bush, Gloria
Tessaria sericea	Arrow-weed
Tillandsia recurvata	Ballmoss, Gallitos, Heno Pequeño
Viguiera laciniata	Ragged-Leaf Goldeneye
Viscainoa geniculata	Guayacán
Yucca valida	Datilillo
Zinnia acerosa	Desert Zinnia
Zizyphus obtusifolia	Graythorn

MAMMALS

Ammospermophilus harrisii	Harris' Ground Squirrel
A. leucurus	White-tailed Antelope Ground Squirrel
Antilocapra americana sonorensis	Sonoran Pronghorn
Bassariscus astutus	Ring-tailed Cat
Canis latrans	Coyote
Dicotyles tajacu	Peccary, Javelina
Dipodomys deserti	Desert Kangaroo Rat
D. merriami	Merriam's Kangaroo Rat
D. peninsularis peninsularis	Vizcaino Desert Kangaroo Rat
Equus asinus	Feral Burro
Lepus californicus	Black-tailed Jackrabbit
Macrotis californicus	California Leaf-nosed Bat
Myotis californicus	California Myotis
Neotoma albigula	White-throated Woodrat
Odocoileus hemionus crooki	Desert Mule Deer
Perognathus amplus	Arizona Pocket Mouse
P. baileyi	Bailey's Pocket Mouse
P. formosus	Long-tailed Pocket Mouse
P. penicillatus	Desert Pocket Mouse
Peromyscus eremicus	Cactus Mouse
P. eremicus eremicus	Arizona Cactus Mouse
Spermophilus tereticaudus	Round-tailed Ground Squirrel
Sylvilagus auduboni	Desert Cottontail
Urocyon cinereoargenteus	Gray Fox
Vulpes macrotus	Kit Fox

BIRDS

Amphispiza bilineata	Black-chinned Sparrow
Athene cunicularia	Burrowing Owl
Auriparus flaviceps	Verdin
Calypte costae	Costa's Hummingbird
Campylorhynchus brunneicapillus	Cactus Wren
Cardinalis sinuatus	Pyrrhuloxia
Chordeiles acutipennis	Lesser Nighthawk
Colaptes auratus	Gilded Flicker
C. chrysoides	—
Geococcyx californianus	Roadrunner
Lophortyx californicus	California Quail
L. gambeli	Gambel Quail
L. douglassii	Elegant Quail
Melanerpes uropygialis	Gila Woodpecker
Micrathene whitneyi	Elf Owl
Myiarchus tyrannulus	Wied's Crested Flycatcher
Parabuteo unicinctus	Harris' Hawk
Phainopepla nitens	Phainopepla
Picoides scalaris	Ladder-backed Woodpecker
Polioptila melanura	Black-tailed Gnatcatcher
Scardafella inca	Inca Dove
Toxostoma bendirei	Bendire's Thrasher
T. curvirostra	Curve-billed Thrasher
T. lecontei	LeConte's Thrasher
Zenaida asiatica	White-winged Dove
Z. macroura	Mourning Dove

AMPHIBIANS AND REPTILES

Arizona elegans	Glossy Snake
A. elegans eburnata	Desert Glossy Snake
A. elegans noctivaga	Arizona Glossy Snake
Bufo retiformis	Sonoran Green Toad
Callisaurus draconoides	Zebratail Lizard
Chilomeniscus cinctus	Banded Sand Snake
Chionactis occipitalis	Western Shovelnose Snake
Cnemidophorus hyperythrus	Orangethroat Lizard
C. tigris gracilis	Southern Whiptail
C. tigris multiscutatus	Coastal Whiptail
C. tigris tigris	Western Whiptail
Coleonyx variegatus	Banded Gecko
Crotalus atrox	Western Diamondback Rattlesnake
C. cerastes	Sidewinder
C. ruber	Red Diamond Rattlesnake
C. scutulatus	Mohave Rattlesnake
C. tigris	Tiger Rattlesnake
Dipsosaurus dorsalis	Desert Iguana
Gopherus agassizi	Desert Tortoise
H. suspectum	Gila Monster
H. suspectum suspectum	Reticulated Gila Monster
Lichanura trivirgata	Rosy Boa
Micruroides euryxanthus	Arizona Coral Snake
Phrynosoma m'calli	Flat-tail Horned Lizard
P. platyrhinos calidiarum	Southern Desert Horned Lizard
P. solare	Regal Horned Lizard
Phyllorhynchus decurtatus	Spotted Leaf-nose Snake
Salvadora hexalepis	Western Patchnose Snake
Sauromalus obesus	Chuckwalla
Sceloporus magister	Desert Spiny Lizard
Sonora semiannulata	Western Ground Snake
Terrapene ornata	Ornate Box Turtle
Uma inornata	—
U. notata	Fringe-toed Lizard
Urosaurus graciosus	Brush Lizard
U. microscutatus	Small-scaled Lizard
U. ornatus	Tree Lizard

Wetlands

Arctic-Boreal Wetlands
Subalpine Wetlands

Scientific Name	Common Name
PLANTS	
Carex spp.	sedges
Cyperus spp.	flatsedges
Eleocharis spp.	spike rushes
Glyceria spp.	mannagrasses
G. borealis	Northern Mannagrass
Juncus spp.	rushes
Luzula spp.	wood rushes
Picea pungens	Blue Spruce
Populus tremuloides	Quaking Aspen
Potentilla fruticosa	Shrubby Cinquefoil
Ribes spp.	currants
Rubus spp.	raspberries
Salix bebbiana	Bebb Willow
S. irrorata	—
S. monticola	Serviceberry Willow
S. scouleriana	Scouler Willow
Sambucus racemosa	Red Elderberry
Scirpus spp.	bulrushes, tules
S. pallidus	Round-stem Bulrush
MAMMALS	
Microtus spp.	voles
Sorex palustris	Water Shrew
Zapus princeps	Western Jumping Mouse
BIRDS	
Anas platyrhynchos	Mallard
Fulica americana	American Coot
Melospiza lincolnii	Lincoln's Sparrow
Oporornis tolmiei	MacGillivray's Warbler

Podiceps caspicus	Eared Grebe
Podilymbus podiceps	Pied-billed Grebe
Porzana carolina	Sora
Zonotrichia leucophrys	White-crowned Sparrow

AMPHIBIANS AND REPTILES

Ambystoma tigrinum nebulosum	Tiger Salamander
Hyla spp.	treefrogs
H. eximia	Mountain Treefrog
H. regilla	Pacific Treefrog
Pseudacris triseriata	Striped Chorus Frog

FISHES

Salmo clarki	Cutthroat Trout
S. apache	Arizona Trout
S. gairdneri	Rainbow Trout
S. gilae	Gila Trout
Salvelinus fontinalis	Brook Trout

Cold-temperate Wetlands
Montane Riparian Wetlands

Scientific Name	Common Name

PLANTS

Abies concolor	White Fir
Acer glabrum	Rocky Mountain Maple
A. grandidentatum	Bigtooth Maple
A. macrophyllum	Bigleaf Maple
A. negundo	Box Elder
Alnus oblongifolia	Arizona Alder
A. rhombifolia	White Alder
A. tenuifolia	Thinleaf Alder
Betula occidentalis	Water Birch
Blepharoneuron tricholepsis	Hairy Dropseed
Cornus stolonifera	Red-osier Dogwood
Crataegus spp.	hawthorns
Parthenocissus vitacea	Virginia Creeper
Pinus ponderosa	Ponderosa Pine
Populus angustifolia	Narrowleaf Cottonwood
P. tremuloides	Quaking Aspen
P. trichocarpa	Black Cottonwood
Prunus americana	American Plum
P. emarginata	Bitter Cherry
Quercus gambelii	Gambel Oak
Rhus glabra	Smooth Sumac
Robinia neomexicana	New Mexican Locust
Salix spp.	willows
S. irrorata	—
S. lasiolepis	Arroyo Willow
S. scouleriana	Scouler Willow
Sambucus glauca	Blueberry Elder

MAMMALS

Castor canadensis	Beaver
Microtus spp.	voles
Odocoileus virginianus	White-tailed Deer
Procyon lotor	Raccoon

BIRDS

Cinclus mexicanus	Water Ouzel, Dipper
Dendroica petechia	Yellow Warbler
Empidonax difficilis	Western Flycatcher
Icteria virens	Yellow-breasted Chat
Megaceryle alcyon	Belted Kingfisher
Meleagris gallopavo	Wild Turkey
Passerina amoena	Lazuli Bunting
Pheucticus melanocephalus	Black-headed Grosbeak
Selasphorus platycercus	Broad-tailed Hummingbird
Vireo gilvus	Warbling Vireo

AMPHIBIANS AND REPTILES

Ambystoma rosaceum	Tarahumara Salamander

Gerrhonotus kingi	Arizona Alligator Lizard
Hyla arenicolor	Canyon Treefrog
Hyla eximia	Mountain Treefrog
H. regilla	Pacific Treefrog
Rana mucosa	Mountain Yellow-legged Frog
Thamnophis couchi	Western Aquatic Garter Snake
T. cyrtopsis	Blackheaded Garter Snake
T. elegans	Western Terrestrial Garter Snake
T. eques	Mexican Garter Snake
T. rufipunctatus	Narrowhead Garter Snake

FISHES

Lepidomeda spp.	spinedaces
Notemigonus crysoleucus	Golden Shiner
Rhinichthys cataractae	Longnose Dace
R. osculus	Speckled Dace
Salmo chrysogaster	Mexican Golden Trout
S. apache	Arizona Trout
S. gairdneri	Rainbow Trout
S. gilae	Gila Trout
S. trutta	European Brown Trout

Plains and Great Basin Riparian Wetlands

Scientific Name	Common Name

PLANTS

Alhagi camelorum	Camelthorn
Elaeagnus angustifolia	Russian Olive
Populus deltoides spp. sargentii	Plains Cottonwood
P. wislizenii	Rio Grande Cottonwood
Salix amygdaloides	Peachleaf Willow
S. exigua	Coyote Willow, Sandbar Willow
Tamarix chinensis	Salt Cedar

BIRDS

Catharus fuscescens	Veery
Dumatella carolinensis	Catbird
Icteria virens	Yellow-breasted Chat
Melanerpes erythrocephalus	Red-headed Woodpecker
Pica pica	Black-billed Magpie
Sayornis phoebe	Eastern Phoebe
Setophaga ruticilla	American Redstart

AMPHIBIANS AND REPTILES

Bufo woodhousei	Woodhouse's Toad
Rana pipiens	Leopard Frog
Scaphiopus bombifrons	Plains Spadefoot
S. hammondi	Western Spadefoot
S. intermontanus	Great Basin Spadefoot
Thamnophis radix	Plains Garter Snake

FISHES

Camptostoma anomalum	Plains Stoneroller
Catostomus commersoni	White Sucker
Crenichthys baileyi	White River Springfish
Cyprinodon pecosensis	Pecos Pupfish
Fundulus kansae	Plains Killifish
F. zebrinus	Rio Grande Killifish
Gila pandora	Pecos Chub
G. robusta jordani	Round-tail Chub
G. robusta robusta	Round-tail Chub
G. robusta seminuda	Round-tail Chub
Hybognathus placitus	Plains Minnow
Lepidomeda albivallis	White River Spinedace
L. altivelis	Pahranagat Spinedace
L. mollispinis mollispinis	Virgin River Spinedace
L. mollispinis pratensis	—
L. vittata	Little Colorado Spinedace
Moapa coriacea	Moapa Dace
Notropis stramineus	Sand Shiner
Pantosteus clarki	Desert Mountain-sucker
P. discobolus	Blue-head Mountain-sucker
P. plebeius	Rio Grande Mountain-sucker

Pimephales promelas	Fathead Minnow
Platygobio gracilis	Flathead Chub
Rhinichthys osculus	Speckled Dace
Semotilus atromaculatus	Creek Chub

Montane, Plains and Great Basin Marshlands

Scientific Name	Common Name
PLANTS	
Carex spp.	sedges
Chara spp.	muskgrasses
Distichilis spicata	Saltgrass
Eleocharis spp.	spike rushes
Elodea spp.	water-weeds
Glyceria spp.	mannagrasses
Juncus spp.	rushes
Myriophyllum spicatum	Water Milfoil
Nitella spp.	stoneworts
Potamogeton spp.	pond weeds
Scirpus acutus	Hardstem Rush, Great Bulrush
S. americanus	Three-square
Typha latifolia	Cattail
MAMMALS	
Mustela vison	Mink
Ondatra zibethicus	Muskrat
Zapus princeps	Western Jumping Mouse
BIRDS	
Agelaius phoeniceus	Red-winged Blackbird
Anas acuta	Pintail
A. cyanoptera	Cinnamon Teal
A. platyrhynchos	Mallard
Aythya americana	Redhead
Botaurus lentiginosus	American Bittern
Cistothorus palustris	Long-billed Marsh Wren
Geothlypis trichas	Common Yellowthroat
Grus canadensis	Sandhill Crane
Oxyura jamaicensis	Ruddy Duck
Porzana carolina	Sora
Rallus limicola	Virginia Rail
Xanthocephalus xanthocephalus	Yellow-headed Blackbird
FISHES	
Esox lucius	Northern Pike
Lepomis cyanellus	Green Sunfish
Notemigonus chrysoleucus	Golden Shiner

Warm-temperate Wetlands
Interior and Californian Riparian Deciduous Forests and Woodlands

Scientific Name	Common Name
PLANTS	
Acer grandidentatum	Bigtooth Maple
A. negundo	Box Elder
Ailanthus altissima	Tree-of-heaven
Alnus oblongifolia	Arizona Alder
A. rhombifolia	White Alder
Catalpa bignonioides	Catalpa
Celtis reticulata	Netleaf Hackberry
Cupressus arizonica	Arizona Cypress
Fraxinus pennsylvanica var. *velutina*	Velvet Ash
Juglans major	Arizona Black Walnut, Nogal
Juniperus deppeana	Alligatorbark Juniper
Maclura pomifera	Osage-orange
Morus microphylla	Texas Mulberry
Pinus ponderosa	Ponderosa Pine, Arizona Pine

Platanus sp.	sycamore, aliso
P. racemosa	California Sycamore
P. wrightii	Arizona Sycamore
Populus angustifolia	Narrowleaf Cottonwood
P. fremontii	Fremont Cottonwood
P. fremontii var. *mesotae*	—
Prunus spp.	cherries
Pteridium aquilinum	Bracken Fern
Quercus agrifolia	Coast Live Oak
Q. arizonica	Arizona White Oak
Q. chrysolepis	Canyon Live Oak
Q. emoryi	Emory Oak
Q. gambelii	Gambel Oak
Rhus glabra	Smooth Sumac
R. toxicodendron	Poison-ivy, Poison-oak
Rubus vitifolius	California Blackberry
Salix bonplandiana	Bonpland Willow
S. exigua	Coyote Willow, Sandbar Willow
S. gooddingii	Goodding Willow
S. nigra	Black Willow
Sambucus mexicana	Mexican Elder
Sapindus saponaria var. *drummondii*	Western Soapberry
Tamarix chinensis	Salt Cedar
Vitis arizonica	Canyon Grape

MAMMALS	
Bassariscus astutus	Ringtailed Cat
Didelphis marsupialis	Opossum
Lasiurus borealis	Red Bat
Lutra canadensis	River Otter
Mephitis spp.	skunks
Myotis spp.	myotises
Odocoileus virginianus	White-tailed Deer
Pipistrellus hesperus	Western Pipistrelle
Sciurus arizonensis	Arizona Gray Squirrel
S. griseus	Western Gray Squirrel
S. nayaritensis	Apache Fox Squirrel, Nayarit Squirrel
S. niger	Fox Squirrel
Spilogale putorius	Eastern Spotted Skunk
Thomomys spp.	pocket gophers
Ursus americanus	Black Bear

BIRDS	
Amazilia verticalis	Violet-crowned Hummingbird
Buteo albonotatus	Zone-tailed Hawk
Buteogallus anthracinus	Black Hawk
Calothorax lucifer	Lucifer Hummingbird
Coccyzus americanus	Yellow-billed Cuckoo
Cynanthus latirostris	Broad-billed Hummingbird
Dendroica petechia	Yellow Warbler
Elanus leucurus	White-tailed Hawk
Icterus bullocki	Bullock's Oriole
Ictinia misisippiensis	Mississippi Kite
Lampornis clemenciae	Blue-throated Hummingbird
Mylodynastes luteiventris	Sulphur-bellied Flycatcher
Petrochelidon pyrrhonota	Cliff Swallow
Piranga rubra	Summer Tanager
Platypsaris aglaiae	Rose-throated Becard
Trogon elegans	Coppery-tailed Trogon

AMPHIBIANS AND REPTILES	
Ambystoma rosaceum	Tarahumara Salamander
Anniella pulchra	California Legless Lizard
Batrochoceps attenuatus	California Slender Salamander
B. pacificus	Pacific Slender Salamander
Bufo boreas holophilus	California toad
Diadophis punctatus	Ringneck Snake
Elaphe triaspis	Green Rat Snake
Ensatina eschscholtzi	Ensatina
Gerrhonotus kingi	Arizona Alligator Lizard
G. multicarinatus	Southern Alligator Lizard
Hyla arenicolor	Canyon Treefrog
H. cadaverina	California Treefrog
H. regilla	Pacific Treefrog
H. wrightorum	Arizona Treefrog
Lampropeltis spp.	kingsnakes

Oxybelis aeneus	Vine Snake
Rana pipiens	Leopard Frog
R. tarahumarae	Tarahumara Frog
Tantilla wilcoxi wilcoxi	Huachuca Blackhead Snake
Taricha torosa	California Newt

FISHES

Agosia chrysogaster	Longfin Dace
Campostoma ornatum	Mexican Stoneroller
Catostomus bernardini	Yaqui Sucker
C. conchos	Conchos Sucker
C. insignis	Sonoran Sucker
Codoma ornata	Ornate Minnow
Fundulus parvipinnis	California Killifish
Gasterosteus aculeatus	Threespine Stickleback
Gila nigrescens	Chihuahua Chub
G. orcutti	Arroyo Chub
G. pulchra	Mesa del Norte Chub
G. robusta robusta	Round-tail Chub
Meda fulgida	Spikedace
Notropis chihuahua	Chihuahuan Shiner
N. formosus	Beautiful Shiner
Pantosteus clarki	Desert Mountain-sucker
P. plebeius	Rio Grande Mountain-sucker
P. santaanae	Santa Ana Mountain-sucker
Pimephales promelas	Fathead Minnow
Poeciliopsis occidentalis	Sonoran Topminnow
Rhinichthys cataractae	Longnose Dace
R. osculus	Speckled Dace
Tiaroga cobitis	Loach Minnow

Riparian Scrublands

| Scientific Name | Common Name |

PLANTS

Acacia greggii	Cat-claw Acacia, Uña de Gato
A. farnesiana	Sweet Acacia, Huisache
A. rigidula	—
Arundo donax	Giant Reed
Aster spinosus	Aster
Baccharis salicifolia	Seepwillow
B. sarothroides	Desert Broom
Cephalanthus occidentalis	Buttonbush
Cercidium texanum	Texas Paloverde
Chilopsis linearis	Desert-willow
Cynodon dactylon	Bermuda Grass
Distichlis spicata var. stricta	Desert Saltgrass
Equisetum spp.	horsetails, scouring rushes
Heliotropium curassavicum	Heliotrope
Hymenoclea spp.	burrobrushes
Nicotiana glauca	Tree Tobacco
Phragmites australis	Common Reed
Pluchea camphorata	Camphor-weed
Prosopis glandulosa	Honey Mesquite
P. pubescens	Screwbean Mesquite
Salix gooddingii	Goodding Willow
S. interior	Sandbar Willow, Taray
S. nigra	Black Willow
Tamarix chinensis	Salt Cedar
Verbesina encelioides	Cowpen Daisy

MAMMALS

Castor canadensis	Beaver
Perognathus penicillatus	Desert Pocket Mouse
Peromyscus leucopus	White-footed Mouse
Procyon lotor	Raccoon
Sigmodon hispidus	Hispid Cotton Rat

BIRDS

Auriparus flaviceps	Verdin
Phainopepla nitens	Phainopepla
Polioptila melanura	Black-tailed Gnatcatcher
Sayornis nigricans	Black Phoebe
Toxostoma dorsale	Crissal Thrasher
Vermivora luciae	Lucy's Warbler

AMPHIBIANS AND REPTILES

| Bufo punctatus | Red-spotted Toad |

B. woodhousei	Woodhouse's Toad
Chrysemys scripta	Pond Slider
Natrix erythrogaster	Plain-bellied Water Snake
Scaphiopus hammondi	Western Spadefoot
Trionyx spiniferus emoryi	Spiny Softshell Turtle
Uta stansburiana	Side-blotched Lizard

FISHES

Astyanax mexicanus	Mexican Tetra
Campostoma ornatum	Mexican Stoneroller
Carpiodes carpio	River Carpsucker
Cycleptus elongatus	Blue Sucker
Cyprinodon eximius	Conchos Pupfish
Dionda episcopa	Roundnose Minnow
Gambusia affinis	Mosquitofish
Hybopis aestivalis	Speckled Chub
Ictalurus furcatus	Blue Catfish
I. punctatus	Channel Catfish
Ictiobus spp.	buffalofish
Lepomis spp.	sunfishes
Notropis braytoni	Tamaulipas Shiner
N. chihuahua	Chihuahuan Shiner
N. lutrensis	Red Shiner
Pimephales promelas	Fathead Minnow

Warm-temperate Maritime and Interior Marshlands

| Scientific Name | Common Name |

PLANTS

Allenrolfea occidentalis	Pickleweed, Iodine Bush
Anemopsis californica	Yerba Mansa, Lizard-tail
Atriplex patula	Saltbush
Bacopa monnieri	Water Hyssop
Batis maritima	Batis, Saltwort
Berula erecta	Water Parsnip
Carex pringlei	—
Cuscuta salina	Dodder
Distichlis spicata	Saltgrass
Eleocharis caribaea	Spike Rush
E. cellulosa	Spike Rush
E. rostellata	Spike Rush
Fimbristylis thermalis	—
Frankenia grandiflora	Alkali-heath
Fuirena simplex	Umbrella-grass
Heliotropium curassavicum	Heliotrope
Jaumea carnosa	
Juncus cooperi	Rush
Limonium californicum	Sea-lavender
Ludwigia octovalvis	Water Primrose
Monanthochloë littoralis	—
Najas marina	Holly-leaf Naiad
Nitrophila occidentalis	—
Nymphaea ampla	Waterlily
Phragmites australis	Common Reed, Carrizo
Potamogeton nodosus	
P. pectinatus	Sago Pondweed
Ruppia maritima	Widgeon-grass
Salicornia bigelovii	pickleweed
S. subterminalis	pickleweed
S. virginica	pickleweed
Schoenus nigricans	
Scirpus spp.	bulrushes, tules
S. americanus	Three-square
S. californicus	Giant Bulrush
S. maritimus var. macrostachyus	Salt-marsh Bulrush
Setaria geniculata	
Spartina foliosa	Cordgrass
S. spartinae	Cordgrass
Suaeda californica	California Seepweed
Tamarix chinensis	Salt Cedar
Triglochin maritima	Arrowgrass
Typha angustifolia	Cattail
Utricularia obtusa	Bladderwort
Zannichellia palustris	Common Pondmat

MAMMALS

| Dipodomys deserti | Desert Kangaroo Rat |

Microtus pennsylvanicus — Meadow Vole
Neotoma lepida — Desert Woodrat
Reithrodontomys megalotis — Harvest Mouse
Sorex vagrans — Vagrant Shrew
Spermophilus tereticaudus — Round-tailed Ground Squirrel

BIRDS

Anas acuta — Pintail
A. americana — American Widgeon
A. crecca — Green-winged Teal
A. cyanoptera — Cinnamon Teal
A. discors — Blue-winged Teal
A. platyrhynchos diazi — Mexican Duck
Branta nigricans — Black Brant
Calidris melanotos — Pectoral Sandpiper
C. minutilla — Least Sandpiper
Catotrophorus semipalmatus — Willet
Chen caerulescens — Snow Goose
Cistothorus palustris — Long-billed Marsh Wren
Grus canadensis — Sandhill Crane
Himantopus mexicanus — Black-necked Stilt
Laterallus jamaicensis — Black Rail
Limnodromus griseus — Short-billed Dowitcher
L. scolopaceus — Long-billed Dowitcher
Limosa fedoa — Marbled Godwit
Numenius americanus — Long-billed Curlew
N. phaeopus — Whimbrel
Nycticorax nycticorax — Black-crowned Night Heron
Oxyura jamaicensis — Ruddy Duck
Rallus longirostra levipes — Light-footed Clapper Rail
Recurvirostra americana — American Avocet
Spatula clypeata — Shoveler
Totanus flavipes — Lesser Yellowlegs
T. melanoleucus — Greater Yellowlegs
Tringa solitaria — Solitary Sandpiper

AMPHIBIANS AND REPTILES

Ambystoma tigrinum nebulosum — Tiger Salamander
Bufo cognatus — Great Plains Toad
B. debilis — Green Toad
Chrysemys scripta taylori — Slider
Clemmys marmorata — Western Pond Turtle
Gerrhonotus kingi — Arizona Alligator Lizard
Scaphiopus spp. — spadefoot toads
Terrapene coahuila — Coahuila Box Turtle
Thamnophis spp. — garter snakes
T. elegans — Western terrestrial Garter Snake
T. sirtalis — Common Garter Snake
Trionyx ater — Softshell Turtle

FISHES

Catostomus fumeiventris — Owens Sucker
Clevelandia ios — Arrow Goby
Crenichthys baileyi — White River Springfish
C. nevadae — Railroad Valley Springfish
Cymatogaster aggregata — Shiner Perch
Cyprinodon bovinus — Leon Spring Pupfish
C. diabolis — Devil's Hole Pupfish
C. elegans — Comanche Springs Pupfish
C. fontinalis — Carbonaria Pupfish
C. macularius — Desert Pupfish
C. milleri — Cottonball Marsh Pupfish
C. nevadensis — Amargosa Pupfish
C. radiosus — Owens Pupfish
C. rubrofluviatilis — Red River Pupfish
C. salinus — Salt Creek Pupfish
Empetrichthys latos — Pahrump Poolfish
E. merriami — Ash Meadows Poolfish
Eucyclogobius newberryi — Tidewater Goby
Fundulus parvipinnis — California Killifish
Gambusia spp. — mosquitofishes
Gasterosteus aculeatus — Threespine Stickleback
Gila bicolor mohavensis — Mohave Chub
G. bicolor snyderi — Owens Tui Chub
G. nigrescens — Chihuahua Chub
Gillichthys mirabilis — Longjaw Mudsucker
Moapa coriacea — Moapa Dace
Platichthys stellatus — Starry Flounder
Rhinichthys osculus — Speckled Dace

Californian Maritime Strands

Scientific Name	Common Name

PLANTS

Abronia maritima — Sand Verbena
A. umbellata — Sand Verbena
Ambrosia chamissonis — —
Ammophila arenaria — Beachgrass
Atriplex leocophylla — Saltbush
Cakile maritima — Sea-rocket
Calystegia soldanella — Beach Morning Glory
Haplopappus ericoides — Mock Heather
H. venetus — Goldenweed
Lupinus chamissonis —
Mesembryanthemum crystallinum — Ice Plant, Flor-de-sol
M. chilense —
Monanthochloë littoralis — —
Oenothera cheiranthifolia — Evening-primrose

MAMMALS

Arctocephalus townsendi — Guadalupe Fur Seal
Mirounga angustirostris — Northern Elephant Seal
Phoca vitulina — Harbor Seal
Zalophus californianus — California Sea Lion

BIRDS

Charadrius alexandrinus — Snowy Plover
Endomychura hypoleuca — Xantus' Murrelet
Gymnogyps californianus — California Condor
Haematopus bachmani — Black Oystercatcher
Larus occidentalis — Western Gull
Phalacrocorax auritis — Double-crested Cormorant
P. pencillatus — Brant's Cormorant
Sterna albifrons — Least Tern
Thalasseus elegans — Elegant Tern

FISHES

Acanthogobius flavemanos — Yellowfin Goby
Atherinopsis californiensis — Jack Smelt
Clevelandia ios — Arrow Goby
Clinocottus analis — Woolly Sculpin
Cymatogaster aggregata — Shiner Perch
Cynoscion nobilis — White Seabass
C. xanthulus — Orangemouth Corvina
Eucinostomus argenteus — Silver Mojarra
Gibbonsia spp. — clinnid blennies
Gillichthys mirabilis — Longjaw Mudsucker
Girella nigricans — Opaleye
Gobieosox meandricus — Clingfish
Hyperprosopon ellipticum — Silver Surfperch
Hypsoblennius gentilis — Bay Blenny
H. giberti — Rockpool Blenny
H. jenkinsi — Mussel Blenny
Hysterocarpus traski — Tule Perch
Leuresthes tenuis — Grunion
Lythrypnus dalli — Bluebanded Goby
L. zebra — Zebra Goby
Oligocottus rubellio — Rosy Sculpin
Quietula y-cauda — Shadow Goby
Scorpaenichthys marmoratus — Cabezon
Sebastes auriculatus — Gopher Rockfish
S. chrysomelas — Black-and-yellow Rockfish
Strongylura exilis — Needlefish
Typhlogobius californiensis — Blind Goby
Umbrina roncador — Yellowfin Croaker
Xererpes fucorum — Rockweed Gurmel

Warm-temperate Interior Strands

Scientific Name	Common Name

PLANTS

Allenrolfea occidentalis — Pickleweed, Iodine Bush
Allionia incarnata — Windmills, Trailing Four-o'clock
Atriplex acanthocarpa — Saltbush
Atriplex obovata —
Baccharis salicifolia — Seepwillow
Bouteloua karwinskii —

Distichlis spicata	Saltgrass	Olneya tesota	Ironwood, Palofierro
Eragrostis obtusifolia	Halophytic Lovegrass	Panicum obtusum	Vine-mesquite Grass
Heliotropium curassavicum	Heliotrope	Platanus wrightii	Arizona Sycamore, Aliso
Hoffmanseggia glauca	Hog-potato, Carrote-de-ratou	P. dimorpha	—
Leptochloa fascicularis	—	Prosopis pubescens	Screwbean Mesquite
L. uninervia	—	P. velutina	Velvet Mesquite
Panicum spp.	panic grasses	Salix gooddingii var. variabilis	—
Phragmites australis	Common Reed	Sambucus mexicana	Mexican Elder
Populus fremontii var. macdougali	Cottonwood, Alamo	Sarcostemma spp.	climbing milkweeds
Potamogeton pectinatus	Sago Pondweed	Schismus barbatus	Mediterranean Grass
Ruppia maritima	Widgeon-grass	Sisymbrium irio	Yellow Rocket
Salicornia rubra	Pickleweed	Suaeda torreyana	Desert Seepweed, Quelite Salado
Salsola iberica	Russian Thistle	Tamarix chinensis	Salt Cedar
Sesuvium verrucosum	Sea-purslane	Vitis arizonica	Canyon Grape
Sporobolus airoides	Alkali Sacaton	Zizyphus obtusifolia	Graythorn, Crucifixion Thorn
S. wrightii	Wright Sacaton		
Suaeda spp.	seepweeds, sea blights		
S. fruticosa	—		
S. jacoensis	—		
S. palmerii	—		
Tamarix chinensis	Salt Cedar		
Tidestromia lanuginosa	—		
Zannichellia palustris	Common Pondmat		

MAMMALS

Lepus californicus	Black-tailed Jackrabbit
Procyon lotor	Raccoon

BIRDS

Actitis macularia	Spotted Sandpiper
Amphispiza bilineata	Black-throated Sparrow
Branta canadensis	Canada Goose
Buteogallus anthracinus	Black Hawk
Charadrius alexandrinus	Snowy Plover
C. vociferus	Killdeer
Grus canadensis	Sandhill Crane
Haliaeetus leucocephalus	Bald Eagle
Recurvirostra americana	American Avocet
Sayornis nigricans	Black Phoebe
Stelgidopteryx ruficollis	Rough-winged Swallow
Zenaida macroura	Mourning Dove

AMPHIBIANS AND REPTILES

Bufo microscaphus	Southwestern Toad
B. woodhousei	Woodhouse's Toad
Cnemidophorus spp.	whiptails
Masticophis flagellum	Coachwhip
Rana pipiens	Leopard Frog
Sceloporus magister	Desert Spiny Lizard
Thamnophis spp.	garter snakes
Trionyx spiniferus emoryi	Spiny Softshell Turtle
Urosaurus ornatus	Tree Lizard

FISHES

Cyprinodon macularius	Desert Pupfish

Tropical-subtropical Wetlands
Sonoran Riparian Deciduous Forests and Woodlands

Scientific Name	Common Name

PLANTS

Acacia greggii	Cat-claw Acacia, Uña de Gato
Amaranthus palmeri	Careless-weed
Atriplex lentiformis	Lens Scale, Quail Bush
A. polycarpa	All Scale, Cattle Spinach
Bromus rubens	Red Brome, Foxtail Chess
Celtis reticulata	Netleaf Hackberry
Cercidium floridum	Blue Palo Verde
Cucurbita spp.	gourds
Cynodon dactylon	Bermuda Grass
Erodium cicutarium	Filaree, Heron's Bill
Fraxinus pennsylvanica var. velutina	Velvet Ash, Fresno
Lycium andersonii	Anderson Thornbush
L. berlandieri	Cilindrillo, Barchata
L. fremontii	Fremont Thornbush

MAMMALS

Castor canadensis	Beaver
Eptesicus fuscus	Big Brown Bat
Felis onca	Jaguar
Lasionycteris noctivagans	Silver-haired Bat
Perognathus penicillatus	Desert Pocket Mouse
Procyon lotor	Raccoon

BIRDS

Ardea herodias	Great Blue Heron
Buteo nitidus	Gray Hawk
Cardinalis cardinalis	Cardinal
C. sinuatus	Pyrrhuloxia
Coccyzus americanus	Yellow-billed Cuckoo
Haliaeetus leucocephalus	Bald Eagle
Ictinia misisippiensis	Mississippi Kite
Passerina versicolor	Varied Bunting
Phainopepla nitens	Phainopepla
Pipilo aberti	Abert's Towhee
Platypsaris aglaiae	Rose-throated Becard
Pyrocephalus rubinus	Vermilion Flycatcher
Tyrannus crassirostris	Thick-billed Kingbird
Vermivora luciae	Lucy's Warbler
Zenaida asiatica	White-winged Dove
Z. macroura	Mourning Dove

AMPHIBIANS AND REPTILES

Cnemidophorus spp.	whiptails
Rana pipiens	Leopard Frog
Sceloporus clarki	Clark's Spiny Lizard
Urosaurus ornatus	Tree Lizard

FISHES

Agosia chrysogaster	Longfin Dace
Anisotremus davidsoni	Sargo
Bairdiella icistius	Bairdiella
Campostoma ornatum	Mexican Stoneroller
Catostomus latipinnis	Flannelmouth Sucker
C. wigginsi	Opata Sucker
Chaenobryttus gulosus	Warmouth
Cynoscion xanthulus	Orangemouth Corvina
Cyprinodon macularius	Desert Pupfish
Cyprinus carpio	Carp
Dorosoma petenense	Threadfin Shad
Elops affinis	Machete
Gambusia affinis	Mosquitofish
Gila ditaenia	Sonora Chub
G. elegans	Bonytail Chub
G. purpurea	Yaqui Chub
G. robusta robusta	Round-tail Chub
Gillichthys mirabilis	Longjaw Mudsucker
Lepomis cyanellus	Green Sunfish
L. macrochirus	Bluegill
L. microlophus	Redear Sunfish
Micropterus dolomieui	Smallmouth Bass
M. salmoides	Largemouth Bass
Morone saxatilis	Striped Bass
Mugil cephalus	Striped Mullet
Notropis lutrensis	Red Shiner
Plagopterus argentissimus	Woundfin
Poecilia latipinna	Sailfin Molly
P. mexicana	Shortfin Molly
Poeciliopsis gracilis	—
P. occidentalis	Sonoran Topminnow

Ptychocheilus lucius	Colorado Squawfish
Pylodictis olivaris	Flathead Catfish
Tilapia aurea	Blue Tilapia
T. mossambica	Mozambique Tilapia
T. zilli	Redbelly Tilapia
Xiphophorus helleri	Green Swordtail
X. variatus	Variable Platyfish
Xyrauchen texanus	Razorback Sucker

Sonoran Oasis Forest and Woodlands

Scientific Name	Common Name

PLANTS

Acacia greggi	Cat-claw Acacia, Uña de Gato
Ambrosia ambrosioides	Bursage
Atriplex spp.	saltbushes
Baccharis salicifolia	Seepwillow
B. sarothroides	Desert Broom
Carex ultra	sedge
Cercidium floridum	Blue Paloverde
Cynodon dactylon	Bermuda Grass
Erythea aculeata	—
E. armata	Bluefan Palm
Hymenoclea monogyra	Burrobrush
Phoenix dactylifera	Date Palm
Phragmites australis	Common Reed, Carrizo
Populus fremontii	Fremont Cottonwood
Prosopis velutina	Velvet Mesquite
Sabal uresana	Mexican Blue Palm
Tamarix chinensis	Salt Cedar
Tessaria sericea	Arrow-weed
Washingtonia filifera	Washington Fan Palm
W. robusta	Sky Duster Palm
Zizyphus obtusifolia	Graythorn

MAMMALS

Dasypterus egaxanthinus	Western Yellow Bat

Sinaloan Riparian Evergreen Forest and Woodland

Scientific Name	Common Name

PLANTS

Arrabidaea litoralis	Bejuco Vaquero
Bacopa monnieri	Water Hyssop
Begonia spp.	begonias
Celtis iguanea	Garabato
Clethra lanata	Jicarillo
Erythea aculeata	—
Eustoma exaltatum	Catchfly Gentian
Ficus spp.	figs
Fuirena simplex	Umbrella-grass
Gouania mexicana	Guirote de Palo
Gratiola brevifolia	Sticky Hedge-hyssop
Guazuma ulmifolia	Guacimilla
Hechtia spp.	Guapilla
Heteranthera limosa	Mud Plantain
Magnolia schiediana	—
Marsdenia edulis	Talayote
Oncidium cebolleta	orchid
Oreopanax salvinii	Mano de Leon
Pisonia capitata	Garabato Prieto
Pithecellobium spp.	ebonies
Platanus sp.	sycamore
Populus dimorpha	Cottonwood
Prosopis juliflora	Mesquite
Quercus spp.	Oaks
Rotala ramosior	Tooth-cup
Sabal uresana	Mexican Blue Palm
Salix gooddingii	Goodding Willow
Samolus ebracteatus	Water-pimpernel, Brookweed
Sartwellia mexicana	—
Sassafridium macrophyllum	Laurelón
Selaginella sp.	Resurrection Plant
Sesbania sesban	—
Stanhopea spp.	—

Taxodium mucronatum	Montezuma Bald Cypress, Ahuehuete
Tillandsia inflata	—
Vallesia glabra	Cacarahue, Otatave
Vitex mollis	Obalamo, Uralama

MAMMALS

Dipodomys spp.	kangaroo rats
Felis onca	Jaguar
F. pardalis	Ocelot
Nasua nasua	Coati
Perognathus spp.	pocket mice
Sigmodon hispidus	Hispid Cotton Rat
S. minimus	Cotton Rat

BIRDS

Amazilia beryllina	Berylline Hummingbird
Amazonia finschi	Lilac-crowned Parrot
Ara militaris	Military Macaw
Aratinga holochlora	Green Parakeet
Archilochus alexandri	Black-chinned Hummingbird
Calocitta formosa	Magpie Jay
Dendrocygna autumnalis	Black-bellied Tree Duck
Forpus cyanopygius	Blue-rumped Parrot
Melanerpes uropygialis	Gila Woodpecker
Momotus mexicanus	Russet-crowned Motmot
Trogon elegans	Coppery-tailed Trogon

AMPHIBIANS AND REPTILES

Bufo marinus	Giant Toad
Chrysemys picta	Painted Turtle
Constrictor constrictor	Boa Constrictor
Drymarchon corias	Indigo Snake
Kinosternon alamosae	Mud Turtle
K. hirtipes	Mexican Mud Turtle
K. integrum	Mud Turtle
Natrix valida	Water Snake
Oxybelis aeneus	Vine Snake
Pachymedusa dacnicolor	Treefrog
Pseudemys scripta mayae	Pond Turtle
Rana pipiens	Leopard Frog
Rhinoclemmys pulcherrima	Pond Turtle
Tantilla planiceps yaquiae	Western Blackhead Snake

FISHES

Agonostomus monticola	Mountain Mullet
Agosia chrysogaster	Longfin Dace
Awaous transandeanus	Goby
Campostoma ornatum	Mexican Stoneroller
Carpiodes carpio	River Carpsucker
Castostomus bernardini	Yaqui Sucker
Cichlasoma beani	Sinaloan Cichlid
Cyprinus carpio	Carp
Dorosoma smithi	Pacific Shad
Gila robusta robusta	Round-tail Chub
Ictalurus pricei	Yaqui Catfish
Lepomis cyanellus	Creen Sunfish
L. macrochirus	Bluegill
L. microlophus	Redear Sunfish
Micropterus salmoides	Largemouth Bass
Notropis formosus	Beautiful Shiner
Poeciliopsis monacha-occidentalis	—
P. prolifica	—
Poxomis annularis	White Crappie

Sonoran Riparian Scrubland

Scientific Name	Common Name

PLANTS

Acacia greggii	Cat-claw Acacia, Uña de Gato
Allenrolfea occidentalis	Pickleweed, Iodine Bush
Atriplex lentiformis	Lens Scale, Quail Bush
A. polycarpa	All Scale, Cattle Spinach
Baccharis salicifolia	Seepwillow
B. sarothroides	Desert Broom
B. viminea	Mule Fat
Celtis pallida	Desert Hackberry
Chilopsis linearis	Desert Willow

Hymenoclea monogyra	Burrobrush
Lycium brevipes	Tomatillo
Pluchea camphorata	Camphor-weed
P. purpurescens	Marsh Fleabane, Stinkweed
Prosopis glandulosa var. *torreyana*	Western Honey Mesquite
P. juliflora	Mesquite
P. pubescens	Screwbean Mesquite
Suaeda torreyana	Desert Seepweed, Quelite Salado
Tamarix chinensis	Salt Cedar
T. aphylla	Athel
Tessaria sericea	Arrow-weed

MAMMALS

Sylvilagus auduboni	Desert Cottontail

BIRDS

Lophortyx gambelii	Gambel's Quail
Pipilo aberti	Abert's Towhee
P. fuscus	Brown Towhee
Polioptila melanura	Black-tailed Gnatcatcher
Rallus longirosta yumaensis	Yuma Clapper Rail
Sayornis saya	Say's Phoebe
Toxostoma dorsale	Crissal Thrasher
Zenaida asiatica	White-winged Dove
Z. macroura	Mourning Dove

Sinaloan Maritime (Tidal) Scrublands

Scientific Name	Common Name

PLANTS

Avicennia germinans	Black Mangrove
Laguncularia racemosa	White Mangrove
Rhizophora mangle	Red Mangrove

BIRDS

Ajaia ajaja	Roseate Spoonbill
Anhinga anhinga	Anhinga
Coccyzus minor	Mangrove Cuckoo
Dendroica erithachorides	Mangrove Warbler
Iridoprocne albilinea	Mangrove Swallow
Mycteria americana	Wood Stork
Rallus longirostra	Clapper Rail
Tigrisoma mexicanum	Tiger Bittern
Zenaida asiatica	White-winged Dove

FISHES

Epinephelus itajaro	Giant Jewfish
Lutjanus spp.	snappers

Sonoran and Sinaloan Interior Marshlands

Scientific Name	Common Name

PLANTS

Atriplex lentiformis	Lens Scale, Quail Bush
Bacopa monnieri	Water Hyssop
Ceratophyllum demersum	Common Hornwort
Cynodon dactylon	Bermuda Grass
Cyperus erythrorhizos	—
C. strigosus	—
Distichlis spicata	Saltgrass
Eleocharis spp.	spike rushes
E. caribaea	—
E. parvula	—
Hydrocotyle verticillata	Pennywort
Juncus spp.	rushes
Lemna spp.	duckweeds
Leptochloa uninervia	—
Mentha spicata	Spearmint
Myriophyllum brasilense	Parrot-feather
M. spicatum	Water Milfoil
Najas guadalupensis	Water Nymph
N. marina	Holly-leaf Naiad
Paspalum dilatatum	Dallis Grass

Phragmites australis	Common Reed, Carrizo
Polygonum fusiforme	—
Potamogeton foliosus	—
P. pectinatus	Sago Pondweed
Prosopis juliflora	Mesquite
Scirpus americanus	Three-square
S. californicus	Giant Bulrush
S. maritimus var. *paludosus*	Salt-marsh Bulrush
Tamarix chinensis	Salt Cedar
Tessaria sericea	Arrow-weed
Typha dominguensis	Cattail
Utricularia spp.	bladderworts
Zannichellia palustris	Common Pondmat

MAMMALS

Ondatra zibethicus	Muskrat

BIRDS

Agelaius phoeniceus	Red-winged Blackbird
Dendrocygna bicolor	Fulvous Whistling Duck
Egretta thula	Snowy Egret
Fulica americana	American Coot
Geothypis trichas	Common Yellowthroat
Ixobrychus exilis	Least Bittern
Nycticorax nycticorax	Black-crowned Night Heron
Podiceps dominicus	Least Grebe
Porphyrula martinica	Purple Gallinule
Porzana carolina	Sora
Rallus longirostra yumaensis	Yuma Clapper Rail

AMPHIBIANS AND REPTILES

Bufo alvarius	Colorado River Toad
B. marinus	Giant Toad
Kinosternon sonoriense	Sonoran Mud Turtle
Rana catesbeiana	Bullfrog
Thamnophis marcianus	Checkered Garter Snake
Trionyx spiniferus emoryi	Spiny Softshell Turtle

FISHES

Cyprinodon macularis	Desert Pupfish
Poecilia latipinna	Sailfin Molly
P. mexicana	Shortfin Molly
Poeciliopsis occidentalis	Sonoran Topminnow
Tilapia spp.	mouthbrooders

Sonoran Maritime Strand

Scientific Name	Common Name

PLANTS

Abronia maritima	Sand Verbena
A. villosa	Sand Verbena
Allenrolfea occidentalis	Pickleweed, Iodine Bush
Amaranthus watsoni	—
Astragalus magdalenae	Loco Weed
Atriplex barclayana	Saltbush
Batis maritima	Batis, Saltwort
Croton californicus	—
Dicoria canescens	—
Distichlis spicata	Saltgrass
Eucnide rupestris	—
Euphorbia leucophylla	—
Ficus palmeri	Anabá, Zalate
F. petiolaris	Higuera, Higuerón
Frankenia spp.	alkali-heaths
Helianthus niveus	Sunflower
Hofmeisteria crassifolia	—
H. fasciculata	—
Jouvea pilosa	—
Limonium californicum	Sea-lavender
Maytenus phyllanthoides	Sweet Mangel, Mangle Dulce
Mesembryanthemum crystallinum	Ice Plant, Flor-de-Sol
Monanthochloë littoralis	—
Nicotiana trigonophylla	Desert Tobacco, Tabaquillo
Pleurocoronis laphamioides	—
Salicornia spp.	glassworts
Suaeda spp.	seepweeds, sea blights

BIRDS

Calidris spp.	sandpipers
Charadrius alexandrinus	Snowy Plover
C. wilsonia	Wilson's Plover
Falco peregrinus	Peregrine Falcon
Haematopus pallitus	American Oystercatcher
Halocyptena microsoma	Least Storm Petrel
Larus atricilla	Laughing Gull
L. heermanni	Heermann's Gull
L. occidentalis	Western Gull
Limnodromus griseus	Short-billed Dowitcher
Numenius americanus	Long-billed Curlew
Oceanodroma melania	Black Storm Petrel
Pandion haliaetus	Osprey
Pelecanus occidentalis	Brown Pelican
Phaethon aethereus	Red-billed Tropicbird
Phalacrocorax auritis	Double-crested Cormorant
Puffinus puffinus	Manx Shearwater
Rynchops nigra	Black Skimmer
Sterna albifrons	Least Tern
Sula dactylatra	Blue-faced Booby
S. leucogaster	Brown Booby
S. nebouxii	Blue-footed Booby
Thalasseus elegans	Elegant Tern
T. maximus	Royal Tern

AMPHIBIANS AND REPTILES

Chelonia mydas	Green Sea-turtle

FISHES

Abudefduf troscheli	Panamic Sargeant Major
Achirus mazatlanus	Flounder
Albula vulpes	Bonefish
Anisotremus davidsoni	Sargo
Bairdiella icistius	Bairdiella
Bodianus diplotaenia	Wrasse
Chaetodon humeralis	butterfly Fish
Cynoscion macdonaldi	Totoaba
Cynoscion parvipinnis	Croaker
Diapterus peruvianus	Mojarra
Epinephelus itajaro	Giant Jewfish
Etropus crossotus	Flounder
Eucinostomus spp.	mojarras
Eupomacentrus rectifraenum	Cortez Damselfish
Gerres cinereus	Mojarra
Gillichthys mirabilis	Longjaw Mudsucker
Girella simplicidens	Gulf Opaleye
Gobieosox pinninger	Clingfish

Gobionellis sagittula	Longtail Goby
Gobisoma chiquita	Sonoran Goby
Gymnothorax castaneus	Moray Eel
Haemulon sexfasciatum	Grunt
Halichoeres dispilus	Wrasse
H. nicholsi	Wrasse
Heniochus nigrirostris	Butterfly Fish
Holacanthus passer	Angelfish
Hypsoblennius gentilis	Bay Blenny
Leuresthes sardina	Gulf Grunion
Malacoctenus gigas	Blenny
Microlepidotus momatus	Grunt
Mugil cephalus	Striped Mullet
M. curema	Mullet
Muraena lentiginosa	Moray Eel
Mycteroperca jordani	Baja Grouper
Paraclinus sini	Blenny
Paralabrax maculatofasciatus	Spotted Sand Bass
Pomacanthus zonipectus	Angelfish
Pomadasys branicki	Grunt
Prionurus punctatus	Yellowtail Surgeonfish
Quietula guaymasiae	Guaymas Goby
Scarus perrico	Bumphead Parrotfish
Symphurus melanorum	Flounder
Thalassoma lucansanum	Wrasse
Tomicodon humeralis	—
Urolophus halleri	Stingray

Tropical-Subtropical Interior Strands

Scientific Name	Common Name

PLANTS

Amaranthus palmeri	Careless-weed
Baccharis salicifolia	Seepwillow
Datura spp.	thorn apples
Helianthus spp.	sunflowers
Nicotiana glauca	Tree Tobacco
Polypogon monspeliensis	Rabbit Foot Grass
Rumex spp.	docks
Solanum spp.	nightshades
Xanthium strumarium	Common Cocklebur

FISHES

Agosia chrysogaster	Longfin Dace
Poeciliopsis occidentalis	Sonoran Topminnow

124.6 Sinaloan Deciduous Forest (cont.)

BIRDS

Amazona albifrons	White-fronted Parrot, Perico Frentiblanco
Calocitta formosa colliei	Black-throated Magpie Jay, Urraca Hermosa
Ciccaba virgata	Mottled Owl, Mochuelo café
Columba flavirostris	Red-billed Pigeon, Paloma Morada
Cyanocorax beecheii	Purplish-backed Jay, Queisque
Dryocopus lineatus	Lineated Woodpecker, Carpintero Real
Euthlypis lachrymosa	Fan-tailed Warbler, Pavito Amarillo
Forpus cyanopygius	Mexican Parrotlet, Blue-rumped Parrotlet, Catarinita
Geranospiza caerulescens	Crane Hawk, Gavilán Zancón
Heliomaster constantii	Plain-capped Starthroat, Chupamirto Ocotero
Icterus pustulatus	Streak-backed Oriole, Calandria de Fuego
Momotus mexicanus	Russet-crowned Motmot, Pajáro Reloj, Pajáro Cu
Ortalis wagleri	Rufous-bellied Chachalaca, Chachalaca
Parula pitiayumi	Tropical Parula, Verdín Espalda Verde
Pheucticus chrysopeplus	Yellow Grosbeak, Piquigrueso Amarillo
Piaya cayana	Squirrel Cuckoo, Vaquero, Guaco
Thryothorus felix	Happy Wren, Saltapared Cluequita
Turdus rufopalliatus	Rufous-backed Robin, Primavera Chivillo

AMPHIBIANS AND REPTILES

Agkistrodon bilineatus	Mexican Cottonmouth, Cantāl, Pichiguaté
Anolis nebuloides	Lagarto, Cachura
Boa constrictor	Boa Constrictor, Corrua
Crotalus basiliscus	Mexican West-coast Rattlesnake, Cascabel
Drymarchon corais	Mexican Indigo Snake, Culebra Azul
Heloderma horridum	Mexican Beaded Lizard, Escorpión

Iguana iguana	Green Iguana, Iguana	*Pseudosicimia hiltoni*	Hilton's Hook-nosed Snake, Culebra
Imantodes geministrata	Blunt-headed Tree Snake	*Rhinochelmys pulcherrima*	Mexican Wood Turtle
Lampropeltus triangulum nelsoni	Sinaloan Milk Snake, Corál	*Sceloporus nelsoni*	Lagarto, Cachura
Leptodactylus melanonotus	Mexican Little Frog, Ranita	*Sonora aemula*	Sinaloan Ground Snake, Coralillo
Leptophis diplotropis	Parrot Snake, Culebra Verde	*Sympholis lippiens*	Fat Snake
Micrurus distans	Neotropical Coral Snake, Corál	*Terrapene nelsoni*	Sinaloan Box Turtle
Pachymedusa dacnicolor	Mexican Giant Tree Frog, Green Tree Frog, Rana Verde	*Trimorphodon tau*	Mexican Lyre Snake
Phylodactylus homolepidurus	Mexican Leaf-toed Gecko, Salamaquesa		